Springer Series in Computational Neuroscience

Volume 6

For further volumes:
http://www.springer.com/series/8164

Nathan W. Schultheiss • Astrid A. Prinz
Robert J. Butera
Editors

Phase Response Curves in Neuroscience

Theory, Experiment, and Analysis

Springer

Editors
Nathan W. Schultheiss
Boston University
2 Cummington Street, RM 109
Boston, Massachusetts 02215
USA
nschult@bu.edu

Astrid A. Prinz
Department of Biology
Emory University
1510 Clifton Road
Atlanta, Georgia 30322
USA
astrid.prinz@emory.edu

Robert J. Butera
School of Electrical and Computer Engineering
Georgia Institute of Technology
Laboratory of Neuroengineering
313 Ferst Drive
Atlanta, Georgia 30332
USA
rbutera@gatech.edu

Organization for Computational Neurosciences: http://www.cnsorg.org.

ISBN 978-1-4614-0738-6 e-ISBN 978-1-4614-0739-3
DOI 10.1007/978-1-4614-0739-3
Springer New York Dordrecht Heidelberg London

Library of Congress Control Number: 2011943621

Printed on acid-free paper

Springer is part of Springer Science+Business Media (www.springer.com)

Preface

The two decades leading up to this year's twentieth annual Computational Neuroscience conference (CNS) have seen a dramatic upswing in applications of quantitative and analytical methods taken from mathematics, physics, and engineering (among others) to the traditionally more biological approaches to Neuroscience. Much of the progress in the Computational Neurosciences, as in the broader field of Neuroscience, has taken the form of advancements in our understanding of neural systems at two key levels: the cellular processes underlying the dynamic electrical and chemical behaviors of individual neurons and the complex interactions among neurons in networks of varying composition and size, ranging from two reciprocally connected neurons, to detailed local microcircuitry, to large scale networks of thousands or more. One of the most difficult challenges, however, has been (and remains) to bridge the cellular and network levels of computation, i.e., to identify and understand how the properties of individual neurons contribute to the behaviors of functional networks underlying perception, motor performance, memory, and cognition. Given that neurons, like people, communicate with and influence one another through a variety of means, this problem is quite a bit like relating the individual personalities of two or more people to the interactions between them; or more generally, it is like relating the psychology of individuals to the sociology of a community.

One of the most fruitful means of addressing the interface between cellular and network computation has been the application of phase response analysis to neuronal systems. Neuronal phase response curves (PRCs) describe the pattern of shifts in the timing of action potentials (spikes) that are caused by inputs to a neuron arriving at different times within that neuron's spike cycle. The degree to which an input can affect spike timing depends not only on the properties of the neuron but also on the characteristics of the input, and the relationship between the PRCs of individual neurons and the behavior of a neuronal network additionally depends on the connectivity structure within the network. Consequently, many of the complexities of computation at the cellular and network levels are embodied in the variety of applications of phase response analyses to neuronal systems. This book provides a cross section of the considerable body of work by many of the prominent theoreticians and experimentalists in the Computational Neurosciences which make use of PRCs to further our understanding of neurons and networks, more generally, the brain, and more abstractly, ourselves. Part 1 introduces the theoretical underpinnings of phase response analysis and presents the central concepts and context for the rest of the book; Part 2 surveys techniques for estimating neuronal phase response curves and many of the technical considerations necessary to do so; Part 3 presents many of the key investigations relating the phase response properties of neurons to their cellular characteristics; and finally, the chapters in Part 4 illustrate how phase response curves can be used to understand and predict patterning of network activity in neuronal systems.

To make this text exciting and accessible to a diverse audience, the contributors to this book were asked to write "across the aisle," so-to-speak, such that the more theoretical or "mathy" authors considered more biologically-minded readers in preparing their contributions, and vice versa. Although this text generally proceeds from more theoretical to more applied topics, and major themes are partitioned into the book's four major parts, readers are not expected to move purely linearly through the content from start to finish. Rather, we encourage readers to familiarize themselves with the general concepts and perspectives and then move from one chapter to another as curiosity and perhaps relevance to their own interests dictate.

We, the editors, dedicate this volume to our mentors, in particular among them Drs. Dieter Jaeger, Eve Marder, Jack Byrne, John Clark, Ron Calabrese, and Terry Blumenthal, and to our families.

Contents

Contributors

Srisairam Achuthan Neuroscience Center of Excellence, LSU Health Sciences Center, New Orleans, LA, USA, sairam.achuthan@gmail.com

Kazuyuki Aihara Aihara Complexity Modelling Project, ERATO, Japan Science and Technology Agency (JST), Japan

Institute of Industrial Science, The University of Tokyo, Japan

Toru Aonishi Interdisciplinary Graduate School of Science and Engineering, Tokyo Institute of Technology, Yokohama, Japan, aonishi@dis.titech.ac.jp

G. Bard Ermentrout Department of Mathematics, University of Pittsburgh, Pittsburgh, PA, USA, bard@pitt.edu

Bryce Beverlin II Department of Physics, University of Minnesota, Minneapolis, MN, USA, beverlin@physics.umn.edu

Victoria Booth Departments of Mathematics and Anesthesiology, University of Michigan Ann Arbor, MI, USA, vbooth@umich.edu

Amitabha Bose Department of Mathematical Sciences, New Jersey Institute of Technology, Newark, NJ, USA

School of Physical Sciences, Jawaharlal Nehru University, New Delhi, India

Robert Butera Georgia Institute of Technology, Atlanta, GA, USA, rbutera@gatech.edu

Carmen C. Canavier Neuroscience Center of Excellence, LSU Health Sciences Center, New Orleans, LA, USA, ccanav@lsuhsc.edu

Lakshmi Chandrasekaran Neuroscience Center of Excellence, LSU Health Sciences Center, New Orleans, LA, USA, chandrasekaran.lakshmi@gmail.com

Christian G. Fink Department of Physics, University of Michigan Ann Arbor, MI, USA, tcgfink@umich.edu

Roberto F. Galán Case Western Reserve University, School of Medicine, Department of Neurosciences, 10900 Euclid Avenue, Cleveland, OH, USA, rfgalan@case.edu

Leon Glass McGill University, Montreal, QC, Canada, glass@cnd.mcgill.ca

Nathan W. Gouwens Department of Physiology, Development and Neuroscience, University of Cambridge, Cambridge, UK, CB2 3EG, ngouwens@gmail.com

Boris S. Gutkin Group for Neural Theory, Départment d'Études Cognitives, École Normale Supérieure, Paris, France

Group for Neural Theory, Laboratoire de Neuroscience Cognitive, INSERM U960, Paris, France

CNRS, Campus Gérard-Mégie, 3 rue Michel-Ange - F-75794 Paris cedex 16, France, boris.gutkin@ens.fr

David Hansel Neurophysique et Physiologie, CNRS UMR8119, Université Paris Descartes, 45 rue des Saint-Pères 75270 Paris, France, david.hansel@univ-paris5.fr

Anmo J. Kim Department of Electrical Engineering, Columbia University, New York, NY, USA, ak2263@columbia.edu

Trine Krogh-Madsen Weill Cornell Medical College, New York, NY, USA, trk2002@med.cornell.edu

Aurel A. Lazar Department of Electrical Engineering, Columbia University, New York, NY, USA, aurel@ee.columbia.edu

Máté Lengyel Computational and Biological Learning Lab, Department of Engineering, University of Cambridge, Cambridge, UK, m.lengyel@eng.cam.ac.uk

Timothy J. Lewis Department of Mathematics, One Shields Ave, University of California Davis, CA, USA, tjlewis@ucdavis.edu

Germán Mato Centro Atómico Bariloche and Instituto Balseiro (CNEA and CONICET), San Carlos de Bariloche, Argentina, matog@cab.cnea.gov.ar

Farzan Nadim Department of Mathematical Sciences, New Jersey Institute of Technology, Newark, NJ, USA

Department of Biological Sciences, Rutgers University, Newark, NJ, USA, farzan@njit.edu

Theoden Netoff Department of Biomedical Engineering, University of Minnesota, Minneapolis, MN, USA, tnetoff@umn.edu

Sorinel Adrian Oprisan Department of Physics and Astronomy, College of Charleston, 66 George Street, Charleston, SC, USA, oprisans@cofc.edu

Keisuke Ota Brain Science Institute, RIKEN, Wako, Japan, k-ota@brain.riken.jp

Carlos A. Paladini UTSA Neurosciences Institute, University of Texas at San Antonio, One UTSA Circle, San Antonio, TX, USA, carlos.paladini@utsa.edu

Benjamin Pfeuty PhLAM, CNRS UMR8523, Université Lille I, F-59655 Villeneuve d'Ascq, France, pfeuty@phlam.univ-lille1.fr

Daniel G. Polhamus UTSA Neurosciences Institute, University of Texas at San Antonio, One UTSA Circle, San Antonio, TX, USA, danp@metrumrg.com

Michiel W.H. Remme Group for Neural Theory, Département d'Études Cognitives, École Normale Supérieure, Paris, France

Current address: Institute for Theoretical Biology, Humboldt-Universität zu Berlin, Berlin, Germany, michiel.remme@hu-berlin.de

Hugh P.C. Robinson Department of Physiology, Development and Neuroscience, University of Cambridge, Cambridge, UK, CB2 3EG, hpcr@cam.ac.uk

Nathan W. Schultheiss Center for Memory and Brain, Psychology Department, Boston University, Boston, MA, USA, nschult@bu.edu

Michael A. Schwemmer Department of Mathematics, One Shields Ave, University of California Davis, CA, USA, mschwemm@math.princeton.edu

Frances K. Skinner Toronto Western Research Institute, University Health Network and University of Toronto, Ontario, Canada, frances.skinner@gmail.com

Klaus M. Stiefel Theoretical and Experimental Neurobiology Unit, Okinawa Institute of Science and Technology, Okinawa, Japan, stiefel@oist.jp

Takashi Tateno Graduate School of Engineering Science, Osaka University, Japan

PRESTO, Japan Science and Technology Agency, 4–1–8 Honcho Kawaguchi, Saitama, Japan, tateno@fbs.osaka-u.ac.jp

Kunichika Tsumoto Aihara Complexity Modelling Project, ERATO, Japan Science and Technology Agency (JST), Japan

Institute of Industrial Science, The University of Tokyo, Japan

Charles J. Wilson UTSA Neurosciences Institute, University of Texas at San Antonio, One UTSA Circle, San Antonio, TX, USA, charles.wilson@utsa.edu

Hugo Zeberg Department of Physiology, Development and Neuroscience, University of Cambridge, Cambridge, UK, CB2 3EG, hugo.zeberg@gmail.com

Shunbing Zhao Department of Biological Sciences, Rutgers University, Newark, NJ, USA, shunbing@gmail.com

Michal Zochowski Department of Physics, University of Michigan Ann Arbor, MI, USA

Biophysics Program, University of Michigan Ann Arbor, MI, USA, michalz@umich.edu

Part I
Foundations of Phase Response Analysis

Introduction

The first section of this text provides an overview of the basic principles of applying phase response analysis to the study of neurons and neuronal networks. Each chapter describes general strategies by which phase response curves can be used for the prediction of phase-locked states among neuronal oscillators. Chapter 1 by Schwemmer and Lewis details the theory of weakly coupled oscillators. This theory entails the reduction of high dimensional descriptions of neurons to phase equations, and the authors describe three approaches for obtaining such phase model descriptions: the "seat of the pants" approach, the geometric approach, and the singular perturbation approach. Chapter 2 by Krogh–Madsen and colleagues presents the topological approach to phase response analysis. This is the earliest developed method of phase response analysis. It simply assumes that biological oscillations are stable limit cycles and considers the effects of perturbations of a neuron's spiking limit cycle within the surrounding phase space. Chapter 3 by Remme and colleagues combines cable theory, used to describe the dynamics of voltage spread in the dendritic processes of neurons, with weak coupling theory to characterize the interactions of dendritic oscillations in different regions of a neuron's dendritic tree. This chapter takes a broader view of the integrative properties of individual neurons by considering the contribution of dendrites to neuronal computation which will be addressed further in later chapters. Finally, in Chap. 4, Canavier and Achuthan introduce analyses of pulse-coupled networks wherein the constituent neuronal oscillators interact at discrete times, perturbing one another away from their respective limit cycles. This approach makes use of maps to describe the evolution of individual neurons' phases across network cycles and to predict stable periodic firing modes. In some cases perturbations among pulse-coupled networks can be significant, violating the weak-coupling assumptions described in the first chapter. Thus, the conditions or systems that can be analyzed with weak-coupling and pulse-coupling methods are relatively distinct.

Chapter 1
The Theory of Weakly Coupled Oscillators

Michael A. Schwemmer and Timothy J. Lewis

Abstract This chapter focuses on the application of phase response curves (PRCs) in predicting the phase locking behavior in networks of periodically oscillating neurons using the theory of weakly coupled oscillators. The theory of weakly coupled oscillators can be used to predict phase-locking in neuronal networks with any form of coupling. As the name suggests, the coupling between cells must be sufficiently weak for these predictions to be quantitatively accurate. This implies that the coupling can only have small effects on neuronal dynamics over any given cycle. However, these small effects can accumulate over many cycles and lead to phase locking in the neuronal network. The theory of weak coupling allows one to reduce the dynamics of each neuron, which could be of very high dimension, to a single differential equation describing the phase of the neuron.

The main goal of this chapter is to explain how a weakly coupled neuronal network is reduced to its phase model description. Three different ways to derive the phase equations are presented, each providing different insight into the underlying dynamics of phase response properties and phase-locking dynamics. The technique is illustrated for a weakly coupled pair of identical neurons. We then show how the phase model for a pair of cells can be extended to include weak heterogeneity and small amplitude noise. Lastly, we outline two mathematical techniques for analyzing large networks of weakly coupled neurons.

1 Introduction

A phase response curve (PRC) (Winfree 1980) of an oscillating neuron measures the phase shifts in response to stimuli delivered at different times in its cycle. PRCs are often used to predict the phase-locking behavior in networks of neurons

M.A. Schwemmer • T.J. Lewis (✉)
Department of Mathematics, One Shields Ave, University of California, Davis, CA 95616, USA
e-mail: mschwemm@math.princeton.edu; tjlewis@ucdavis.edu

N.W. Schultheiss et al. (eds.), *Phase Response Curves in Neuroscience: Theory, Experiment, and Analysis*, Springer Series in Computational Neuroscience 6,
DOI 10.1007/978-1-4614-0739-3_1, © Springer Science+Business Media, LLC 2012

and to understand the mechanisms that underlie this behavior. There are two main techniques for doing this. Each of these techniques requires a different kind of PRC, and each is valid in a different limiting case. One approach uses PRCs to reduce neuronal dynamics to firing time maps, e.g., (Ermentrout and Kopell 1998; Guevara et al. 1986; Goel and Ermentrout 2002; Mirollo and Strogatz 1990; Netoff et al. 2005b; Oprisan et al. 2004). The second approach uses PRCs to obtain a set of differential equations for the phases of each neuron in the network.

For the derivation of the firing time maps, the stimuli used to generate the PRC should be similar to the input that the neuron actually receives in the network, i.e., a facsimile of a synaptic current or conductance. The firing time map technique can allow one to predict phase locking for moderately strong coupling, but it has the limitation that the neuron must quickly return to its normal firing cycle before subsequent input arrives. Typically, this implies that input to a neuron must be sufficiently brief and that there is only a single input to a neuron each cycle. The derivation and applications of these firing time maps are discussed in Chap. 4.

This chapter focuses on the second technique, which is often referred to as the theory of weakly coupled oscillators (Ermentrout and Kopell 1984; Kuramoto 1984; Neu 1979). The theory of weakly coupled oscillators can be used to predict phase locking in neuronal networks with any form of coupling, but as the name suggests, the coupling between cells must be sufficiently "weak" for these predictions to be quantitatively accurate. This implies that the coupling can only have small effects on neuronal dynamics over any given period. However, these small effects can accumulate over time and lead to phase locking in the neuronal network. The theory of weak coupling allows one to reduce the dynamics of each neuron, which could be of very high dimension, to a single differential equation describing the phase of the neuron. These "phase equations" take the form of a convolution of the input to the neuron via coupling and the neuron's infinitesimal PRC (iPRC). The iPRC measures the response to a small brief (δ-function-like) perturbation and acts like an impulse response function or Green's function for the oscillating neurons. Through the dimension reduction and exploiting the form of the phase equations, the theory of weakly coupled oscillators provides a way to identify phase-locked states and understand the mechanisms that underlie them.

The main goal of this chapter is to explain how a weakly coupled neuronal network is reduced to its phase model description. Three different ways to derive the phase equations are presented, each providing different insight into the underlying dynamics of phase response properties and phase-locking dynamics. The first derivation (the "Seat-of-the-Pants" derivation in Sect. 3) is the most accessible. It captures the essence of the theory of weak coupling and only requires the reader to know some basic concepts from dynamical system theory and have a good understanding of what it means for a system to behave linearly. The second derivation (The Geometric Approach in Sect. 4) is a little more mathematically sophisticated and provides deeper insight into the phase response dynamics of neurons. To make this second derivation more accessible, we tie all concepts back to the explanations in the first derivation. The third derivation (The Singular

Perturbation Approach in Sect. 5) is the most mathematically abstract but it provides the cleanest derivation of the phase equations. It also explicitly shows that the iPRC can be computed as a solution of the "adjoint" equations.

During these three explanations of the theory of weak coupling, the phase model is derived for a pair of coupled neurons to illustrate the reduction technique. The later sections (Sects. 6 and 7) briefly discuss extensions of the phase model to include heterogeneity, noise, and large networks of neurons.

For more mathematically detailed discussions of the theory of weakly coupled oscillators, we direct the reader to (Ermentrout and Kopell 1984; Hoppensteadt and Izhikevich 1997; Kuramoto 1984; Neu 1979).

2 Neuronal Models and Reduction to a Phase Model

2.1 *General Form of Neuronal Network Models*

The general form of a single or multicompartmental Hodgkin–Huxley-type neuronal model (Hodgkin and Huxley 1952) is

$$\frac{\mathrm{d}X}{\mathrm{d}t} = F(X), \tag{1.1}$$

where X is a N-dimensional state variable vector containing the membrane potential(s) and gating variables[1], and $F(X)$ is a vector function describing the rate of change of the variables in time. For the Hodgkin–Huxley (HH) model (Hodgkin and Huxley 1952), $X = [V, m, h, n]^{\mathrm{T}}$ and

$$F(X) = \begin{bmatrix} \frac{1}{C}(-g_{\mathrm{Na}}m^3h(V - E_{\mathrm{Na}}) - g_K n^4(V - E_K) - g_L(V - E_L) + I) \\ \dfrac{m_\infty(V) - m}{\tau_m(V)} \\ \dfrac{h_\infty(V) - h}{\tau_h(V)} \\ \dfrac{n_\infty(V) - n}{\tau_n(V)}, \end{bmatrix}, \tag{1.2}$$

In this chapter, we assume that the isolated model neuron (1.1) exhibits stable T-periodic firing (e.g., top trace of Fig. 1.2). In the language of dynamical systems, we assume that the model has an asymptotically stable T-periodic *limit cycle*. These oscillations could be either due to intrinsic conductances or induced by applied current.

[1]The gating variables could be for ionic membrane conductances in the neuron, as well as those describing the output of chemical synapses.

A pair of coupled model neurons is described by

$$\frac{dX_1}{dt} = F(X_1) + \varepsilon I(X_1, X_2) \tag{1.3}$$

$$\frac{dX_2}{dt} = F(X_2) + \varepsilon I(X_2, X_1), \tag{1.4}$$

where $I(X_1, X_2)$ is a vector function describing the coupling between the two neurons, and ε scales the magnitude of the coupling term. Typically, in models of neuronal networks, cells are only coupled through the voltage (V) equation. For example, a pair of electrically coupled HH neurons would have the coupling term

$$I(X_1, X_2) = \begin{bmatrix} \frac{1}{C}\left(g_C(V_2 - V_1)\right) \\ 0 \\ 0 \\ 0 \end{bmatrix}. \tag{1.5}$$

where g_C is the coupling conductance of the electrical synapse (see Chap. 14).

2.2 *Phase Models, the G-Function, and Phase Locking*

The power of the theory of weakly coupled oscillators is that it reduces the dynamics of each neuronal oscillator in a network to single phase equation that describes the rate of change of its relative phase, ϕ_j. The phase model corresponding to the pair of coupled neurons (1.3)–(1.4) is of the form

$$\frac{d\phi_1}{dt} = \varepsilon H(\phi_2 - \phi_1) \tag{1.6}$$

$$\frac{d\phi_2}{dt} = \varepsilon H(-(\phi_2 - \phi_1)). \tag{1.7}$$

The following sections present three different ways of deriving the function H, which is often called the interaction function.

Subtracting the phase equation for cell 1 from that of cell 2, the dynamics can be further reduced to a single equation that governs the evolution of the phase difference between the cells, $\phi = \phi_2 - \phi_1$

$$\frac{d\phi}{dt} = \varepsilon(H(-\phi) - H(\phi)) = \varepsilon G(\phi). \tag{1.8}$$

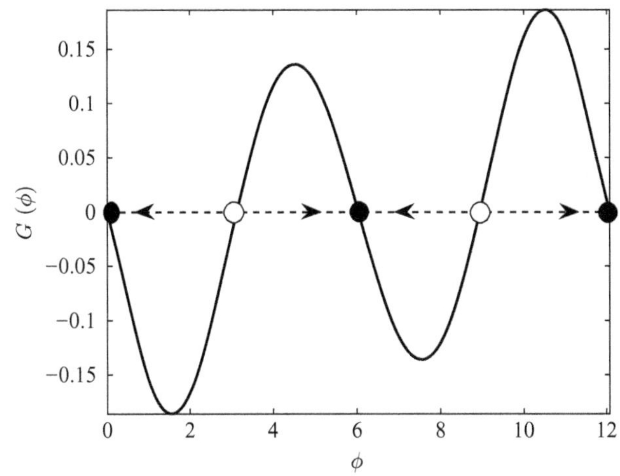

Fig. 1.1 *Example G function.* The G function for two model Fast–Spiking (FS) interneurons (Erisir et al. 1999) coupled with gap junctions on the distal ends of their passive dendrites is plotted. The arrows show the direction of the trajectories for the system. This system has four steady state solutions $\phi_S = 0, T$ (synchrony), $\phi_{\mathrm{AP}} = T/2$ (antiphase), and two other nonsynchronous states. One can see that synchrony and antiphase are stable steady states for this system (*filled in circles*) while the two other nonsynchronous solutions are unstable (*open circles*). Thus, depending on the initial conditions, the two neurons will fire synchronously or in antiphase

In the case of a pair of coupled Hodgkin–Huxley neurons (as described above), the number of equations in the system is reduced from the original 8 describing the dynamics of the voltage and gating variables to a single equation. The reduction method can also be readily applied to multicompartment model neurons, e.g., (Lewis and Rinzel 2004; Zahid and Skinner 2009), which can render a significantly larger dimension reduction. In fact, the method has been applied to real neurons as well, e.g., (Mancilla et al. 2007).

Note that the function $G(\phi)$ or "G-function" can be used to easily determine the phase-locking behavior of the coupled neurons. The zeros of the G-function, ϕ^*, are the steady state phase differences between the two cells. For example, if $G(0) = 0$, this implies that the synchronous solution is a steady state of the system. To determine the stability of the steady state note that when $G(\phi) > 0$, ϕ will increase and when $G(\phi) < 0$, ϕ will decrease. Therefore, if the derivative of G is positive at a steady state ($G'(\phi^*) > 0$), then the steady state is unstable. Similarly, if the derivative of G is negative at a steady state ($G'(\phi^*) < 0$), then the steady state is stable. Figure 1.1 shows an example G-function for two coupled identical cells. Note that this system has 4 steady states corresponding to $\phi = 0, T$ (synchrony), $\phi = T/2$ (antiphase), and two other nonsynchronous states. It is also clearly seen that $\phi = 0, T$ and $\phi = T/2$ are stable steady states and the other nonsynchronous states are unstable. Thus, the two cells in this system exhibit bistability, and they will either synchronize their firing or fire in antiphase depending upon the initial conditions.

In Sects. 3, 4, and 5, we present three different ways of derive the interaction function H and therefore the G-function. These derivations make several approximations that require the coupling between neurons to be sufficiently weak. "Sufficiently weak" implies that the neurons' intrinsic dynamics dominate the effects due to coupling at each point in the periodic cycle, i.e., during the periodic oscillations, $|F(X_j(t))|$ should be an order of magnitude greater than $|\varepsilon I(X_1(t), X_2(t))|$. However, it is important to point out that, even though the phase models quantitatively capture the dynamics of the full system for sufficiently small ε, it is often the case that they can also capture the qualitative behavior for moderate coupling strengths (Lewis and Rinzel 2003; Netoff et al. 2005a).

3 A "Seat-of-the-Pants" Approach

This section will describe perhaps the most intuitive way of deriving the phase model for a pair of coupled neurons (Lewis and Rinzel 2003). The approach highlights the key aspect of the theory of weakly coupled oscillators, which is that neurons behave linearly in response to small perturbations and therefore obey the principle of superposition.

3.1 Defining Phase

T-periodic firing of a model neuronal oscillator (1.1) corresponds to repeated circulation around an asymptotically stable T-periodic limit cycle, i.e., a closed orbit in state space X. We will denote this T-periodic limit cycle solution as $X_{\mathrm{LC}}(t)$. The phase of a neuron is a measure of the time that has elapsed as the neuron's moves around its periodic orbit, starting from an arbitrary reference point in the cycle. We define the phase of the periodically firing neuron j at time t to be

$$\theta_j(t) = (t + \phi_j) \mod T, \tag{1.9}$$

where $\theta_j = 0$ is set to be at the peak of the neurons' spike (Fig. 1.2).[2] The constant ϕ_j, which is referred to as the *relative phase* of the jth neuron, is determined by the position of the neuron on the limit cycle at time $t = 0$. Note that each phase of the neuron corresponds to a unique position on the cell's T-periodic limit cycle, and any solution of the uncoupled neuron model that is on the limit cycle can be expressed as

$$X_j(t) = X_{\mathrm{LC}}(\theta_j(t)) = X_{\mathrm{LC}}(t + \phi_j). \tag{1.10}$$

[2]Phase is often normalized by the period T or by $T/2\pi$, so that $0 \le \theta < 1$ or $0 \le \theta < 2\pi$ respectively. Here, we do not normalize phase and take $0 \le \theta < T$.

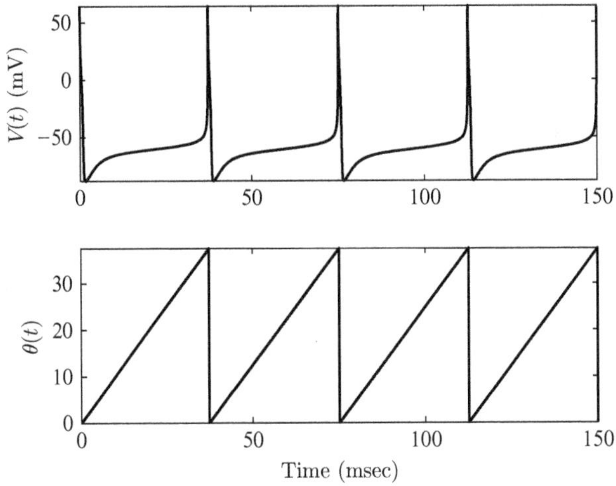

Fig. 1.2 *Phase*. (**a**) Voltage trace for the Fast-Spiking interneuron model from Erisir et al. (1999) with $I_{\rm appl} = 35\ \mu$A/cm^2 showing T-periodic firing. (**b**) The phase $\theta(t)$ of these oscillations increases linearly from 0 to T, and we have assumed that zero phase occurs at the peak of the voltage spike

When a neuron is perturbed by coupling current from other neurons or by any other external stimulus, its dynamics no longer exactly adhere to the limit cycle, and the exact correspondence of time to phase (1.9) is no longer valid. However, when perturbations are sufficiently weak, the neuron's intrinsic dynamics are dominant. This ensures that the perturbed system remains close to the limit cycle and the interspike intervals are close to the intrinsic period T. Therefore, we can approximate the solution of neuron j by $X_j(t) \simeq X_{\rm LC}(t + \phi_j(t))$, where the relative phase ϕ_j is now a function of time t. Over each cycle of the oscillations, the weak perturbations to the neurons produce only small changes in ϕ_j. These changes are negligible over a single cycle, but they can slowly accumulate over many cycles and produce substantial effects on the relative firing times of the neurons.

The goal now is to understand how the relative phase $\phi_j(t)$ of the coupled neurons evolves slowly in time. To do this, we first consider the response of a neuron to small abrupt current pulses.

3.2 *The Infinitesimal Phase Response Curve*

Suppose that a small brief square current pulse of amplitude εI_0 and duration Δt is delivered to a neuron when it is at phase θ^*. This small, brief current pulse causes the membrane potential to abruptly increase by $\delta V \simeq \varepsilon I_0 \Delta t / C$, i.e., the change in voltage will approximately equal the total charge delivered to the cell by

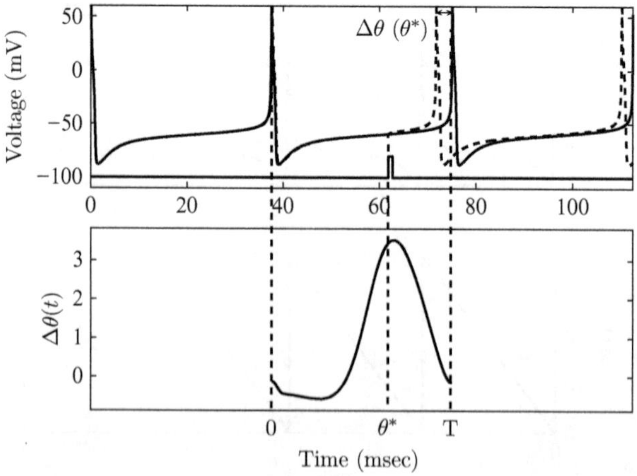

Fig. 1.3 *Measuring the Phase Response Curve from Neurons.* The voltage trace and corresponding PRC is shown for the same FS model neuron from Fig. 1.2. The PRC is measured from a periodically firing neuron by delivering small current pulses at every point, θ^*, along its cycle and measuring the subsequent change in period, $\Delta\theta$, caused by the current pulse

the stimulus, $\varepsilon I_0 \Delta t$, divided by the capacitance of the neuron, C. In general, this perturbation can cause the cell to fire sooner (phase advance) or later (phase delay) than it would have fired without the perturbation. The magnitude and sign of this *phase shift* depends on the amplitude and duration of the stimulus, as well as the phase in the oscillation at which the stimulus was delivered, θ^*. This relationship is quantified by the Phase Response Curve (PRC), which gives the phase shift $\Delta\phi$ as a function of the phase θ^* for a fixed $\varepsilon I_0 \Delta t$ (Fig. 1.3).

For sufficiently small and brief stimuli, the neuron will respond in a linear fashion, and the PRC will scale linearly with the magnitude of the current stimulus

$$\Delta\phi(\theta^*) \simeq Z_V(\theta^*)\,\delta V = Z_V(\theta^*)\left(\frac{1}{C}\varepsilon I_0 \Delta t\right), \qquad 0 \le \theta^* < T, \tag{1.11}$$

where $Z_V(\theta^*)$ describes the proportional phase shift as a function of the phase of the stimulus. The function $Z_V(\theta)$ is known as the infinitesimal phase response curve (iPRC) or the phase-dependent sensitivity function for voltage perturbations. The iPRC $Z_V(\theta)$ quantifies the normalized phase shift due to an infinitesimally small δ-function-like voltage perturbation delivered at any given phase on the limit cycle.

3.3 The Phase Model for a Pair of Weakly Coupled Cells

Now we can reconsider the pair of weakly coupled neuronal oscillators (1.3)–(1.4). Recall that, because the coupling is weak, the neurons' intrinsic dynamics dominate the dynamics of the coupled-cell system, and $X_j(t) \simeq X_{\mathrm{LC}}(\theta_j(t)) = X_{\mathrm{LC}}(t + \phi_j(t))$ for $j = 1, 2$. This assumes that the coupling current can only affect the speed at which cells move around their limit cycle and does not affect the amplitude of the oscillations. Thus, the effects of the coupling are entirely captured in the slow time dynamics of the relative phases of the cells $\phi_j(t)$.

The assumption of weak coupling also ensures that the perturbations to the neurons are sufficiently small so that the neurons respond linearly to the coupling current. That is, (i) the small phase shifts of the neurons due to the presence of the coupling current for a brief time Δt can be approximated using the iPRC (1.11), and (ii) these small phase shifts in response to the coupling current sum linearly (i.e., the principle of superposition holds). Therefore, by (1.11), the phase shift due to the coupling current from t to $t + \Delta t$ is

$$\begin{aligned}\Delta\phi_j(t) &= \phi_j(t + \Delta t) - \phi_j(t)\\ &\simeq Z_V(\theta_j(t))\,(\varepsilon I(X_j(t), X_k(t)))\Delta t.\\ &= Z_V(t + \phi_j(t))\,\big(\varepsilon I(X_{\mathrm{LC}}(t + \phi_j(t)), X_{\mathrm{LC}}(t + \phi_k(t)))\big)\,\Delta t. \qquad (1.12)\end{aligned}$$

By dividing the above equation by Δt and taking the limit as $\Delta t \to 0$, we obtain a system of differential equations that govern the evolution of the relative phases of the two neurons

$$\frac{\mathrm{d}\phi_j}{\mathrm{d}t} = \varepsilon\, Z_V(t + \phi_j)\, I(X_{\mathrm{LC}}(t + \phi_j), X_{\mathrm{LC}}(t + \phi_k)), \quad j, k = 1, 2;\ j \neq k. \quad (1.13)$$

Note that, by integrating this system of differential equations to find the solution $\phi_j(t)$, we are assuming that phase shifts in response to the coupling current sum linearly.

The explicit time dependence on the right-hand side of (1.13) can be eliminated by "averaging" over the period T. Note that $Z_V(t)$ and $X_{\mathrm{LC}}(t)$ are T-periodic functions, and the scaling of the right-hand side of (1.13) by the small parameter ε indicates that changes in the relative phases ϕ_j occur on a much slower timescale than T. Therefore, we can integrate the right-hand side over the full period T holding the values of ϕ_j constant to find the average rate of change of ϕ_j over a cycle. Thus, we obtain equations that approximate the slow time evolution of the relative phases ϕ_j,

$$\begin{aligned}\frac{\mathrm{d}\phi_j}{\mathrm{d}t} &= \varepsilon \frac{1}{T}\int_0^T Z_V(\tilde{t})\,\big(I(X_{\mathrm{LC}}(\tilde{t}), X_{\mathrm{LC}}(\tilde{t} + \phi_k - \phi_j))\big)\,\mathrm{d}\tilde{t}\\ &= \varepsilon H(\phi_k - \phi_j), \quad j, k = 1, 2;\ j \neq k, \qquad (1.14)\end{aligned}$$

i.e., the relative phases ϕ_j are assumed to be constant with respect to the integral over T in $\tilde{t}$, but they vary in t. This averaging process is made rigorous by averaging theory (see Ermentrout and Kopell 1991; Guckenheimer and Holmes 1983).

We have reduced the dynamics of a pair of weakly coupled neuronal oscillators to an autonomous system of two differential equations describing the phases of the neurons and therefore finished the first derivation of the equations for a pair of weakly coupled neurons.[3] Note that the above derivation can be easily altered to obtain the phase model of a neuronal oscillator subjected to T-periodic external forcing as well. The crux of the derivation was identifying the iPRC and exploiting the approximately linear behavior of the system in response to weak inputs. In fact, it is useful to note that the interaction function H takes the form of a convolution of the iPRC and the coupling current, i.e., the input to the neuron. Therefore, one can think of the iPRC of an oscillator as acting like an impulse response function or Green's function.

3.3.1 Averaging Theory

Averaging theory (see Ermentrout and Kopell 1991; Guckenheimer and Holmes 1983) states that there is a change of variables that maps solutions of

$$\frac{\mathrm{d}\phi}{\mathrm{d}\tilde{t}} = \varepsilon g(\phi, \tilde{t}), \tag{1.15}$$

where $g(\phi, \tilde{t})$ is a T-periodic function in ϕ and $\tilde{t}$, to solutions of

$$\frac{\mathrm{d}\varphi}{\mathrm{d}t} = \varepsilon \bar{g}(\varphi) + \mathcal{O}(\varepsilon^2), \tag{1.16}$$

where

$$\bar{g}(\varphi) = \frac{1}{T}\int_0^T g(\varphi, \tilde{t})\mathrm{d}\tilde{t}, \tag{1.17}$$

and $\mathcal{O}(\varepsilon^2)$ is Landau's "Big O" notation, which represents terms that either have a scaling factor of ε^2 or go to zero at the same rate as ε^2 goes to zero as ε goes to zero.

4 A Geometric Approach

In this section, we describe a geometric approach to the theory of weakly coupled oscillators originally introduced by Kuramoto (1984). The main asset of this approach is that it gives a beautiful geometric interpretation of the iPRC and deepens our understanding of the underlying mechanisms of the phase response properties of neurons.

---

[3]Note that this reduction is not valid when T is of the same order of magnitude as the timescale for the changes due to the weak coupling interactions (e.g., close to a SNIC bifurcation), however an alternative dimension reduction can be performed in this case (Ermentrout 1996).

4.1 The One-to-One Map Between Points on the Limit Cycle and Phase

Consider again a model neuron (1.1) that has a stable T-periodic limit cycle solution $X_{\rm LC}(t)$ such that the neuron exhibits a T-periodic firing pattern (e.g., top trace of Fig. 1.2). Recall that the phase of the oscillator along its limit cycle is defined as $\theta(t) = (t + \phi) \mod T$, where the relative phase ϕ is a constant that is determined by the initial conditions. Note that there is a one-to-one correspondence between phase and each point on the limit cycle. That is, the limit cycle solution takes phase to a unique point on the cycle, $X = X_{\rm LC}(\theta)$, and its inverse maps each point on the limit cycle to a unique phase, $\theta = X_{\rm LC}^{-1}(X) = \Phi(X)$.

Note that it follows immediately from the definition of phase (1.9) that the rate of change of phase in time along the limit cycle is equal to 1, i.e., $\frac{d\theta}{dt} = 1$. Therefore, if we differentiate the map $\Phi(X)$ with respect to time using the chain rule for vector functions, we obtain the following useful relationship

$$\frac{d\theta}{dt} = \nabla_X \Phi(X_{\rm LC}(t)) \cdot \frac{dX_{\rm LC}}{dt} = \nabla_X \Phi(X_{\rm LC}(t)) \cdot F(X_{\rm LC}(t))) = 1, \qquad (1.18)$$

where $\nabla_X \Phi$ is the gradient of the map $\Phi(X)$ with respect to the vector of the neuron's state variables $X = (x_1, x_2, \ldots, x_N)$

$$\nabla_X \Phi(X) = \left[\left(\frac{\partial \Phi}{\partial x_1}, \frac{\partial \Phi}{\partial x_2}, \ldots, \frac{\partial \Phi}{\partial x_N} \right)\Bigg|_X \right]^{\rm T}. \qquad (1.19)$$

(We have defined the gradient as a column vector for notational reasons).

4.2 Asymptotic Phase and the Infinitesimal Phase Response Curve

The map $\theta = \Phi(X)$ is well defined for all points X on the limit cycle. We can extend the domain of $\Phi(X)$ to points off the limit cycle by defining *asymptotic phase*. If X_0 is a point on the limit cycle and Y_0 is a point in a neighborhood of the limit cycle[4], then we say that Y_0 has the same asymptotic phase as X_0 if $||X(t; X_0) - X(t; Y_0)|| \to 0$ as $t \to \infty$. This means that the solution starting at the initial point Y_0 off the limit cycle converges to the solution starting at the point X_0 on the limit cycle as time goes to infinity. Therefore, $\Phi(Y_0) = \Phi(X_0)$. The set of

[4]In fact, the point Y_0 can be anywhere in the basin of attraction of the limit cycle.

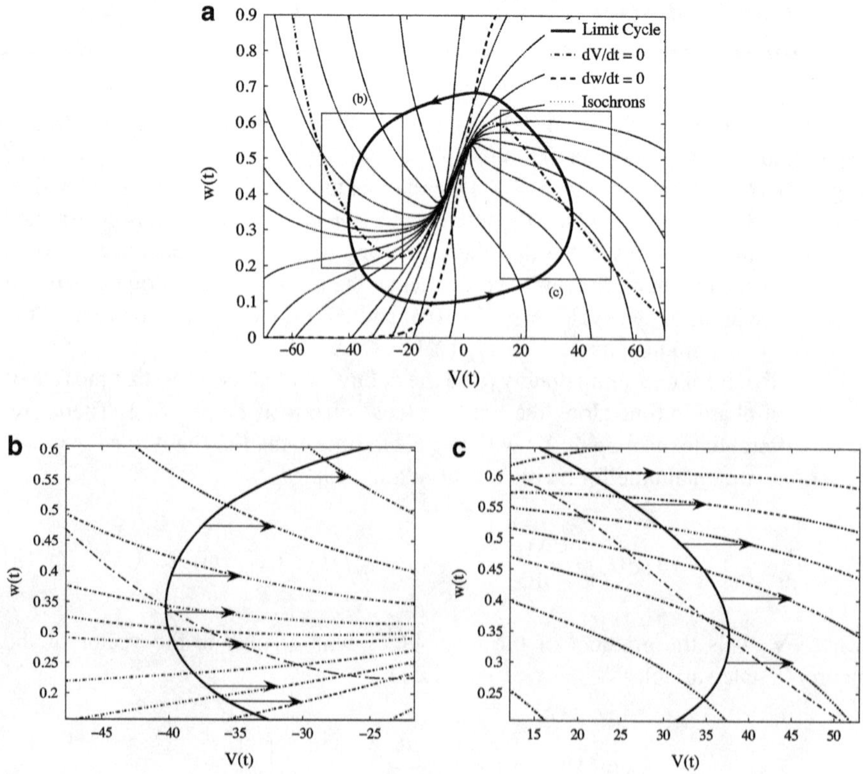

Fig. 1.4 *Example Isochron Structure*. (**a**) The limit cycle and isochron structure for the Morris–Lecar neuron (Morris and Lecar 1981) is plotted along with the nullclines for the system. (**b**) Blow up of a region on the left-hand side of the limit cycle showing how the same strength perturbation in the voltage direction can cause different phase delays or phase advances. (**c**) Blow up of a region on the right-hand side of the limit cycle showing also that the same size voltage perturbation can cause phase advances of different sizes

all points off the limit cycle that have the same asymptotic phase as the point X_0 on the limit cycle is known as the *isochron* (Winfree 1980) for phase $\theta = \Phi(X_0)$. Figure 1.4 shows some isochrons around the limit cycle for the Morris–Lecar neuron (Morris and Lecar 1981). It is important to note that the figure only plots isochrons for a few phases and that *every* point on the limit cycle has a corresponding isochron.

Equipped with the concept of asymptotic phase, we can now show that the iPRC is in fact the gradient of the phase map $\nabla_X \Phi(X_{LC}(t))$ by considering the following phase resetting "experiment". Suppose that, at time t, the neuron is on the limit cycle in state $X(t) = X_{LC}(\theta^*)$ with corresponding phase $\theta^* = \Phi(X(t))$. At this time, it receives a small abrupt external perturbation εU, where ε is the magnitude of the perturbation and U is the unit vector in the direction of the perturbation in state space. Immediately after the perturbation, the neuron is in the state $X_{\mathrm{LC}}(\theta^*) + \varepsilon U$, and its new asymptotic phase is $\tilde{\theta} = \Phi(X_{\mathrm{LC}}(\theta^*) + \varepsilon U)$. Using Taylor series,

$$\tilde{\theta} = \Phi(X_{\mathrm{LC}}(\theta^*) + \varepsilon U) = \Phi(X_{\mathrm{LC}}(\theta^*)) + \nabla_X \Phi(X_{\mathrm{LC}}(\theta^*)) \cdot (\varepsilon U) + \mathcal{O}(\varepsilon^2). \quad (1.20)$$

Keeping only the linear term (i.e., $\mathcal{O}(\varepsilon)$ term), the phase shift of the neuron as a function of the phase θ^* at which it received the εU perturbation is given by

$$\Delta\phi(\theta^*) = \tilde{\theta} - \theta^* \simeq \nabla_X \Phi(X_{\mathrm{LC}}(\theta^*)) \cdot (\varepsilon U). \quad (1.21)$$

As was done in Sect. 3.2, we normalize the phase shift by the magnitude of the stimulus,

$$\frac{\Delta\phi(\theta^*)}{\varepsilon} \simeq \nabla_X \Phi(X_{\mathrm{LC}}(\theta^*)) \cdot U = Z(\theta^*) \cdot U. \quad (1.22)$$

Note that $Z(\theta) = \nabla_X \Phi(X_{\mathrm{LC}}(\theta))$ is the iPRC. It quantifies the normalized phase shift due to a small delta-function-like perturbation delivered at any given on the limit cycle. As was the case for the iPRC Z_V derived in the previous section [see (1.11)], $\nabla_X \Phi(X_{\mathrm{LC}}(\theta))$ captures only the linear response of the neuron and is quantitatively accurate only for sufficiently small perturbations. However, unlike Z_V, $\nabla_X \Phi(X_{\mathrm{LC}}(\theta))$ captures the response to perturbations in any direction in state space and not only in one variable (e.g., the membrane potential). That is, $\nabla_X \Phi(X_{\mathrm{LC}}(\theta))$ is the vector iPRC; its components are the iPRCs for every variable in the system (see Fig. 1.5).

In the typical case of a single-compartment HH model neuron subject to an applied current pulse (which perturbs only the membrane potential), the perturbation would be of the form $\varepsilon U = (u, 0, 0, \ldots, 0)$ where x_1 is the membrane potential V. By (1.20), the phase shift is

$$\Delta\phi(\theta) = \frac{\partial \Phi}{\partial V}(X_{\mathrm{LC}}(\theta))\; u = Z_V(\theta)\; u, \quad (1.23)$$

which is the same as (1.11) derived in the previous section.

With the understanding that $\nabla_X \Phi(X_{\mathrm{LC}}(t))$ is the vector iPRC, we now derive the phase model for two weakly coupled neurons.

4.3 A Pair of Weakly Coupled Oscillators

Now consider the system of weakly coupled neurons (1.3)–(1.4). We can use the map Φ to take the variables $X_1(t)$ and $X_2(t)$ to their corresponding asymptotic phase, i.e., $\theta_j(t) = \Phi(X_j(t))$ for $j = 1, 2$. By the chain rule, we obtain the change in phase with respect to time

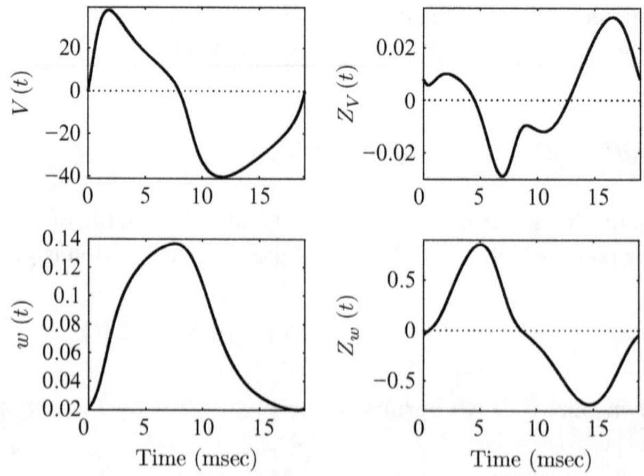

Fig. 1.5 *iPRCs for the Morris–Lecar Neuron.* The voltage, $V(t)$ and channel, $w(t)$, components of the limit cycle for the same Morris–Lecar neuron as in Fig. 1.4 are plotted along with their corresponding iPRCs. Note that the shape of voltage iPRC can be inferred from the insets of Fig. 1.4. For example, the isochronal structure in Fig. 1.4c reveals that perturbations in the voltage component will cause phase advances when the voltage is ~30 to 38 mV

$$\begin{aligned}\frac{d\theta_j}{dt} &= \nabla_X \Phi(X_j(t)) \cdot \frac{dX_j}{dt} \\ &= \nabla_X \Phi(X_j(t)) \cdot \left[F(X_j(t)) + \varepsilon I(X_j(t), X_k(t))\right] \\ &= \nabla_X \Phi(X_j(t)) \cdot F(X_j(t)) + \nabla_X \Phi(X_j(t)) \cdot \left[\varepsilon I(X_j(t), X_k(t))\right] \\ &= 1 + \varepsilon \nabla_X \Phi(X_j(t)) \cdot I(X_j(t), X_k(t)), \end{aligned} \tag{1.24}$$

where we have used the "useful" relation (1.18). Note that the above equations are exact. However, in order to solve the equations for $\theta_j(t)$, we would already have to know the full solutions $X_1(t)$ and $X_2(t)$, in which case you wouldn't need to reduce the system to a phase model. Therefore, we exploit that fact that ε is small and make the approximation $X_j(t) \sim X_{LC}(\theta_j(t)) = X_{LC}(t + \phi_j(t))$, i.e., the coupling is assumed to be weak enough so that it does not affect the amplitude of the limit cycle, but it can affect the rate at which the neuron moves around its limit cycle. By making this approximation in (1.24) and making the change of variables $\theta_j(t) = t + \phi_j(t)$, we obtain the equations for the evolution of the relative phases of the two neurons

$$\frac{d\phi_j}{dt} = \varepsilon \nabla_X \Phi(X_{LC}(t + \phi_j(t))) \cdot I(X_{LC}(t + \phi_j(t)), X_{LC}(t + \phi_k(t))). \tag{1.25}$$

Note that these equations are the vector versions of (1.13) with the iPRC written as $\nabla_X \Phi(X_{LC}(t))$. As described in the previous section, we can average these equations over the period T to eliminate the explicit time dependence and obtain the phase model for the pair of coupled neurons

$$\frac{\mathrm{d}\phi_j}{\mathrm{d}t} = \varepsilon \frac{1}{T}\int_0^T \nabla_X \Phi(X_{\mathrm{LC}}(\tilde{t})) \cdot I(X_{\mathrm{LC}}(\tilde{t}), X_{\mathrm{LC}}(\tilde{t} + (\phi_k - \phi_j)))\mathrm{d}\tilde{t} = \varepsilon H(\phi_k - \phi_j). \tag{1.26}$$

Note that while the above approach to deriving the phase equations provides substantial insight into the geometry of the neuronal phase response dynamics, it does not provide a computational method to compute the iPRC for model neurons, i.e., we still must directly measure the iPRC using extensive numerical simulations as described in the previous section.

5 A Singular Perturbation Approach

In this section, we describe the singular perturbation approach to derive the theory of weakly coupled oscillators. This systematic approach was developed by Malkin (1949; 1956), Neu (1979), and Ermentrout and Kopell (1984). The major practical asset of this approach is that it provides a simple method to compute iPRCs for model neurons.

Consider again the system of weakly coupled neurons (1.3)–(1.4). We assume that the isolated neurons have asymptotically stable T-periodic limit cycle solutions $X_{\mathrm{LC}}(t)$ and that coupling is weak (i.e., ε is small). As previously stated, the weak coupling has small effects on the dynamics of the neurons. On the timescale of a single cycle, these effects are negligible. However, the effects can slowly accumulate on a much slower timescale and have a substantial influence on the relative firing times of the neurons. We can exploit the differences in these two timescales and use the method of multiple scales to derive the phase model.

First, we define a "fast time" $t_f = t$, which is on the timescale of the period of the isolated neuronal oscillator, and a "slow time" $t_s = \varepsilon t$, which is on the timescale that the coupling affects the dynamics of the neurons. Time, t, is thus a function of both the fast and slow times, i.e., $t = f(t_f, t_s)$. By the chain rule, $\frac{\mathrm{d}}{\mathrm{d}t} = \frac{\partial}{\partial t_f} + \varepsilon\frac{\partial}{\partial t_s}$. We then assume that solutions $X_1(t)$ and $X_2(t)$ can be expressed as power series in ε that are dependent both on t_f and t_s,

$$X_j(t) = X_j^0(t_f, t_s) + \varepsilon X_j^1(t_f, t_s) + \mathcal{O}(\varepsilon^2), \quad j = 1, 2.$$

Substituting these expansions into (1.3)–(1.4) yields

$$\begin{aligned}\frac{\partial X_j^0}{\partial t_f} + \varepsilon\frac{\partial X_j^0}{\partial t_s} + \varepsilon\frac{\partial X_j^1}{\partial t_f} + \mathcal{O}(\varepsilon^2) &= F(X_j^0 + \varepsilon X_j^1 + \mathcal{O}(\varepsilon^2)) \\ &\quad + \varepsilon I(X_j^0 + \varepsilon X_j^1 + \mathcal{O}(\varepsilon^2), X_k^0 + \varepsilon X_k^1 + \mathcal{O}(\varepsilon^2)), \\ &\quad j,k = 1,2;\ j \neq k. \end{aligned} \tag{1.27}$$

Using Taylor series to expand the vector functions F and I in terms of ε, we obtain

$$F(X_j^0 + \varepsilon X_j^1 + \mathcal{O}(\varepsilon^2)) = F(X_j^0) + \varepsilon DF(X_j^0)X_j^1 + \mathcal{O}(\varepsilon^2) \tag{1.28}$$

$$\varepsilon I(X_j^0 + \varepsilon X_j^1 + \mathcal{O}(\varepsilon^2), X_k^0 + \varepsilon X_k^1 + \mathcal{O}(\varepsilon^2)) = \varepsilon I(X_j^0, X_k^0) + \mathcal{O}(\varepsilon^2), \tag{1.29}$$

where $DF(X_j^0)$ is the Jacobian, i.e., matrix of partial derivatives, of the vector function $F(X_j)$ evaluated at X_j^0. We then plug these expressions into (1.27), collect like terms of ε, and equate the coefficients of like terms.[5]

The leading order ($\mathcal{O}(1)$) terms yield

$$\frac{\partial X_j^0}{\partial t_f} = F(X_j^0), \quad j = 1, 2. \tag{1.30}$$

These are the equations that describe the dynamics of the uncoupled cells. Thus, to leading order, each cell exhibits the T-periodic limit cycle solution $X_j^0(t_f, t_s) = X_{\mathrm{LC}}(t_f + \phi_j(t_s))$. Note that (1.30) implies that the relative phase ϕ_j is constant in t_f, but it can still evolve on the slow timescale t_s.

Substituting the solutions for the leading order equations (and shifting t_f appropriately), the $\mathcal{O}(\varepsilon)$ terms of (1.27) yield

$$\mathcal{L}X_j^1 \equiv \frac{\partial X_j^1}{\partial t_f} - DF(X_{\mathrm{LC}}(t_f))X_j^1 = I(X_{\mathrm{LC}}(t_f), X_{\mathrm{LC}}(t_f - (\phi_j(t_s) - \phi_k(t_s)))) \\ - X'_{\mathrm{LC}}(t_f)\frac{\mathrm{d}\phi_j}{\mathrm{d}t_s}. \tag{1.31}$$

To simplify notation, we have defined the linear operator $\mathcal{L}X \equiv \frac{\partial X}{\partial t_f} - DF(X_{\mathrm{LC}}(t_f))X$, which acts on a T-periodic domain and is therefore bounded. Note that (1.31) is a linear differential equation with T-periodic coefficients. In order for our power series solutions for $X_1(t)$ and $X_2(t)$ to exist, a solution to (1.31) must exist. Therefore, we need to find conditions that guarantee the existence of a solution to (1.31), i.e., conditions that ensure that the right-hand side of (1.31) is in the range of the operator $\mathcal{L}$. The Fredholm Alternative explicitly provides us with these conditions.

Theorem 1 (Fredholm Alternative). Suppose that

$$(*) \qquad \mathcal{L}x = \frac{\mathrm{d}x}{\mathrm{d}t} + A(t)x = f(t); \;\; x \in \mathbb{R}^N,$$

where the matrix $A(t)$ and the vector function $f(t)$ are continuous and T-periodic. Then, there is a continuous T-periodic solution $x(t)$ to (*) if and only if

$$(**) \qquad \frac{1}{T}\int_0^T Z(t) \cdot f(t)\mathrm{d}t = 0,$$

[5] Because the equation should hold for arbitrary ε, coefficients of like terms must be equal.

for each continuous T-periodic solution, $Z(t)$, to the adjoint problem

$$\mathcal{L}^* x = -\frac{\mathrm{d}Z}{\mathrm{d}t} + \{A(t)\}^{\mathrm{T}} Z = 0.$$

where $\{A(t)^T\}$ is the transpose of the matrix $A(t)$.

In the notation of the above theorem,

$$A(t) = -DF(X_{\mathrm{LC}}(t_f)) \quad \text{and} \quad f(t) = I(X_{\mathrm{LC}}(t_f), X_{\mathrm{LC}}(t_f - (\phi_j(t_s) - \phi_k(t_s)))) \\ -X'_{\mathrm{LC}}(t_f)\frac{\mathrm{d}\phi_j}{\mathrm{d}t_s}.$$

Thus, the solvability condition (**) requires that

$$\frac{1}{T}\int_0^T Z(t_f)\cdot\left[I(X_{\mathrm{LC}}(t_f), X_{\mathrm{LC}}(t_f - (\phi_j(t_s) - \phi_k(t_s)))) - X'_{\mathrm{LC}}(t_f)\frac{\mathrm{d}\phi_j}{\mathrm{d}t_s}\right]\mathrm{d}t_f = 0 \tag{1.32}$$

where Z is a T-periodic solution of the adjoint equation

$$\mathcal{L}^* Z = -\frac{\partial Z}{\partial t_f} - DF(X_{\mathrm{LC}}(t_f))^T Z = 0. \tag{1.33}$$

Rearranging (1.32),

$$\frac{\mathrm{d}\phi_j}{\mathrm{d}t_s} = \frac{1}{T}\int_0^T Z(t_f)\cdot\left[I(X_{\mathrm{LC}}(t_f), X_{\mathrm{LC}}(t_f - (\phi_j(t_s) - \phi_k(t_s))))\right]\mathrm{d}t_f \tag{1.34}$$

where we have normalized $Z(t_f)$ by

$$\frac{1}{T}\int_0^T Z(t_f)\cdot[X'_{\mathrm{LC}}(t_f)]\mathrm{d}t_f = \frac{1}{T}\int_0^T Z(t_f)\cdot F(X_{\mathrm{LC}}(t_f))\mathrm{d}t_f = 1. \tag{1.35}$$

This normalization of $Z(t_f)$ is equivalent to setting $Z(0)\cdot X'_{\mathrm{LC}}(0) = Z(0)\cdot F(X'_{\mathrm{LC}}(0)) = 1$, because $Z(t)\cdot X'_{\mathrm{LC}}(t)$ is a constant (see below).

Finally, recalling that $t_s = \varepsilon t$ and $t_f = t$, we obtain the phase model for the pair of coupled neurons

$$\frac{\mathrm{d}\phi_j}{\mathrm{d}t} = \varepsilon\frac{1}{T}\int_0^T Z(\tilde{t})\cdot[I\left(X_{\mathrm{LC}}(\tilde{t}), X_{\mathrm{LC}}(\tilde{t} - (\phi_j - \phi_k))\right)]\mathrm{d}\tilde{t} = \varepsilon H(\phi_k - \phi_j), \tag{1.36}$$

By comparing these phase equations with those derived in the previous sections, it is clear that the appropriately normalized solution to the adjoint equations $Z(t)$ is the iPRC of the neuronal oscillator.

5.1 A Note on the Normalization of $Z(t)$

$$\begin{aligned}\frac{\mathrm{d}}{\mathrm{d}t}\left[Z(t)\cdot F(X_{\mathrm{LC}}(t))\right] &= \frac{\mathrm{d}Z}{\mathrm{d}t}\cdot F(X_{\mathrm{LC}}(t)) + Z(t)\cdot\frac{\mathrm{d}}{\mathrm{d}t}\left[F(X_{\mathrm{LC}}(t))\right]\\ &= (-DF(X_{\mathrm{LC}}(t))^T Z)\cdot F(X_{\mathrm{LC}}(t))\\ &\quad + Z(t)\cdot\left(DF(X_{\mathrm{LC}}(t))X'_{\mathrm{LC}}(t)\right)\\ &= -Z(t)\cdot(DF(X_{\mathrm{LC}}(t))F(X_{\mathrm{LC}}(t)))\\ &\quad + Z(t)\cdot(DF(X_{\mathrm{LC}}(t))F(X_{\mathrm{LC}}(t)))\\ &= 0.\end{aligned}$$

This implies that $Z(t)\cdot F(X_{\mathrm{LC}}(t))$ is a constant. The integral form of the normalization of $Z(t)$ (1.35) implies that this constant is 1. Thus, $Z(t)\cdot F(X_{\mathrm{LC}}(t)) = Z(t)\cdot X'_{\mathrm{LC}}(t) = 1$ for all t, including $t = 0$.

5.2 Adjoints and Gradients

The intrepid reader who has trudged their way through the preceding three sections may be wondering if there is a direct way to relate the gradient of the phase map $\nabla_X\Phi(X_{\mathrm{LC}}(t))$ to solution of the adjoint equation $Z(t)$. Here, we present a direct proof that $\nabla_X\Phi(X_{\mathrm{LC}}(t))$ satisfies the adjoint equation (1.33) and the normalization condition (1.35) (Brown et al. 2004).

Consider again the system of differential equations for an isolated neuronal oscillator (1.1) that has an asymptotically stable T-periodic limit cycle solution $X_{\mathrm{LC}}(t)$. Suppose that $X(t) = X_{\mathrm{LC}}(t+\phi)$ is a solution of this system that is on the limit cycle, which starts at point $X(0) = X_{\mathrm{LC}}(\phi)$. Further suppose that $Y(t) = X_{\mathrm{LC}}(t+\phi) + p(t)$ is a solution that starts at from the initial condition $Y(0) = X_{\mathrm{LC}}(\phi) + p(0)$, where $p(0)$ is small in magnitude. Because this initial perturbation $p(0)$ is small and the limit cycle is stable, (i) $p(t)$ remains small and, to $\mathcal{O}(|p|)$, $p(t)$ satisfies the linearized system

$$\frac{\mathrm{d}p}{\mathrm{d}t} = DF(X_{\mathrm{LC}}(t+\phi))p, \tag{1.37}$$

and (ii) the phase difference between the two solutions is

$$\Delta\phi = \Phi(X_{\mathrm{LC}}(t+\phi)+p(t)) - \Phi(X_{\mathrm{LC}}(t+\phi)) = \nabla_X\Phi(X_{\mathrm{LC}}(t+\phi))\cdot p(t) + \mathcal{O}(|p|^2) \tag{1.38}$$

Furthermore, while the asymptotic phases of the solutions evolve in time, the phase difference between the solutions $\Delta\phi$ remains constant. Therefore, by differentiating equation (1.38), we see that to $\mathcal{O}(|p|)$

$$\begin{aligned}
0 &= \frac{\mathrm{d}}{\mathrm{d}t}\left[\nabla_X \Phi(X_{\mathrm{LC}}(t+\phi)) \cdot p(t)\right] \\
&= \frac{\mathrm{d}}{\mathrm{d}t}\left[\nabla_X \Phi(X_{\mathrm{LC}}(t+\phi))\right] \cdot p(t) + \nabla_X \Phi(X_{\mathrm{LC}}(t+\phi)) \cdot \frac{\mathrm{d}p}{\mathrm{d}t} \\
&= \frac{\mathrm{d}}{\mathrm{d}t}\left[\nabla_X \Phi(X_{\mathrm{LC}}(t+\phi))\right] \cdot p(t) + \nabla_X \Phi(X_{\mathrm{LC}}(t+\phi)) \cdot (DF(X_{\mathrm{LC}}(t+\phi))p(t)) \\
&= \frac{\mathrm{d}}{\mathrm{d}t}\left[\nabla_X \Phi(X_{\mathrm{LC}}(t+\phi))\right] \cdot p(t) + (DF(X_{\mathrm{LC}}(t+\phi))^T \nabla_X \Phi(X_{\mathrm{LC}}(t+\phi))) \cdot p(t) \\
&= \left\{\frac{\mathrm{d}}{\mathrm{d}t}\left[\nabla_X \Phi(X_{\mathrm{LC}}(t+\phi))\right] + DF(X_{\mathrm{LC}}(t+\phi))^T\,(\nabla_X \Phi(X_{\mathrm{LC}}(t+\phi)))\right\} \cdot p(t).
\end{aligned}$$

Because p is arbitrary, the above argument implies that $\nabla_X \Phi(X_{\mathrm{LC}}(t))$ solves the adjoint equation (1.33). The normalization condition simply follows from the definition of the phase map [see (1.18)], i.e.,

$$\frac{\mathrm{d}\theta}{\mathrm{d}t} = \nabla_X \Phi(X_{\mathrm{LC}}(t)) \cdot X'_{\mathrm{LC}}(t) = 1. \tag{1.39}$$

5.3 *Computing the PRC Using the Adjoint method*

As stated in this beginning of this section, the major practical asset of the singular perturbation approach is that it provides a simple method to compute the iPRC for model neurons. Specifically, the iPRC is a T-period solution to

$$\frac{\mathrm{d}Z}{\mathrm{d}t} = -DF(X_{\mathrm{LC}}(t))^T Z \tag{1.40}$$

subject to the normalization constraint

$$Z(0) \cdot X'_{\mathrm{LC}}(0) = 1. \tag{1.41}$$

This equation is the adjoint equation for the isolated model neuron (1.1) linearized around the limit cycle solution $X_{\mathrm{LC}}(t)$.

In practice, the solution to (1.40) is found by integrating the equation backward in time (Williams and Bowtell 1997). The adjoint system has the opposite stability of the original system (1.1), which has an asymptotically stable T-periodic limit cycle solution. Thus, we integrate backward in time from an arbitrary initial condition so as to dampen out the transients and arrive at the (unstable) periodic solution of (1.40). To obtain the iPRC, we normalize the periodic solution using (1.41).

This algorithm is automated in the software package XPPAUT (Ermentrout 2002), which is available for free on Bard Ermentrout's webpage www.math.pitt.edu/~bard/bardware/.

6 Extensions of Phase Models for Pairs of Coupled Cells

Up to this point, we have been dealing solely with pairs of identical oscillators that are weakly coupled. In this section, we show how the phase reduction technique can be extended to incorporate weak heterogeneity and weak noise.

6.1 *Weak Heterogeneity*

Suppose that the following system

$$\frac{\mathrm{d}X_j}{\mathrm{d}t} = F_j(X_j) + \varepsilon I(X_k, X_j) = F(X_j) + \varepsilon \left[f_j(X_j) + I(X_k, X_j) \right] \tag{1.42}$$

describes two weakly coupled neuronal oscillators (note that the vector functions $F_j(X_j)$ are now specific to each neuron). If the two neurons are weakly heterogeneous, then their underlying limit cycles are equivalent up to an $\mathcal{O}(\varepsilon)$ difference. That is, $F_j(X_j) = F(X_j) + \varepsilon f_j(X_j)$, where $f_j(X_j)$ is a vector function that captures the $\mathcal{O}(\varepsilon)$ differences in the dynamics of cell 1 and cell 2 from the function $F(X_j)$. These differences may occur in various places such as the value of the neurons' leakage conductances, the applied currents, or the leakage reversal potentials, etc.

As in the previous sections, (1.42) can be reduced to the phase model

$$\begin{aligned}\frac{\mathrm{d}\phi_j}{\mathrm{d}t} &= \varepsilon \left(\frac{1}{T} \int_0^T Z(\tilde{t}) \cdot \left[f_j(X_{\mathrm{LC}}(\tilde{t})) + I(X_{\mathrm{LC}}(\tilde{t}), X_{\mathrm{LC}}(\tilde{t} + \phi_k - \phi_j)) \right] \mathrm{d}\tilde{t} \right) \\ &= \varepsilon \omega_j + \varepsilon H(\phi_k - \phi_j),\end{aligned} \tag{1.43}$$

where $\omega_j = \frac{1}{T} \int_0^T Z(\tilde{t}) \cdot f_j(X_{\mathrm{LC}}(\tilde{t}))\mathrm{d}\tilde{t}$ represents the difference in the intrinsic frequencies of each neuron caused by the presence of the weak heterogeneity. If we now let $\phi = \phi_2 - \phi_1$, we obtain

$$\begin{aligned}\frac{\mathrm{d}\phi}{\mathrm{d}t} &= \varepsilon(H(-\phi) - H(\phi) + \Delta\omega) \\ &= \varepsilon(G(\phi) + \Delta\omega),\end{aligned} \tag{1.44}$$

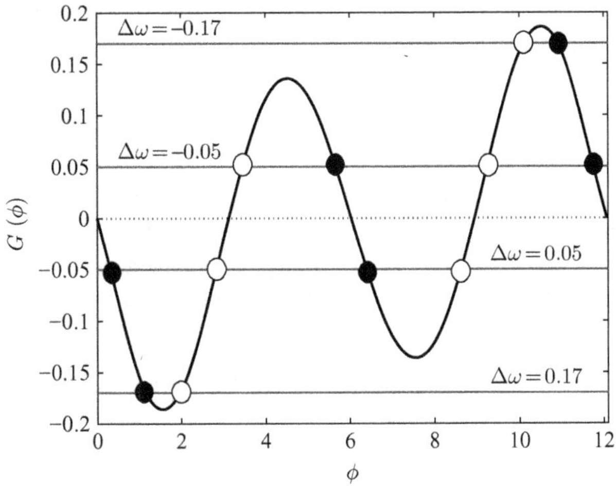

Fig. 1.6 *Example G Function with Varying Heterogeneity*. Example of varying levels of heterogeneity with the same G function as in Fig. 1.1. One can see that the addition of any level of heterogeneity will cause the stable steady-state phase-locked states to move to away from the synchronous and antiphase states to nonsynchronous phase-locked states. Furthermore, if the heterogeneity is large enough, the stable steady state phase-locked states will disappear completely through saddle node bifurcations

where $\Delta\omega = \omega_2 - \omega_1$. The fixed points of (1.44) are given by $G(\phi) = -\Delta\omega$. The addition of the heterogeneity changes the phase-locking properties of the neurons. For example, suppose that in the absence of heterogeneity ($\Delta\omega = 0$) our G function is the same as in Fig. 1.1, in which the synchronous solution, $\phi_S = 0$, and the antiphase solution, ϕ_{AP}, are stable. Once heterogeneity is added, the effect will be to move the neurons away from either firing in synchrony or anti-phase to a constant non-synchronous phase shift, as in Fig. 1.6. For example, if neuron 1 is faster than neuron 2, then $\Delta\omega < 0$ and the stable steady state phase-locked values of ϕ will be shifted to left of synchrony and to the left of anti-phase, as is seen in Fig. 1.6 when $\Delta\omega = -0.5$. Thus, the neurons will still be phase-locked, but in a nonsynchronous state that will either be to the left of synchronous state or to the left of the antiphase state depending on the initial conditions. Furthermore, if $\Delta\omega$ is decreased further, saddle node bifurcations occur in which a stable and unstable fixed point collide and annihilate each other. In this case, the model predicts that the neurons will not phase-lock but will drift in and out of phase.

6.2 *Weakly Coupled Neurons with Noise*

In this section, we show how two weakly coupled neurons with additive white noise in the voltage component can be analyzed using a probability density approach (Kuramoto 1984; Pfeuty et al. 2005).

The following set of differential equations represent two weakly heterogeneous neurons being perturbed with additive noise

$$\frac{\mathrm{d}X_j}{\mathrm{d}t} = F_j(X_j) + \varepsilon I(X_k, X_j) + \delta N_j(t), \quad i, j = 1, 2; i \neq j, \tag{1.45}$$

where δ scales the noise term to ensure that it is $\mathcal{O}(\varepsilon)$. The term $N_j(t)$ is a vector with Gaussian white noise, $\xi_j(t)$, with zero mean and unit variance (i.e., $\langle \xi_j(t) \rangle = 0$ and $\langle \xi_j(t)\xi_j(t') \rangle = \delta(t - t')$) in the voltage component, and zeros in the other variable components. In this case, the system can be mapped to the phase model

$$\frac{\mathrm{d}\phi_j}{\mathrm{d}t} = \varepsilon(\omega_j + H(\phi_k - \phi_j)) + \delta\sigma_\phi \xi_j(t), \tag{1.46}$$

where the term $\sigma_\phi = \left(\frac{1}{T}\int_0^T [Z(\tilde{t})]^2 \mathrm{d}\tilde{t}\right)^{1/2}$ comes from averaging the noisy phase equations (Kuramoto 1984). If we now let $\phi = \phi_2 - \phi_1$, we arrive at

$$\frac{\mathrm{d}\phi}{\mathrm{d}t} = \varepsilon(\Delta\omega + (H(-\phi) - H(\phi))) + \delta\sigma_\phi \sqrt{2}\eta(t), \tag{1.47}$$

where $\Delta\omega = \omega_2 - \omega_1$ and $\sqrt{2}\eta(t) = \xi_2(t) - \xi_1(t)$ where $\eta(t)$ is Gaussian white noise with zero mean and unit variance.

The nonlinear Langevin equation (1.47) corresponds to the Fokker–Planck equation (Risken 1989; Stratonovich 1967; Van Kampen 1981)

$$\frac{\partial \rho}{\partial t}(\phi, t) = -\frac{\partial}{\partial \phi}[\varepsilon(\Delta\omega + G(\phi))\rho(\phi, t)] + (\delta\sigma_\phi)^2 \frac{\partial^2 \rho}{\partial \phi^2}(\phi, t), \tag{1.48}$$

where $\rho(\phi, t)\ \Delta\phi$ is the probability that the neurons have a phase difference between ϕ and $\phi + \Delta\phi$ at time t, where $\Delta\phi$ is small. The steady-state $\left(\frac{\partial \rho}{\partial t} = 0\right)$ solution of (1.48) is

$$\rho(\phi) = \frac{1}{N} \mathrm{e}^{M(\phi)} \left[\frac{\mathrm{e}^{-\alpha T \Delta\omega} - 1}{\int_0^T \mathrm{e}^{-M(\bar{\phi})} \mathrm{d}\bar{\phi}} \int_0^\phi \mathrm{e}^{-M(\bar{\phi})} \mathrm{d}\bar{\phi} + 1 \right], \tag{1.49}$$

where

$$M(\phi) = \alpha \int_0^\phi (\Delta\omega + G(\bar{\phi})) \mathrm{d}\bar{\phi}, \tag{1.50}$$

N is a normalization factor so that $\int_0^T \rho(\phi)\mathrm{d}\phi = 1$, and $\alpha = \frac{\varepsilon}{\delta^2 \sigma_\phi^2}$ represents the ratio of the strength of the coupling to the variance of the noise.

The steady-state solution $\rho(\phi)$ gives the distribution of the phase differences between the two neurons ϕ as time goes to infinity. Pfeuty et al. (2005) showed that

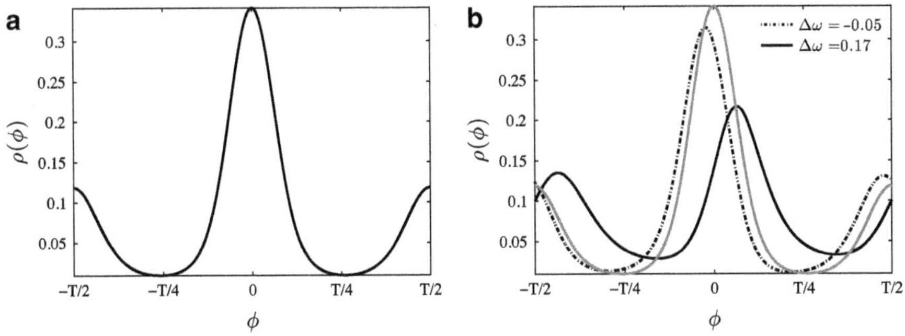

Fig. 1.7 *The Steady-State Phase Difference Distribution* $\rho(\phi)$ *is the Cross-Correlogram for the Two Neurons.* (**a**) Cross-correlogram for the G function given in Fig. 1.1 with $\alpha = 10$. Note that ϕ ranges from $-T/2$ to $T/2$. The cross-correlogram has two peaks corresponding to the synchronous and antiphase phase-locked states. This is due to the fact that in the noiseless system, synchrony and antiphase were the only stable steady states. (**b**) Cross-correlograms for two levels of heterogeneity from Fig. 1.6. The cross-correlogram from (a) is plotted as the light solid line for comparison. The peaks in the cross-correlogram have shifted to correspond with the stable nonsynchronous steady-states in Fig. 1.6

spike-train cross-correlogram of the two neurons is equivalent to the steady state distribution (1.49) for small ε. Figure 1.7a shows the cross-correlogram for two identical neurons ($\Delta\omega = 0$) using the G function from Fig. 1.1. One can see that there is a large peak in the distribution around the synchronous solution ($\phi_S = 0$), and a smaller peak around the antiphase solution ($\phi_{AP} = T/2$). Thus, the presence of the noise works to smear out the probability distribution around the stable steady-states of the noiseless system.

If heterogeneity is added to the G function as in Fig. 1.6, one would expect that the peaks of the cross-correlogram would shift accordingly so as to correspond to the stable steady states of the noiseless system. Figure 1.7b shows that this is indeed the case. If $\Delta\omega < 0$ ($\Delta\omega > 0$), the stable steady states of the noiseless system shift to the left (right) of synchrony and to the left (right) of antiphase, thus causing the peaks of the cross-correlogram to shift left (right) as well. If we were to increase (decrease) the noise, i.e., decrease (increase) α, then we would see that the variance of the peaks around the stable steady states becomes larger (smaller), according to (1.49).

7 Networks of Weakly Coupled Neurons

In this section, we extend the phase model description to examine networks of weakly coupled neuronal oscillators.

Suppose we have a one spatial dimension network of M weakly coupled and weakly heterogeneous neurons

$$\frac{dX_i}{dt} = F_i(X_i) + \frac{\varepsilon}{M_0}\sum_{j=1}^{M} s_{ij} I(X_j, X_i), \quad i = 1, ..., M; \tag{1.51}$$

where $S = \{s_{ij}\}$ is the connectivity matrix of the network, M_0 is the maximum number of cells that any neuron is connected to and the factor of $\frac{1}{M_0}$ ensures that the perturbation from the coupling is $\mathcal{O}(\varepsilon)$. As before, this system can be reduced to the phase model

$$\frac{d\phi_i}{dt} = \omega_i + \frac{\varepsilon}{M_0}\sum_{j=1}^{M} s_{ij} H(\phi_j - \phi_i), \quad i = 1, ..., M. \tag{1.52}$$

The connectivity matrix, S, can be utilized to examine the effects of network topology on the phase-locking behavior of the network. For example, if we wanted to examine the activity of a network in which each neuron is connected to every other neuron, i.e., all-to-all coupling, then

$$s_{ij} = 1, \quad i, j = 1, ..., M. \tag{1.53}$$

Because of the nonlinear nature of (1.52), analytic solutions normally cannot be found. Furthermore, it can be quite difficult to analyze for large numbers of neurons. Fortunately, there exist two approaches to simplifying (1.52) so that mathematical analysis can be utilized, which is not to say that simulating the system (1.52) is not useful. Depending upon the type of interaction function that is used, various types of interesting phase-locking behavior can be seen, such as total synchrony, traveling oscillatory waves, or, in two spatial dimensional networks, spiral waves, and target patterns, e.g. (Ermentrout and Kleinfeld 2001; Kuramoto 1984).

A useful method of determining the level of synchrony for the network (1.52) is the so-called Kuramoto synchronization index (Kuramoto 1984)

$$r e^{2\pi\sqrt{-1}\psi/T} = \frac{1}{M}\sum_{j=1}^{M} e^{2\pi\sqrt{-1}\phi_j/T}, \tag{1.54}$$

where ψ is the average phase of the network, and r is the level of synchrony of the network. This index maps the phases, ϕ_j, to vectors in the complex plane and then averages them. Thus, if the neurons are in synchrony, the corresponding vectors will all be pointing in the same direction and r will be equal to one. The less synchronous the network is, the smaller the value of r.

In the following two sections, we briefly outline two different mathematical techniques for analyzing these phase oscillator networks in the limit as M goes to infinity.

7.1 Population Density Method

A powerful method to analyze large networks of all-to-all coupled phase oscillators was introduced by Strogatz and Mirollo (1991) where they considered the so-called Kuramoto model with additive white noise

$$\frac{\mathrm{d}\phi_i}{\mathrm{d}t} = \omega_i + \frac{\varepsilon}{M}\sum_{j=1}^{M} H(\phi_j - \phi_i) + \sigma\xi(t), \tag{1.55}$$

where the interaction function is a simple sine function, i.e., $H(\phi) = \sin(\phi)$. A large body of work has been focused on analyzing the Kuramoto model as it is the simplest model for describing the onset of synchronization in populations of coupled oscillators (Acebrón et al. 2005; Strogatz 2000). However, in this section, we will examine the case where $H(\phi)$ is a general T-periodic function.

The idea behind the approach of (Strogatz and Mirollo 1991) is to derive the Fokker–Planck equation for (1.55) in the limit as $M \to \infty$, i.e., the number of neurons in the network is infinite. As a first step, note that by equating real and imaginary parts in (1.54) we arrive at the following useful relations

$$r\cos(2\pi(\psi - \phi_i)/T) = \frac{1}{M}\sum_{j=1}^{M}\cos(2\pi(\phi_j - \phi_i)/T) \tag{1.56}$$

$$r\sin(2\pi(\psi - \phi_i)/T) = \frac{1}{M}\sum_{j=1}^{M}\sin(2\pi(\phi_j - \phi_i)/T). \tag{1.57}$$

Next, we note that since $H(\phi)$ is T-periodic, we can represent it as a Fourier series

$$H(\phi_j - \phi_i) = \frac{1}{T}\sum_{n=0}^{\infty} a_n\cos(2\pi n(\phi_j - \phi_i)/T) + b_n\sin(2\pi n(\phi_j - \phi_i)/T). \tag{1.58}$$

Recognizing that (1.56) and (1.57) are averages of the functions cosine and sine, respectively, over the phases of the oscillators, we see that, in the limit as M goes to infinity (Neltner et al. 2000; Strogatz and Mirollo 1991)

$$ra_n\cos(2\pi n(\psi_n - \phi)/T) = a_n\int_{-\infty}^{\infty}\int_0^T g(\omega)\rho(\tilde{\phi},\omega,t)\cos(2\pi n(\tilde{\phi} - \phi)/T)\mathrm{d}\tilde{\phi}\mathrm{d}\omega \tag{1.59}$$

$$rb_n\sin(2\pi n(\psi_n - \phi)/T) = b_n\int_{-\infty}^{\infty}\int_0^T g(\omega)\rho(\tilde{\phi},\omega,t)\sin(2\pi n(\tilde{\phi} - \phi)/T)\mathrm{d}\tilde{\phi}\mathrm{d}\omega, \tag{1.60}$$

where we have used the Fourier coefficients of $H(\phi_j - \phi_i)$. $\rho(\phi, \omega, t)$ is the probability density of oscillators with intrinsic frequency ω and phase ϕ at time t, and $g(\omega)$ is the density function for the distribution of the frequencies of the oscillators. Furthermore, $g(\omega)$ also satisfies $\int_{-\infty}^{\infty} g(\omega)\mathrm{d}\omega = 1$. With all this in mind, we can now rewrite the infinite M approximation of (1.55)

$$\frac{\mathrm{d}\phi}{\mathrm{d}t} = \omega + \varepsilon\frac{1}{T}\sum_{n=0}^{\infty}[ra_n \cos(2\pi n(\psi_n - \phi)/T) + rb_n \sin(2\pi n(\psi_n - \phi)/T)] + \sigma\xi(t). \tag{1.61}$$

The above nonlinear Langevin equation corresponds to the Fokker–Planck equation

$$\frac{\partial \rho}{\partial t}(\phi, \omega, t) = -\frac{\partial}{\partial \phi}[J(\phi, t)\rho(\phi, \omega, t)] + \frac{\sigma^2}{2}\frac{\partial^2 \rho}{\partial \phi^2}(\phi, \omega, t), \tag{1.62}$$

with

$$J(\phi, t) = \omega + \varepsilon\frac{1}{T}\sum_{n=0}^{\infty}[ra_n \cos(2\pi n(\psi_n - \phi)/T) + rb_n \sin(2\pi n(\psi_n - \phi)/T)], \tag{1.63}$$

and $\int_0^T \rho(\phi, \omega, t)\mathrm{d}\phi = 1$ and $\rho(\phi, \omega, t) = \rho(\phi + T, \omega, t)$. Equation (1.62) tells us how the fraction of oscillators with phase ϕ and frequency ω evolves with time. Note that (1.62) has the trivial solution $\rho_0(\phi, \omega, t) = \frac{1}{T}$, which corresponds to the incoherent state in which the phases of the neurons are uniformly distributed between 0 and T.

To study the onset of synchronization in these networks, Strogatz and Mirollo (1991) and others, e.g. (Neltner et al. 2000), linearized equation (1.62) around the incoherent state, ρ_0, in order to determine its stability. They were able to prove that below a certain value of ε, the incoherent state is neutrally stable and then loses stability at some critical value $\varepsilon = \varepsilon_C$. After this point, the network becomes more and more synchronous as ε is increased.

7.2 *Continuum Limit*

Although the population density approach is a powerful method for analyzing the phase-locking dynamics of neuronal networks, it is limited by the fact that it does not take into account spatial effects of neuronal networks. An alternative approach to analyzing (1.52) in the large M limit that takes into account spatial effects is to assume that the network of neuronal oscillators forms a spatial continuum (Bressloff and Coombes 1997; Crook et al. 1997; Ermentrout 1985).

Suppose that we have a one-dimensional array of neurons in which the jth neuron occupies the position $x_j = j\Delta x$ where Δx is the spacing between the neurons. Further suppose that the connectivity matrix is defined by $S = \{s_{ij}\} = W(|x_j - x_i|)$, where $W(|x|) \to 0$ as $|x| \to \infty$ and $\sum_{j=-\infty}^{\infty} W(x_j)\Delta x = 1$. For example, the spatial connectivity matrix could correspond to a Gaussian function, $W(|x_j - x_i|) = \mathrm{e}^{-\frac{|x_j - x_i|^2}{2\sigma^2}}$, so that closer neurons have more strongly coupled to each other than to neurons that are further apart. We can now rewrite (1.52) as

$$\frac{\mathrm{d}\phi}{\mathrm{d}t}(x_i, t) = \omega(x_i) + \varepsilon \sum_{j=-\infty}^{\infty} \left[W(|x_j - x_i|)\, \Delta x\, H\left(\phi(x_j, t) - \phi(x_i, t)\right)\right], \quad (1.64)$$

where $\phi(x_i, t) = \phi_i(t)$, $\omega(x_i) = \omega_i$ and we have taken $1/M = \Delta x$. By taking the limit of $\Delta x \to 0$ ($M \to \infty$) in (1.64), we arrive at the continuum phase model

$$\frac{\partial \phi}{\partial t}(x, t) = \omega(x) + \varepsilon \int_{-\infty}^{\infty} W(|x - \bar{x}|)\, H(\phi(\bar{x}, t) - \phi(x, t))\, \mathrm{d}\bar{x}, \qquad (1.65)$$

where $\phi(x, t)$ is the phase of the oscillator at position x and time t. Note that this continuum phase model can be modified to account for finite spatial domains (Ermentrout 1992) and to include multiple spatial dimensions.

Various authors have utilized this continuum approach to prove results about the stability of the synchrony and traveling wave solutions of (1.65) (Bressloff and Coombes 1997; Crook et al. 1997; Ermentrout 1985, 1992). For example, Crook et al. (1997) were able to prove that presence of axonal delay in synaptic transmission between neurons can cause the onset of traveling wave solutions. This is due to the presence of axonal delay which encourages larger phase shifts between neurons that are further apart in space. Similarly, Bressloff and Coombes (1997) derived the continuum phase model for a network of integrate-and-fire neurons coupled with excitatory synapses on their passive dendrites. Using this model, they were able to show that long range excitatory coupling can cause the system to undergo a bifurcation from the synchronous state to traveling oscillatory waves. For a rigorous mathematical treatment of the existence and stability results for general continuum and discrete phase model neuronal networks, we direct the reader to Ermentrout (1992).

8 Summary

- The infinitesimal PRC (iPRC) of a neuron measures its sensitivity to infinitesimally small perturbations at every point along its cycle.
- The theory of weak coupling utilizes the iPRC to reduce the complexity of neuronal network to consideration of a single phase variable for every neuron.
- The theory is valid only when the perturbations to the neuron, from coupling or an external source, is sufficiently "weak" so that the neuron's intrinsic dynamics dominate the influence of the coupling. This implies that coupling does not cause the neuron's firing period to differ greatly from its unperturbed cycle.

- For two weakly coupled neurons, the theory allows one to reduce the dynamics to consideration of a single equation describing how the phase difference of the two oscillators changes in time. This allows for the prediction of the phase-locking behavior of the cell pair through simple analysis of the phase difference equation.
- The theory of weak coupling can be extended to incorporate effects from weak heterogeneity and weak noise.

Acknowledgements This work was supported by the National Science Foundation under grants DMS-09211039 and DMS-0518022.

References

Acebrón, J., Bonilla, L., Vicénte, C., Ritort, F., and Spigler, R. (2005). The kuramoto model: A simple paradigm for synchronization phenomena. *Rev. Mod. Phys.*, 77:137–185.

Bressloff, P. and Coombes, S. (1997). Synchrony in an array of integrate-and-fire neurons with dendritic structure. *Phys. Rev. Lett.*, 78:4665–4668.

Brown, E., Moehlis, J., and Holmes, P. (2004). On the phase reduction and response dynamics of neural oscillator populations. *Neural Comp.*, 16:673–715.

Crook, S., Ermentrout, G., Vanier, M., and Bower, J. (1997). The role of axonal delay in the synchronization of networks of coupled cortical oscillators. *J. Comp. Neurosci.*, 4:161–172.

Erisir, A., Lau, D., Rudy, B., and Leonard, C. (1999). Function of specific k+ channels in sustained high-frequency firing of fast-spiking neocortical interneurons. *J. Neurophysiol.*, 82:2476–2489.

Ermentrout, B. (2002). *Simulating, Analyzing, and Animating Dynamical Systems: A Guide to XPPAUT for Researchers and Students*. SIAM.

Ermentrout, G. (1985). The behavior of rings of coupled oscillators. *J. Math. Biology*, 23:55–74.

Ermentrout, G. (1992). Stable periodic solutions to discrete and continuum arrays of weakly coupled nonlinear oscillators. *SIAM J. Appl. Math.*, 52(6):1665–1687.

Ermentrout, G. (1996). Type 1 membranes, phase resetting curves, and synchrony. *Neural Computation*, 8:1979–1001.

Ermentrout, G. and Kleinfeld, D. (2001). Traveling electrical waves in cortex: Insights from phase dynamics and speculation on a computational role. *Neuron*, 29:33–44.

Ermentrout, G. and Kopell, N. (1984). Frequency plateaus in a chain of weakly coupled oscillators, i. *SIAM J. Math. Anal.*, 15(2):215–237.

Ermentrout, G. and Kopell, N. (1991). Multiple pulse interactions and averaging in systems of coupled neural oscillators. *J. Math. Bio.*, 29:33–44.

Ermentrout, G. and Kopell, N. (1998). Fine structure of neural spiking and synchronization in the presence of conduction delays. *Proc. Nat. Acad. of Sci.*, 95(3):1259–1264.

Goel, P. and Ermentrout, G. (2002). Synchrony, stability, and firing patterns in pulse-coupled oscillators. *Physica D*, 163:191–216.

Guckenheimer, J. and Holmes, P. (1983). *Nonlinear Oscillations, Dynamical Systems, and Bifurcations of Vector Fields*. Springer, NY.

Guevara, M. R., Shrier, A., and Glass, L. (1986). Phase resetting of spontaneously beating embryonic ventricular heart cell aggregates. *Am J Physiol Heart Circ Physiol*, 251(6): H1298–1305.

Hodgkin, A. and Huxley, A. (1952). A quantitative description of membrane current and its application to conduction and excitation in nerve. *J. Physiol.*, 117:500–544.

Hoppensteadt, F. C. and Izhikevich, E. M. (1997). *Weakly Connected Neural Networks*. Springer, New York.

Kuramoto, Y. (1984). *Chemical Oscillations, Waves, and Turbulence*. Springer, Berlin.

Lewis, T. and Rinzel, J. (2003). Dynamics of spiking neurons connected by both inhibitory and electrical coupling. *J. Comp. Neurosci.*, 14:283–309.

Lewis, T. and Rinzel, J. (2004). Dendritic effects in networks of electrically coupled fast-spiking interneurons. *Neurocomputing*, 58-60:145–150.

Malkin, I. (1949). *Methods of Poincare and Liapunov in Theory of Non-Linear Oscillations*. Gostexizdat, Moscow.

Malkin, I. (1956). *Some Problems in Nonlinear Oscillation Theory*. Gostexizdat, Moscow.

Mancilla, J., Lewis, T., Pinto, D., Rinzel, J., and Connors, B. (2007). Synchronization of electrically coupled pairs of inhibitory interneurons in neocortex. *J. Neurosci.*, 27(8): 2058–2073.

Mirollo, R. and Strogatz, S. (1990). Synchronization of pulse-coupled biological oscillators. *SIAM J. Applied Math.*, 50(6):1645–1662.

Morris, C. and Lecar, H. (1981). Voltage oscillations in the barnacle giant muscle fiber. *Biophys J*, 35:193–213.

Neltner, L., Hansel, D., Mato, G., and Meunier, C. (2000). Synchrony in heterogeneous networks of spiking neurons. *Neural Comp.*, 12:1607–1641.

Netoff, T., Acker, C., Bettencourt, J., and White, J. (2005a). Beyond two-cell networks: experimental measurement of neuronal responses to multiple synaptic inputs. *J. Comput. Neurosci.*, 18:287–295.

Netoff, T., Banks, M., Dorval, A., Acker, C., Haas, J., Kopell, N., and White, J. (2005b). Synchronization of hybrid neuronal networks of the hippocampal formation strongly coupled. *J. Neurophysiol.*, 93:1197–1208.

Neu, J. (1979). Coupled chemical oscillators. *SIAM J. Appl. Math.*, 37(2):307–315.

Oprisan, S., Prinz, A., and Canavier, C. (2004). Phase resetting and phase locking in hybrid circuits of one model and one biological neuron. *Biophys. J.*, 87:2283–2298.

Pfeuty, B., Mato, G., Golomb, D., and Hansel, D. (2005). The combined effects of inhibitory and electrical synapses in synchrony. *Neural Computation*, 17:633–670.

Risken, H. (1989). *The Fokker–Planck Equation: Methods of Solution and Applications*. Springer, NY.

Stratonovich, R. (1967). *Topics in the Theory of Random Noise*. Gordon and Breach, NY.

Strogatz, S. (2000). From kuramoto to crawford: Exploring the onset of synchronization in populations of coupled oscillators. *Physica D*, 143:1–20.

Strogatz, S. and Mirollo, R. (1991). Stability of incoherence in a population of coupled oscillators. *J. Stat. Physics*, 63:613–635.

Van Kampen, N. (1981). *Stochastic Processes in Physics and Chemistry*. Amsterdam: Elsevier Science.

Williams, T. and Bowtell, G. (1997). The calculation of frequency-shift functions for chains of coupled oscillators, with application to a network model of the lamprey locomotor pattern generator. *J. Comput. Neurosci.*, 4:47–55.

Winfree, A. T. (1980). *The Geometry of Biological Time*. Springer, NY.

Zahid, T. and Skinner, F. (2009). Predicting synchronous and asynchronous network groupings of hippocampal interneurons coupled with dendritic gap junctions. *Brain Research*, 1262: 115–129.

Chapter 2
Phase Resetting Neural Oscillators: Topological Theory Versus the Real World

Trine Krogh-Madsen, Robert Butera, G. Bard Ermentrout, and Leon Glass

Abstract Biological oscillations, despite their vast range of periodicities, can be described mathematically by stable limit cycles. Therefore, a general theory characterizing the effects of perturbations applied to such limit cycles allows predictions of qualitative features of a particular oscillation subject to perturbation. In this chapter, we summarize this topological approach and discuss ways in which the theory breaks down, mainly for neuronal and cardiac oscillators. In particular, we describe experimental and computational studies that demonstrate apparent discontinuities in the response to perturbations, and others where there is not a rapid return to the limit cycle following a perturbation. Finally, we discuss differences between the topological and the excitability-type descriptions of neuronal oscillators.

1 Introduction

Biological systems display stable oscillations. These oscillations have periods that range over many orders of magnitude: milliseconds in the case of some neural oscillations; seconds for cardiac and respiratory rhythms; days for circadian and

T. Krogh-Madsen (✉)
Weill Cornell Medical College, New York, NY, USA
e-mail: trk2002@med.cornell.edu

R. Butera
Georgia Institute of Technology, Atlanta, GA, USA
e-mail: rbutera@gatech.edu

G.B. Ermentrout
Department of Mathematics, University of Pittsburgh, Pittsburgh, PA, USA
e-mail: bard@pitt.edu

L. Glass
McGill University, Montreal, QC, Canada
e-mail: glass@cnd.mcgill.ca

N.W. Schultheiss et al. (eds.), *Phase Response Curves in Neuroscience: Theory, Experiment, and Analysis*, Springer Series in Computational Neuroscience 6,
DOI 10.1007/978-1-4614-0739-3_2,

menstrual rhythms; a year for the rhythms underlying hibernation. However, to fulfill its physiological role, any particular biological oscillation must be controlled by various feedbacks, and it must appropriately interact with the rich dynamics generated from other cells and organs that might affect it.

The fundamental concept underlying the mathematical study of biological rhythms is that they can be associated with stable limit cycle oscillations in some appropriate dynamical system – typically an ordinary, partial or time delay differential equation. A stable limit cycle is a periodic cycle in state space that will be reestablished in the limit of $t \to \infty$ for all initial conditions in the neighborhood of the cycle. The set of points in state space that attract to the limit cycle is called its *basin of attraction*. Since many biological rhythms, such as the heartbeat or respiration, must be robust to perturbation, the basin of attraction of biological oscillations may often be quite large, and amenable to experimental study by delivering appropriate perturbations. Further, because of this presumed underlying mathematical structure, based solely on mathematical analysis, all biological rhythms will be expected to display some similar characteristics (Canavier 2006; Glass and Mackey 1998; Winfree 2000).

In particular, Winfree described basic features that could be theoretically predicted during phase resetting experiments in which perturbations were delivered to biological oscillations under controlled circumstances that vary the phase and amplitude of the delivery of the perturbation. Subsequent studies by others, particularly in the field of neurophysiology, have carried out further theoretical and experimental studies of resetting biological oscillations, which in some cases have used similar terminology to that introduced by Winfree to reflect different concepts. Winfree championed the notion that the underlying limit cycle structure of biological oscillations will lead to certain universal characteristics amenable to classification and mathematical analysis (Glass and Mackey 1998; Winfree 2000). Our point in the following is to examine the robust predictions of the topological theory of resetting of limit cycle oscillations in light of subsequent experimental and theoretical studies. In Sect. 2, we briefly present the main predictions of the topological theory with regard to resetting and show how in the limit of very fast return to the limit cycle following a perturbation, the effects of periodic forcing can be mathematically analyzed from iteration of circle maps determined from resetting experiments. In Sect. 3, we describe two ways in which the theory in Sect. 2 breaks down. First, we show circumstances in which the resetting curves differ in a fundamental sense from the topological predictions. Then, we describe experimental and theoretical studies that demonstrate circumstances in which there is not a rapid convergence to a limit cycle, so that iteration of a circle map cannot be used to predict dynamics. In Sect. 4, we discuss the main concepts of phase oscillator models and weak coupling as commonly used in the computational neuroscience community.

2 Topology of Phase Resetting

A very large number of investigators, coming from different disciplines, have investigated the resetting of biological oscillators experimentally. Since in many instances, investigators were not aware of similar questions being asked in different areas, there are a great many different notations and conventions for describing phase resetting. In the following, we present the topological theory of phase resetting as introduced by Winfree, and present appropriate terminology and notation. This topological perspective on phase-resetting predates most applications of phase-response theory in neuroscience, and has been more commonly used in cardiac electrophysiology. Furthermore, the notions of both strong vs. weak resetting within this topological theory are quite distinct from notions of coupling strength using other theoretical frameworks in computational neuroscience, which we will discuss in Sect. 4.

In general, phase resetting occurs when the dynamics of an oscillator is shifted in time. This can be accomplished when a stimulus transiently perturbs the state point away from the stable limit cycle associated with the oscillation, followed by the asymptotic return of the state point to the limit cycle. An example of such a resetting experiment is shown in Fig. 2.1. The unperturbed limit cycle oscillation (here of a model of spontaneous cardiac pacemaking activity) is of period T_0. All points within this cycle can be assigned a phase; we assign it values from 0 to 1 and set the fiducial point where the phase is zero to be on the upstroke of the action potential.

A stimulus delivered at a time t_c after the previous action potential upstroke is therefore applied at a phase

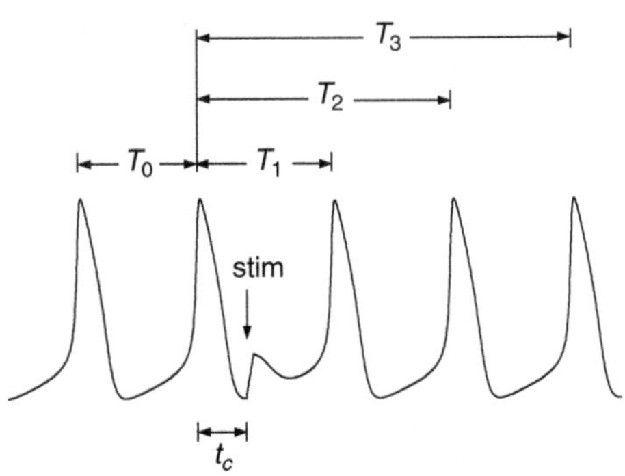

Fig. 2.1 Notation for resetting protocol in a schematic cardiac pacemaker cell. A depolarizing stimulus is given at a coupling interval t_c after the crossing of 0 mV on the upstroke of the second action potential. The (old) phase of the cycle at which the stimulus is delivered is t_c/T_0. This stimulus causes a delay, with the first perturbed cycle length, T_1, being larger than the unperturbed period, T_0

$$\phi = \frac{t_c}{T_0}, \tag{2.1}$$

which we term the *old phase*.

Following the perturbation by the applied stimulus, the asymptotic return to the limit cycle can be measured from the timing of the subsequent action potential upstrokes, i.e., $T_1, T_2, \ldots$. Mathematically, this return occurs in the limit $i \to \infty$. In general, the state point will return to a point on the limit cycle that is different from the point it would have been had there been no perturbation. From the timing of the subsequent action potential upstrokes, we quantify this as the *new phase*:

$$g(\phi) = \phi + \frac{iT_0 - T_i}{T_0} \pmod 1, \quad i \to \infty. \tag{2.2}$$

The degree to which the value of $T_i - T_{i-1}$ differs from T_0 depends on the rate at which the rhythm returns to the limit cycle oscillation. Although many use the change of phase (sometimes designated $\Delta\phi$) to measure the resetting, since both the new and old phases are represented by points on the circle the concept of advance and delay is not always well defined – this is particularly the case when a single stimulus can cause a pause greater than the intrinsic cycle length before the cycle is reestablished. Further, if there is a transient longer than the intrinsic cycle length until the asymptotic behavior is established the timing of the next beat following a stimulus cannot be used to accurately determine the resetting. However, for the sake of easier comparison to the neural literature, in many plots we will show both the phase response curve (also called the phase resetting curve), which plots $\Delta\phi$ vs. ϕ, and the phase transition curve, which plots $g(\phi)$ vs. ϕ. The relationship between these two functions is

$$g(\phi) = \phi + \Delta\phi. \tag{2.3}$$

Figure 2.2 shows resetting in a two-dimensional system called the Poincaré oscillator (also called the radial isochron clock or the Andronov–Hopf oscillator). This system is governed by the equations

$$\frac{\mathrm{d}\theta}{\mathrm{d}t} = 1, \tag{2.4}$$

$$\frac{\mathrm{d}r}{\mathrm{d}t} = kr(1 - r^2), \tag{2.5}$$

where k is a positive parameter. The system has an asymptotically stable limit cycle with $|r| = 1$ (black circle in Fig. 2.2) and an unstable fixed point at $r = 0$ (cross). This unstable fixed point is the only point outside of the basin of attraction of the stable limit cycle. The rate of convergence to the limit cycle is determined by k. Let $x(t = 0)$, $x'(t = 0)$ be the initial conditions of a point on the cycle and a point not on the cycle, respectively, and $x(t), x'(t)$ be the coordinates of the trajectories at time t. Then, if $\lim_{t\to\infty} d[x(t), x'(t)] = 0$, where d is the Euclidean distance, we say that the latent, or asymptotic, phase of $x'(0)$ is the same as the phase of $x(0)$.

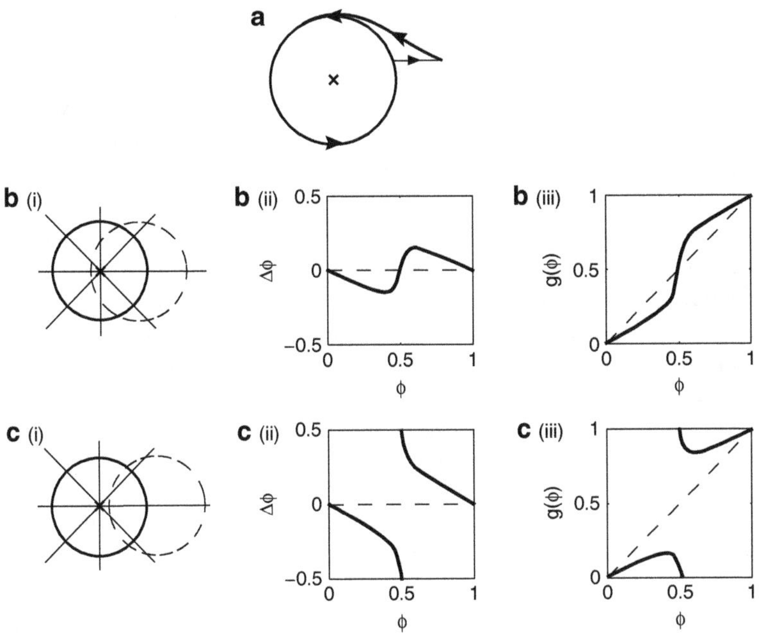

Fig. 2.2 Phase resetting of the Poincare oscillator. (**a**): Example of phase-resetting experiment with $k = 2.0$. The state point is removed from the limit cycle by a perturbation of amplitude b in the horizontal direction (*thin line*), and then converges back to the limit cycle (thick trajectory). "x" designates the unstable fixed point. (**b**): Type 1 resetting ($b = 0.8$). (**i**) Isochrons (*thin lines*) and shifted cycle (*dashed curve*); (**ii**) phase response curve; and (**iii**) phase transition curve. (**c**): Same as (**b**), but for type 0 resetting ($b = 1.2$)

A *W-isochron* consists of all points with the same latent phase (for example, see radial lines in Fig. 2.2b(i)). Thus, every point in the basin of attraction of a limit cycle, lies on an isochron (Winfree 2000).

A phase resetting experiment probes the effect of stimuli delivered at different phases of the cycle. For example, the dashed curves in Figs. 2.2b(i) and c(i) shows the *shifted cycle*, which consists of the locus of points resulting from stimuli delivered at all phases of the cycle. The stimulus delivered in Fig. 2.2b is a weak stimulus, whereas the stimulus in Fig. 2.2c is strong, resulting in a larger shift. The resulting phase transition curves in Fig. 2.2b(iii) and c(iii), can be thought of as circle maps, i.e., they map a circle into itself; g: $S^1 \rightarrow S^1$. If continuous, a circle map can be characterized topologically by its winding number, i.e., the number of times that $g(\phi)$ wraps around the unit circle as ϕ goes around the circle once. Geometrically, the winding number equals the average slope of $g(\phi)$. In Fig. 2.2b(iii) the winding number is 1. Indeed, for very weak stimuli, $g(\phi) \approx \phi$ by continuity, such that all points fall close to the diagonal and the winding number is 1, leading to "weak" or "type 1" resetting. However, in this example, for stronger stimuli the winding number of $g(\phi)$ is 0, Fig. 2.2c(iii). Unlike weak resetting, where the shifted cycle intersects all the W-isochrons, if the stimulus is of sufficiently strong amplitude to move the shifted cycle to a location where it no longer intersects all W-isochrons, degree 0 resetting will occur (Winfree 2000). The curve $g(\phi)$ is still continuous however, as the phase is defined using *modulus* 1. The example here shows a typical situation for cardiac pacemaker systems subjected to a depolarizing stimulus: an early stimulus causes a delay, while a late stimulus leads to an advance in the rhythm.

If the winding number is 1 for weak stimuli and 0 for strong stimuli, then there must be an intermediate strength (or range of strengths) that results in the state point being perturbed to a location outside of the basin of attraction of the stable limit cycle. In this example, this happens when the shifted cycle intersects the unstable fixed point within the stable limit cycle. Hence, for the old phase corresponding to that intersection, the oscillation is abolished, the new phase is undefined, and the phase transition curve discontinuous.

An important property of the phase transition curve is the Continuity Theorem (Gedeon and Glass 1998; Guckenheimer 1975). It states that if a stimulus of a given amplitude delivered at any phase of a limit-cycle oscillation leaves the state point within the basin of attraction of that asymptotically stable limit cycle, then the phase transition curve will be continuous for that stimulus amplitude. Similarly, if the phase transition curve is truly discontinuous there must be a stimulus phase (or range of phases) that will result in the state point being perturbed outside of the basin of attraction of the limit cycle. In the next section, we discuss situations where apparent discontinuities occur in phase resetting.

Although the determination of properties of the phase resetting curve of biological oscillators has a certain intrinsic interest, perhaps the major reason for investigating phase resetting is to understand the interaction of a biological oscillator with inputs from other oscillators or the environment. In the simplest circumstance there is a brief perturbation delivered to an oscillator, for example, as delivered by a stimulus from a microelectrode or from an action potential from another nerve cell. In that case, and assuming a rapid return to the limit cycle following a perturbation, the phase transition curve can be used to predict the effects of a train of stimuli. Indeed, if a series of periodic stimuli is delivered to a limit cycle oscillator with phase transition curve $g(\phi)$, then successive phases are given by

$$\phi_{j+1} = g(\phi_j) + \tau \ (\text{mod}\ 1), \tag{2.6}$$

where $\tau = t_s/T_0$, and t_s is the interval between successive stimuli and the phase transition curve g depends on the strength of the stimuli (Guevara and Glass 1982; Guevara et al. 1983).

This equation thus shows how the circle map defining the phase transition curve may be iterated to predict the effect of periodic forcing. Indeed, since the resulting Eq. (2.6) involves a circle map with two parameters, corresponding to the period and strength of the stimuli, there has been significant analysis of the periodic stimulation of biological oscillators in the context of theoretical studies of the bifurcation of circle maps (Arnol'd 1965; Glass and Mackey 1998). In particular, following early studies on a spiking neuron by Perkel and colleagues (1964), there were descriptions of experimentally observed dynamics such as complex bifurcations and chaos in cardiac (Guevara et al. 1981) and neuronal oscillators (Kaplan et al. 1996; Matsumoto et al. 1984). In some cases, the one dimensional circle map, Eq. (2.6), is able to predict the dynamics under periodic stimulation as frequency and amplitude change (Guevara et al. 1981, 1983).

3 Discrepancies Between Theoretical Predictions and Experimental and Computational Results

Although the topological theory of phase resetting reviewed in the previous section has a certain simplicity and elegance, current experimental and theoretical studies of resetting and entrainment of biological (especially neural) oscillators are largely focused in other directions. We can think of three possible reasons.

Experimental studies of resetting and entrainment using the theory are tedious and require experimental systems with low noise, comparatively short period, and a sufficiently long experimental duration. Any real experimental system will be subjected to noise, and in some cases, there can be significant fluctuations in the period of biological oscillators. If the period is not constant, then there is necessarily uncertainty in measuring both the phase of the stimulus and the resetting induced by the stimulus. To carry out resetting experiments using a variety of amplitudes of stimulus requires very stable oscillations and sufficiently fast oscillations that repeated resetting experiments can be carried out. Furthermore, higher resolution phase response curves require a longer duration of experiment, which is often limited for acute single cell *in vitro* experimentation. There is an extensive literature on resetting circadian rhythms, but resetting experiments take very long times and are difficult due to the difficulty of measuring the phase of the oscillation. In neuroscience, it is commonly observed that experimental phase response curves from invertebrate neurons (Pinsker 1977; Preyer and Butera 2005; Sieling et al. 2009) are much less noisy than those from mammalian neurons (Mancilla et al. 2007; Netoff et al. 2005).

The topological predictions about resetting oscillations are of limited interest to most biologists and are not particularly relevant to understanding biological oscillations. Although the topological analysis makes robust predictions about continuity and winding numbers of phase resetting, these predictions have not proven useful for scientists studying mechanisms of oscillations. Indeed, there have been few studies that relate the ionic mechanisms to the phase response curve itself (Canavier et al. 1997; Demir et al. 1997; Kunysz et al. 1995). (See also Part 3 of this book, particularly Chaps. 9, 12, and 13.) Recently, the use of the phase response curve to infer predictions about synchrony has been an active area of study in neuroscience, particularly using dynamic clamp techniques (Butera et al. 2001;

Dorval et al. 2001; Sharp et al. 1993) to study controlled interactions among pairs of cells (Bem et al. 2005; Mancilla et al. 2007; Netoff et al. 2005; Sharp et al. 1996).

Actual properties of real biological oscillators and the ways in which they interact differ substantially from the simple properties outlined in Sect. 2. The basic theory above assume very brief stimuli. Although there are certainly circumstances that correspond to this paradigm in experimental systems, the relevance in networks of interacting cells is less clear. Further, there really has been very little systematic testing of the predictions in Sect. 2, even assuming very brief stimuli. However, in the remainder of this section we discuss two ways in which the predictions break down: experimentally measured resetting curves may not be continuous except for low level stimuli; and relaxation rates to the limit cycle may be sufficiently slow to preclude rapid return to the limit cycle for realistic stimuli.

3.1 Discontinuities in Phase Resetting

According to the topological theory discussed in Sect. 2, the phase transition curve will be continuous for any stimulus amplitudes that does not lead to a transition outside the basin of attraction of the limit cycle. However, in many studies, the phase transition curve is very steep or even discontinuous for some combinations of stimulus amplitude and stimulus timing, but only in some of these situations was the oscillation abolished (Clay et al. 1984; Glass and Winfree 1984; Guevara et al. 1986; Vanmeerwijk et al. 1984). In this section, we reconcile this observation with the theoretical analysis.

Apparent discontinuities. In experiments on a cardiac pacemaker preparation, fine probing of the dynamics close to the transition between delay and advance of the next action potential showed that for a certain value of the coupling interval, either advance or delay occurred in different trials (Guevara et al. 1986). A slightly smaller (larger) value of the coupling interval caused only delays (advances) in multiple trials [Fig. 4 of Guevara et al. (1986)]. Further, this separation was maintained during several subsequent cycles. Thus, if one were to construct the phase transition curve for this preparation, it would be bi-valued at one value of the old phase (the one corresponding to the coupling interval where both advance and delay were seen) and these two values would be separated by a large gap. Clearly, this curve would be discontinuous. However, no annihilation of the intrinsic rhythm was seen, contradicting the Continuity Theorem.

The abrupt transition between advance and delay of the oscillation is related to the "all-or-none" dynamics typical of neuronal and some cardiac cells due to the fast positive feedback dynamics of the sodium current generating the action potential upstroke. Using a simplified three-variable model, Krogh–Madsen et al. showed that between the trajectories that lead to "all" or "none" responses lies a family of exceptionally sensitive solutions that secure the continuity of the phase transition curve (Fig. 2.3, Krogh-Madsen et al. (2004)). These so-called canard solutions hug the unstable branch of the slow manifold (Fig. 2.3c) before diverging

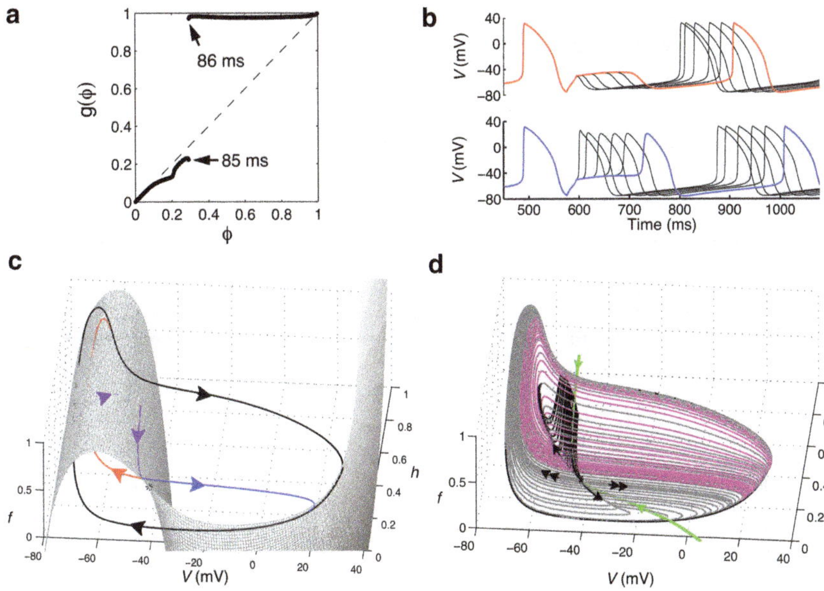

Fig. 2.3 (**a**): Discontinuous phase transition curve with t_c sampled every 1.0 ms. (**b**): Voltage waveforms close to the abrupt transition between delay and advance. Red trace obtained for t_c = 85.336376727283 ms, blue trace is for 85.336376727284 ms. (**c**): Trajectories in phase space travel along the slow manifold (gray structure, defined as $dV/dt = 0$). (**d**): Quasi-threshold structure (black, defined as the stable manifold of the trajectories following the slowest direction away from the fixed point). (**c**) and (**d**) are reprinted with modifications from Krogh-Madsen et al. (2004) with permission

off a quasi-threshold structure (Fig. 2.3d). Because these solutions are so sensitive, a tiny amount of noise such as that stemming from the stochastic gating of even a single ion channel, can provide sufficient perturbation to convert the "all" to the "none" response and vice versa (Krogh-Madsen et al. 2004). Similar sensitivity can occur in the neuronal Hodgkin-Huxley model (Clay 1977).

Type 1 to type 0 transition. The transition from type 1 to type 0 resetting is well defined theoretically. It may be direct as in the example in Fig. 2.2, where for one exact value of the stimulus amplitude, the shifted cycle crosses the unstable fixed point causing annihilation of the oscillation and an undefined topological degree. A direct transition has been observed in a model of a cardiac pacemaker cell with a slow action potential upstroke (Guevara and Jongsma 1990). Alternatively, the transition may be indirect if there is a range of intermediate stimulus amplitudes for which the state point is taken outside of the basin of attraction of the limit cycle.

However, in experimental and even computational studies, apparent discontinuities can make it very difficult to distinguish between type 1 and type 0 topologies (Krogh-Madsen et al. 2004). An example is shown in Fig. 2.4a (same model as in Fig. 2.3), where t_c was sampled in intervals of 1 ms. For a low stimulus amplitude (−30 pA), the phase transition curve is clearly of type 1 and with a very large

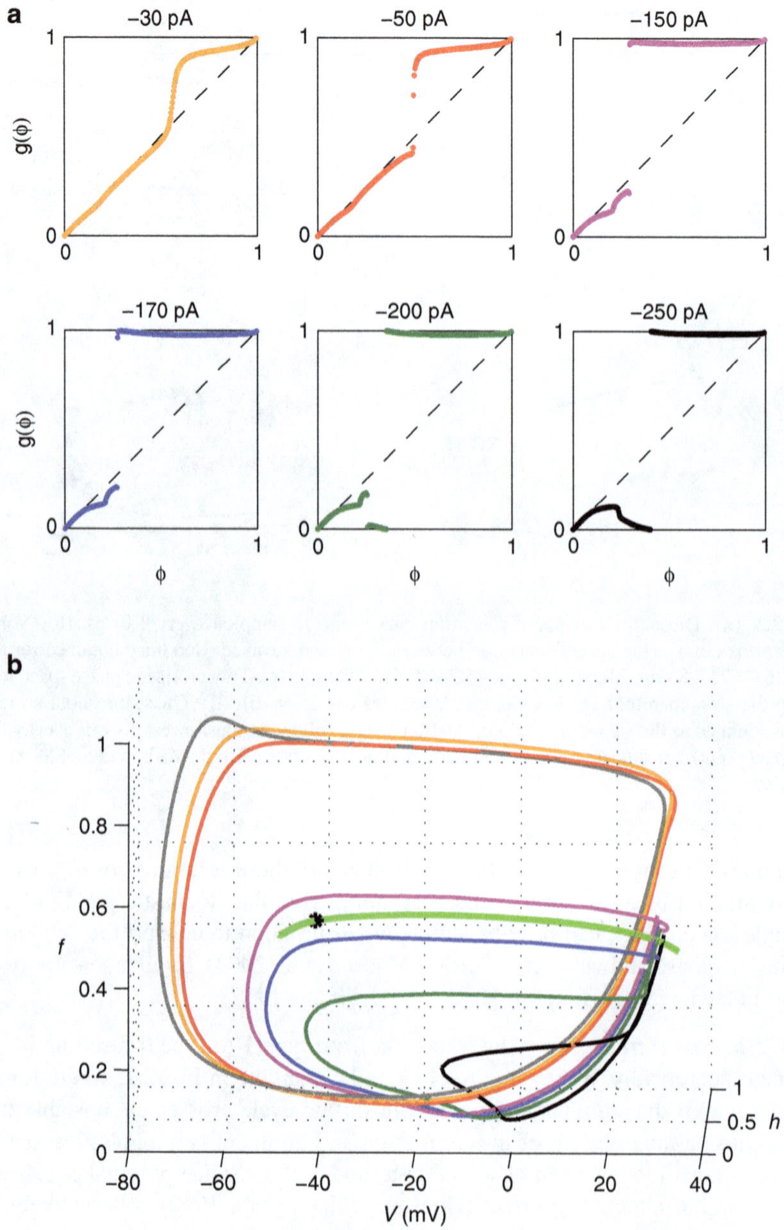

Fig. 2.4 Type 1 to type 0 transition. (**a**): Phase transition curves for a range of stimulus amplitudes. (**b**): Location of shifted cycles corresponding to the curves in (**a**) relative to the stable manifold of the fixed point (light green)

stimulus amplitude (–250 pA), it is evidently of type 0. However, for some of the intermediate stimulus amplitudes (in particular −150 and −170) the degree is indeterminate.

In a noise-free model, the location of the transition may be determined by much finer sampling of the stimulus timing. Alternatively, in a model where one has access to all the variables, the shifted cycle is readily computed and the topological transition determined from the location of a range of shifted cycles relative to the phaseless set. In our example, the transition from type 1 to type 0 resetting occurs for a stimulus amplitude between −150 and −170 pA (Fig. 2.4b). However, due to intrinsic noise in experimental preparations, this inherent difficulty in assigning a topological degree for intermediate stimulus amplitudes persists (Guevara et al. 1986; Vanmeerwijk et al. 1984). Noise is particularly prevalent in the phase response curves of many mammalian spiking neurons (Galan et al. 2005; Netoff et al. 2005).

There are only a limited number of studies examining the existence of a type 1 to type 0 transition in single neurons. In one study where both type 1 and type 0 resetting was found by systematic variation of the stimulus amplitude, the two types occurred for both excitatory and inhibitory inputs to a bursting neuron (Prinz et al. 2003). Further, the type transition was hypothesized to occur in spiking neurons also (Prinz et al. 2003). Except for this case, the published phase response curves from spiking neurons of which we are aware [e.g., Cui et al. (2009); Galan et al. (2005); Mancilla et al. (2007); Netoff et al. (2005); Perkel et al. (1964); Preyer and Butera (2005)] all appear to be type 1. It is quite likely that the experimental rationale never required a sufficiently strong input to elicit a type 1 to type 0 transition. Conversely, the published phase response curves of bursting neurons (Pinsker 1977; Sieling et al. 2009) in response to inhibitory input almost always appear to be type 0. This response is also supported by modeling studies (Demir et al. 1997). In the case of these bursting neurons, this type 0 response is due to an all-or-none like effect (see Sect. 3.1) – inhibitory inputs during a burst (when the neuron is spiking) either have a negligible effect on the burst trajectory (and thus the phase) or cause an immediate hyperpolarization to a different portion of the limit cycle. A type 1 topology in bursting neurons in response to excitatory input is only possible for extremely weak inputs, and is difficult to achieve in practice. These phenomena are discussed in (Sieling et al. 2009). In essence, the existence and nature of the type 1 to type 0 transition has not been extensively studied in spiking or bursting neurons and is an open area of investigation.

3.2 Finite Relaxation Times and the Limits of the Circle Map

One of the important applications of resetting experiments is the use of the resetting curves to determine dynamics set up between an oscillator and the environment (including other oscillations). The notions of weak interactions have played an important role in the development and application of the infinitesimal resetting curve (see Sect. 4). However, a major conclusion of Sect. 2 is that if the phase transition curve $g(\phi)$ is known, and a stimulation protocol is adapted, then the effects of the stimulation can be determined provided there is a rapid return to the limit cycle. For example, Eq. (2.6) can be used to predict the effects of periodic stimulation. Indeed, work from several groups has shown the applicability of this approach to predict the effects of periodic stimulation of cardiac tissue (Guevara et al. 1981). However, there have been limited experimental and theoretical studies that have explored the consequences if the time to relax back to the limit cycle is not fast compared to the cycle length.

Perhaps the most direct way to explore the notion of instantaneous relaxation back to the limit cycle is to deliver stimuli at a fixed time delay after a fiducial event. By definition, Eq. (2.1), the phase of a stimulus is t_c/T_0, so provided there is rapid relaxation to the limit cycle, the effect of a stimulus will only depend on the phase of the stimulus and not on the previous history. From an experimental perspective, it is straightforward to deliver a stimulus at fixed delays and to determine the effects. We mention several examples.

Lewis et al. determined the effects of stimulation of the superior laryngeal nerve (SLN) on the respiratory rhythm in anesthetized decerebrate cats (Lewis et al. 1989). These experiments were initially undertaken to apply the theory of Sect. 2 to measure the resetting of the respiratory rhythm by SLN stimulation, and then to predict the effects of periodic SLN stimulation. Normally, a pulse of SLN activity delivered at a time delay of 200 ms after the onset of phrenic activity (associated with the onset of inspiration) would lead to a termination of the inspiratory phase. However, in the course of doing these experiments, it became clear that the SLN stimulation had long lasting effects, and that consequently Eq. (2.6) would not hold. The SLN stimuli were delivered every phrenic burst, or after some specified fixed number of phrenic bursts. The stimulus would either shorten or prolong the respiratory cycle. If the stimulus was given every 8 cycles, each stimulus would lead to a shortening of the respiratory cycle, but if the stimulus was given at every 1st, 2nd, or 4th cycle, some stimuli would lead to shortening of the cycle, whereas others would lead to lengthening of the cycle (Fig. 2.5).

Kunysz et al. carried out similar experiments on spontaneously beating aggregates of chick heart cells, but used a depolarizing current that would normally induce an action potential if given after a sufficiently long delay after an action potential (Kunysz et al. 1997). If each stimulus induced a new action potential, then there would be a resulting burst of activity, at a faster rate than normal, a finding that could provide a physiological basis for some types of rapid heart rhythms (tachycardias). However, over a range of delays, each stimulus would not give rise to an action potential, so that after a short burst of activity the stimulus would not be effective, and there would be a long delay until there would again be a spontaneous action potential. Thus, this would lead to bursts of activity and could also provide a potential physiological mechanism for paroxysmal tachycardias that are observed clinically.

Another approach for this situation is to quantify the effects of the perturbation on subsequent limit cycles (Oprisan and Canavier 2004; Reyes and Fetz 1993). This approach defines each limit cycle by measurement of spike times and separates

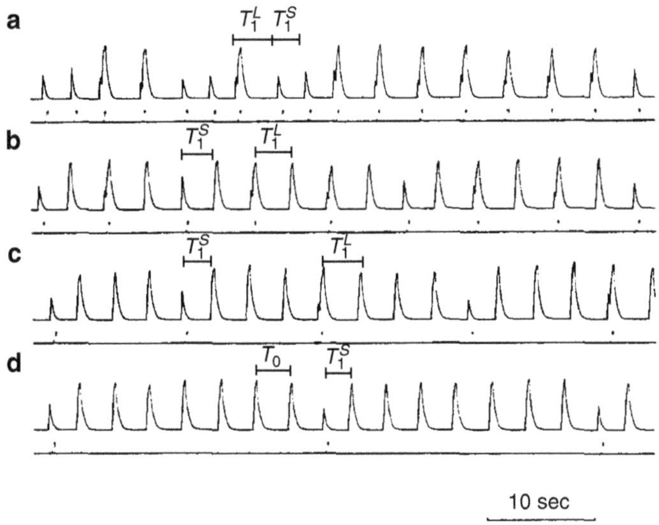

Fig. 2.5 Phrenic nerve activity during SLN stimulation. (**a**): Stimulus train delivered every cycle; (**b**): every 2 cycles; (**c**): every 4 cycles; (**d**): every 8 cycles. T_0, T_1^S, and T_1^L mark examples of the intrinsic period, stimulus-induced cycle shortening, and stimulus-induced cycle lengthening, respectively. Reproduced from Lewis et al. (1989) with permission

the effect of the stimulus into a first order phase response curve (effect on first limit cycle), second order phase response curve (effect on second period after the limit cycle with the perturbation), and so forth. Oprisan et al. (2004) showed that incorporating second order resetting improved the prediction of the synchronous state between a bursting neuron and a bursting neuron model synaptically coupled via a dynamic clamp.

More recently, Cui et al. have explored similar effects associated with adaptation of spontaneously spiking neurons in *Aplysia* (Cui et al. 2009). While the neurons display adaptation that occurs over several cycles, their method was to characterize a phase response curve calculated by repetitive stimulation of the neuron after a fixed delay from phase 0. After a sufficient number of cycles the phase relationship between the delayed stimulus and the spike oscillation stabilized. The rationale for this approach is that adapting neurons participating in a rhythm-generating network frequently lock into stable phase relationships among multiple oscillators, with an input arriving at approximately the same phase each cycle. Thus this method of characterizing a phase response curve represents the phase response curve of the neuron under a "steady-state" level of adaptation.

From a theoretical perspective, there has been some work. Given the Poincaré oscillator with periodic or fixed delay stimulation, extremely interesting dynamics can be obtained by simply assuming a finite relaxation time to the limit cycle. For example, a detailed mathematical study of the entrainment rhythms for different relaxation rates would be warranted, but we know of only one study at some selected relaxation rates (Glass and Sun 1994). Similarly, stimulation of the Poincaré oscillator at a fixed phase can lead to chaotic dynamics and interesting bifurcations as the phase of the stimulus is changed (Lewis et al. 1987).

4 Topological Analysis of Neural Models Near the Saddle Node Bifurcation

Although some studies of pulse coupling of neural oscillators use phase response curves in a similar fashion to those used in the present chapter (also see Chap. 4), most previous studies of resetting of neural oscillators have not been strongly influenced by the topological approach. Rather, studies of resetting neural oscillators emphasize the classification "class I" and "class II" oscillators (in this chapter, we use the terminology of type 1/0 for the topological classification and class I/II for the traditional neural classification). This classification was initially based on experimental observations of different classes of excitability by Alan Hodgkin (1948) and has subsequently been extended by many [see Izhikevich (2000, 2007)]. In class I neurons the firing frequency codes an input intensity, whereas in class II neurons the firing frequency is largely independent of a (suprathreshold) input. There are many examples of published phase response curves from single repetitive spiking neurons that are predominantly class I and also apparently type 1 [e.g., Cui et al. (2009); Galan et al. (2005); Mancilla et al. (2007); Netoff et al. (2005); Perkel et al. (1964); Preyer and Butera (2005)]. Consideration of the class I/II neurons from a dynamical perspective by Ermentrout, Izhikevich and others (Rinzel and Ermentrout (1989); Izhikevich (2000, 2007)) has led to the recognition that class I neurons may arise from a saddle node bifurcation on an invariant circle, whereas class II neurons may more likely arise from an Andronov–Hopf bifurcation. The concept of class I/II neurons has been particularly fruitful in analysis of weak coupling which is assumed when using phase-oscillator model approaches to studying interacting oscillations (Ermentrout and Kopell 1990).

In the current theoretical neurophysiological literature, there is often a common statement "class I neurons will exhibit a phase-advance, whereas class II neurons exhibit both advance and delay" [e.g., Abouzeid and Ermentrout (2009)]. However, based on the topological description in Sect. 2, continuity of the underlying dynamical system ensures the existence of isochrons corresponding to all real values of the new phase between 0 and 1 (or 0 and 2π if measured in radians). Hence, the topological approach predicts both delay and advance, making the above assertion regarding class I neurons appear at odds with the topological description. We thought it worthwhile to analyze the source of the apparent discrepancies. In order to carry out the computations, we first assume an equation for a model oscillator that displays a saddle-node bifurcation on an invariant cycle, and consider a parameter value "close" to the bifurcation point.

We analyze the equation

$$\frac{d\theta}{dt} = 1 - ar\cos(\theta - \Phi_{SN}), \tag{2.7}$$

$$\frac{dr}{dt} = r(1 - r^2). \tag{2.8}$$

For $a < 1$ there is a stable limit cycle with $|r| = 1$ and when $a = 1$ there is a saddle node bifurcation on the invariant circle at Φ_{SN}. Again, stimuli of amplitude b are applied to perturb the state point in the direction of $\theta = 0$. We show the results of three simulations of resetting curves with $a = 0.97$, representing a value that is near the saddle node bifurcation. Notice that when $a = 0$ we have the Poincaré oscillator discussed in Sect. 2 with an oscillation period of 2π. When $a = 0.97$ the period increases to ≈ 25.85 reflecting the slowing of the angular derivative near the point Φ_{SN}.

Figure 2.6a shows the phase transition and phase response curve curves for $b = 0.1$ and $\Phi_{SN} = 3\pi/2$. In this case the phase response curve is almost entirely positive. This arises from (a) the slowing of the trajectory near the point $\Phi_{SN} = 3\pi/2$; and (b) the observation that stimuli near the phase $\phi = 3\pi/2$ in the Poincaré oscillator will lead to positive phase response curve (see Fig. 2.2). However, there is also a region in which the phase response curve is negative, as would be expected from the topological analysis. Because of isochron clustering, the region of delay in the phase response curve is small and might easily be missed or even be unmeasurable in experiments. This situation represents the typical situation in most neuron models, where the location of saddle node bifurcation, the orientation of the limit cycle trajectory, and the direction of the perturbation occurs in a region of phase space where stimuli generally lead to a phase advance. Others have observed that the phase response curve of class I neurons models appears to have at least a small portion of the limit cycle where the phase response curve is negative (Oprisan et al. 2002; Rinzel and Ermentrout 1989).

Now consider the situation when $\Phi_{SN} = \pi/2$ (Fig. 2.6b). In contrast to the situation in Fig. 2.6a, the phase response curve is now almost entirely negative. The explanation for this behavior is clear and it is a consequence of a slowing of the trajectory near a point in phase space that will lead to a negative phase response curve. Once again, the phase response curve is bimodal, though now most of the phase response curve is negative and only a small region is positive.

For completeness, in Fig. 2.6c we illustrate the phase transition curve and phase response curve for $a = 0.97$, $\Phi_{SN} = 3\pi/2$, and $b = 1.1$ for which we have topological type 0 resetting.

In conclusion, these examples show that (a) limit cycles that arise as a consequence of a saddle node bifurcation will show bimodal phase response curves and (b) the relative ranges over which the curves are advancing or delaying depends on the location of the incipient bifurcation relative to the fiducial point and the imposed perturbation.

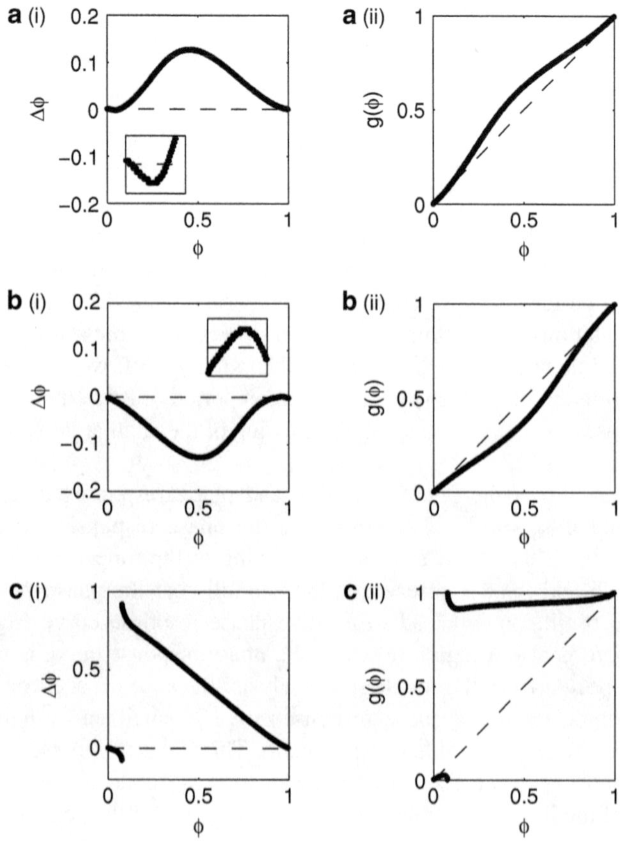

Fig. 2.6 Phase response curves (*left*) and phase transition curves (*right*) for the simple model close to the SNIC Eq. (2.7). (**a**) (**i** and **ii**): $b = 0.1$, $\Phi_{\mathrm{SN}} = 3\pi/2$; (**b**) (**i** and **ii**): $b = 0.1$, $\Phi_{\mathrm{SN}} = \pi/2$; (**c**) (**i** and **ii**): $b = 1.1$, $\Phi_{\mathrm{SN}} = 3\pi/2$; where b is the stimulus amplitude. Inserts in (**a**) (**i**) and (**b**) (**i**) show zoom-ins of negative and positive going parts of the phase response curves. Axes range is 0.1 in the horizontal direction and 0.06 in the vertical direction

5 Summary and Conclusions

In summary, we have reviewed the general theory of topological phase resetting and shown how it can be applied to experimental analysis. The method is straightforward to apply to experimental data and relies upon testable assumptions. These methods are similar to the pulse-coupled approaches used for neuronal oscillators as described in Chap. 4. Although some of the statements that have been made with regard to phase resetting appear to contradict the topological theory, there is no inconsistency between the two approaches. In particular, phase response curves will in general be bimodal for both class I and class II neurons. However, in the limit when one is infinitesimally close to the bifurcation point, the phase response curve can be positive for class I neurons. In addition, the assertion that class I neurons will display principally positive phase response curves, results both from the occurrence of a stable limit cycle near a saddle node bifurcation *and* the actual spatial location of the incipient saddle-node in the phase space. In typical mathematical models of neurons, the saddle node bifurcation is located in a region of the phase space where a depolarization would lead to an action potential sooner than expected in the absence of a perturbation causing a positive phase response curve.

The use of phase oscillator models and the assumption of weak coupling (Ermentrout and Kopell 1990) is an alternative approach for studying the phase response of neuronal oscillators. The advantage of such theory is the ability to separate the input waveform from the intrinsic phase response properties of the neuron. A disadvantage of this theory is that the estimation of phase response curves using this method is less straightforward [e.g., Galan et al. (2005); Preyer and Butera (2005)] and requires a degree of numerical estimation, optimization, or deconvolution. A type 0 phase response curve is not possible within the paradigm of weak coupling, as the assumptions of weak coupling are inclusive of (and more stringent than) the assumptions of type 1 resetting described herein. These assumptions include that the perturbation is weak and relatively close to the limit cycle. Topological theory may be a useful adjunct for weak coupling in situations where sufficiently strong inputs occur such that the weak coupling assumptions are violated. We are currently investigating more completely the relationship between the influence of the bifurcation (class I or II) on the topology of the phase space and its implications on topological phase resetting.

Acknowledgments T. Krogh-Madsen acknowledges the National Institutes of Health for financial support (R01-HL094620-01, PI: Christini). R. Butera acknowledges support from the National Institutes of Health (R01-HL08886, PI: Butera; R01-NS054281, PI: Canavier; R01-RR020115, PI: Christini). G. B. Ermentrout acknowledges support from NSF DMS 817131. L. Glass thanks NSERC for financial support.

References

Abouzeid, A., & Ermentrout, G. B. (2009). Type-II phase resetting curve is optimal for stochastic synchrony. *Physical Review E*, 80:011911.

Arnol'd, V. I. (1965). Small denominators. I. mappings of the circumference onto itself. *American Mathematical Society Translations*, 46:213–284.

Bem, T., Feuvre, Y. L., Rinzel, J., & Meyrand, P. (2005). Electrical coupling induces bistability of rhythms in networks of inhibitory spiking neurons. *European Journal of Neuroscience*, 22:2661–8.

Butera, R. J., Wilson, C. G., DelNegro, C. A., & Smith, J. C. (2001). A methodology for achieving high-speed rates of artificial conductance injection in electrically excitable biological cells. *IEEE Transactions on Biomedical Engineering*, 48:1460–1470.

Canavier, C. C. (2006). Phase response curve. *Scholarpedia*, 1(12):1332. Retrieved March 16, 2010, from http://www.scholarpedia.org/article/Phase_response_curve

Canavier, C. C., Butera, R. J., Dror, R. O., Baxter, D. A., Clark, J. W., & Byrne, J. H. (1997). Phase response characteristics of model neurons determine which patterns are expressed in a ring circuit model of gait generation. *Biological Cybernetics*, 77:367–380.

Clay, J. R. (1977). Monte Carlo simulation of membrane noise: An analysis of fluctuations in graded excitation of nerve membrane. *Journal of Theoretical Biology*, 64:671–680.

Clay, J. R., Guevara, M. R., & Shrier, A. (1984). Phase resetting of the rhythmic activity of embryonic heart cell aggregates. experiment and theory. *Biophysical Journal*, 45:699–714.

Cui, J., Canavier, C. C., & Butera, R. J. (2009). Functional phase response curves: A method for understanding synchronization of adapting neurons. *Journal of Neurophysiology*, 102:387–98.

Demir, S. S., Butera, R. J., Defranceschi, A. A., Clark, J. W., & Byrne, J. H. (1997). Phase-sensitivity and entrainment in a modeled bursting neuron. *Biophysical Journal*, 72:579–594.

Dorval, A. D., Christini, D. J., and White, J. A. (2001). Real-time linux dynamic clamp: A fast and flexible way to construct virtual on channels in living cells. *Annals of Biomedical Engineering*, 29:897–907.

Ermentrout, G. B., & Kopell, N. (1990). Oscillator death in systems of coupled neural oscillators. *SIAM Journal of Applied Mathematics*, 50:125–146.

Galan, R. F., Ermentrout, G. B., & Urban, N. N. (2005). Efficient estimation of phase-resetting curves in real neurons and its significance for neural-network modeling. *Physical Review Letters*, 94:158101.

Gedeon, T., & Glass, L. (1998). Continuity of resetting curves for FitzHugh-Nagumo equations on the circle. In *Fields Institute Communications, Differential Equations with Applications to Biology* (pp. 225–236). New York: Springer.

Glass, L., & Mackey, M. C. (1998). *From clocks to chaos: The rhythms of life*. Princeton: Princeton University Press.

Glass, L., & Sun, J. (1994). Periodic forcing of a limit cycle oscillator: Fixed points, Arnold tongues, and the global organization of bifurcations. *Physical Review E*, 50:5077–5084.

Glass, L., & Winfree, A. T. (1984). Discontinuities in phase resetting experiments. *American Journal of Physiology*, 15:R251–R258.

Guckenheimer, J. (1975). Isochrones and phaseless sets. *Journal of Mathematical Biology*, 1:259–273.

Guevara, M. R., & Glass, L. (1982). Phase locking, periodic doubling bifurcations and chaos in a mathematical model of a periodically driven oscillator: A theory for the entrainment of biological oscillators and the generation of cardiac dysrhythmias. *Journal of Mathematical Biology*, 14:1–23.

Guevara, M. R., Glass, L., Mackey, M. C., & Shrier, A. (1983). Chaos in neurobiology. *IEEE Transactions on Systems, Man, and Cybernetics*, SMC–13:790–798.

Guevara, M. R., Glass, L., & Shrier, A. (1981). Phase locking, period-doubling bifurcations, and irregular dynamics in periodically stimulated cardiac cells. *Science*, 214:1350–1353.

Guevara, M. R., & Jongsma, H. J. (1990). Phase resetting in a model of sinoatrial nodal membrane: Ionic and topological aspects. *American Journal of Physiology*, 258:H734–H737.

Guevara, M. R., Shrier, A., & Glass, L. (1986). Phase resetting of spontaneously beating embryonic ventricular heart cell aggregates. *American Journal of Physiology*, 251:H1298–H1305.

Hodgkin, A. L. (1948). The local electric changes associated with repetitive action in a non-medullated axon. *Journal of Physiology*, 107:165–81.

Izhikevich, E. M. (2000). Neural excitability, spiking, and bursting. *International Journal of Bifurcation and Chaos*, 10:1171–1266.

Izhikevich, E. (2007). *Dynamical Systems in Neuroscience: The Geometry of Excitability and Bursting*. Cambridge, Massachusetts: M.I.T. Press.

Kaplan, D. T., Clay, J. R., Manning, T., Glass, L., Guevara, M. R., & Shrier, A. (1996). Subthreshold dynamics in periodically stimulated squid giant axons. *Physical Review Letters*, 76:4074–4077.

Krogh-Madsen, T., Glass, L., Doedel, E. J., & Guevara, M. R. (2004). Apparent discontinuities in the phase-resetting response of cardiac pacemakers. *Journal of Theoretical Biology*, 230: 499–517.

Kunysz, A. M., Munk, A. A., & Shrier, A. (1995). Phase resetting and dynamics in isolated atrioventricular nodal cell clusters. *Chaos*, 5:184–192.

Kunysz, A. M., Shrier, A., & Glass, L. (1997). Bursting behavior during fixed-delay stimulation of spontaneously beating chick heart cell aggregates. *American Journal of Physiology*, 273:C331–C346.

Lewis, J., Bachoo, M., Glass, L., & Polosa, C. (1987). Complex dynamics resulting from repeated stimulation of nonlinear oscillators at a fixed phase. *Physics Letters A*, 125:119–122.

Lewis, J., Bachoo, M., Polosa, C., & Glass, L. (1989). The effects of superior laryngeal nerve stimulation on the respiratory rhythm: Phase resetting and aftereffects. *Brain Research*, 517: 44–50.

Mancilla, J. G., Lewis, T. J., Pinto, D. J., Rinzel, J., & Connors, B. W. (2007). Synchronization of electrically coupled pairs of inhibitory interneurons in neocortex. *Journal of Neuroscience*, 27:2058–73.

Matsumoto, G., Aihara, K., Ichikawa, M., & Tasaki, A. (1984). Periodic and nonperiodic responses of membrane potentials in squid giant axons during sinusoidal current stimulation. *Journal of Theoretical Neurobiology*, 3:1–14.

Netoff, T. I., Acker, C. D., Bettencourt, J. C., & White, J. A. (2005). Beyond two-cell networks: Experimental measurement of neuronal responses to multiple synaptic inputs. *Journal of Computational Neuroscience*, 18:287–95.

Oprisan, S., & Canavier, C. (2004). Phase resetting and phase locking in hybrid circuits of one model and one biological neuron. *Biophysical Journal*, 87:2283–2298.

Oprisan, S., Prinz, A., & Canavier, C. (2002). The influence of limit cycle topology on the phase resetting curve. *Neural Computation*, 14:1027–1057.

Perkel, D. H., Schulman, J. H., Bullock, T. H., Moore, G. P., & Segundo, J. P. (1964). Pacemaker neurons: Effects of regularly spaced synaptic input. *Science*, 145:61–63.

Pinsker, H. M. (1977). Aplysia bursting neurons as endogenous oscillators. I. Phase-response curves for pulsed inhibitory synaptic input. *Journal of Neurophysiology*, 40:527–43.

Preyer, A. J., & Butera, R. J. (2005). Neuronal oscillators in aplysia californica that demonstrate weak coupling in vitro. *Physical Review Letters*, 95:138103.

Prinz, A. A., Thirumalai, V., & Marder, E. (2003). The functional consequences of changes in the strength and duration of synaptic inputs to oscillatory neurons. *Journal of Neuroscience*, 23:943–954.

Reyes, A. D., & Fetz, E. E. (1993). Two modes of interspike interval shortening by brief transient depolarizations in cat neocortical neurons. *Journal of Neurophysiology*, 69:1661–1672.

Rinzel, J., & Ermentrout, G. B. (1989). Analysis of neural excitability and oscillations. In C. Koch and I. Segev, editors, *Methods in neuronal modeling*. Cambridge, MA: MIT Press.

Sharp, A. A., O'Neil, M. B., Abbott, L. F., & Marder, E. (1993). Dynamic clamp: Computer-generated conductances in real neurons. *Journal of Neurophysiology*, 69:992–5.

Sharp, A. A., Skinner, F. K., & Marder, E. (1996). Mechanisms of oscillation in dynamic clamp constructed two-cell half-center circuits. *Journal of Neurophysiology*, 76:867–83.

Sieling, F. H., Canavier, C. C., & Prinz, A. A. (2009). Predictions of phase-locking in excitatory hybrid networks: Excitation does not promote phase-locking in pattern-generating networks as reliably as inhibition. *Journal of Neurophysiology*, 102:69–84.

Vanmeerwijk, W. P. M., Debruin, G., van Ginneken, A. C. G., Vanhartevelt, J., Jongsma, H. J., Kruyt, E. W., Scott, S. S., & Ypey, D. L. (1984). Phase resetting properties of cardiac pacemaker cells. *The Journal of General Physiology*, 83:613–629.

Winfree, A. T. (2000). *The Geometry of Biological Time*. (Second Edition). New York: Springer.

Chapter 3
A Theoretical Framework for the Dynamics of Multiple Intrinsic Oscillators in Single Neurons

Michiel W.H. Remme, Máté Lengyel, and Boris S. Gutkin

Abstract The dendritic tree contributes significantly to the elementary computations a neuron performs while converting its synaptic inputs into action potential output. Traditionally, these computations have been characterized as both temporally and spatially localized. Under this account, neurons compute near-instantaneous mappings from their current input to their current output, brought about by somatic summation of dendritic contributions that are generated in functionally segregated compartments. However, recent evidence about the presence of oscillations in dendrites suggests a qualitatively different mode of operation: the instantaneous phase of such oscillations can depend on a long history of inputs, and, under appropriate conditions, even dendritic oscillators that are remote may interact through synchronization. Here, we develop a mathematical framework to analyze the interactions of local dendritic oscillations, and the way these interactions influence single cell computations. Combining weakly coupled oscillator methods with cable theoretic arguments, we derive phase-locking states for multiple oscillating dendritic compartments. We characterize how the phase-locking properties depend on key parameters of the oscillating dendrite: the electrotonic properties of the

M.W.H. Remme (✉)
Group for Neural Theory, Département d'Études Cognitives, École Normale Supérieure, Paris, France

Current address: Institute for Theoretical Biology, Humboldt-Universität zu Berlin, Berlin, Germany
e-mail: michiel.remme@hu-berlin.de

M. Lengyel
Computational and Biological Learning Lab, Department of Engineering, University of Cambridge, Cambridge, UK
e-mail: m.lengyel@eng.cam.ac.uk

B.S. Gutkin
Group for Neural Theory, Département d'Études Cognitives, École Normale Supérieure, Paris, France
e-mail: boris.gutkin@ens.fr

N.W. Schultheiss et al. (eds.), *Phase Response Curves in Neuroscience: Theory, Experiment, and Analysis*, Springer Series in Computational Neuroscience 6,
DOI 10.1007/978-1-4614-0739-3_3, © Springer Science+Business Media, LLC 2012

(active) dendritic segment, and the intrinsic properties of the dendritic oscillators. As a direct consequence, we show how input to the dendrites can modulate phase-locking behavior and hence global dendritic coherence. In turn, dendritic coherence is able to gate the integration and propagation of synaptic signals to the soma, ultimately leading to an effective control of somatic spike generation. Our results suggest that dendritic oscillations enable the dendritic tree to operate on more global temporal and spatial scales than previously thought; notably that local dendritic activity may be a mechanism for generating on-going whole-cell voltage oscillations.

1 Introduction

The dendritic tree contributes significantly to the elementary computations a neuron can perform, both by its intricate morphology and its composition of voltage-gated ionic conductances (Stuart et al. 2007). Such active conductances can underlie a wide variety of dynamical behaviors such as ongoing oscillations of the dendritic membrane potential, both sub- and supra-threshold. Membrane potential oscillations have been demonstrated in various types of neurons. Prominent intrinsic subthreshold oscillations have been found in stellate cells from entorhinal cortex layer 2 (Alonso and Llinás 1989; Alonso and Klink 1993), neurons from the frontal cortex (Gutfreund et al. 1995), neurons from the amygdala complex (Pape et al. 1998; Sanhueza and Bacigalupo 2005), and pyramidal cells and interneurons from the hippocampal CA1 area (Leung and Yim 1991; Chapman and Lacaille 1999).

Although these membrane potential oscillations are normally recorded at the soma and thus are considered to be of somatic origin, several lines of evidence suggest dendritic loci of generation. First, many of the conductances thought to underlie the generation of such oscillations reside predominantly in the dendrites, sometimes specifically in the distal parts of the dendritic tree. For example, in the apical dendrites of hippocampal CA1 pyramidal neurons, the density of I_{h} increases strongly with distance from the soma (Magee 1998), and reaches very high values in the thin distal branches (Lörincz et al. 2002). Second, several studies have suggested the existence of clusters of ionic conductances that are responsible for the generation of dendritic spikes (Llinás and Sugimori 1980). While most of the direct electrophysiological evidence regards excitable behavior, demonstrating the generation of dendritic spikes in response to sufficient levels of depolarization, mathematical analysis has shown that neural membranes exhibiting excitability can readily pass to oscillatory regimes in an input-dependent manner (e.g., see Rinzel and Ermentrout 1998). Third, in several cases, oscillations have been directly recorded in dendrites. For example, recordings from hippocampal CA1 pyramidal neurons have demonstrated ongoing oscillations in the dendrites that include repetitive dendritic spikes, presumably involving Ca^{2+} currents (Kamondi et al. 1998). Furthermore, significant intrinsic dendritic oscillations have been observed in several neuronal preparations that depended on the interplay between the nonlinear properties of NMDA synaptic receptors and intrinsic voltage-dependent currents (Moore et al. 1999; Placantonakis and Welsh 2001). Crucially, while the onset of these oscillations was conditional on the activation of the NMDA synapses, the oscillations themselves were produced by mechanisms that were intrinsic to the postsynaptic cell and not by periodically structured synaptic inputs. Since NMDA receptors are largely localized on dendritic spines, and are hence electrotonically removed from the soma, these data may also argue for a nonuniform and local dendritic generation of membrane potential oscillations. Taken together, these experimental results suggest that dendritic trees can function as oscillators, perhaps conditional on the level of background depolarization or the presence of neuromodulators (Yoshida and Alonso 2007), while leaving open the question whether global cell-wide voltage oscillations could result from local dendritic mechanisms that are intrinsic even to distal dendrites and hence perhaps only weakly coupled to the soma electrotonically.

Indeed, multiple intrinsic dendritic oscillators have been proposed to underlie the recently discovered intricate firing pattern of entorhinal grid cells (O'Keefe and Burgess 2005; Burgess et al. 2007; Giocomo et al. 2007). This influential model suggests that the functional responses of entorhinal neurons recorded in behaving animals are a direct consequence of the generation of independent oscillations that are intrinsic to individual dendrites. Hence, this model presupposes the existence of multiple oscillators that are integrated at the soma, leading to the questions of how such dendritic oscillators may interact with the soma and with each other, and what sorts of collective behaviors the electrotonic structure of the dendritic tree might impose on the oscillations.

Here, we focus on the dynamics of such interacting oscillators and their impact on signal propagation in single neurons, using mathematical analysis corroborated by numerical simulations of biophysical models. We treat the dendritic tree of a neuron as a network of oscillators coupled by stretches of relatively less active cable. This prompts us to combine two analytical methods: weakly coupled oscillator theory and cable theory. The theory of weakly coupled oscillators has been extensively used previously to study synchronization of multiple oscillators residing in separate cells interacting through synapses or gap junctions (Izhikevich 2007). In this framework, the response of oscillators to perturbations is described by their infinitesimal phase response curves. This greatly simplifies this complex dynamical system and allows for an analytical treatment of the phase-locking behavior of the coupled oscillators. Since we focus on intradendritic oscillators, which are continuously coupled via the membrane voltage, we use cable theory (Rall 1967) to compute the perturbations via which the oscillators interact.

Part of the results of this chapter have been published in Remme et al. (2009).

2 Methods

We first develop a theory for the behavior of a dendritic tree that contains multiple intrinsic oscillators and then use this framework to gain understanding of how such a tree would behave dynamically and hence control the neuron's output depending on the input. In order to develop the mathematical framework, we begin by considering

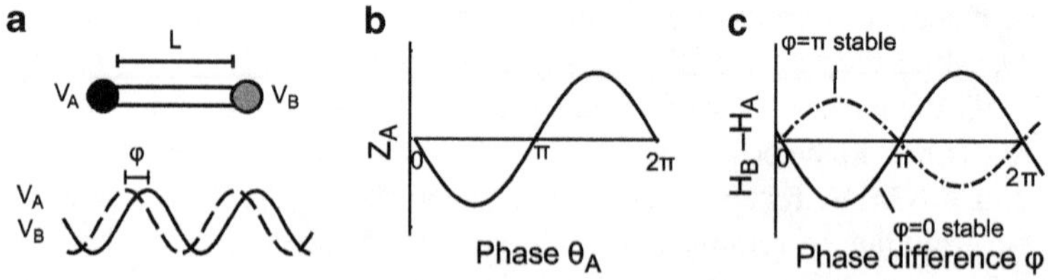

Fig. 3.1 Minimal system of two oscillators coupled via a dendritic cable. (**a**) The oscillators with voltage trajectories $V_A(t)$ and $V_B(t)$ and phase difference ϕ determine the membrane potential at the ends of a cable with electrotonic length L. (**b**) Example of a phase response function $Z_A(\theta_A)$ giving the phase shift of oscillator A as a perturbation at phase θ_A. (**c**) The stable phase-locked solution is determined by $\dot{\phi} = 0$ and $\frac{d}{d\phi}\dot{\phi} < 0$ and is either at $\phi = 0$ (e.g. for the *solid curve*) or at $\phi = \pi$ (e.g. for the *dash-dotted curve*)

a minimal setup of two cable-coupled oscillators. As we will see, even this setup is too complicated for direct analytical treatment; hence, we will go through a number of reduction steps which we sketch out below.

We study the behavior of a system of two oscillators with period T being connected via an active (though not intrinsically oscillating) dendritic cable with length constant λ and membrane time constant τ. The cable also expresses a voltage-dependent conductance with a gating variable $m(x,t)$ with activation function $m_\infty(V)$ and time constant τ_m (in milliseconds). The equations governing the membrane potential $V(x,t)$ and the gating variable $m(x,t)$ along the cable (excluding the oscillators) are

$$\tau \frac{\partial}{\partial t} V(x,t) = \lambda^2 \frac{\partial^2}{\partial x^2} V(x,t) - (V(x,t) - E_L) - \gamma_m\, m(x,t)\,(V(x,t) - E_m)$$

$$\tau_m \frac{\partial}{\partial t} m(x,t) = m_\infty(V(x,t)) - m(x,t), \tag{3.1}$$

where E_L is the leak reversal potential, E_m is the reversal potential of the active current, and γ_m is the ratio of the maximal conductance of the active current to the leak conductance. The two oscillators A and B are located at the ends of the cable at $x = 0$ and $x = l$, separated by an electrotonic distance $L = l/\lambda$ (Fig. 3.1a). The two oscillators form the periodically forced end conditions of the cable:

$$V(0,t) = V_A(t),$$

$$V(l,t) = V_B(t) \tag{3.2}$$

with $V_A(t)$ and $V_B(t)$ being the voltage traces of the two oscillators A and B that evolve according to

$$C_m \frac{d}{dt} V_A(t) = -g_L(V_A(t) - E_L) - I_A(V_A(t), \mathbf{m}_A(t)) - \varepsilon\, p_A(t),$$

$$C_m \frac{d}{dt} V_B(t) = -g_L(V_B(t) - E_L) - I_B(V_B(t), \mathbf{m}_B(t)) - \varepsilon\, p_B(t), \tag{3.3}$$

where C_m is the membrane capacitance (in μF/cm^2), g_L is the leak conductance (in mS/cm^2), $I_{A,B}(t)$ summarizes the voltage-dependent membrane currents generating the oscillations with the vector of gating variables $\mathbf{m}_{A,B}(t)$ given by standard kinetic equations. The terms $\varepsilon\, p_{A,B}(t)$ describe the perturbing currents that each oscillator receives from the cable and are proportional to $\frac{\partial}{\partial x}V(0,t)$ and $\frac{\partial}{\partial x}V(l,t)$. In the weak coupling framework ε is typically a small parameter, implying that the currents in the stretch of cell membrane that generate the intrinsic oscillations are much stronger than the perturbing currents that arrive from the dendritic cable, hence, $|\varepsilon\, p_{A,B}(t)|/|g_L(V_{A,B}(t) - E_L) + I_{A,B}(t)| \ll 1$. For a cable with diameter d (in cm) and oscillators that are described as a single isopotential compartment with membrane surface area A (in cm^2), we have $\varepsilon = \pi d^2/4R_i\, A$, where R_i is the intracellular resistivity of the dendritic cable (in kΩ cm).

The two oscillators described by (3.3) form the boundary conditions (3.2) for the cable (3.1). In turn, the cable yields the current flux through its ends into (and thereby perturbing) the two oscillators: the terms $\varepsilon\, p_{A,B}(t)$ in (3.3). It is clear that it is next to impossible to solve (3.1)–(3.3) directly. However, we will use a number of reductions to arrive at a phase description of the system that is simple enough to handle analytically. This allows us to derive interaction functions for the two oscillators, describing how much they perturb each other through the dendrite depending on their phases. We then use these interaction functions to determine the stable phase relationship between the oscillators for different parameters, i.e., the properties of the cable and the type of oscillators (see for similar approaches Crook et al. 1998; Bressloff 1999; Goldberg et al. 2006).

We begin by observing that the oscillators from (3.3) can be reduced to a phase description (Izhikevich 2007). The phases θ_A and θ_B (in radians) describe the state of each oscillator. The dynamics of the phases are then described by

$$\dot{\theta}_A = \frac{2\pi}{T} + \varepsilon\, Z_A(t) p_A(t),$$
$$\dot{\theta}_B = \frac{2\pi}{T} + \varepsilon\, Z_B(t) p_B(t). \tag{3.4}$$

Here, the first term in the right hand side of each equation is the natural frequency of each oscillator and the second term describes the interaction between the oscillators. The crux of the analysis is thus to derive this function which we do explicitly in Appendix.

The interaction between the two oscillators depends on two factors: the intrinsic properties of the oscillators, as reflected by their infinitesimal phase response curves $Z_{A,B}(t)$, and the perturbations $p_{A,B}(t)$ to each oscillator via the cable. The infinitesimal phase response curve of an oscillator describes the phase shift induced by a perturbation delivered at a given phase (Fig. 3.1b). It can be determined using standard methods (Izhikevich 2007). The perturbations to the oscillators come from solving (3.1) with the oscillators described by (3.3) as the boundary conditions described by (3.2). For the active cable, this task can be greatly simplified if we consider a quasi-active approximation of the cable, and if we realize that the cable should behave periodically. The former can be done by linearizing the cable (3.1) about the voltage to which the cable would relax if it was not driven by the oscillators (Sabah and Leibovic 1969; Koch 1984). Under such approximations, the active properties of the dendritic cable can be summarized by a single parameter, μ, which can be derived from its basic biophysical properties (see Appendix). The sign of μ indicates whether the active conductance that is present in the cable is regenerative ($\mu < 0$), restorative ($\mu > 0$), or passive ($\mu = 0$) (see also Goldberg et al. 2006). A regenerative current will amplify perturbations (e.g., a persistent sodium current I_{NaP}), while a restorative current actively counteracts such perturbations (e.g., the hyperpolarization activated inward current I_{h}).

Since the solution to the cable equation with periodically forced end conditions is also periodic, it depends only on the difference of the phases of the two oscillators $\phi = \theta_{\mathrm{B}}(t) - \theta_{\mathrm{A}}(t)$. The dynamics of ϕ is the central object of our interest. Assuming that the oscillator interactions via the cable are relatively weak, we can obtain the interaction functions $H_{\mathrm{A}}(\phi)$ and $H_{\mathrm{B}}(\phi)$ (see Izhikevich 2007 and Appendix). These describe the change in the oscillators' phases as a function of the phase difference (Fig. 3.1c). Now the phase difference between the oscillators evolves, on a slower timescale, as

$$\dot{\phi} = \varepsilon \, (H_{\mathrm{B}}(\phi) - H_{\mathrm{A}}(\phi)) \tag{3.5}$$

It is easy to see that phase-locked states for our dendrite can be identified as values of ϕ where $\dot{\phi} = 0$. The derivative of $\dot{\phi}$ with respect to ϕ gives the stability of such states (negative implies stable, positive unstable).

3 Results

3.1 Phase-Locking Dynamics of Specific Oscillator Models

The interaction functions and hence the phase-locking dynamics depend critically on the biophysics of the oscillators considered. We now turn to illustrating our analysis for two different oscillator types: one that generates action potentials and the other a model for subthreshold oscillations.

As a first example we analyze the phase-locking for the type II Morris–Lecar neural oscillator (Morris and Lecar 1981). The voltage trace and the phase response function of this oscillator are plotted in Fig. 3.2a for one oscillation cycle, starting at the peak of the voltage trace. With this, we can compute the bifurcation diagram and determine the stable and unstable phase-locked solutions as a function of the electrotonic distance L, one key parameter in determining the phase-locking dynamics (Fig. 3.2b). The dendrite shows a bistable region where both the in-phase and the antiphase solution are stable (around $L \sim 1.65$). For smaller L, the in-phase solution is stable. As the electrotonic separation between the oscillators approaches $L = 4$, there is also a sharp transition from a stable anti-phase to a stable in-phase solution.

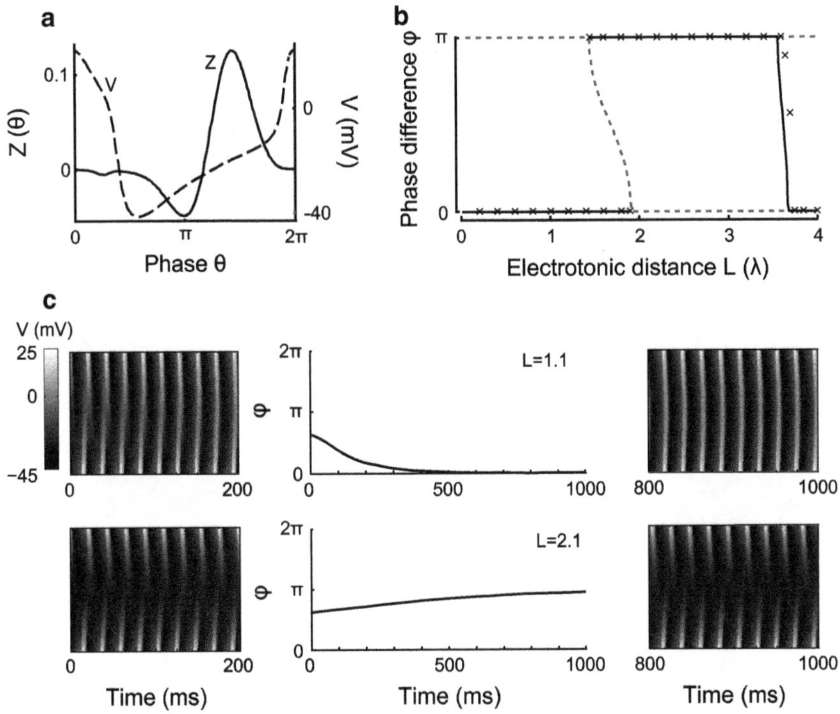

Fig. 3.2 Phase-locking of two Morris–Lecar type II oscillators. (**a**) Voltage trajectory (*dashed line*) and phase response function (*solid line*) of the Morris–Lecar type II oscillator with period $T = 21$ ms. (**b**) Bifurcation diagram showing the stable (*solid black line*) and unstable (*dashed gray line*) phase-locked solutions as a function of L. *Cross marks* give the stable phase difference determined with numerical simulations. (**c**) The *middle two panels* show simulations of the phase difference dynamics for $L = 1.1$ (*top*) and $L = 2.1$ (*bottom*). Space-time plots of the membrane potential along the dendritic cable are plotted for the first 200 ms (*left*) and for the final 200 ms (*right*) of the two simulations

Using numerical simulations of (3.1)–(3.3), we can demonstrate the dynamics of the phase difference between the two Morris–Lecar oscillators, as well as the membrane potential dynamics along the cable. Figure 3.2c illustrates these dynamics when the oscillators are separated by an electrotonic distance of $L = 1.1$ (top panels) or $L = 2.1$ (bottom panels). The oscillators start out with a phase difference of $\phi = 2\pi/3$. As expected from the bifurcation diagram in Fig. 3.2b, the two oscillators move to the in-phase configuration $\phi = 0$ when $L = 1.1$, synchronizing the voltage oscillations along the cable. When $L = 2.1$ the two oscillators settle in the antiphase solution $\phi = \pi$, producing large voltage gradients along the cable.

As a second example, we determine the phase locking for a model of subthreshold oscillations in entorhinal stellate cells under both passive and active cable coupling. These oscillations are thought to arise from an interaction between a

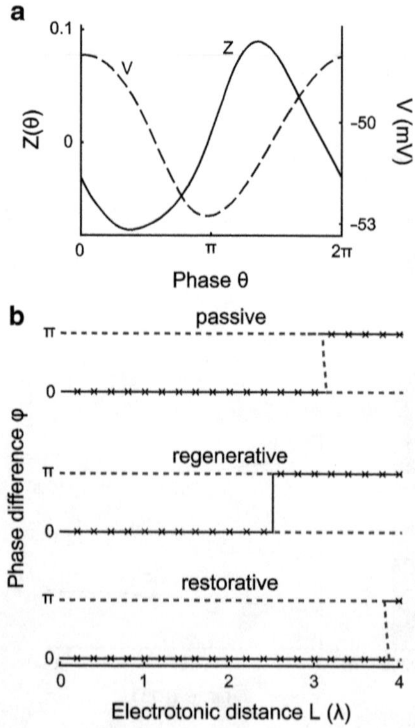

Fig. 3.3 Phase locking behavior of subthreshold oscillators. The oscillations are generated by interactions between I_{NaP} and I_h (see Appendix). (**a**) Voltage trajectory (*solid line*) and phase response function (*dashed line*) of the oscillator. (**b**) Bifurcation diagrams showing the stable (*solid black lines*) and unstable (*dashed gray lines*) phase-locked solutions as a function of L for a passive cable (*top*), a cable with a regenerative current (*middle*), and a cable with a restorative current (*bottom*). The restorative current I_h and regenerative current I_{NaP} are inserted in the cable using, respectively, $\mu = -1.35$, $\gamma_R = 1.1$ and $\tau_m = 1$ ms, and $\mu = 0.84$, $\gamma_R = 1.21$ and $\tau_m = 52.3$ ms. *Cross marks* in the bifurcation diagrams give the stable phase difference determined with numerical simulations of the nonlinear system

persistent sodium current I_{NaP} and a hyperpolarization-activated inward current I_h (see Appendix). Both the voltage trajectory and the phase response function are close to a sinusoid (Fig. 3.3a). The bifurcation diagrams (Fig. 3.3b) are shown for two oscillators coupled via a passive cable (top), a cable with a regenerative current (middle), and a cable with a restorative current (bottom). The regenerative current makes the transition between in-phase and antiphase solutions to occur for smaller L, compared to passive cable coupling. In contrast, adding the restorative current to the cable causes the transition to occur at larger L, making the synchronous phase-locked solution stable up to $L \sim 3.8$.

Numerical simulations of (3.1)–(3.3) agree very well with the theoretical predictions of the phase-locking, both for the type II Morris-Lecar oscillators (Fig. 3.2) and the subthreshold oscillators (Fig. 3.3), when using the maximal ε that still allows for oscillations. Larger values of ε lead to such strong interaction currents that the oscillations are annihilated. Numerical simulations of (3.1)–(3.3) using voltage-dependent cable currents match exactly with the predictions of the weak coupling analysis (bottom two panels in Fig. 3.3b), thereby also emphasizing the validity of using linearized descriptions of those active currents in our analytical framework.

3.2 *Multiple Oscillators: Chains and Branched Structures*

So far, we have focused on a minimal configuration of two oscillators connected by a cable. However, our analysis can be easily extended to predict phase locking of a chain of oscillators. This follows since the phase-locking behavior only depends on each neighboring pair of oscillators. Figure 3.4a shows numerical simulations of a chain of three oscillators, using the same Morris–Lecar model as in Fig. 3.2. The two pairs are separated by a passive dendritic cable of either $L = 1.1$ (top panel) or $L = 2.1$ (bottom panel). The phase-locked solutions follow from the bifurcation diagram in Fig. 3.2b: the three oscillators move into an in-phase solution for $L = 1.1$, whereas for $L = 2.1$ each neighboring pair of oscillators moves into the anti-phase solution.

Our framework also allows us to understand phase locking in a branched cable structure. Hence, we examined the phase difference dynamics of a triangular configuration of three Morris–Lecar oscillators (Fig. 3.4b). In this situation, each oscillator is separated from the other two oscillators by a passive dendritic cable with electrotonic length $L = 1.1$ (top panel) or $L = 2.1$ (bottom panel). For $L = 1.1$, all three oscillators synchronize. When $L = 2.1$, we expect from the bifurcation diagram in Fig. 3.2c that the oscillators go into antiphase. However, as we have three mutually coupled oscillators, two pairs of antiphase locked oscillators would lead to an in-phase configuration of the final pair of oscillators. The bifurcation diagram shows that the in-phase configuration is unstable. We see from the simulation that the system settles into the solution closest to the antiphase solution, which is a phase difference of $2\pi/3$ between each pair of oscillators.

3.3 *Dendritic Phase-Locked States: Controlled by Inputs and Read Out with Spikes*

Above, we developed a framework for analyzing the behavior of local oscillators embedded in the dendritic tree. Now we turn to the question of how such oscillating dendrites respond to inputs and impact the output of the neuron. We will show that the external synaptic input can control the phase-locked configuration of the dendritic oscillators and that this phase-locked configuration can then be transmitted through patterning of the cell's action potentials. We give several salient illustrative examples using a model with a branched oscillating dendritic tree and a spike-generating soma. The model consists of a passive branching dendritic compartment

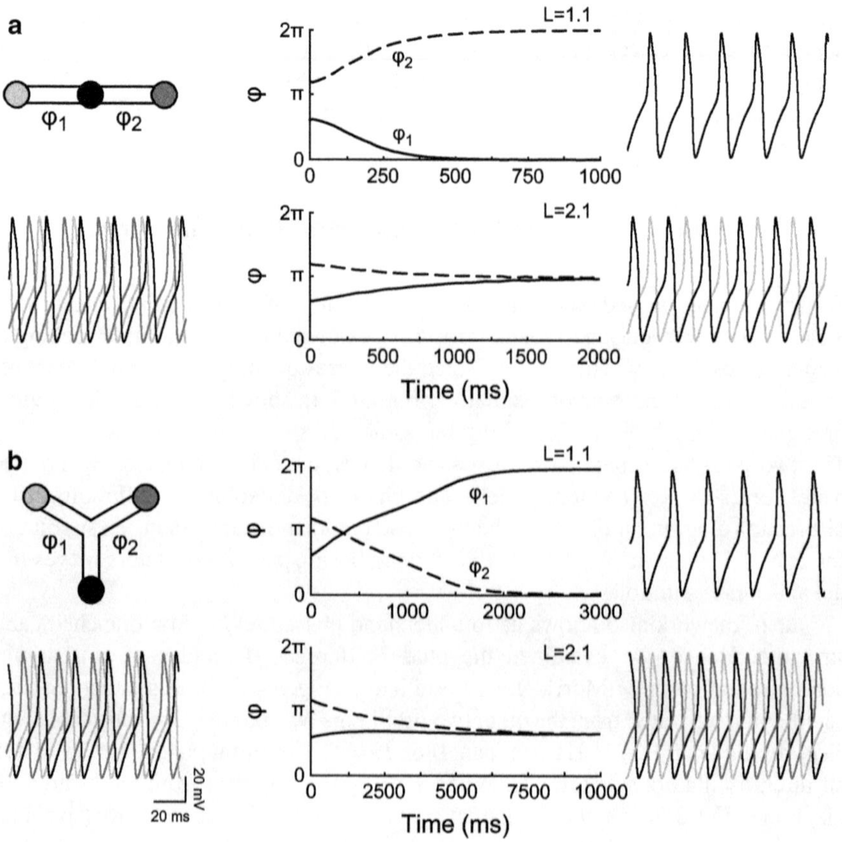

Fig. 3.4 Phase difference dynamics of three oscillators in a chain or a branched configuration. The Morris-Lecar type II oscillators are separated by a passive cable. Panels (**a**) and (**b**) show from *left to right*: a scheme of the model with below it the membrane potential of the oscillators at the start of the simulation; the dynamics of the phase difference ϕ between the oscillators for $L = 1.1$ (*top*) and $L = 2.1$ (*bottom*); and the membrane potential of the oscillators at the end of the simulation. The properties of the Morris-Lecar oscillators and the dendritic cable are as in Fig. 3.2

with two Morris–Lecar type II oscillators at its two distal ends and an excitable soma that, for simplicity, we describe with an integrate and fire mechanism (Fig. 3.5a).

Above, we showed that the dendritic tree can be in a phase-locking regime where two stable phase-locked states coexist (see Fig. 3.2b). In such a bistable regime, well-timed inputs to one or more dendritic oscillators can switch the locking between in-phase and anti-phase. Clearly, the membrane potential fluctuations at the soma depend on whether the dendritic oscillators are synchronized or not. In our model, they are largest in amplitude when the dendritic oscillators are in-phase. The soma can show this difference with its spiking pattern when such large amplitude fluctuations are supra-threshold, while smaller fluctuations (e.g., with asynchronous oscillators) are not.

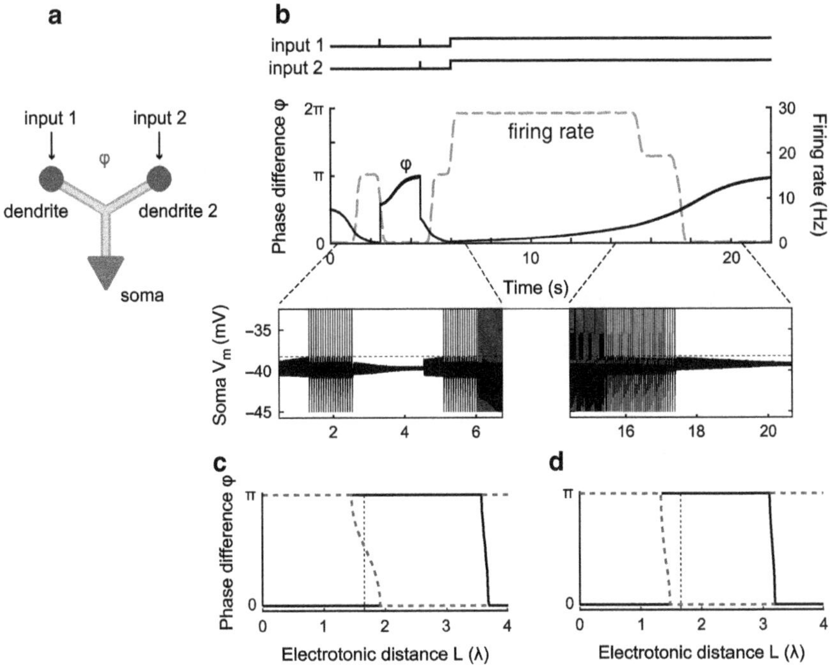

Fig. 3.5 Changing the phase-locked solution of dendritic oscillators with external input and detection of the phase-locked state with an excitable soma. (**a**) Schematic drawing showing the configuration of two dendritic Morris-Lecar type II oscillators and a spike-generating soma. All are separated by a passive cable with electrotonic length $L = 1.65$ and $\tau = 20$ ms. (**b**) From *top to bottom* are shown the inputs to the two dendritic oscillators, the phase difference dynamics (*solid black line*) and somatic firing rate (*dashed gray line*), and the somatic membrane potential V_m (*solid line*) with the spike threshold (*dotted line*). Note that the spikes have been cut off in order to show the subthreshold membrane potential. (**c–d**) Bifurcation diagrams describing the phase-locked solutions up to $t = 6$ s (C, see also Fig. 3.2b) and after $t = 6$ s (D) with *dotted line* at $L = 1.65$ giving the electrotonic distance between the dendritic oscillators

In Fig. 3.5, we illustrate the above mechanism. The initial parameters are such that both the in-phase and antiphase state of the dendritic oscillators are stable (vertical dotted line in Fig. 3.5c). Oscillators starting from an initial phase difference $\phi = \pi/4$ move into the synchronous phase-locked state (solid black line in Fig. 3.5b, top panel). This consequently leads to repetitive somatic spiking (Fig. 3.5b, top panel shows firing rate and bottom panel shows zooms of somatic voltage). A brief depolarizing current pulse to one of the oscillators moves them into the antisynchronous state and the somatic spiking ceases. A subsequent synchronous current pulse to both dendritic oscillators can switch them back into the synchronous state and hence restart the spiking. Note that all the stimuli here are excitatory, yet depending on their timing, they can have a net excitatory or inhibitory effects on the cell's spiking.

Another mechanism by which inputs to the dendrites can affect the phase-locked state is by changing the input amplitude and thereby the oscillator frequency. In Fig. 3.5b at time $t = 6$ s we increase the amplitude of the current input impinging on the oscillators which causes the system to move out of the bistable regime. The synchronized state loses stability and the oscillators gradually move into anti-phase locking. As a result, the soma stops spiking (at time $t \sim 17$ s). Note that the electrotonic separation between the oscillators remains constant (vertical dotted line in Fig. 3.5d) but that the bifurcation diagram itself changes. In turn, a decrease in the excitatory input would reinstate spiking. Hence, this mechanism allows the cell to encode an inverse of the input amplitude, or the inverse of the excitatory input rate.

4 Discussion

We studied the dynamics of dendrites that show intrinsic oscillations due to active voltage-dependent currents that present strong spatial inhomogeneities, hence leading to discrete oscillatory segments. We developed an analytical framework to describe and understand the behavior of interacting dendritic oscillators and their impact on signal propagation within a neuron. A major focus was to understand when the oscillators within the dendrite would lock and hence the whole dendritic tree would act as a single oscillatory unit.

Using the weakly coupled oscillator framework, we have identified the requirements for the various phase-locking regimes of the dendritic oscillators. A central parameter determining the phase-locked solutions is the electrotonic distance between the oscillators. This distance determines how strongly the dendritic cable filters the interactions between the oscillators. In particular, dendritic coupling introduces a delay in the oscillator interactions. The time it takes for the state of one oscillator to perturb another oscillator increase with electrotonic distance, and thereby effectively shifts the phase response function (see also Goldberg et al. 2006). For tightly coupled oscillators, the synchronous solution is stable, and the antiphase solution is unstable. However, at a certain electrotonic distance, the phase response functions of the oscillators are shifted in time such that the synchronous solution loses stability and the antiphase solution becomes stable. Hence, the phase locking of oscillators alternates between synchronized solutions and anti-phase solutions as a function of the electrotonic distance.

The analysis also reveals how the phase-locking is affected by the presence of voltage-dependent conductances in the cable connecting the oscillators. Using the quasi-active approximation of the cable (Sabah and Leibovic 1969; Koch 1984), we find that the dependence of the stable phase-locked solution on the electrotonic distance is amplified by regenerative conductances (i.e., ionic conductances that amplify a voltage perturbation), whereas it is counteracted by restorative conductances (i.e., ionic conductances that counteract voltage perturbations) (see also Goldberg et al. 2006). It should be noted that the linearization of the active conductances in the dendrites is appropriate for small amplitude oscillations in the dendrite and is therefore in general a better approximation for subthreshold oscillations than for spiking oscillators. The effects of active conductances in the connecting cable are explored in more detail in Remme et al. (2009), where we show that the above results hold in general.

The mathematical approach that we used, builds on several studies which focused on the interaction between two neurons with repetitively spiking somata that interact via inputs at the dendrites (Crook et al. 1998; Bressloff 1999; Goldberg et al. 2006). A crucial difference with these studies is that rather than coupling via discrete synaptic events, we treat continuous coupling between the oscillators via the current-conducting cables. One consequence of the continuous coupling is that one needs both the phase response function and the voltage trajectory of the oscillators in order to compute the interaction functions and ultimately the phase-locked solutions. Both the voltage trajectory of an oscillator and its phase response function can be determined numerically from a model of an oscillator and, at least in principle, also experimentally (see, e.g., Galán et al. 2005).

In the final section of our study, we demonstrated how inputs to the dendritic tree can set the phase-locked state and how in turn the phase-locked configuration can control somatic spike generation. The first can for instance be accomplished by changing the frequency of the oscillators with the external input. The soma can subsequently detect the amplitude of the membrane potential fluctuations since this is affected by the phase-locked configuration. The time scale at which the dendritic oscillators move from one solution to another is set by the strength of the interactions between the oscillators. This time scale can be much longer than that of the different components of the system, e.g., the membrane time constant or the period of the oscillators. In this way, the phase difference between the oscillators can function as a memory. Related ideas have been previously discussed by Huhn et al. (2005). We also showed that in the bistable phase-locked regime the state of the dendrites is easily set by transient inputs and "read-out" by the soma. This also can endow the neuron with a memory since brief external inputs can switch the neuron from a spiking to a quiescent mode and vice versa. Interestingly, we showed that both the turn-on and turn-off signals (inputs) can be excitatory, their final effects defined by their timing.

The focus of our report is complementary to that of a recent theoretical study of the subthreshold oscillations in the dendrites of mesencephalic dopaminergic neurons (Medvedev et al. 2003). As these cells do not show any indication of distinct dendritic oscillators, the whole cell was modeled as one continuous oscillator with gradients in oscillator properties along the dendrites. Moreover, since there were no distinct oscillators, in their analysis Medvedev and colleagues assumed strong voltage coupling between neighboring compartments, enforcing synchronized oscillations throughout the cell. In contrast, our approach assumed weak coupling between the dendritic oscillators. This would not be appropriate for a spatially continuous oscillator. However, it is not possible to state in general at what precise electrotonic distance between two oscillators the weak coupling assumption becomes valid, since it depends on the strength of the interaction currents with respect to the intrinsic currents of the oscillators. However, our numerical simulations for a dendritic cable without the assumption of weak coupling, show that the phase-locking behavior of Morris-Lecar oscillators is consistent with weak coupling.

We have set up an analytical framework for studying interacting dendritic oscillators. This opens up a wide range of questions that were outside the scope of the present study. For example, we focused our analysis on identical oscillators, while it is likely that dendritic oscillators will vary in their properties throughout the dendritic tree. For example, the diameter of the dendrites, which typically becomes smaller with increasing distance from the soma, can affect the intrinsic frequency of the oscillators. A gradient in the frequency of distinct oscillators is likely to lead to more complex phenomena such as traveling waves (see, e.g., Kopell and Ermentrout 1986).

A major focus of our study was to explore how local dendritic mechanisms may lead to oscillations expressed globally in the cell and hence visible at the soma, for example, in somatic intracellular recordings. Our analysis showed that even electrotonically far removed dendritic oscillators can lead to voltage oscillations that significantly affect the soma voltage and hence spike generation. This suggests several experimentally testable predictions. In one possible experiment, one can take advantage of imperfect space clamp in a electrotonically extended neuron. As a proof of principle, in a neuron where the oscillations are generated distally in the dendritic tree, voltage clamping the soma would not block such oscillations, and these should be seen in the current necessary to hold the somatic potential. In fact, results from (Moore et al. 1999) point in this direction, where in chick spinal cord neuron NMDA-dependent intrinsic oscillations were not blocked by somatic voltage clamp. A further prediction stems from the weak coupling between active dendrites. If active oscillations, such as periodically generated dendritic spikes, are generated in different segments of the dendritic tree, our analysis predicts that such spikes should interact and should exist in a stable phase-locked configuration, e.g., synchrony. Hence, should one of the dendritic segments be phase-shifted, such perturbation should propagate to the other segment (the other segment should be phase reset), the dendritic spikes should return to the phase-locked configuration, and the time scale of this return should be relatively long and determined by the electrotonic distance between the active segments. While difficult, such experiments are possible using the multiple dendritic recording techniques, such as those developed by Davie et al. (2006) in Purkinje cells.

A recent model for the grid field properties of the entorhinal cortex layer II stellate cells (O'Keefe and Burgess 2005; Burgess et al. 2007; Hasselmo 2007) relies precisely on the ingredients considered in the present study. The model assumes that different dendritic branches emanating from the soma of these cells function as distinct oscillators. The oscillations are modulated by external inputs and the interference of the oscillators eventually determines the somatic spiking. Crucially, the model assumes that the dendritic oscillators operate independently. At a first glance, our results appear to argue against this: the various oscillators should phase lock (hence, lose their independence) even when the mutual coupling is weak. However, in principle, the locking may be slower than the behavioral time scale, allowing the oscillators to act quasi-independently on the behavioral time scale.

Our analysis provides the appropriate framework to examine these issues: the scaling of locking in time and the biophysical implementation of grid-field formation via dendritic oscillators (see Remme et al. 2010).

Above, we studied relatively simple cell geometries; however, these form basic building blocks for more complex dendritic trees. Thus our framework should be valid for understanding global voltage oscillations in more realistic models of spatially extended cells. We would like to emphasize at this point that our general framework should also hold when – in addition to the distinct oscillators distributed throughout the dendritic tree – also the soma is regarded as an oscillator. These and other issues will be addressed in future publications.

The framework we have developed, builds on the extensive mathematical theory of coupled oscillators and nestles nicely below the complexity of full compartmental models of neuronal dendritic trees. Yet our framework is sufficiently powerful and clear to both take into account certain key aspects of the dendritic tree structure and to be amenable to theoretical analysis of the dynamics of active dendrites and the computational function of such dendritic structures. These remain an active focus for further investigations.

Appendix

Interaction Functions for Two Weakly Coupled Dendritic Oscillators

In order to determine the perturbations $\varepsilon\, p_{\mathrm{A,B}}(t)$ in (3.3), we need to solve (3.1) with the boundary conditions from (3.2). To do so, we linearize (3.1) about the membrane potential V_{R} to which the cable would relax if it was not driven by the oscillators, yielding the quasi-active approximation for the cable (Sabah and Leibovic 1969; Koch 1984). This approximation is appropriate as long as the voltage fluctuations around V_{R} are sufficiently small. We define $U(x,t)$ as the difference between the oscillating solution and the resting membrane potential V_{R}, i.e., $U(x,t) \equiv V(x,t) - V_{\mathrm{R}}$ and we define $w(x,t)$ analogously as $w(x,t) \equiv m(x,t) - m_{\infty}(V_{\mathrm{R}})$. The equations describing the quasi-active cable now read

$$\begin{aligned}\tau \frac{\partial}{\partial t} U(x,t) &= \lambda^2 \frac{\partial^2}{\partial x^2} U(x,t) - \gamma_{\mathrm{R}}\, U(x,t) - \gamma_{\mathrm{m}}(V_{\mathrm{R}} - E_{\mathrm{m}})\, w(x,t) \\ \tau_{\mathrm{m}} \frac{\partial}{\partial t} w(x,t) &= \frac{\partial}{\partial V} m_{\infty}(V_{\mathrm{R}})\, U(x,t) - w(x,t),\end{aligned} \tag{3.6}$$

where $\gamma_{\mathrm{R}} = 1 + \gamma_{\mathrm{m}} m_{\infty}(V_{\mathrm{R}})$ is the total membrane conductance of the cable at V_{R} divided by the cable's membrane leak conductance.

The oscillators determine the voltage of the cable at $x = 0$ and $x = l$. These voltages would need to be computed by solving the full system of equations for the dynamics of each oscillator; however, since we consider weak coupling (meaning that the trajectories are only weakly perturbed by the cable currents) we can make use of the fact that the trajectories are periodic. Hence, we expand U_A and U_B in a Fourier series, allowing for a possible phase difference ϕ (in radians) between the oscillators:

$$U(0,t) = U_\mathrm{A}(t) = \sum_{n=-\infty}^{\infty} \widetilde{U}_n^\mathrm{A}\, \mathrm{e}^{\mathrm{i}\omega_n t},$$

$$U(l,t) = U_\mathrm{B}\left(t + \phi\tfrac{T}{2\pi}\right) = \sum_{n=-\infty}^{\infty} \widetilde{U}_n^\mathrm{B}\, \mathrm{e}^{\mathrm{i}(\omega_n t + n\phi)}, \tag{3.7}$$

where $\omega_n = n\,2\pi/T$, T is the intrinsic oscillator period, and membrane voltages U_A and U_B (in mV) are measured relative to V_R.

The solution of the cable (3.6) will also be periodic and we can write the equation in the frequency domain as

$$\lambda^2 \frac{d^2}{dx^2}\widetilde{U}_n(x) - \left(\gamma_\mathrm{R} + \frac{\mu}{1+(\omega_n\tau_\mathrm{m})^2} + i\omega_n\left(\tau - \frac{\mu\,\tau_\mathrm{m}}{1+(\omega_n\tau_\mathrm{m})^2}\right)\right)\widetilde{U}_n(x) = 0. \tag{3.8}$$

Using the boundary conditions defined by (3.7) yields the solution:

$$U(x,t) = \sum_{n=-\infty}^{\infty} \mathrm{e}^{\mathrm{i}\omega_n t}\, \widetilde{U}_n^\mathrm{A} \frac{\sinh(b_n(L - x/\lambda))}{\sinh(b_n L)} + \sum_{n=-\infty}^{\infty} \mathrm{e}^{\mathrm{i}(\omega_n t + n\phi)}\, \widetilde{U}_n^\mathrm{B} \frac{\sinh(b_n x/\lambda)}{\sinh(b_n L)}, \tag{3.9}$$

where

$$b_n = \sqrt{\gamma_\mathrm{R} + \frac{\mu}{1+(\omega_n\tau_\mathrm{m})^2} + i\omega_n\left(\tau - \frac{\mu\,\tau_\mathrm{m}}{1+(\omega_n\tau_\mathrm{m})^2}\right)} \tag{3.10}$$

with $\mu = \gamma_\mathrm{m}(V_\mathrm{R} - E_\mathrm{m})\frac{\partial}{\partial V} m_\infty(V_\mathrm{R})$. The parameter μ determines whether the active conductance that is present in the cable is regenerative ($\mu < 0$), meaning that perturbations are amplified (e.g., a persistent sodium current), or restorative ($\mu > 0$), meaning that the active conductance counteracts perturbations from V_R (e.g., the hyperpolarization activated inward current). As mentioned above, the perturbations that the oscillators receive from the cable is proportional to the derivative of the voltage with respect to x. For the oscillator at $x = 0$, the perturbation from the cable is

$$\begin{aligned} p_\mathrm{A}(t;\phi) &= \frac{\partial}{\partial x} U(0,t) \\ &= \frac{1}{\lambda}\sum_{n=-\infty}^{\infty} \mathrm{e}^{\mathrm{i}(\omega_n t + n\phi)}\, \widetilde{U}_n^\mathrm{B} \frac{b_n}{\sinh(b_n L)} - \frac{1}{\lambda}\sum_{n=-\infty}^{\infty} \mathrm{e}^{\mathrm{i}\omega_n t}\, \widetilde{U}_n^\mathrm{A}\, b_n \coth(b_n L) \\ &= \frac{1}{\lambda}\sum_{n=-\infty}^{\infty} \mathrm{e}^{\mathrm{i}\omega_n t} \frac{b_n}{\sinh(b_n L)} \left(\widetilde{U}_n^\mathrm{B}\,\mathrm{e}^{\mathrm{i}n\phi} - \widetilde{U}_n^\mathrm{A}\cosh(b_n L)\right). \end{aligned} \tag{3.11}$$

The perturbation from the cable at $x = l$ can be derived in the same way.

We have now derived the perturbations that an oscillator receives depending on the phase difference ϕ between the oscillators. In order to complete our analysis, we also need to compute how these perturbations act back on the phases of the two oscillators and thus on the phase difference. Each of the oscillators is described explicitly by a system of equations determining the dynamics of its voltage (3.3). However, if we assume that the periodic solutions of such a system of equations are sufficiently attractive and the coupling is sufficiently weak we can write an equivalent phase model (see Izhikevich 2007). The phases of the two dendritic oscillators, $\theta_A(t)$ and $\theta_B(t)$ (in radians), evolve as

$$\dot{\theta}_A = \frac{2\pi}{T} + \varepsilon\, Z_A(\theta_A)\, p_A\left(\theta\frac{T}{2\pi};\phi\right),$$

$$\dot{\theta}_B = \frac{2\pi}{T} + \varepsilon\, Z_B(\theta_B)\, p_B\left(\theta\frac{T}{2\pi};\phi\right), \tag{3.12}$$

where $\frac{2\pi}{T}$ is the intrinsic oscillator frequency. The second term describes the effect of the cable on the phase. $Z_{A,B}(\theta)$ are the infinitesimal phase response functions of the respective oscillators and describe how much their phases are advanced or delayed in response to an infinitesimally small and short perturbation.

Since we consider weak interactions between the oscillators, ϕ changes slowly with respect to the oscillation period. Therefore, we can average the interaction between the oscillators (i.e., the products $Z_A\, p_A$ and $Z_B\, p_B$ in (3.12)) over a cycle and obtain the interaction functions $H_{A,B}(\phi)$. $H_A(\phi)$ describes the average effect on the phase of oscillator A over one cycle as a function of ϕ:

$$H_A(\phi) = \frac{1}{2\pi}\int_0^{2\pi} Z_A(\theta)\, p_A\left(\theta\frac{T}{2\pi};\phi\right)\, d\theta \tag{3.13}$$

with p_A given by (3.11). The interaction function $H_B(\phi)$ can be determined analogously. Note that with identical oscillators, we have $H_B(\phi) = H_A(-\phi)$.

Oscillator Models

The equations for the Morris–Lecar type II oscillator (Morris and Lecar 1981) with parameters as in Ermentrout (1996) read

$$C_m\frac{dV}{dt} = -g_L(V - E_L) - g_w\, w(V - E_w) - g_m\, m_\infty(V)(V - E_m) + I$$

$$\frac{dw}{dt} = \varphi\frac{w_\infty(V) - w}{\tau_w(V)}. \tag{3.14}$$

The Morris–Lecar type II oscillator uses $C_m = 1\,\mu F/cm^2$, $g_L = 0.5\,mS/cm^2$, $g_w = 2\,mS/cm^2$, $g_m = 1.1\,mS/cm^2$, $E_L = -50\,mV$, $E_w = -70\,mV$, $E_m = 100\,mV$, $\varphi = 0.2$, $I = 25\,\mu A/cm^2$, and where $m_\infty(V) = \frac{1}{2}[1 + \tanh((V + 1)/15)]$, $w_\infty(V) = \frac{1}{2}[1 + \tanh(V/30)]$, and $\tau_w(V) = 1/\cosh(V/60)$.

The oscillatory dynamics of the subthreshold oscillator emerge from the interaction between the persistent sodium current I_{NaP} and the hyperpolarization activated inward current I_h. The current descriptions are based on the data from Dickson et al. (2000) and Fransén et al. (2004). The dynamics of I_h are described by a single gating variable $w(t)$ with activation function $w_\infty(V)$ and time constant $\tau_w(V)/\varphi$ (in milliseconds). The voltage-dependent activation of I_{NaP} is described by $m_\infty(V)$ and is instantaneous. The equations are the same as for the Morris–Lecar type II oscillator with $C_m = 1\,\mu F/cm^2$, $g_L = 0.3\,mS/cm^2$, $g_w = 1.5\,mS/cm^2$, $g_m = 0.076\,mS/cm^2$, $E_L = -69\,mV$, $E_w = -20\,mV$, $E_m = 48\,mV$, $\varphi = 0.014$, $I = 0.9\,\mu A/cm^2$, and where $m_\infty(V) = \frac{1}{2}[1 + \tanh((V + 48.7)/8.8)]$, $w_\infty(V) = \frac{1}{2}[1 + \tanh((V + 74.2)/-14.4)]$, and $\tau_w(V) = 1/\cosh((V + 74.2)/-28.8)$.

References

Alonso, A. and Klink, R. (1993). Differential electroresponsiveness of stellate and pyramidal-like cells of medial entorhinal cortex layer II. *J Neurophysiol*, 70(1):128–43.

Alonso, A. and Llinás, R. (1989). Subthreshold Na^+-dependent theta-like rhythmicity in stellate cells of entorhinal cortex layer II. *Nature*, 342(6246):175–7.

Bressloff, P. C. (1999). Resonantlike synchronization and bursting in a model of pulse-coupled neurons with active dendrites. *J Comput Neurosci*, 6(3):237–49.

Burgess, N., Barry, C., and O'Keefe, J. (2007). An oscillatory interference model of grid cell firing. *Hippocampus*, 17(9):801–12.

Chapman, C. A. and Lacaille, J. C. (1999). Intrinsic theta-frequency membrane potential oscillations in hippocampal CA1 interneurons of stratum lacunosum-moleculare. *J Neurophysiol*, 81(3):1296–307.

Crook, S., Ermentrout, G., and Bower, J. (1998). Dendritic and synaptic effects in systems of coupled cortical oscillators. *J Comput Neurosci*, 5(3):315–29.

Davie, J. T., Kole, M. H. P., Letzkus, J. J., Rancz, E. A., Spruston, N., Stuart, G. J., and Häusser, M. (2006). Dendritic patch-clamp recording. *Nat Protoc*, 1(3):1235–47.

Dickson, C., Magistretti, J., Shalinsky, M., Fransén, E., Hasselmo, M., and Alonso, A. (2000). Properties and role of I_h in the pacing of subthreshold oscillations in entorhinal cortex layer II neurons. *J Neurophysiol*, 83(5):2562–79.

Ermentrout, G. B. (1996). Type I membranes, phase resetting curves, and synchrony. *Neural Comp*, 8(5):979–1001.

Fransén, E., Alonso, A., Dickson, C. T., Magistretti, J., and Hasselmo, M. E. (2004). Ionic mechanisms in the generation of subthreshold oscillations and action potential clustering in entorhinal layer II stellate neurons. *Hippocampus*, 14(3):368–84.

Galán, R. F., Ermentrout, G. B., and Urban, N. N. (2005). Efficient estimation of phase-resetting curves in real neurons and its significance for neural-network modeling. *Phys Rev Lett*, 94(15):158101.

Giocomo, L., Zilli, E., Fransén, E., and Hasselmo, M. (2007). Temporal frequency of subthreshold oscillations scales with entorhinal grid cell field spacing. *Science*, 315(5819):1719–22.

Goldberg, J., Deister, C., and Wilson, C. (2006). Response properties and synchronization of rhythmically firing dendritic neurons. *J Neurophysiol*, 97(1):208–19.

Gutfreund, Y., Yarom, Y., and Segev, I. (1995). Subthreshold oscillations and resonant frequency in guinea-pig cortical neurons: physiology and modelling. *J Physiol*, 483 (Pt 3):621–40.

Hasselmo, M. E. (2007). Arc length coding by interference of theta frequency oscillations may underlie context-dependent hippocampal unit data and episodic memory function. *Learn Mem*, 14(11):782–94.

Huhn, Z., Orbán, G., Érdi, P., and Lengyel, M. (2005). Theta oscillation-coupled dendritic spiking integrates inputs on a long time scale. *Hippocampus*, 15(7):950–62.

Izhikevich, E. (2007). *Dynamical Systems in Neuroscience: The Geometry of Excitability And Bursting*. Cambridge, MA: MIT Press.

Kamondi, A., Acsády, L., Wang, X.-J., and Buzsáki, G. (1998). Theta oscillations in somata and dendrites of hippocampal pyramidal cells *in vivo*: activity-dependent phase-precession of action potentials. *Hippocampus*, 8(3):244–61.

Koch, C. (1984). Cable theory in neurons with active, linearized membranes. *Biol Cyber*, 50(1): 15–33.

Kopell, N. and Ermentrout, G. B. (1986). Symmetry and phaselocking in chains of weakly coupled oscillators. *Comm Pure Appl Math*, 39(5):623–660.

Leung, L. W. and Yim, C. Y. (1991). Intrinsic membrane potential oscillations in hippocampal neurons *in vitro*. *Brain Res*, 553(2):261–74.

Llinás, R. R. and Sugimori, M. (1980). Electrophysiological properties of *in vitro* Purkinje cell dendrites in mammalian cerebellar slices. *J Physiol*, 305:197–213.

Lörincz, A., Notomi, T., Tamás, G., Shigemoto, R., and Nusser, Z. (2002). Polarized and compartment-dependent distribution of HCN1 in pyramidal cell dendrites. *Nat Neurosci*, 5(11):1185–93.

Magee, J. C. (1998). Dendritic hyperpolarization-activated currents modify the integrative properties of hippocampal CA1 pyramidal neurons. *J Neurosci*, 18(19):7613–24.

Medvedev, G. S., Wilson, C. J., Callaway, J. C., and Kopell, N. J. (2003). Dendritic synchrony and transient dynamics in a coupled oscillator model of the dopaminergic neuron. *J Comput Neurosci*, 15(1):53–69.

Moore, L. E., Chub, N., Tabak, J., and O'Donovan, M. (1999). NMDA-induced dendritic oscillations during a soma voltage clamp of chick spinal neurons. *J Neurosci*, 19(19):8271–80.

Morris, C. and Lecar, H. (1981). Voltage oscillations in the barnacle giant muscle fiber. *Biophys J*, 35(1):193–213.

O'Keefe, J. and Burgess, N. (2005). Dual phase and rate coding in hippocampal place cells: theoretical significance and relationship to entorhinal grid cells. *Hippocampus*, 15(7): 853–66.

Pape, H. C., Paré, D., and Driesang, R. B. (1998). Two types of intrinsic oscillations in neurons of the lateral and basolateral nuclei of the amygdala. *J Neurophysiol*, 79(1):205–16.

Placantonakis, D. and Welsh, J. (2001). Two distinct oscillatory states determined by the NMDA receptor in rat inferior olive. *J Physiol*, 534(Pt 1):123–40.

Rall, W. (1967). Distinguishing theoretical synaptic potentials computed for different soma-dendritic distribution of synaptic inputs. *J Neurophysiol*, 30:1138–1168.

Remme, M. W. H., Lengyel, M., and Gutkin, B. S. (2009). The role of ongoing dendritic oscillations in single-neuron dynamics. *PLoS Comput Biol*, 5(9):e1000493.

Remme, M. W. H., Lengyel, M., and Gutkin, B. S. (2010). Democracy-independence trade-off in oscillating dendrites and its implications for grid cells. *Neuron*, 66(3):429–437.

Rinzel, J. and Ermentrout, B. (1998). *Analysis of neural excitability and oscillations. In: Koch C and Segev I, editors. Methods in neuronal modeling*, pages 251–291. MIT Press, Cambridge (MA).

Sabah, N. H. and Leibovic, K. N. (1969). Subthreshold oscillatory responses of the Hodgkin-Huxley cable model for the squid giant axon. *Biophys J*, 9(10):1206–22.

Sanhueza, M. and Bacigalupo, J. (2005). Intrinsic subthreshold oscillations of the membrane potential in pyramidal neurons of the olfactory amygdala. *Eur J Neurosci*, 22(7):1618–26.

Stuart, G., Spruston, N., and Häusser, M. (2007). *Dendrites*. Oxford University Press, New York (NY), 2nd edition.

Yoshida, M. and Alonso, A. (2007). Cell-type specific modulation of intrinsic firing properties and subthreshold membrane oscillations by the M(Kv7)-current in neurons of the entorhinal cortex. *J Neurophysiol*, 98(5):2779–94.

Chapter 4
History of the Application of the Phase Resetting Curve to Neurons Coupled in a Pulsatile Manner

Carmen C. Canavier and Srisairam Achuthan

Abstract Phase resetting curves (PRCs) keep track of how much an input advances or delays the next spike in an oscillatory neuron depending upon where in the cycle the input is applied. In this chapter, a history of the application of PRC theory to circuits of pulse-coupled neurons is presented. In these circuits, the interactions between oscillators take the form of brief pulses such that the effect of one input dies out before the next is received. In some approaches, a specific form is assumed for the interactions between oscillators, but a more general approach is to formulate the problem assuming a PRC that is measured using a perturbation that approximates the input received in the intact network. In general, this approach requires that a circuit architecture and a specific firing pattern be assumed. This allows the construction of a map from one discrete event, such as the phase at which an input is received, to the next event. The fixed points of these maps correspond to periodic firing modes and are relatively easy to locate and analyze for stability compared to analyzing the original network directly. Alternatively, maps based on the PRC have been constructed that do not presuppose a firing order. Specific circuits that have been analyzed under the assumption of pulsatile coupling include one to one lockings in a periodically forced oscillator or an oscillator forced at a fixed delay after a threshold event, two bidirectionally coupled oscillators with and without delays, a unidirectional N-ring of oscillators, and N all-to-all networks. The results obtained provide insights regarding synchronization tendencies.

C.C. Canavier (✉)
Neuroscience Center of Excellence, LSU Health Sciences Center, New Orleans, LA 70112, USA
e-mail: ccanav@lsuhsc.edu

S. Achuthan
Neuroscience Center of Excellence, LSU Health Sciences Center, New Orleans, LA 70112, USA
e-mail: sairam.achuthan@gmail.com

N.W. Schultheiss et al. (eds.), *Phase Response Curves in Neuroscience: Theory, Experiment, and Analysis*, Springer Series in Computational Neuroscience 6,
DOI 10.1007/978-1-4614-0739-3_4,

1 Introduction

The rhythmic activity of cardiac pacemaker cells, pancreatic cells, and neural circuits that underlie breathing and locomotion as well as the synchronized flashings of populations of fireflies (Winfree 1967; Peskin 1975; Kuramoto 1984; Buck 1988; Winfree 1980; de la Iglesia et al. 2004; Yuste et al. 2005; Buzsaki 2006) are some of the well-known biological examples of self-sustaining oscillatory patterns. Each such pattern is thought to emerge from the interactions between constituent oscillators coupled via pulse-like signals. Oscillators that interact with each other at discrete times such that their coupling duration is brief relative to cycle length are referred to as pulse-coupled oscillators. Phase-resetting theory has been used quite extensively to study the firing activity of pulse-coupled neural oscillators (Perkel et al. 1964; Peskin 1975; Winfree 1980; Glass and Mackey 1988; Mirollo and Strogatz 1990; Canavier et al. 1997; Oprisan and Canavier 2001; Goel and Ermentrout 2002; Strogat 2003; Oprisan et al. 2004; Achuthan and Canavier 2009; Canavier et al. 2009).

Phase-resetting theory was developed in the context of limit cycle oscillators (Glass and Mackey 1988; Winfree 1980) under the assumption that external perturbations affect only the phase and not the amplitude of the oscillation. A limit cycle oscillator can be visualized as tracing out a closed orbit on every cycle in some appropriate state space. At any time point within the cycle, a number between 0 and 1 called the phase represents the position of the oscillator along the limit cycle. Phase is computed as modulo 1 and is measured from an arbitrary reference point. In a neural context, this reference point is defined as the spike or burst onset and is assigned a phase of zero. An input can be visualized as displacing the trajectory along the limit cycle (Glass and Mackey 1988; Winfree 1980), with a phase advance causing the trajectory to skip ahead on the limit cycle and a delay causing it to jump backwards. This has the effect of either shortening or lengthening the cycle in which the perturbation occurs.

There are no completely general theoretical approaches to use the phase-resetting framework described above to analyze phase locking in networks (Ermentrout and Chow 2002). There are, however, two special cases that are amenable to analysis: one assumes that the coupling is weak (Hoppensteadt and Izhikevich 1997), and other assumes that it is pulsatile. The weakly coupled oscillator approach is described in Chap. 1 of this volume, and the pulsatile approach is described here. The main difference between the approaches is that in the pulse-coupled method, the effect of any specific pulsatile input received in the circuit must be known or measured, whereas in weak coupling only the response to an infinitesimal current pulse needs to be known in order to estimate the effect of any arbitrary coupling waveform. Weak coupling between oscillators in a network implies that many cycles are required for the oscillators to converge to a phase-locked mode such as synchrony, and therefore may not be appropriate for the analysis of circuits in which synchronization must occur quickly, such as cortical circuits that mediate binding (Singer 1999) or central pattern generators (CPGs) that synchronize their activity quickly (Yuste et al. 2005). Pulsatile coupling can accommodate large

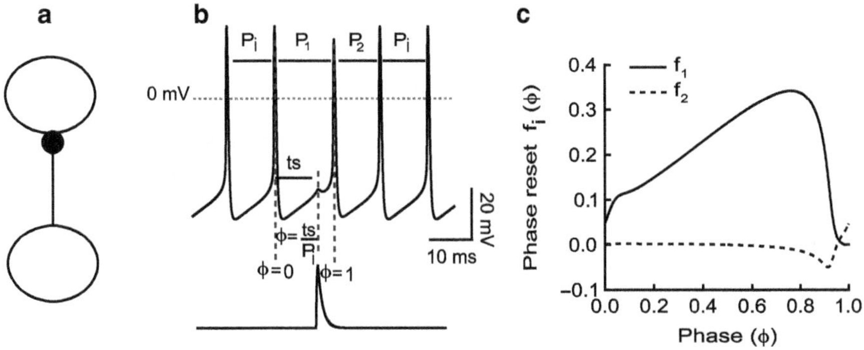

Fig. 4.1 Phase-resetting curve. (**a**) Open loop configuration. (**b**) The voltage waveform represents a regular spiking neuron with intrinsic period P_{i}. The lower trace corresponds to the postsynaptic conductance resulting from a spike in a presynaptic neuron. The normalized change in cycle period length as a result of the perturbation received at a phase $\phi = ts/P_{\mathrm{i}}$ is the phase resetting. P_1 represents the period of the cycle in which the stimulus is received. The following cycle period is represented by P_2. (**c**) Example of first order ($f_1(\phi)$, *solid line*) and second order ($f_2(\phi)$, *dashed line*) phase resetting curves

excursions from the limit cycle and large amounts of resetting in a relatively short time that would invalidate the weak coupling assumptions. In such cases, the system cannot be reduced to the relative phases between oscillators as in the weak coupling assumptions, but instead the precise timing information of threshold events such as spikes and bursts must be retained using maps.

For applications with pulsatile coupling, the relevant PRC is either estimated or measured (see Fig. 4.1) directly with the stimulus waveform that each neuron in the network receives (Canavier et al. 1997, 1999; Canavier 2005; Luo et al. 2004; Achuthan and Canavier 2011). The PRC can be measured in an open loop configuration (see Fig. 4.1a) such that it receives input from the neuron(s) that will be presynaptic to it in the closed loop network configuration, but is isolated in that it does not affect the driving presynaptic neuron. In the example shown, this input is the synaptic conductance waveform (lower trace Fig. 4.1b) that results from an action potential in the presynaptic neuron, such that the PRC is equivalent to the spike time response curve (Acker et al. 2003; Pervouchine et al. 2006). Although spiking neurons are used in this illustration, the same procedure applies to bursting neurons except that burst onset rather than spike onset is assigned a phase of zero, and the perturbation in conductance results from a presynaptic burst rather than a single action potential. In the open loop condition (i.e., when there is only a single unidirectional coupling) the phase at which a stimulus is received is $\phi = ts/P_{\mathrm{i}}$, where P_{i} is the intrinsic period and ts is the time between the last action potential in the model neuron for which the PRC is being generated and the action potential initiation in the presynatic model neuron. The normalized change in the length of the cycle containing the perturbation and the one following it are called the first order resetting, $f_1(\phi) = (P_1 - P_{\mathrm{i}})/P_{\mathrm{i}}$, and second order resetting, $f_2(\phi) = (P_2 - P_{\mathrm{i}})/P_{\mathrm{i}}$, respectively. P_1 is the length of the cycle containing the perturbation and P_2 is the length of the subsequent cycle (see Fig. 4.1a,b). Plotting the resetting with this convention, often used by biologists, causes a delay to be positive and an advance negative, in contrast to the opposite convention often used by mathematicians.

The assumption of pulsatile coupling allows the time evolution of the phases in a network to be described by maps from their values in one cycle to their values in the next cycle. The assumptions required in order to generate maps based on the PRC are that (1) each neuron in the network can be represented as a limit cycle oscillator, (2) the trajectory of each neuron returns to its limit cycle between inputs rendering the coupling effectively pulsatile, and (3) that effect of each pulsatile interaction on the phase of the oscillator is known. In many cases, this is satisfied provided that the open loop PRC described in Fig. 4.1 still applies in the intact network despite feedback effects. In order to predict network activity using the PRCs, we must either (1) assume a firing pattern and solve for all possible sets of phases that satisfy the periodicity constraints imposed by that firing pattern, or (2) pick a set of initial phases, identify the next firing time, update the phases, and repeat (Achuthan and Canavier 2009; Netoff et al. 2005; Canavier et al. 1999; Sieling et al. 2009). In this chapter, we focus on the first approach.

2 Application of the PRC to a Periodically Forced Oscillator

Perkel et al. (1964) measured the phase resetting function, $f(\phi)$, in invertebrate pacemaker neurons. Perkel et al. considered only first order resetting. The analysis is generalized here to include second order resetting. Pacemakers fire action potentials spontaneously and repetitively in a periodic manner. Perkel et al. induced precisely timed inhibitory or excitatory synaptic potentials in the pacemaker cell by evoking an action potential in an excitatory or inhibitory presynaptic neuron. They used the PRC to predict stable m:1 entrainment of the pacemaker neuron by a periodic train of evoked synaptic potentials. In a 1:1 phase-locked mode (see Fig. 4.2), the forcing input arrives at exactly the same phase in the forced neuron on each cycle, and we refer to this phase as the equilibrium phase, ϕ^*. The period of the forcing neuron, P_f, must equal the intrinsic period of the forced neuron, P_i, plus any phase resetting at the equilibrium phase, ϕ^*, of the perturbation: $P_f = P_i(1 + f_1(\phi^*) + f_2(\phi^*))$. For m:1 lockings in which only every mth cycle contains a perturbation the appropriate expression is: $P_f = (m - 1)P_i + P_i(1 + f_1(\phi^*) + f_2(\phi^*))$. Therefore, all possible m:1 lockings can be determined from the PRC alone by finding which phases, if any, satisfy the equality. Perkel et al. also determined which lockings were stable by using a map based on the PRC. The stimulus time $ts[n]$ indicates the time elapsed between an action potential in the pacemaker and the arrival of a presynaptic potential due to the periodic forcing. The first order resetting occurs in the interval between the receipt of an input and the next spike, whereas the second order resetting is assumed to be contained in the stimulus

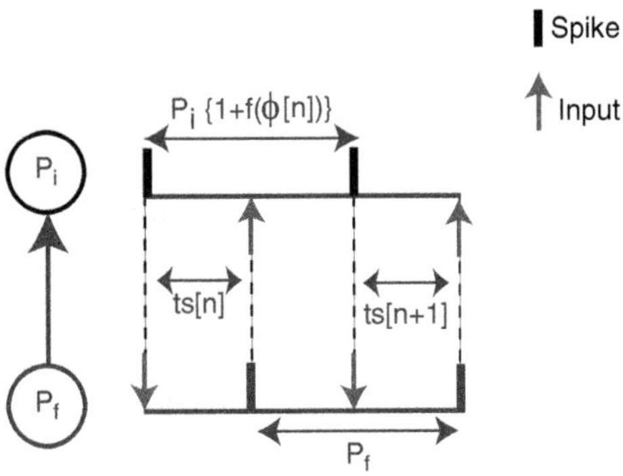

Fig. 4.2 Stability analysis for periodically forced oscillator. Stimulus times (ts) in the nth and $(n + 1)$st cycles in a phase-locked mode with no delays is shown here. The *black vertical bars* indicate firing times and the *vertical gray arrows* indicate neuronal inputs arriving at a phase ϕ $(= ts/P_\mathrm{i})$. P_i represents the period of the forced oscillator and P_f represents the period of the forcing oscillator

interval $t_\mathrm{s}[n] = P_\mathrm{i}(\phi[n]+f_2(\phi[n-1]))$, where $\phi[n]$ stands for the phase of the forced neuron in the nth cycle. Note that the definition of the stimulus interval is different for repetitive perturbations (Fig. 4.2) compared to that for a single perturbation (Fig. 4.1) because prior perturbations have to be considered. For 1:1 forcing, the map is generated as follows:

$$P_\mathrm{i}\{1 + f_1(\phi[n]) + f_2(\phi[n-1])\} + ts[n+1] = ts[n] + P_\mathrm{f}. \tag{4.1}$$

If we rewrite P_f, and the stimulus intervals in terms of the phase as described above, divide by P_i and rearrange so that terms involving phase on the earlier cycle appear only on the right, we obtain:

$$\phi[n+1] = \phi[n] + f_1(\phi^*) - f_1(\phi[n]) + f_2(\phi^*) - f_2(\phi[n]). \tag{4.2}$$

The above equation is a nonlinear return map which gives the phase on one cycle in terms of the phase on the previous cycle (note the $[n-1]$ terms drop out). We can linearize the PRCs in the neighborhood of the presumed fixed point ϕ^* using $f_i(\phi[n]) = f_i(\phi^*) + f_i'(\phi^*)\Delta\phi[n]$, where the perturbation from the fixed point is given by $\Delta\phi[n] = \phi[n] - \phi^*$ for $i = 1$ and 2. $f_i'(\phi^*)$ is the slope of the ith order PRC at the locking point ϕ^*. Substituting in (4.2) followed by simplification produces the linearized map:

$$\Delta\phi[n+1] = \{1 - f_1'(\phi^*) - f_2'(\phi^*)\}\Delta\phi[n]. \tag{4.3}$$

If $0 < f_1'(\phi^*) + f_2'(\phi^*) < 2$, then $\Delta\phi[n+1]$ goes to zero and the locking at ϕ^* is stable. Often $f_2'(\phi^*)$ is negligible.

3 Application of the PRC to Two Coupled Oscillators

The dynamics of two reciprocally pulsed-coupled oscillators has been studied quite extensively (Peskin 1975; Dror et al. 1999; Oprisan et al. 2004). Peskin (1975) considered synchronization in a network of two mutually excitatory pulse-coupled leaky integrate and fire neural oscillators.

$$\mathrm{d}V/\mathrm{d}t = -\gamma V(t) + S_0 + S(t). \tag{4.4}$$

The oscillators were characterized by a state variable $V(t)$ with $0 \leq V(t) \leq 1$ and satisfying (4.4) where γ and S_0 were considered to be intrinsic properties of the oscillators and $S(t)$ an external stimulus. At any given instant of time, the effect of pulsatile coupling due to the firing of an oscillator was to advance the voltage of the second oscillator instantaneously so as to become more depolarized by a fixed amount (ε). Therefore,

$$V_i(t) = 1 \rightarrow V_j(t) = V_j(t) + \varepsilon \quad \text{for} \quad j \neq i. \tag{4.5}$$

This implicitly defines a phase resetting curve because if a perturbation is received at time t, the next spike will be advanced by an amount that can be calculated using the explicit solution for the differential equation that describes the leaky integrate and fire oscillator. Therefore,

$$f_1(\phi) = \ln(1 - \gamma\varepsilon S_0 \mathrm{e}^{A\phi})/A, \tag{4.6}$$

where $A = \ln(S_0/(S_0 - \gamma))$. Peskin derived a return map assuming the two oscillators take turns firing. The maps gives the dependence of the phase $\phi[n+1]$ of one oscillator immediately after its partner fired in cycle $n+1$ upon the phase $\phi[n]$ of the other oscillator immediately after its partner fired in cycle n. This mapping was proven to have a unique, unstable fixed point at a phase of 0.5. This fixed point repelled trajectories toward synchronization at a phase of zero or one. Once the oscillators synchronize, the coupling term drops out because each neuron is already at threshold when its partner fires; hence, synchronization results for every set of initial conditions.

Mirollo and Strogatz (1990) generalized the results obtained by Peskin (1975) to any two coupled oscillators in which the state variable (i.e., the membrane potential V) is a smooth monotonically increasing and concave down function, as it is for the leaky integrator and fire model neuron. The pulse coupling when one oscillator fired depolarized the other by an amount ε or pulled it up to the firing threshold, whichever was less.

$$V_i(t) = 1 \rightarrow V_j(t) = \min(1, V_j(t) + \varepsilon) \quad \text{for all} \quad j \neq i.$$

Again, this implicitly defines a phase-resetting curve, allowing a return map to be formulated in terms of the phase of the nonfiring oscillator immediately after its

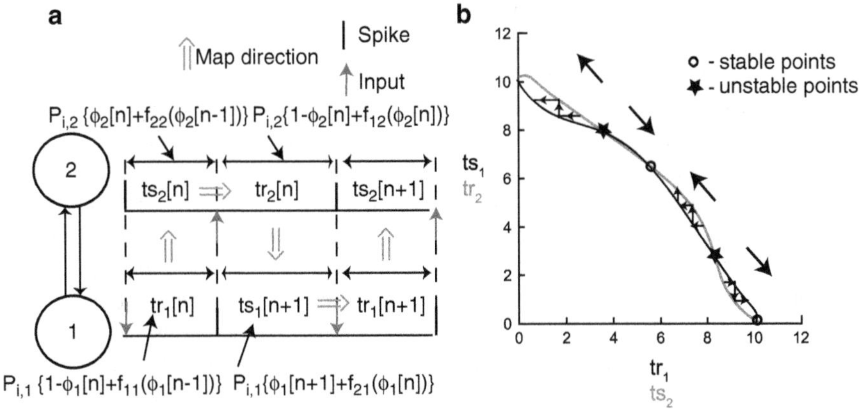

Fig. 4.3 Existence and stability analysis for two bidirectionally pulse-coupled oscillators. (**a**) Firing map for 1:1 phase locking. The two oscillators are labeled as 1 and 2, respectively. Stimulus times in a phase-locked mode are shown here. The *black vertical bars* indicate firing times and the *gray vertical arrows* indicate neuronal inputs at phase $\phi_i (= ts/P_{i,j})$. $P_{i,j}$, f_{1k}, and f_{2k} represent the intrinsic period, the first order resetting and the second order resetting of the kth oscillator, respectively. The cycles of the firing pattern define a map on the plane containing the ts–tr curves. (**b**) Graphical method for determining the fixed points for the firing map given in (**a**). The intersections of the two curves (*black* and *gray*) indicate intervals at which the two periodicity criteria $tr1 = g(ts_1)$ and $ts_2 = g^{-1}(tr_2)$ are satisfied. The circle and the cross symbols indicate the stable and unstable fixed points of the discrete map given in (**a**). The stair steps indicate the direction of flow of the map

partner fired. This mapping was proven to have a unique fixed point which was unstable, based on the assumption of concavity. Once again the firing is globally repelled toward synchrony. The disappearance of the coupling term allowed the synchronized state to exist.

For identical oscillators, synchrony is always a solution. The criteria given below (Dror et al. 1999 and Oprisan et al. 2004) are formulated for oscillators that are not necessarily identical. The existence criteria may be obtained from the algebraic relationships between the intervals at steady state, whereas for the stability criteria a mapping similar to the one presented in the section on forced oscillators is required (Fig. 4.3a). The recovery interval (tr) is defined as the time elapsed between when a neuron receives an input and when it next fires. The stimulus interval (ts) in a network is the time between when a neuron fires and when it receives an input. By definition, $tr1[n] = ts2[n]$ and $tr2[n] = ts1[n+1]$. The next step is to use the assumption of pulsatile coupling in order to define the intervals in terms of the phase and the phase resetting. Although we assume that the effect of one input has dissipated by the time the next is received, if an input is the last one received in a cycle (and in this example there is only one input per cycle), then we do not need to assume that the effect has dissipated by the time the next spike or burst is generated because the total resetting is divided into two parts. The first order resetting is contained in the recovery interval $tr = P_i(1 - \phi + f_1(\phi))$, whereas the second order resetting is contained in the stimulus interval $ts = P_i(\phi + f_2(\phi))$. Note that the definition of the stimulus interval is different in the closed and open loop configuration because in contrast to the open loop, in the closed loop prior perturbations have to be considered. We next assume a steady locking point in which the neurons receive an input at the steady state phases ϕ_1^* and ϕ_2^*. Since the steady state values of both tr and ts depend only on phase, tr for each neuron can be considered a function of ts, $tr = g(ts)$. When phase-locking occurs, $tr1 = ts2$ and $tr2 = ts1$, so the intersection of the $tr1 = g(ts1)$ and $ts2 = g^{-1}(tr2)$ curves (Fig. 4.3b) yields the fixed point(s) of the discrete map given in Fig. 4.3a. The stability of the fixed point can be calculated using discrete linear systems theory if one assumes a small perturbation from the fixed point, and further assumes that the PRCs are linear in a small neighborhood about the fixed point. For 1:1 phase locking, the periodicity criteria together with the definition of stimulus interval (ts) and recovery interval (tr) define a map on the plane containing the ts–tr curves (see Fig. 4.3a). The direction of flow of the map as illustrated in Fig. 4.3b converges at a stable fixed point and diverges from an unstable fixed point of the map (Achuthan et al. 2010a). The map based on the PRCs for two pulse coupled oscillators firing in an alternating firing pattern as shown in Fig. 4.3a is generated as follows:

$$P_{i,1}(1 - \phi_1[n] + f_{11}(\phi_1[n])) = P_{i,2}(\phi_2[n] + f_{22}(\phi_2[n-1])), \qquad (4.7)$$

$$P_{i,2}(1 - \phi_2[n] + f_{12}(\phi_2[n])) = P_{i,1}(\phi_1[n+1] + f_{21}(\phi_1[n])). \qquad (4.8)$$

Here, $f_{1i}(\phi_i[n])$ and $f_{2i}(\phi_i[n])$ correspond to the first order and second order phase resetting of the ith oscillator in response to the input received in the form of a pulse from the other oscillator, respectively. In the neighborhood of $\phi_i{}^*$, $f_{1i}(\phi_i[n]) = f_{1i}(\phi_i{}^*) + f_{1i}{}'(\phi_i{}^*)\ \Delta\phi_i[n]$ and $f_{2i}(\phi_i[n]) = f_{2i}(\phi_i{}^*) + f_{2i}{}'(\phi_i{}^*)\ \Delta\phi_i[n]$ where $\Delta\phi_i[n] = \phi_i[n] - \phi_i{}^*$. $f_{1i}{}'(\phi_i{}^*)$ and $f_{2i}{}'(\phi_i{}^*)$ are the slopes of the first order and second order PRC at the locking point $\phi_i{}^*$. Substituting for ϕ_1 and ϕ_2 (in terms of the perturbations) in (4.7) and (4.8) and simplifying produces: $\Delta[k+1] = M\Delta[k]$ where $\Delta[k] = [\Delta\phi_1[k], \Delta\phi_2[k]]^T$ with M as the following matrix:

$$\begin{pmatrix} f'_{21} & \dfrac{P_{0,2}}{P_{0,1}}\left(f'_{21} - 1\right) \\ \dfrac{-P_{0,1}}{P_{0,2}} f'_{21}\left(f'_{11} - 1\right) & \left(f'_{11} - 1\right)\left(f'_{12} - 1\right) - f'_{22} \end{pmatrix}.$$

The eigenvalues of M are computed from the following characteristic equation:

$$\lambda^2 - \lambda[(1 - f'_{11}(\phi^*{}_1))(1 - f'_{12}(\phi^*{}_2)) - f'_{21}(\phi^*{}_1) - f'_{22}(\phi^*{}_2)] + f'_{21}(\phi^*{}_1) f'_{22}(\phi^*{}_2) = 0. \qquad (4.9)$$

The above derivation can be applied to exact synchrony if it is applied twice, at phases of 0^+ and 1^- for each oscillator (Oprisan and Canavier 2001; Achuthan and Canavier 2009). This ensures robustness to perturbations from synchrony in which either neuron leads. The leading cluster is assigned a phase of 0 approached from the right and the other a phase of 1 approached from the left. The sum of the first and second order resetting is presumed to have a continuous derivative at 0 and 1, but the first and second order resetting curves themselves do not have a continuous derivative at zero and at one (Goel and Ermentrout 2002). In practice, if the coupling is approximately pulsatile in nature and dies out within one cycle of the oscillation, the first order resetting disappears as a phase of one is approached, from the left, and second order resetting dies out as a phase of zero is approached from the right: $f_{11}(1) = f_{22}(0) = f_{12}(1) = f_{21}(0) = 0$. If the sum is continuous modulo 1 then $f_{1i}(0) + f_{2i}(0) = f_{1i}(1) + f_{2i}(1)$, which together with the previous sentence implies that $f_{1i}(0) = f_{2i}(1)$, and $f'_{1i}(0^+) = f'_{2i}(1^-)$. It is often (but not always[1]) reasonable to assume that if the magnitude of the resetting is small then the slope is quite small as well: $f'_{11}(1^-) = f'_{22}(0^+) = f'_{12}(1^-) = f'_{21}(0^+) = 0$. Thus, in that case, the expression for the eigenvalues from the characteristic equation given above reduces to $\lambda = 1 - f'_{12}(0^+) - f'_{11}(0^+)$ and $\lambda = 1 - f'_{11}(0^+) - f'_{12}(0^+)$. If the absolute value of both eigenvalues is less than one then $\Delta\phi[n+1]$ goes to zero at the stable phase-locked points $\phi^* = (0^+, 1^-)$ and $\phi^* = (1^-, 0^+)$.

Pervouchine et al. (2006) used the spike time response curves (STRCs) and spike time difference maps (STDMs) to analyze the case of two identical, identically coupled oscillators in antiphase with identical short delays (see section below on delays). In our terminology, the spike time difference for two neurons 1 and 2 is $F_{12}(ts) = ts1[n+1] - ts1[n]$, where $ts1[n+1] = \psi_1(\psi_2(ts1[n]))$, and the map ψ_j is simply the response time as a function of the stimulus time: $\psi_j(ts) = P_i + f_{1j}(ts/P_i) - ts$, where P_i is the intrinsic period of the ith neuron and f_{1i} is the first order resetting of the ith neuron. This method produces results identical to the pulse-coupled phase resetting methods presented above except that second order resetting is neglected.

Goel and Ermentrout (2002) obtained analogous results with a slightly different approach, and extended them to exact synchrony in identical oscillators by making additional assumptions. The phase transition curve, $F(\phi) = \phi - f_1(\phi)$, was used to define the return map, $G(\phi) = 1 - F(1 - F(\phi))$. They considered only first order resetting $f_1(\phi)$. Linearizing the map and taking the derivative with respect to ϕ fields:

$$G'(\phi) = (1 - f_1'(\phi))(1 - f'(1 - \phi + f(\phi))). \tag{4.10}$$

They assumed that $f_1(0) = f_1(1) = 0$. It is only true that $f_1(0) = 0$ if there is no second order resetting since for pulsatile coupling $f_1(0) = f_2(1)$. Then the assumption was made that $F(\phi)$ is monotonically increasing so that $F'(\phi) > 0$ and $(1 - F'(\phi)) > 0$. This ensures that the firing order cannot change so that one oscillator cannot fire twice in a row. The map then has a stable fixed point at synchrony if $(1 - f_1'(0^+))(1 - f_1'(1^-)) < 1$. The use of the left and right slopes acknowledges the possibility that the first order PRC is discontinuous at 0 and 1. The

[1] $f_{1j}(1^-) = 0$ does not always imply that $f'_{1j}(1^-) = 0$. For an excitatory input that immediately evokes a threshold event, $f_{1j}(\phi) = \phi - 1$ and $f'_{1j}(1^-) = 1$. In this case, the full expression must be retained.

source of this discontinuity is (Maran and Canavier 2008) the existence of second order resetting (see above). In some cases, second order resetting must be considered in order to get the correct stability results. In these cases, (4.10) is not sufficient and (4.9) must be used.

4 Application of the PRC to a Unidirectional Ring of Coupled Oscillators

Dror et al. (1999) considered the case of a unidirectional ring of coupled oscillators. Each oscillator receives exactly one input per cycle. Stable phase-locked modes must satisfy the following criterion (see Fig. 4.4):

$$P_{i,1}\phi_1{}^* + P_{i,2}\phi_2{}^* + \cdots + P_{i,N-1}\phi_{N-1}{}^* + P_{i,N}\phi_N{}^* = kP_e, \tag{4.11}$$

where $P_{i,j}(1 + f(\phi_i{}^*)) = P_e$ and $t_{si}[n] = P_{i,j}(\phi_i{}^* + \Delta\phi_i[n])$ for $j = 1, 2, \ldots N$ as before and $k \in [0, N-1]$. P_e represents the common entrained period of the N neural oscillators. Each value of k produces a different firing order. $(N + k - 1)$

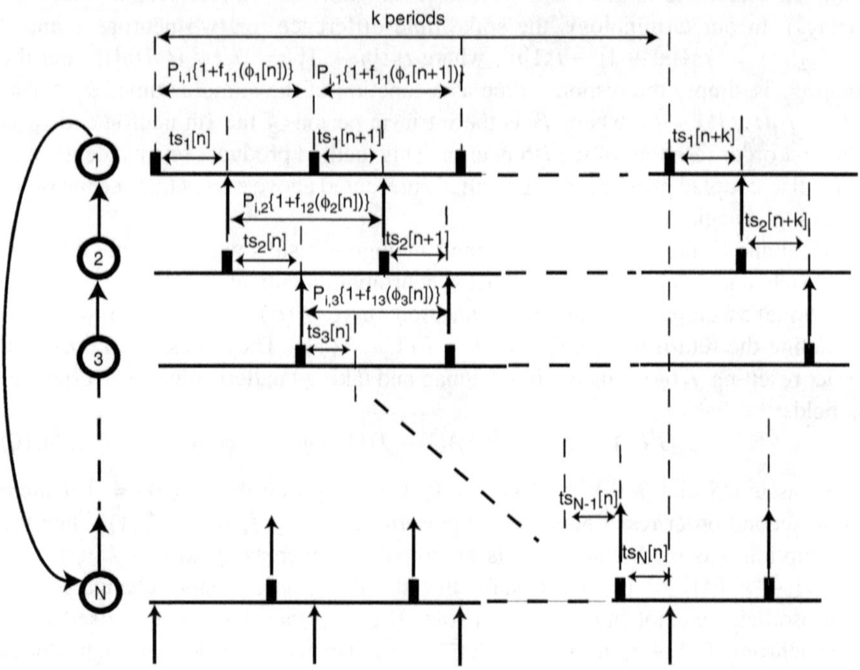

Fig. 4.4 Stability analysis for unidirectional N-ring pulse-coupled oscillators. The analysis depends on k because the sum of $ts_i[n]$ for $i = 1$ to N equals kP_e at steady state, where P_e represents the common entrained period of the N neurons. If the firing time of neuron 1 is perturbed slightly from steady state, the effects of this perturbation will propagate all the way around the ring in k periods

cycles elapse before the feedback due to a spike in a particular oscillator propagates back to that oscillator, similar to the integer k in Foss and Milton (2000) (see the section on pulse-coupled oscillators with delays). Let $\Delta\phi_i[n+1]$ denote the change in the phase of the ith oscillator in the $(n+1)$st cycle. $\Delta\phi_1[n+1]$ in a ring of N coupled oscillators using PRCs is fully described by the $(N-1)$ state variables $\Delta\phi_q[n+1]$ where $q \in [0, N-1]$. Here, unlike the case of the two coupled oscillators, we have: $\Delta\phi_1[n+1] = A\Delta\phi_1[n]$ where A is the Jacobian matrix of the discrete-time system determined by the PRCs (Dror et al. 1999). The matrix A has been worked out by Dror et al. (1999) where time was used instead of the phase variable. The matrix A is unchanged for identical oscillators and is given by Oprisan and Canavier (2001) for heterogeneous rings of three. The matrix A is distinct for each assumed firing order.

In the case of N-ring coupled oscillators, the eigenvalues of the matrix A determine the stability of the system given the values of $f_i(\phi_i{}^*)$ for $i = 1, 2, \ldots N$. If λ_{max} denotes the maximum eigenvalue and if $|\lambda_{max}| < 1$ stability is guaranteed. These existence and stability criteria predict the activity of model neural networks accurately as long as the duration of the coupling is short compared to the period (Canavier et al. 1997, 1999; Luo et al. 2004).

Goel and Ermentrout (2002) used PRCs to study similar geometries such as a bidirectional ring and a chain of N coupled oscillators. The firing order cannot be assumed to be resistant to all possible small perturbations in such a system so the issue of stability for this general case remains as an open question.

5 Application of the PRC to N all-to-all Coupled Oscillators

Mirollo and Strogatz (1990) formulated a map for N all-to-all coupled identical oscillators (i.e., each oscillator coupled to all the others) with assumptions analogous to those for their two neuron circuit, described above. The interaction between the oscillators was defined as before so that when $V_i(t)$ reached 1, it reset to zero and increments the voltage of all the other oscillators by a pulse of fixed amount. Once again, this implicitly defines a phase-resetting curve. They formulated a map of the phases at each firing event, under the assumption that the firing order cannot change due to the form assumed for the resetting and the identical frequencies of each oscillator. As the system evolved in time, oscillators formed groups that fired at the same time. The effect that a group has on other groups is proportional to its size. Larger groups advanced smaller groups that follow them more than they were advanced by the smaller group on each cycle, such that the gap between them decreases on each cycle until the smaller group is brought up to threshold along with the larger group. This continues until a single group remains except in the unlikely event that groups of identical size form and the gaps remain constant. The proof that their map converged to synchrony was performed in two stages. The first part utilized the assumed concavity of $V_i(t)$ to show that for almost all initial conditions, an absorption occurred in finite time because the mode in which all neurons fire sequentially is unstable. The second part of the proof showed that the measure of the sets of initial conditions that lived forever without experiencing the final absorption to synchrony was also zero. Bottani (1995) stated that the only way to stop the progression to synchrony once the first absorption occurs was to settle into a splay mode in which the groups are exactly the same size. Bottani retained the assumption for the form of the coupling, but dispensed with the requirement that $V_i(t)$ be concave down to ensure global synchrony and required only that it be monotonically increasing, provided it was not too convex (i.e., concave up) by showing that for a sufficient large pulse it is unlikely for no absorptions to occur.

Goel and Ermentrout (2002) also studied N all-to-all pulse-coupled identical oscillators in which every oscillator was identically connected to every other oscillator. As in the two neuron case, they assumed that $f_1(0) = 0$ and $f_1(1) = 0$, and that $f_1'(\phi) > 0$, which guaranteed that the oscillators always fired in the same order. The oscillators were indexed in the order of their firing and an $(N - 1)$-dimensional map was formulated. Stability conditions were derived by linearizing the map about its fixed point. The stability conditions were derived based on whether the PRC was continuously differentiable or not. If $f_1'(0^+) = f_1'(1^-)$, then stability was guaranteed if $(1 - f_1'(\phi)) < 1$. Otherwise the following $(N - 1)$ conditions must be satisfied for the stability of the synchronous solution: $(1 - f_1'(0^+))^a$ $(1 - f_1'(1^-))^{N-a} < 1$, where $1 \leq a \leq N$. In the latter case, stability depends upon the size of the network and synchrony can be destabilized for large N. There are two problems with this proof: the structure of the perturbation assumes that no two neurons fire simultaneously as synchrony is approached, but rather coalesce all at once into a single cluster. Therefore, this proof does not guarantee stability in the case of a perturbation when only a single neuron is displaced from synchrony and the remainder forms a large cluster (see Achuthan et al. in this volume). The Goel and Ermentrout proof is actually more useful as a proof for the stability of the splay mode (see Achuthan et al. in this volume). The second problem, as noted in the two oscillator section, is that second order resetting is neglected but is most likely to be important near synchrony, as one or more neurons may receive inputs very late in the cycle.

Achuthan and Canavier (2009) found that the fully synchronous mode was found to lose stability when the PRC generated using $(N - 1)$ simultaneously active inputs becomes too steep. The implication was that the stability of synchrony could be lost by increasing the slope of this PRC either by increasing the network size or the strength of the individual synapses. Clustering solutions are also analyzed in Achuthan et al. in this volume.

6 Pulse-Coupled Oscillators with Delay

A study of pulsatile coupling (Ernst et al. 1995, 1998) extended the results of Peskin (1975) and Mirollo and Strogatz (1990) to the case of two pulse-coupled oscillators reciprocally coupled with delays up to half the intrinsic period. This line of work assumed that the relevant state variable (voltage, for example) was monotonically increasing and concave down (as for a leaky integrator) between instances of reaching firing threshold, and was reset to zero afterward. A specific form for the PRC was assumed in that excitatory coupling was assumed to increase this state variable by a fixed amount or bring it to threshold, whichever was less. In an extension of previous work, inhibition was assumed to decrease the state variable by a fixed amount or reduce it to zero, whichever was greater. For small delays and strong excitatory coupling, at all coupling strengths synchronization with a phase lag equal to the delay was found to be always stable, and for very strong coupling bistability with a variable, smaller, nonzero phase lag occurred. For inhibitory coupling at lower strengths, bistability between the in phase and out of phase solution occurred, and as the coupling strength was increased the out of phase solution disappeared. At the strongest coupling strengths, a transition occurred to synchronization with a variable, smaller, nonzero phase lag. This study stated that the "remarkable" mechanism of zero phase lag synchronization with inhibition was directly related to the form assumed for the state variable that was concave down.

Other lines of research have not assumed a specific form for the PRC and are therefore more general. The proof for feedback to a single neuron at fixed delay from a spike in that neuron follows the proof given by Foss (1999) and Foss and Milton (2000) except that we use terminology consistent with the rest of the chapter and include second order resetting, which was ignored in the earlier proof. Here, curly braces {} group terms with a common multiplier, parentheses () indicate a function, and straight brackets [] enclose the cycle number in a discrete map. The integer k denotes the number of network periods contained in the delay δ. The definitions of the intervals depicted in Fig. 4.5 are as follows. The stimulus interval $ts[n]$ is the interval between the last spike and the receipt of an input in cycle n, and can be written as the sum of the phase $\phi[n]$ at the time the input is received and the second order resetting from the previous cycle $f_2(\phi[n-1])$, multiplied by the intrinsic period P_{i} i.e., $ts[n] = P_{\mathrm{i}}\{\phi[n] + f_2(\phi[n-1])\}$. The response time $tr[n]$ is the interval between the receipt of an input in cycle n and the next spike, and can be written as the interval remaining until the next spike in the absence of a perturbation $P_{\mathrm{i}}\{1-\phi[n]\}$ and the first order resetting in the current cycle times the intrinsic period P_{i}, i.e., $tr[n] = P_{\mathrm{i}}\{1-\phi[n] + f_1(\phi[n])\}$. Therefore, the network period is equal to $ts[n] + tr[n] = P_{\mathrm{i}}\{1 + f_1(\phi[n]) + f_2(\phi[n-1])\}$. As per Fig. 4.5a, the fixed delay is equal to k complete cycles plus stimulus interval $ts[n+k]$. The self-consistency, or periodicity criterion can be written as follows.

$$\begin{aligned}\delta &= \sum_{j=n}^{n+k-1} (ts[j] + tr[j]) + ts[n+k] \\ &= P_0 \sum_{j=0}^{k-1} \{1 + f_1(\varphi[n+j]) + f_2(\varphi[n+j-1)\} \\ &\quad + P_0\{\varphi[n+k] + f_2(\varphi[n+k-1])\}.\end{aligned} \tag{4.12}$$

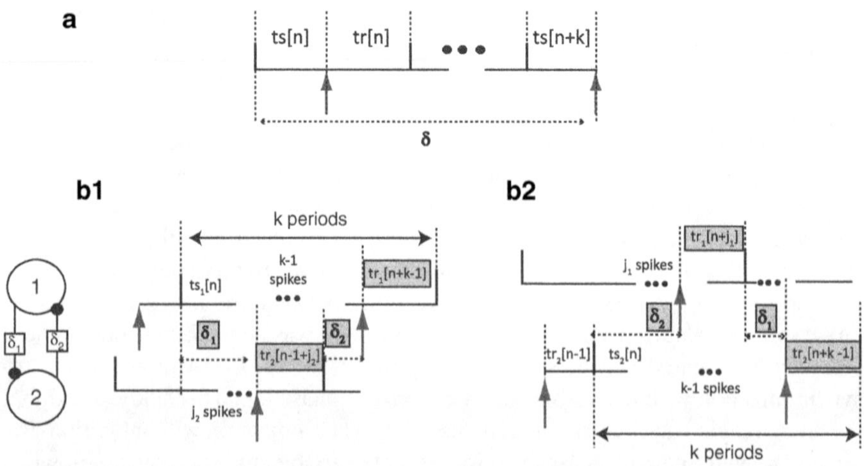

Fig. 4.5 Neural firing time maps with delays of arbitrary lengths. (**a**) Forced oscillator with a feedback input that has a fixed delay (δ) from each time the oscillator fires. (**b**) A circuit of two coupled neurons corresponding to two out of the four possible firing patterns. The delay between a spike in neuron 1 (2) and the receipt of an input by neuron 2 (1) is designated δ_1 (δ_2). The delay between a spike in neuron 1 (2) and the receipt of an input by neuron 2 (1) is designated δ_1 (δ_2). (B1) The quantities highlighted in gray sum to k cycles in neuron 1. The number of spikes that occur in neuron 2 during δ_1 is j_2. (B2) The quantities highlighted in *gray* sum to k cycles in neuron 2. The number of spikes that occur in neuron 1 during δ_2 is j_1. Stimulus intervals (t_{si}) denote the time elapsed between a spike in neuron i and the receipt of the next input by neuron i, whereas recovery intervals (t_{ri}) denote the time elapsed between the receipt of an input by neuron i and the next spike in neuron i. The time intervals are variable and indexed by cycle number n

The values of ϕ that satisfy the equality given above are fixed points of the mapping from one cycle to the next and are denoted as ϕ^*. A stability criterion can be derived by assuming a perturbation from a fixed point such that $\phi[j] = \phi^* + \Delta\phi[j]$, and linearizing the PRCs such that $f_{\mathrm{i}}(\phi[j]) = f_{\mathrm{i}}(\phi^*) + f_{\mathrm{i}}'(\phi^*)\Delta\phi[j]$, where $f_{\mathrm{i}}'(\phi^*)$ indicates the slope of the ith order PRC at the locking point. If these substitutions are made in the equality above, the steady state terms are canceled from each side, and the term for $\phi[n+k]$ is moved to the left hand side, the following linear system is obtained:

$$\varphi[n+k] = -\sum_{j=0}^{k-1}\{f_1'(\varphi[n+j]) + f_2'(\varphi[n+j])\} - f_2'(\varphi[n-1]).$$

The matrix corresponding to the perturbed linear system is as follows.

$$\begin{pmatrix} \Delta\varphi[n+k] \\ \Delta\varphi[n+k-1] \\ \Delta\varphi[n+k-2] \\ \vdots \\ \Delta\varphi[n] \end{pmatrix} = \begin{pmatrix} -m_1 - m_2 & -m_1 - m_2 & \cdots & -m_1 - m_2 & -m_2 \\ 1 & 0 & \cdots & 0 & 0 \\ 0 & 1 & \cdots & 0 & 0 \\ \vdots & \vdots & \vdots & \vdots & \vdots \\ 0 & 0 & \cdots & 1 & 0 \end{pmatrix} \begin{pmatrix} \Delta\varphi[n+k-1] \\ \Delta\varphi[n+k-2] \\ \Delta\varphi[n+k-3] \\ \vdots \\ \Delta\varphi[n-1] \end{pmatrix},$$

where $m_i = f_i'(\phi^*)$. This matrix is the companion matrix for the polynomial expression (Horn and Johnson 1985):

$$\pi(\lambda) = \lambda^{k+1} - \{m_1 + m_2\}\lambda^k - \{m_1 + m_2\}\lambda^{k-1} \ldots - \{m_1 + m_2\}\lambda - m_2.$$

The zeros of the polynomial are the eigenvalues of the matrix, and must all have an absolute value less than one, that is, they must lie within the unit circle, in order for the fixed point of the mapping to be stable, which implies that the periodic locking at a fixed delay is also stable in the system from which the PRC was derived. For the case in which second order resetting is ignored ($m_2 = 0$), Foss (1999) was able to derive a simple bound on the slopes of the first order PRC, by first noting that the zeros cross the unit circle when $m_1 = -1$, then by setting λ to 1 and $\pi(\lambda)$ to 0 in order obtain the lower bound on m_1, such that stability is guaranteed when $-1/k < m_1 < 1$.

Woodman and Canavier (2009) recently derived stability results for all possible firing patterns in two reciprocally pulse-coupled oscillators with delays. In Fig. 4.5b1, it takes k cycles before the firing of a spike in neuron 1 affects the timing of the next spike in neuron 1 via the feedback loop through neuron 2. We define j_2 as the number of spikes in neuron 2 that occur during the delay period δ_1 and j_1 as the number of spikes in neuron 1 that occur during the delay period δ_2. The integer k is equal to $(j_1 + j_2 + 1)$. Based on this figure, we can construct a map based on the stimulus and recovery intervals.

$$ts_1[n+k-1] = tr_2[n+j_2-1] + \delta_1 + \delta_2 - \sum_{i=0}^{k-2} (ts_1[n+i] + tr_1[n+i]),$$

$$ts_2[n+k-1] = tr_1[n+j_1] + \delta_1 + \delta_2 - \sum_{i=0}^{k-2} (ts_2[n+i] + tr_2[n+i]).$$

The quantities on the left side are written in terms of earlier intervals. The summation is only applied for k greater than two. Surprisingly, this exact expression generalizes to all firing patterns (for example, see Fig. 4.5b2). The same techniques that were previously used to write these expressions in terms of the phase resetting, then linearize and derive stability results can be applied here to produce the general results described in Woodman and Canavier (2009).

7 Maps with No Predetermined Firing Order

In addition to the methods based on the existence and stability criteria for an assumed firing pattern, there is another way to predict network activity based on the PRC, which we call here an iterated map (Canavier et al. 1997, 1999; Netoff et al. 2005). Given the current phase of each oscillator (neuron) and its period, one can easily determine which neuron(s) will fire next. Given the phase resetting behavior of each neuron in response to the firing of every other neuron, one can then determine how the phases of each neuron are altered by each firing, so one can determine the future sequence of firing for all time, in the absence of noise. The determination of each firing event is considered one map iteration. The map may produce different firing patterns depending upon how it is initialized because no fixed, predetermined firing order is assumed. The firing intervals can be specified by subtracting the starting phase from the ending phase, adding any resetting presumed to occur during the interval, and multiplying by the intrinsic period. Note that first order resetting is added to the recovery interval, and second order resetting to the next stimulus interval.

Real circuits subject to noise necessitates the use of statistical methods to determine whether there is a locking (Mancilla et al. 2007). Sieling et al. (2009) address the issue of robustness to noise by randomizing the phase resetting in the map. A Gaussian distribution (with truncated tails) was assumed about the average PRC that charaterized the actual noise observed while measuring the PRC. The width of this distribution was adjusted to reflect the full amount of observed noise. Causality dictates that the first order resetting cannot cause the recovery interval to become negative, and second order resetting cannot force the next stimulus interval to become negative, so the noise in the PRCs was clipped to prevent violations of causality. Structural stability was determined by comparing predictions using the upper and lower envelopes of the experimentally derived PRCs with those obtained using the average PRC. If the predictions did not contain the same number of fixed points with the same stability, the case was considered structually unstable. The iterated map was robust to noise in the structurally stable cases and very sensitive to noise in the structurally unstable cases. Therefore, the amount of noise that can be tolerated before one to one phase locking is lost can also be predicted using the PRC under the assumption of pulsatile coupling.

8 Summary

Pulsatile-coupling methods have been applied to a number of circuit architectures in order to derive stability results for phase-locked modes; in the most general cases, the stability depends only on the slope of the resetting curves at the locking points. Since PRCs can be measured for biological neurons, these results can provide insight into the synchronization tendencies of neural networks. In general, pulsatile-coupling methods can be extended to inputs that are not strictly pulsatile under the assumption that the effect of a perturbation has dissipated by the time the next one is received. As explained above, the methods can also tolerate a certain level of noise. The violation of the assumption of a constant firing order is problematic. Firing orders that are not constant from cycle to cycle were considered by Maran and Canavier 2008; Oh and Matveev 2009. In these papers, the leading neuron can switch from cycle to cycle. The assumption of pulsatile coupling may also be violated, especially in networks comprised of oscillators that receive multiple inputs per cycle as opposed to those that receive only a single input per cycle. Slow currents that cause adaptation from cycle to cycle also violate the assumption of pulsatile coupling. One possible solution to this problem is to apply repetitive inputs at a fixed delay rather than just a single pulse (Cui et al. 2009). Nonetheless, the application of phase-resetting methods to systems in which the coupling is neither weak nor pulsatile is very much an open problem.

Acknowledgments This work was supported by NIH grants NS54281 and MH 85387 under the CRCNS program.

References

Achuthan S and Canavier CC (2009). Phase resetting curves determine synchronization, phase-locking, and clustering in networks of neural oscillators. *J. Neurosci.*, 29:5218–5233.

Achuthan S, Cui J, Butera RJ and Canavier CC (2010). Maps based on the phase resetting curve explains spike statistics of coupled neural oscillators observed in the presence of noise. *BMC Neurosci.*, 11(Suppl 1): P18.

Achuthan S and Canavier CC (2011). Synaptic and intrinsic determinants of the phase resetting curve for weak coupling. *J. Comput. Neurosci*, 2:373–390.

Acker CD, Kopell N and White JA (2003). Synchronization of strongly coupled excitatory neurons: relating network behavior to biophysics. *J. Comput. Neurosci.*, 15:71–90.

Bottani S (1995). Pulse-coupled relaxation oscillators: from biological synchronization to self-organized criticality. *Phys. Rev. Lett.*, 74:4189–4192.

Buck J (1988). Synchronous rhythmic flashing of fireflies II. *Quart. Rev. Biol.*, 63:265–289.

Buzsaki G (2006). *Rhythms of the Brain.* Oxford University Press Inc., New York.

Canavier CC, Butera RJ, Dror RO, Baxter DA, Clark JW and Byrne JH (1997). Phase response characteristics of model neurons determine which patterns are expressed in a ring circuit model of gait generation. *Biol. Cybernetics*, 77: 367–380.

Canavier CC, Baxter DA, Clark JW and Byrne JH (1999). Control of multistability in ring circuits of oscillators. *Biol. Cybernetics*, 80: 87–102.

Canavier CC (2005). The application of phase resetting curves to the analysis of pattern generating circuits containing bursting neurons. In: Coombes S. and Bressloff P. (eds.) *Bursting: The Genesis of Rhythm in the Nervous System.* Series in Mathematical Neuroscience, World Scientific, Singapore, 175–200.

Canavier CC, Gurel Kazanci F, and Prinz AA (2009). Phase resetting curves allow for simple and accurate prediction of robust N:1 phase locking for strongly coupled neural oscillators. *Biophys. J.*, 97: 59–73.

Cui J, Canavier CC and Butera RJ (2009). Functional phase response curves: a method for understanding synchronization of adapting neurons. *J. Neurophysiol.*, 102:387–398.

de la Iglesia HO, Cambras T, Schwartz WJ, and Diez-Noguera A (2004). Forced desynchronization of dual circadian oscillators within the rat suprachiasmatic nucleus. *Curr. Biol.*, 14: 796–800.
Dror RO, Canavier CC, Butera RJ, Clark JW and Byrne JH (1999). A mathematical criterion based on phase response curves for stability in a ring of coupled oscillators. *Biol. Cybern.*, 80: 11–23.
Ermentrout GB and Chow CC (2002). Modeling neural oscillations. *Physiol. Behav.*, 77:629–633.
Ernst U, Pawelzik K, and Geisel T (1995). Synchronization induced by temporal delays in pulse-coupled oscillators. *Phys. Rev. Lett.*, 74:1570–1573.
Ernst U, Pawelzik K, and Geisel T (1998). Delay-induced multi-stable synchronization of biological oscillators. *Phys. Rev. E*, 57:2150–2162.
Foss J (1999). Control of multistability in neural feedback systems with delay (Ph.D. thesis) Chicago, IL: The University of Chicago.
Foss J and Milton J (2000). Multistability in recurrent neural loops arising from delay. *J. Neurophysiol.*, 84:975–985.
Glass L and Mackey MC (1988). *From Clocks to Chaos: the Rhythms of Life*. Princeton, N.J. Princeton University Press.
Goel P and Ermentrout GB (2002). Synchrony, stability, and firing patterns in pulse-coupled oscillators. *Physica D*, 163 (3– 4): 191–216.
Hoppensteadt FC and Izhikevich EM (1997). *Weakly Connected Neural Networks.* New York: Springer-Verlag.
Horn RA and Johnson CR (1985). *Matrix Analysis.* Cambridge: Cambridge University Press.
Kuramoto Y (1984). *Chemical Oscillations, Waves and Turbulence*. Springer-Verlag, Berlin.
Luo C, Clark JW Jr, Canavier CC, Baxter DA and Byrne JH (2004) Multimodal behavior in a four neuron ring circuit: mode switching. *IEEE Trans. Biomed. Eng.*, 51:205–218.
Mancilla JG, Lewis TJ, Pinto DJ, Rinzel J and Connors BW (2007). Synchronization of electrically coupled pairs of inhibitory interneurons in neocortex. *J. Neurosci.*, 27:2058–2073.
Maran SK, and Canavier CC (2008). Using phase resetting to predict 1:1 and 2:2 locking in two neuron networks in which firing order is not always preserved. *J Comput. Neurosci.*, 24: 37–55.
Mirollo RE and Strogatz SH (1990). Synchronization of pulse-coupled biological oscillators. *SIAM J. Appl. Math.*, 50:1645–1662.
Netoff T, Banks M, Dorval A, Acker C, Haas J, Kopell N, and White J (2005). Synchronization in hybrid neuronal networks of the hippocampal formation. *J. Neurophysiol.*, 93:1197–1208.
Oh M and Matveev V (2009). Loss of phase-locking in non-weakly coupled inhibitory networks of type-I neurons. *J. Comput. Neurosci.*, 26:303–320.
Oprisan SA and Canavier CC (2001). Stability analysis of rings of pulse-coupled oscillators: the effect of phase resetting in the second cycle after the pulse is important at synchrony and for long pulses. *Differ. Equat. Dyn. Syst.*, 9:242–259.
Oprisan, SA, Prinz, AA, Canavier, CC (2004). Phase resetting and phase locking in hybrid circuits of one model and one biological neuron, *Biophys. J.*, 87: 2283–2298.
Perkel DH, Schulman JH, Bullock TH, Moore GP, and Segundo JP (1964). Pacemaker neurons: effects of regularly spaced synaptic input. *Science*, 145:61–63.
Peskin CS (1975). *Mathematical Aspects of Heart Physiology*, Courant Institute of Mathematical Sciences, New York University, New York, 268–278.
Pervouchine DD, Netoff TI, Rostein HG, White JA, Cunningham MO, Whittington MA and Kopell NJ (2006). Low dimensional maps encoding dynamics in entorhinal cortex and hippocampus. *Neural Comput.*, 18:2617–2650.
Sieling FH, Canavier CC and Prinz AA (2009). Predictions of phase-locking in excitatory hybrid networks: Excitation does not promote phase-locking in pattern generating networks as reliably as inhibition. *J. Neurophys.*, 102:69–84.
Singer W (1999). Neural synchrony: a versatile code for the definition of relations? *Neuron*, 24:49–65.
Strogatz SH (2003). Sync: *How Order Emerges from Chaos in the Universe, Nature, and Daily Life*. New York: Hyperion.
Winfree AT (1967). Biological rhythms and the behavior of populations of coupled oscillators. *J. Theor. Biol.*, 16:15–42.

Winfree AT (1980). *The Geometry of Biological Time*. Springer, New York.
Woodman M and Canavier C (2009). Phase locking of pulse-coupled oscillators with delays is determined by the phase response curve. *Frontiers in Systems Neuroscience. Conference Abstract: Computational and systems neuroscience*. doi: 10.3389/conf.neuro.06.2009.03.139
Yuste R, MacLean JN, Smith J and Lansner A (2005). The cortex as a central pattern generator. *Nat. Rev. Neurosci.*, 6:477–483.

Part II
Estimation of Phase Response Curves

Introduction

Neuronal phase response curves are powerful tools for prediction and interpretation of network dynamics. However, accurate prediction of network behavior depends on knowing the precise shape of the PRC, and a number of factors can complicate measuring neuronal PRCs experimentally. This section of the book outlines considerations when estimating neuronal PRCs and presents strategies for dealing with the challenges inherent to experimental preparations and analysis. The first chapter in Part II provides a thorough overview of these considerations, and the subsequent chapters present more targeted approaches for obtaining accurate PRC estimations while overcoming specific challenges. Building strongly on the weak coupling approach introduced in Part I of the book, Chap. 5 by Netoff and colleagues introduces the basic experimental protocol for obtaining a PRC, considerations for stimulus design, strategies for dealing with nonstationarities in neuronal activity including slow processes such as spike frequency adaptation or drift and noise, and finally presents a method for obtaining the PRC using a white noise stimulus. Chapter 6 by Oprisan uses elements of topological theory introduced in Chap. 2 to derive the PRC by relating geometric displacements within phase space to phase resetting in the temporal domain. Chapter 7 by Polhamus and colleagues describes a common problem in experimental estimation of PRCs where only few trials of stimuli delivered during long interspike intervals are possible and offers a solution using truncated estimators of the firing period rather than the mean firing period across control trials. Finally, Chap. 8 by Ota and Aonishi describes a Bayesian algorithm for estimating PRCs from noisy spike response data.

Introduction

Chapter 5
Experimentally Estimating Phase Response Curves of Neurons: Theoretical and Practical Issues

Theoden Netoff, Michael A. Schwemmer, and Timothy J. Lewis

Abstract Phase response curves (PRCs) characterize response properties of oscillating neurons to pulsetile input and are useful for linking the dynamics of individual neurons to network dynamics. PRCs can be easily computed for model neurons. PRCs can be also measured for real neurons, but there are many issues that complicate this process. Most of these complications arise from the fact that neurons are noisy on several time-scales. There is considerable amount of variation (jitter) in the inter-spike intervals of "periodically" firing neurons. Furthermore, neuronal firing is not stationary on the time scales on which PRCs are usually measured. Other issues include determining the appropriate stimuli to use and how long to wait between stimuli.

In this chapter, we consider many of the complicating factors that arise when generating PRCs for real neurons or "realistic" model neurons. We discuss issues concerning the stimulus waveforms used to generate the PRC and ways to deal with the effects of slow time-scale processes (e.g. spike frequency adaption). We also address issues that are present during PRC data acquisition and discuss fitting "noisy" PRC data to extract the underlying PRC and quantify the stochastic variation of the phase responses. Finally, we describe an alternative method to generate PRCs using small amplitude white noise stimuli.

T. Netoff (✉)
Department of Biomedical Engineering, University of Minnesota, Minneapolis, MN 55455, USA
e-mail: tnetoff@umn.edu

M.A. Schwemmer • T.J. Lewis
Department of Mathematics, University of California, Davis, CA 95616, USA
e-mail: mschwemm@math.princeton.edu; tjlewis@ucdavis.edu

N.W. Schultheiss et al. (eds.), *Phase Response Curves in Neuroscience: Theory, Experiment, and Analysis*, Springer Series in Computational Neuroscience 6,
DOI 10.1007/978-1-4614-0739-3_5, © Springer Science+Business Media, LLC 2012

1 Introduction

A phase response curve (PRC) quantifies the response of a periodically firing neuron to an external stimulus (Fig. 5.1). More specifically, it measures the phase shift of the oscillating neuron as a function of the phase that a stimulus is delivered. The task of generating a PRC for a real neuron seems straightforward: (1) If the neuron is not oscillating, then inject a constant current or apply a neuromodulator to make the neuron fire with the desired period. (2) Deliver a pulsatile stimulus at a particular phase in the neuron's periodic cycle, and measure the subsequent change

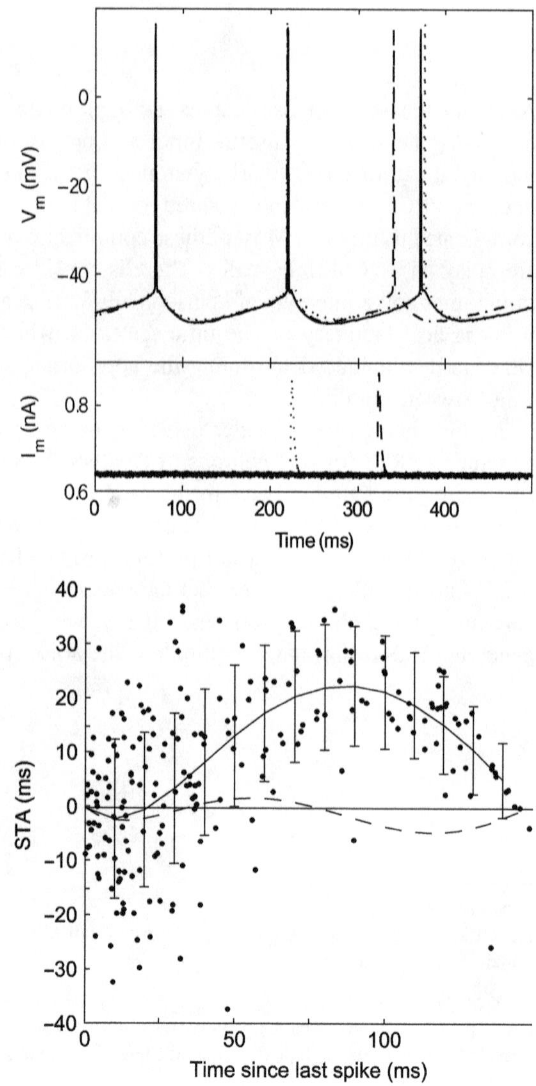

Fig. 5.1 PRC measured from a cortical pyramidal neuron. *Top panel*: voltage from an unperturbed periodically firing neuron (*solid line*); voltage from neuron stimulated with synaptic conductance near the end of the cycle, resulting in an advance of the spike from normal period (*dash-dotted line*); voltage from a neuron stimulated early in the cycle, resulting in slight delay of spike (*dotted line*). *Bottom panel*: spike advance measured as a function of the phase of stimulation. Each dot represents the response to a stimulus. *Solid line* is a function fit to the raw data, estimating the PRC. Error bars indicate the standard deviation at each stimulus phase. The *dashed line* is the second order PRC, i.e. it indicates the effect of the synaptic input at that phase on the period following the stimulated period. Figure adapted from Netoff et al. 2005b

in timing of the next spike (Fig. 5.1 top panel). (3) Repeat these steps for many different phases (Fig. 5.1 bottom panel). However, there are many subtle issues that complicate this seemingly simple process when dealing with real neurons.

Many of the complicating factors in generating PRCs arise from the fact that neurons are inherently noisy on several timescales. There is usually a considerable amount of variation in the interspike intervals of "periodically" firing neurons. This jitter in interspike intervals confounds the change in phase that is due to a stimulus. Furthermore, neuronal firing is typically not stationary over the timescales on which PRCs are measured, and a PRC can change significantly with the firing rate of a neuron. Other important issues that need consideration when constructing PRCs arise from the inherent nonlinearities and slow timescale processes of neuronal dynamics. These issues include determining the appropriate stimuli and deciding how long to wait between stimuli.

In this chapter, we will discuss many of the complicating factors that arise when generating PRCs for real neurons or "realistic" model neurons. In Sects. 2 and 3, we focus on theoretical issues for which noise is not a factor. Section 2 addresses issues around the stimulus waveforms used to generate the PRC. Section 3 discusses the effects of slow timescale processes, such as spike frequency adaption, on the phase response properties of a neuron. In Sects. 4 and 5, we discuss issues that arise when measuring PRCs in "real-world" noisy conditions. Section 4 deals with issues that are present during data acquisition, and Sect. 5 discusses fitting "noisy" PRC data to extract the underlying PRC and quantifying the stochastic variation of the phase responses. Finally, in Sect. 6, we describe an alternative method to generate PRCs using small amplitude white noise stimuli.

2 Choosing an Appropriate Stimulus

PRCs are often used to predict the phase-locking dynamics of coupled neurons, using either spike-time response curve (STRC) maps (e.g., (Canavier 2005; Netoff et al. 2005) also see Chap. 4) or the theory of weakly coupled oscillators (e.g., (Ermentrout & Kopell 1991; Kuramoto 1984); also see Chaps. 1 and 2) as follows:

1. The STRC map approach can be used for networks in which synaptic inputs can be moderately strong but must be sufficiently brief. The limiting assumption of STRC map approach is that the effect of any input to a neuron must be complete before the next input arrives. In this case, the PRC can be used to predict the phase shift due to each synaptic input. Therefore, if one intends to use the STRC map method to predict phase-locking behavior, then PRCs should be generated using a stimulus that approximates the synaptic input in the neuronal circuit under study.
2. The theory of weakly coupled oscillators can be used for completely general coupling but the total coupling current incident on a neuron at any time must be sufficiently small. The limiting assumption of this method is that the effects of the

inputs sum linearly, i.e., the neurons respond to input like a time-dependent linear oscillator. The infinitesimal PRC (iPRC), which is used in the theory of weakly coupled oscillators, can be obtained from any PRC generated with sufficiently small perturbations (so long as the perturbation elicits a "measurable" response). Typically, current-based stimuli that approximate delta functions are used.

As indicated above, the choice of stimulus used to generate a PRC depends on the intended use of the PRC. It also depends on the need for realistic stimuli and ease of implementation. In this section, we will address some of the issues involved in choosing an appropriate stimulus waveform to generate a PRC. For the case of small amplitude stimuli, we will also describe the relationships between PRCs generated with different stimulus waveforms.

2.1 *Stimulus Waveforms*

2.1.1 Current-based Synaptic Input

Perhaps the simplest stimulus waveform used to measure a neuron's PRC is a square pulse of current. Square wave current stimuli are easy to implement in models and in a real neuron, using a waveform generator. A possible drawback is that square wave pulses do not resemble synaptic conductances (however, see Sect. 2.2).

A current stimulus waveform that has a shape similar to realistic synaptic input is an alpha function

$$I_{\rm syn}(t) = S_{\rm max}\frac{1}{\tau_{\rm f}-\tau_{\rm r}}\left(e^{-\frac{t}{\tau_{\rm f}}}-e^{-\frac{t}{\tau_{\rm r}}}\right), \quad t \geq 0,$$

where $S_{\rm max}$ controls the amplitude of the synaptic current, $\tau_{\rm f}$ is the time constant that controls the decay ("fall") of the synaptic current, and $\tau_{\rm r}$ is the time constant that controls the rise time of the synaptic current. Here, $t = 0$ is the time at the onset of each synaptic input. Examples of alpha functions with different coefficients are plotted in Fig. 5.2. The coefficients of the alpha function current stimulus can be chosen in order to fit the postsynaptic potentials (PSPs) measured in neurons of interest.[1] In the neocortex, physiologically reasonable values for the synaptic conductance time constants are $\tau_{\rm f} \approx 1.8$ ms and $\tau_{\rm r} \approx 1.4$ ms for fast excitatory synapses and $\tau_{\rm f} \approx 3.5$ ms and $\tau_{\rm r} \approx 1.0$ ms for fast inhibitory synapses (Cruikshank, Lewis, & Connors 2007). The peak amplitude is synapse specific and depends on the resistance of the neuron. We usually adjust the amplitude of the synaptic current so

[1] The time constants of the synaptic currents will be faster than those for the PSP because the current waveform is filtered due to the RC properties neuronal membrane. To find the time constants of the synaptic currents, one can adjust these time constants until the current stimulus induces a PSP waveform that adequately matches an actual PSP.

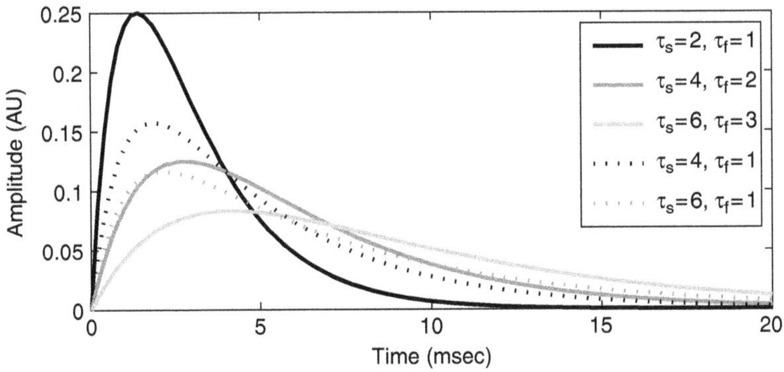

Fig. 5.2 Alpha function current stimuli plotted with different time constants and the same total current (in arbitrary units, AU)

that it elicits a PSP of $\sim$1 mV in amplitude. Figure 5.3 shows PRCs generated using an alpha function current stimulus with a positive S_{max} to simulate an excitatory synaptic input (top) and a negative S_{max} to simulate an inhibitory synapse (bottom).

The shape of the PRC can be significantly affected by the shape of the stimulus waveform used to generate it. PRCs measured from the same neuron using excitatory currents with different time constants are shown in Fig. 5.4. As the synaptic time constants increase, the PRC peak shifts down and to the left. This shift in the PRC is associated with changes in phase-locking of synaptically coupled neurons; simply by slowing the time constants of the synapses, it is possible for a network to transition from synchronous firing to asynchronous firing (Lewis & Rinzel 2003; Netoff et al. 2005; Van Vreeswijk, Abbott, & Ermentrout 1994).

The PRC can also be affected by the magnitude of the stimulus used to generate it. As shown in the example in Fig. 5.5, the peak of the PRC typically shifts up and to the left as the magnitude of excitatory input increases. PRCs for inhibitory pulses are generally flipped compared to those for excitatory pulses, and the peak in the PRC typically shifts down and to the right as the magnitude of inhibitory input increases. For sufficiently small input, the magnitude of the PRC scales approximately linearly with the magnitude of the input. In fact, for sufficiently small input, the changes in the PRC that occur due to changes in the stimulus waveform can be understood in terms of a convolution of the stimulating current with the neuron's so-called infinitesimal PRC. This will be described more fully in Sect. 2.2. Some of the changes in the PRC that occur in response to changes of the stimuli with large magnitudes follow the same trends found for small stimuli; however, other changes are more complicated and involve the nonlinear properties of neurons.

When stimulating a neuron, the maximum a neuron's spike can be advanced is to the time of the stimulus. In other words, the spike cannot be advanced to a time before the stimulus was applied. Therefore, the spike advance is limited by causality. Often we plot the causality limit along with the PRC to show the maximal advance (as shown in Fig. 5.16). When plotting PRCs measured from neurons if the stimulus

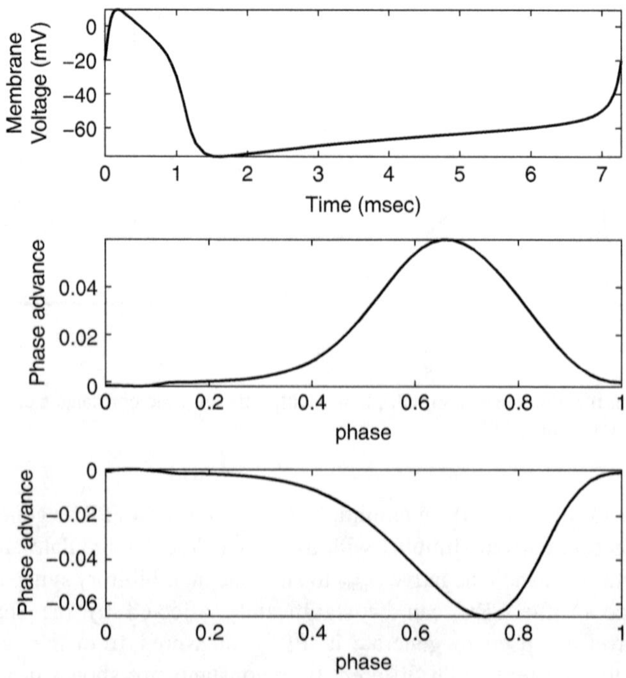

Fig. 5.3 Phase response curves measured with alpha-function current stimuli: [*top*] voltage trace from neuron over one period [*middle*] excitatory stimuli, [*bottom*] inhibitory stimuli. Each plot depicts the spike time advance, in proportion of the period, as a function of stimulus phase. Zero phase and a phase of 1 are defined as voltage crossing of −20 mV. The $I_{Na} + I_K$ model of Izhikevich (2007) was used to model neuronal dynamics with ISI = 15 ms. The synaptic parameters were $\tau_r = 0.25$ ms, $\tau_f = 0.5$ ms, and $S_{max} = 0.04$

was too strong much of the data will hug the causality limit for a significant portion of the phase. This indicates that the stimulus is eliciting an action potential at these phases. If this is the case, we will drop the stimulus strength down. Because each neuron we record has a different resistance, it is not possible to choose one stimulus amplitude that works for every cell. We often have to adjust the amplitude. If the stimulus amplitude is too weak, we find the PRC is indistinguishable from flat.

The line of causality can affect the estimate of the PRC as well. If the neuron is close to the line of causality, it effectively truncates the noise around the PRC. PRCs measured using excitatory synaptic inputs are affected more by the line of causality than those measured with inhibitory synaptic inputs, where the effect is to generally delay the next spike. In Chap. 7 of this book, it will be addressed how the truncation of the noise can affect the estimation of the PRC and how to correct for it.

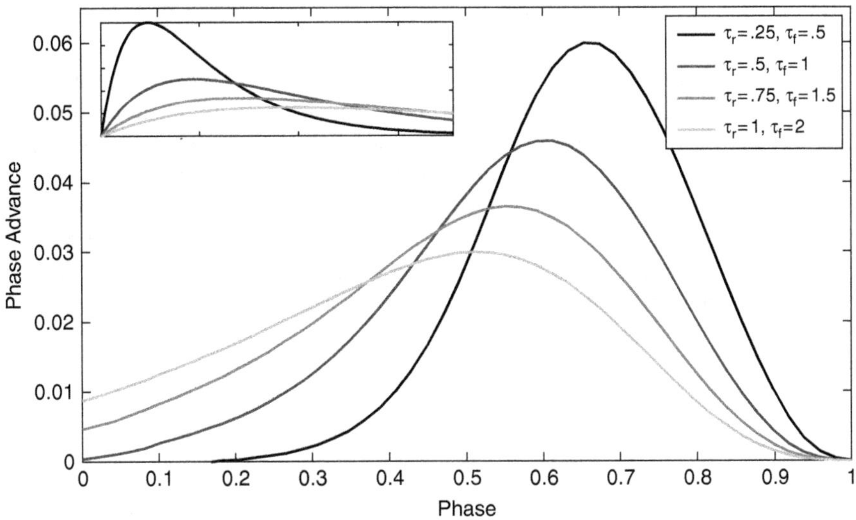

Fig. 5.4 The shape of PRC changes with the shape of the stimulus waveform. Phase response curves are measured with alpha function excitatory current stimuli. *Inset* shows synaptic waveforms as the rise time constants and falling time constants are varied. The $I_{\mathrm{Na}} + I_{\mathrm{K}}$ model of Izhikevich (2007) was used to model neuronal dynamics with a period of 7.2 ms and $S_{\mathrm{max}} = 0.04$

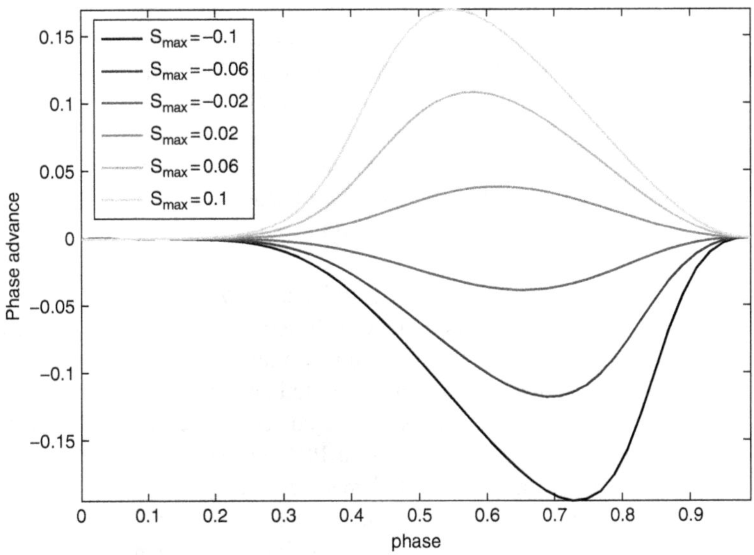

Fig. 5.5 The shape of PRC changes with the magnitude of the stimulus waveform. Phase response curves are measured with alpha-function current stimuli. The $I_{\mathrm{Na}} + I_{\mathrm{K}}$ model of Izhikevich (2007) was used to model neuronal dynamics with ISI = 15 ms. The synaptic time constants were $\tau_{\mathrm{r}} = 0.25$ ms and $t_{\mathrm{f}} = 0.5$ ms

2.1.2 Conductance-based Synaptic Input

When a neurotransmitter is released by a presynaptic cell, it opens ion channels in the postsynaptic cell, evoking a synaptic current carried by the flow of ions through the cell membrane. The synaptic current depends on the number of channels opened (i.e., the activated synaptic *conductance*) and the potential across the membrane. To simulate a more realistic "conductance-based" synapse, an alpha function can be used to describe the synaptic conductance waveform which is then multiplied by synaptic driving force (the difference between the membrane potential V and the reversal potential of the synaptic current E_{syn}) to calculate the synaptic current

$$I_{\text{syn}}(t) = G_{\text{syn}}(t)(E_{\text{syn}} - V(t)), \quad t \geq 0,$$

where the synaptic conductance waveform $G_{\text{syn}}(t)$ is defined to be

$$G_{\text{syn}}(t) = g_{\text{syn}} \frac{1}{\tau_{\text{s}} - \tau_{\text{f}}} \left(\mathrm{e}^{-\frac{t}{\tau_{\text{s}}}} - \mathrm{e}^{-\frac{t}{\tau_{\text{f}}}} \right), \quad t \geq 0.$$

The parameter g_{syn} scales the amplitude of the synaptic conductance. Note that stimulating a neuron with a conductance waveform requires a closed loop feedback system, called a dynamic clamp. The details of a dynamic clamp will be discussed Sect. 4.1.

The synaptic reversal potential is calculated using the Nernst equation if the channels only pass one ion, or the Goldman–Hodgkin–Katz equation if it passes multiple ions (Hille 1992). Excitatory glutamatergic ion channels are cationic, passing some sodium and potassium ions, and therefore the associated excitatory synaptic current reversal potential is usually set near 0 mV. Inhibitory GABAergic ion channels pass mainly chloride or potassium ions, and therefore the associated inhibitory synaptic reversal potential is usually set near −80 mV. The time constants for synaptic conductances are very similar to those previously quoted for synaptic currents.

In Fig. 5.6, the synaptic conductance profiles and corresponding current waveforms for an excitatory synaptic conductance-based input are plotted for different input phases, as illustrated in the Golomb–Amitai model neuron (Golomb & Amitai 1997). As the neuron's membrane potential changes over the cycle, so does the synaptic driving force. Thus, the synaptic current waveform will be different for different input phases. For this reason, a PRC measured with a conductance-based input will be different from a PRC measured using a stimulus with a fixed current waveform. For excitatory conductance-based synaptic inputs, the input current can even reverse direction when the action potential passes through the excitatory synaptic reversal potential. Differences between PRC generated with inhibitory conductance-based input current-based input are more pronounced than for excitatory input because the cell's voltage between action potentials is much closer to the inhibitory reversal potential than for the excitatory reversal potential. This results in larger fractional changes of the driving force (and therefore input current) when compared to excitatory synapses.

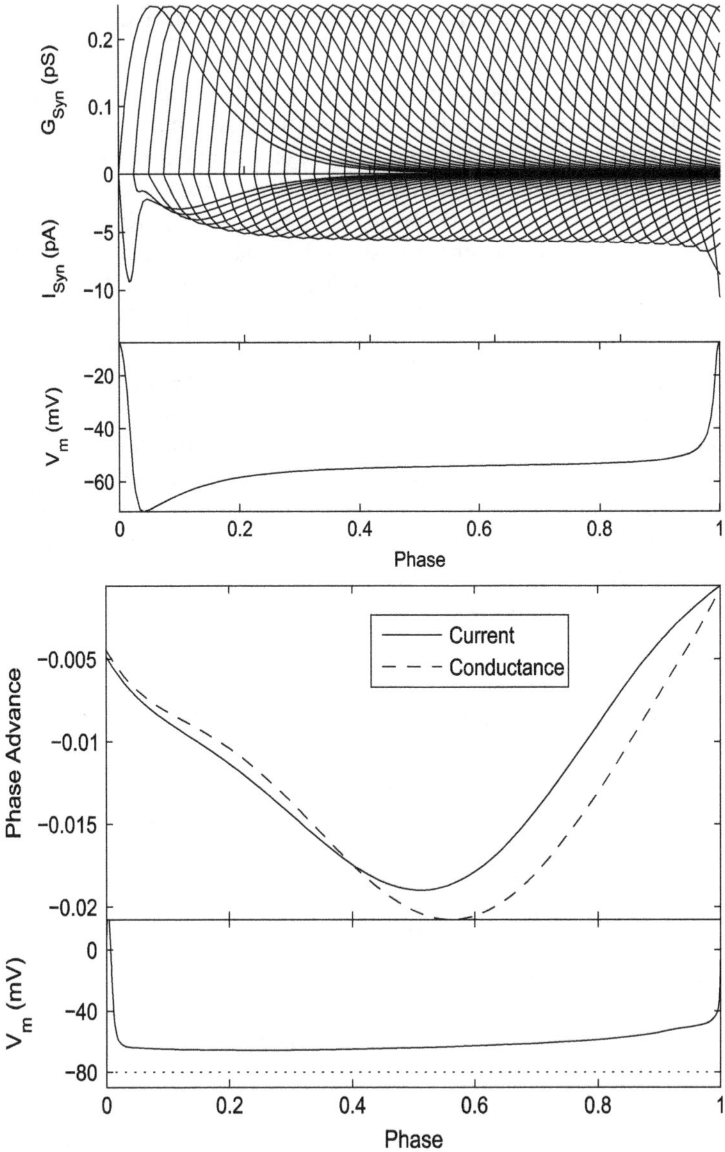

Fig. 5.6 Synaptic input current varies with phase for conductance-based synaptic input. [*Top left panel*] Identical synaptic conductance (G_{syn}) waveforms started at different phases; [*middle left panel*] the corresponding synaptic current waveforms; [*bottom left panel*] the membrane potential used to calculate the synaptic current waveforms. Notice that the synaptic current depends on the membrane potential. [*Right panels*] PRCs measured with inhibitory current input and inhibitory conductance input. The *bottom panel* shows the voltage trace, and the synaptic reversal potential at $-80\,\text{mV}$. To model the current-based waveforms, the synaptic driving force was held constant at 16 mV (i.e., $V - E_{\text{syn}} = -64\,\text{mV} - (-80\,\text{mV})$). The $I_{\text{Na}} + I_{\text{K}}$ model of Izhikevich (2007) was used for figures in the left panels, and the Golomb–Amitai model (1997) was used for figures in the right panels

2.2 The Infinitesimal Phase Response Curve (iPRC)

The infinitesimal phase response curve (iPRC or Z) is a special PRC that directly measures the sensitivity of a neuronal oscillator to small input current[2] at any given phase. The iPRC is used in the theory of weakly coupled oscillators to predict the phase-locking patterns of neuronal oscillators in response to external input or due to network connectivity (Ermentrout & Kopell 1991; Kuramoto 1984) see also Chaps. 1 and 2). For mathematical models, the iPRC can be computed by linearizing the system about the stable limit cycle and solving the corresponding adjoint equations ((Ermentrout & Chow 2002), see also Chap. 1). Equivalently, the iPRC can be constructed by simply generating a PRC in a standard fashion using a small delta function[3] current pulse and then normalizing the phase shifts by the net charge of the pulse (the area of the delta function). More practically, the iPRC can be obtained using any stimulus that approximates a delta function, i.e., any current stimulus that is sufficiently small and brief. Typically, small brief square pulses are used. Note that, for sufficiently small stimuli, the system will behave like a time-dependent linear oscillator, and therefore the iPRC is independent of the net charge of the stimulus that was used. When generating approximations of a real neuron's iPRC, it is useful to generate iPRCs for at least two amplitudes to test for linearity and determine if a sufficiently small stimulus was used.

2.2.1 Relationship Between General PRCs and the iPRC

The iPRC measures the *linear* response of an oscillating neuron (in terms of phase shifts) to small delta-function current pulses. Therefore, it can serve as the impulse response function for the oscillatory system: The phase shift due to a stimulus of arbitrary waveform with sufficiently small amplitude can be obtained by computing the integral of the stimulus weighted by the iPRC. Thus, a PRC of a neuron for any particular current stimulus can be estimated from the "convolution"[4] of the stimulus waveform and the neuron's iPRC

$$\mathrm{PRC}(\phi) \cong \int_0^\infty Z(t + \phi T) I_{\mathrm{stim}}(t)\mathrm{d}t, \tag{5.1}$$

---

[2]In general, the iPRC is equivalent to the gradient of phase with respect to all state variables evaluated at all points along the limit cycle (i.e. it is a vector measuring the sensitivity to perturbations in any variable). However, because neurons are typically only perturbed by currents, the iPRC for neurons is usually taken to be the voltage component of this gradient $\left(\frac{\partial \phi}{\partial V}\right)$ evaluated along the limit cycle.

[3]A delta-function is a pulse with infinite height and zero width with an area of one. Injecting a delta-function current into a cell corresponds to instantaneously injecting a fixed charge into the cell, which results in an instantaneous jump in the cell's membrane potential by a fixed amount.

[4]The definition of a convolution is $g * f(\psi) = \int g(\psi - t) f(t)\mathrm{d}t = \int g(-(t - \psi)) f(t)\mathrm{d}t$, so technically, $PRC(\phi) = Z * I(-\phi T)$.

where $PRC(\phi)$ is the phase shift in response of a neuron with an iPRC $Z(t)$ and a current stimulus of waveform $I_{\mathrm{stim}(t)}$, and ϕ is the phase of the neuron at the onset of the stimulus. Note that (5.1) assumes that the relative phase of the neuron ϕ is a constant over the entire integral. However, because only small stimuli are considered, phase shifts will be small, and thus this assumption is reasonable.

2.2.2 Calculating PRCs from iPRCs

Assuming that the functional forms of the stimulus (as chosen by the experimenter) and the iPRC (as fit to data) are known, an estimate of the PRC can be calculated using (5.1). From a practical standpoint, the interval of integration must be truncated so that the upper limit of the interval is $t_{\max} < \infty$. By discretizing ϕ and t so that $t_j = j\Delta t$ and $\phi_j = j\ \Delta t/T$ with $j = 1 \ldots N$, $\Delta t = t_{\max}/N$, equation becomes

$$\mathrm{PRC}(\phi_j) \cong \sum_{k=0}^{N-1} Z(t_k + \phi_j T) I_{\mathrm{stim}}(t_k)\Delta t. \tag{5.2}$$

(Note that a simple left Reimann sum is used to approximate the integral, but high order numerical integration could be used for greater accuracy). Equation (5.2) can be used to directly compute an approximation of the PRC in the time domain. In this direct calculation, $t_{\max}$ should be chosen sufficiently large to ensure that the effect of the stimulus is almost entirely accounted for. In the case of small pulsatile stimuli, one period of the neuron's oscillation is usually sufficient (i.e., $t_{\max} = T$).

The PRC could also be calculated by solving (5.2) using discrete Fourier transforms (DFTs)

$$\mathrm{PRC}(\phi_j) \cong \frac{1}{N} \sum_{n=0}^{N-1} \hat{Z}_n \hat{I}_{(N-1)-n} \mathrm{e}^{-i2\pi n \phi_j T/t_{\max}} \Delta t, \tag{5.3}$$

where $\hat{Z}_n$ and $\hat{I}_n$ are coefficients of the nth modes of the DFTs of the discretized Z and I_{stim}, as defined by

$$x(t_j) = \sum_{n=0}^{N-1} \hat{x}_n \mathrm{e}^{\frac{i2\pi n t_j}{t_{\max}}}, \quad \hat{x}_n = \frac{1}{N} \sum_{j=0}^{N-1} x(t_j) \mathrm{e}^{-i2\pi n t_j/t_{\max}}. \tag{5.4}$$

Note that, because the DFT assumes that functions are t_{max}-periodic, a $PRC(\phi_j)$ calculated with this method will actually correspond to phase shifts resulting from applying the stimulus I_{stim} $t_{\max}$-periodically. To minimize this confounding effect, $t_{\max}$ should be two or three times the intrinsic period of the neuron for small pulsatile stimuli (i.e., $t_{\max} = 2T$ or $3T$).

2.2.3 Calculating iPRC from the PRC Measured with Current-based Stimuli

If a PRC was measured for a neuron using a stimulus that had a sufficiently small magnitude, then the iPRC of the neuron can be estimated by "deconvolving" the functional form of the PRC with the stimulus waveform, i.e., solving (5.2) for $Z(t_j)$. Deconvolution can be done in the time domain or the frequency-domain. Equation (5.3) shows that the nth mode of the DFTs of the discretized PRC is $\widehat{PRC}_n = \hat{Z}_n \hat{I}_{(N-1)-n} \Delta t$. Therefore, the iPRC can be computed by

$$Z(t_j) \cong \sum_{n=0}^{N-1} \left(\frac{\widehat{PRC}_n}{\hat{I}_{(N-1)-n} \Delta t} \right) e^{i 2\pi n \phi_f T / t_{\max}}. \tag{5.5}$$

We can also directly solve (5.2) for the iPRC, $Z(t_j)$, in the time domain by noting that

$$\begin{aligned} \mathrm{PRC}(\phi_j) &\cong \sum_{k=1}^{N} Z(t_k + \phi_j T) I_{\mathrm{stim}}(t_k) \Delta t, \\ &= \sum_{k=1}^{N} I_{\mathrm{stim}}(t_k - \phi_j T) Z(t_k) \Delta t, \end{aligned} \tag{5.6}$$

which can be written in matrix form as

$$\overline{\mathrm{PRC}} \cong \bar{\bar{I}}_{\mathrm{stim}} \bar{Z}, \tag{5.7}$$

where $\overline{\mathrm{PRC}}$ and $\bar{Z}$ are the vectors representing the discretized PRC and iPRC, respectively, and $\bar{\bar{I}}_{\mathrm{stim}}$ is an $N \times N$ matrix with the j,kth element $I_{\mathrm{stim}}\,(t_k - \phi_j T)$. Therefore, we can find the iPRC $\bar{Z}$ by solving this linear system. Note that this problem will be well posed because all rows of $\bar{\bar{I}}_{\mathrm{stim}}$ are shifts of the other rows.

A related method for measuring the iPRC using a white noise current stimulus will be discussed in Sect. 6.

2.2.4 iPRCs, PRCs, and Conductance-based Stimuli

As inferred from Sect. 2.1.2, the synaptic waveform for conductance-based stimuli is not phase invariant. However, the ideas in previous sections can readily be extended to incorporate conductance-based stimuli. The PRC measured with a synaptic conductance is related to the iPRC by

$$\mathrm{PRC}(\phi) \cong \int_0^{\infty} Z(t + \phi \mathrm{T}) g_{\mathrm{syn}}(t)(E_{\mathrm{syn}} - V(t + \phi T))\mathrm{d}t. \tag{5.8}$$

Assuming that the functional forms of the stimulus, the iPRC and the membrane potential are known, an estimate of the PRC for conductance-based stimuli can be calculated in a similar manner to that described for current-based stimuli in Sect. 2.2.2. That is, the PRC can be computed in the time domain or frequency domain, using

$$\text{PRC}(\phi_j) \cong \sum_{k=0}^{N-1} Z(t_k)[g_{\text{syn}}(t_k - \phi_j T)(E_{\text{syn}} - V(t_k))]\Delta t, \tag{5.9}$$

$$= -\sum_{n=0}^{N-1} \hat{Z}_n \hat{v}_n \hat{g}_{(N-1)-n} \mathrm{e}^{-i2\pi n\phi_f T/t_{\max}} \Delta t, \tag{5.10}$$

where $\hat{v}_n$ and $\hat{g}_n$ are the nth modes of the DFTs of the discretized functions $(V(t) - E_{\text{syn}})$ and $g_{\text{syn}}(t)$. Furthermore, the iPRC can be calculated from the PRC in the frequency domain by noting that the nth mode of the DFTs of the discretized PRC is $\widehat{PRC}_n = -\hat{z}_n \hat{v}_n \hat{g}_{(N-1)-n} \Delta t$ for conductance-based stimuli (see (5.10)), therefore

$$Z(t_j) = -\sum_{n=0}^{N-1} \left(\frac{\widehat{PRC}_{-n}}{\hat{v}_n \hat{g}_{(N-1)-n} \Delta t} \right) \mathrm{e}^{i2\pi n\phi_j T/t_{\max}}. \tag{5.11}$$

When computing the iPRC in the time domain, we can first "deconvolve" (5.8) to find the product $(E_{\text{syn}} - V(t))_{\text{K}} Z(t)$, and then divide out the driving force to find the $Z(t)$. Note that numerical error could be large when $(E_{\text{syn}} - V(t_k))$ is small, therefore care should be taken at these points (e.g., these points could be discarded).

Estimates of the iPRCs for a real neuron that are calculated from PRCs measured with excitatory synaptic conductance in one case and inhibitory synaptic conductance in another are shown in Fig. 5.7 (Netoff, Acker, Bettencourt, & White 2005). While the measured PRCs look dramatically different, the iPRCs are quite similar, indicating that the main difference in the response can be attributed to changes in the synaptic reversal potential. The remaining differences between the estimated iPRCs are likely due to small changes in the state of the neuron, error introduced by fitting the PRCs, and/or the fact that the response of the neuron to the stimuli is not perfectly linear.

3 Dealing with Slow timescale Dynamics in Neurons

Processes that act on relatively slow time scales can endow a neuron with the "memory" of stimuli beyond a single cycle. In fact, a stimulus applied to one cycle is never truly isolated from other inputs. In this section we will address how neuronal memory can affects the phase response properties of a neuron. Specifically, we will discuss how stimuli can affect the cycles following the cycle in which the

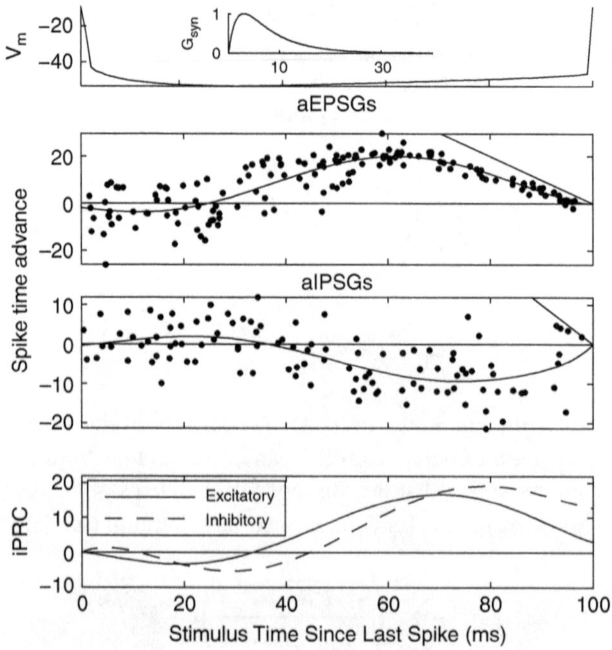

Fig. 5.7 Estimates of the infinitesimal PRC (iPRC) for a pyramidal neuron from CA1 region of the hippocampus as calculated using PRCs. A synaptic conductance stimulus was used to generate PRCs of the neuron, and then the shape of the synaptic waveform was deconvolved from the PRC to estimate the iPRC. (*Top panel*) Voltage trace of neuron over one period. *Inset* is the synaptic conductance waveform. (*Middle two panels*) PRCs measured with excitatory conductances (*upper*) and inhibitory conductances (*lower*). (*Bottom panel*) iPRCs estimated using the excitatory and the inhibitory PRCs. The iPRCs from the two data sets, despite being measured with completely different waveforms, are similar. Figure modified from Netoff et al., 2005a

neuron was stimulated and how to quantify these effects (Sect. 3.1). We also address how the effect of repeated inputs can accumulate over many periods, resulting in accommodation of the firing rate and alteration of the PRC (Sect. 3.2).

3.1 Higher-order PRCs

A stimulus may not only affect the interspike intervals (ISIs) in which it is applied but may also affect the ISIs of the following cycles, although usually to a lesser degree. This can happen in two ways. The first is when the stimulus starts in one cycle but continues into the next cycle. The second is through neuronal memory. For example, a phase shift of a spike during one cycle may result in compensatory changes in the following cycle, or the stimuli may significantly perturb a slow process such as an adaption conductance. Often a large spike advance is followed by a small delay in the next period (Netoff et al. 2005; Oprisan & Canavier 2001).

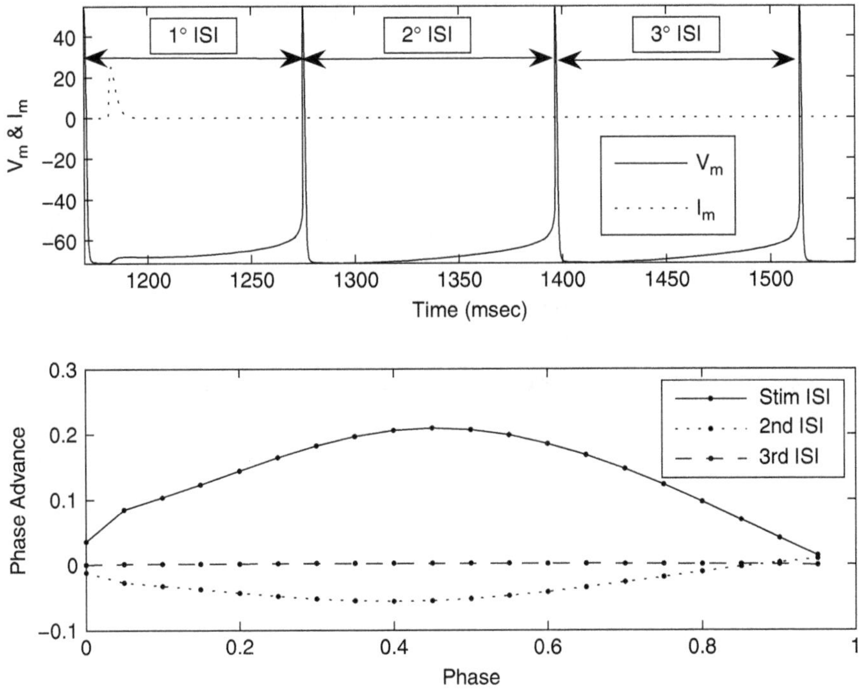

Fig. 5.8 First, second, and third order PRCs. The first order PRC is measured as the change in period of the cycle that the stimulus was applied, while second and third order PRCs are measured from additional phase shifts of spikes in the subsequent cycles. Often the second and third order PRCs are small compared to the first order PRC and are of alternating sign. Simulations were performed using the Golomb–Amitai model (1997)

As mentioned earlier, the PRC represents the phase shifts of the *first* spike following the onset of the stimulus, so the PRC measured this way can be considered the "first order PRC". The additional phase shifts of the second spike (or nth spike) following the onset of the stimulus versus the phase of the stimulus onset is called the "second order PRC" (or nth order PRC). Examples of first, second, and third order PRCs are shown in Fig. 5.8. The higher order PRCs are usually small as compared to the first order PRC, but can have significant implications in predicting network behavior when accounted for (Oprisan & Canavier 2001).

3.2 *Functional PRCs*

Many neurons exhibit significant accommodation when a repeated stimulus is applied. Thus, the shape of the PRC can depend on whether the perturbed cycle is measured before or after the accommodation. Usually, the PRC is measured by applying a single stimulus every few periods, in order to let the neuron recover

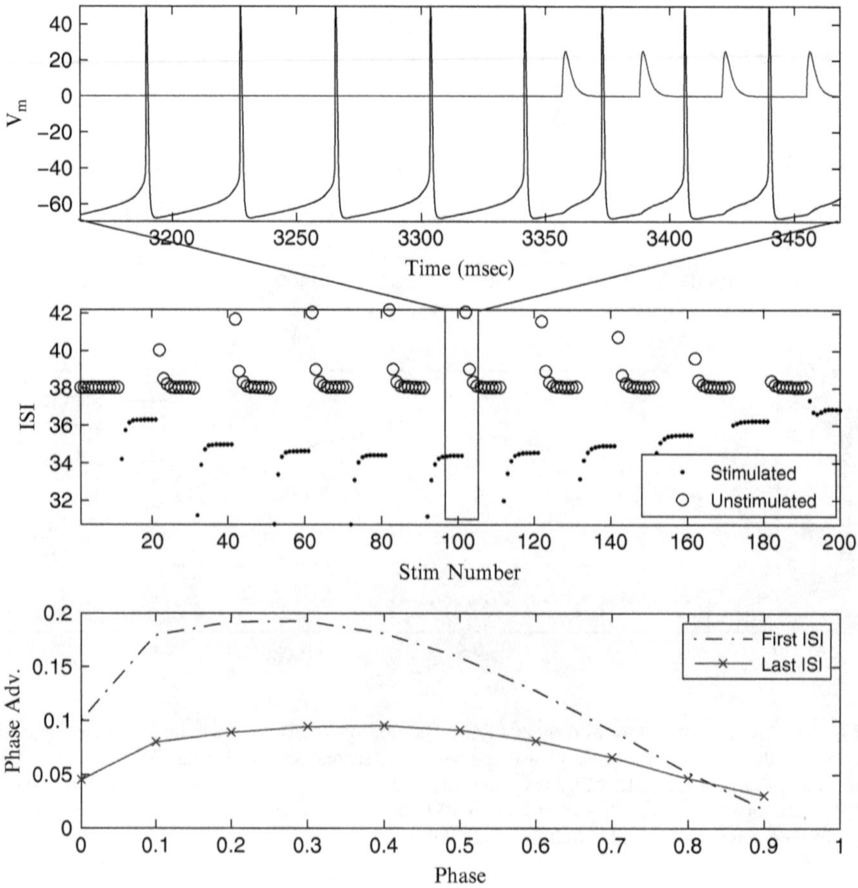

Fig. 5.9 Functional PRC takes accommodation into consideration. The neuron is stimulated at the same phase for many cycles, and the PRC is determined from the average interspike intervals averaged over the last cycles. (*Top trace*) Voltage (in mV) and current for a stimulus applied repeatedly at a fixed phase. Time series taken from one set of stimuli shown in middle panel. (*Middle*) Interspike intervals (ISIs): *circles* represent unstimulated cycles; *dots* are stimulated periods. The phase of the stimulus is systematically varied from the earliest to latest across the stimulus trains. Simulations were performed using the Golomb–Amitai model (1997). (*Bottom*) PRCs without accommodation (calculated from first ISI) and with accommodation (calculated from last ISI)

from the stimulus and return to baseline firing rate. If the stimulus is repeated at each cycle and the same time lag, the neuron may accommodate to the synaptic input by changing the ISI over the first few cycles. One approach to deal with the accommodation is to measure the phase advance after the neuron has accommodated to the input and reached a steady-state response. The phase response curve from the accommodated neuron is termed the functional phase response curve (fPRC) (Cui, Canavier, & Butera 2009). The method is illustrated in Fig. 5.9. The PRC taken from the first stimulus interval looks different from the last train. Under conditions where a neuron may accommodate significantly during network dynamics, the predictions of network phase locking using the fPRC may produce more accurate results than predictions using standard PRCs.

4 Issues in PRC Data Acquisition

On the timescale of a full PRC experiment, the neuron's firing rate can drift significantly. This drift can confound the small phase shifts resulting from the stimuli. "Closed-loop" experimental techniques can be used to counteract this drift and maintain a stable firing rate over the duration of the experiment. In this Sect. 4.1, we introduce the dynamic clamp technique, which enables closed loop experiments (Sect. 4.1), and we describe a method for using the dynamic clamp to control the spike rate in order to reduce firing rate drift over the duration of the experiment (Sect. 4.2). We also show how the dynamic clamp can also be used to choose the phases of stimulation in a quasi-random manner, which can minimize sampling bias (Sect. 4.3).

4.1 Open-loop and Closed-loop Estimation of the PRC

Historically, patch clamp experiments have been done in open loop, where a predetermined stimulus is applied to the neuron and then the neuron's response is measured. With the advent of fast analog-to-digital sampling cards in desktop computers, it has been possible to design experiments that require real-time interactions between the stimulus and the neuron's dynamics in a closed-loop fashion, called a dynamic-clamp (Sharp, O'Neil, Abbott, & Marder 1993).

There are many different real-time systems available for dynamic clamp experiments (Prinz, Abbott, & Marder 2004). We use the Real-Time eXperimental Interface (RTXI) system (Dorval, 2nd, Bettencourt, Netoff, & White, 2007; Dorval, Christini, & White, 2001; Dorval, Bettencourt, Netoff, & White, 2008), which is an open-source dynamic clamp based on real-time Linux. It is freely available to the public for download at http://www.rtxi.org. Modules for controlling the firing rate of the neuron, simulating synapses and measuring the PRC can be downloaded with the RTXI system. The RTXI system is modular, allowing one to write small modules that perform specific tasks and then connect them together to run full sets of experiments. Figure 5.10 illustrates the modules used to generate PRCs experimentally. We note that the modular design makes it relatively easy to replace a synaptic conductance module with a module to trigger a picospritzer to inject neurotransmitters proximal to the dendrite to simulate synapses.

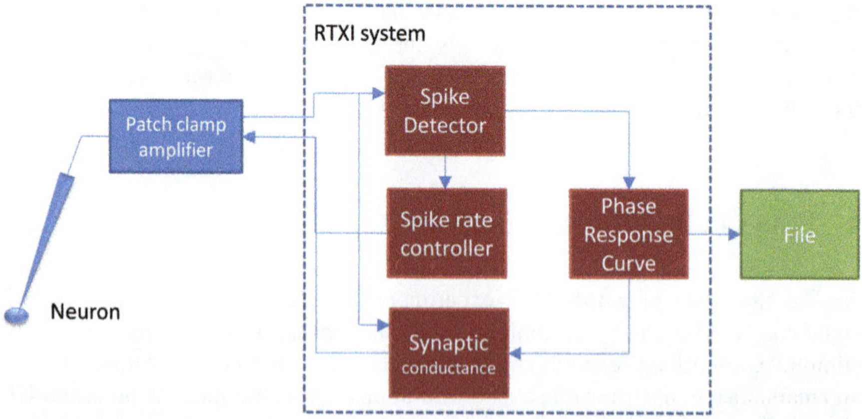

Fig. 5.10 Schematic of RTXI system for measuring phase response curves. Neuron is whole cell patch clamped with glass electrode connected to patch clamp amplifier. Spike detect module identifies occurrence of action potentials. Spike rate controller monitors the inter-spike intervals and adjusts applied current to maintain neuron at user specified firing rate using an ePI closed-loop feedback. PRC module determines the times to stimulate the neuron given the history and sends trigger pulses to the synaptic conductance module. PRC module also records data to file for analysis. Synaptic conductance module simulates synaptic conductances by injecting current waveform that is dependent on time-varying synaptic conductance and voltage of neuron

4.2 Controlling Spike Rate

The PRC measures the neuron's deviation from the natural period due to a stimulus. If the firing rate of the neuron drifts over the time that PRC data is collected, then the measured spike advance caused by the synaptic perturbation will be confounded with the drift in the spike rate. An advantage of closed-loop measurements is that the baseline applied current can be adjusted slightly from cycle to cycle to maintain the firing rate close to a desired frequency.

To maintain the firing rate of the neuron, we developed an event based proportional-integral (ePI) controller (Miranda-Dominguez, Gonia, & Netoff 2010). The baseline current is adjusted immediately after the detection of the spike and only small changes in current are allowed from cycle to cycle. The current at each spike is calculated as $I(n) = K_\mathrm{p}e(n) + K_\mathrm{i}\sum_{j=0}^{n} e(j)$, where $e(n)$ is the difference between the measured inter-spike interval (ISI) on cycle n and the target period, K_p is the proportional constant, and K_i is an integration constant. It is possible to determine optimal values for these coefficients based on the gain of the neuron's response to change in current and the time to settle, which is out of the scope of this chapter and will be published elsewhere. We have found that for most neurons using $K_\mathrm{p} = 6\times10^{-12}$ and $K_\mathrm{i} = 6\times10^{-10}$ works well. Figure 5.11 demonstrates the effects of the spike rate controller. The ISIs are plotted for an open-loop experiment in which a constant current is injected into a neuron and for a closed-loop experiment

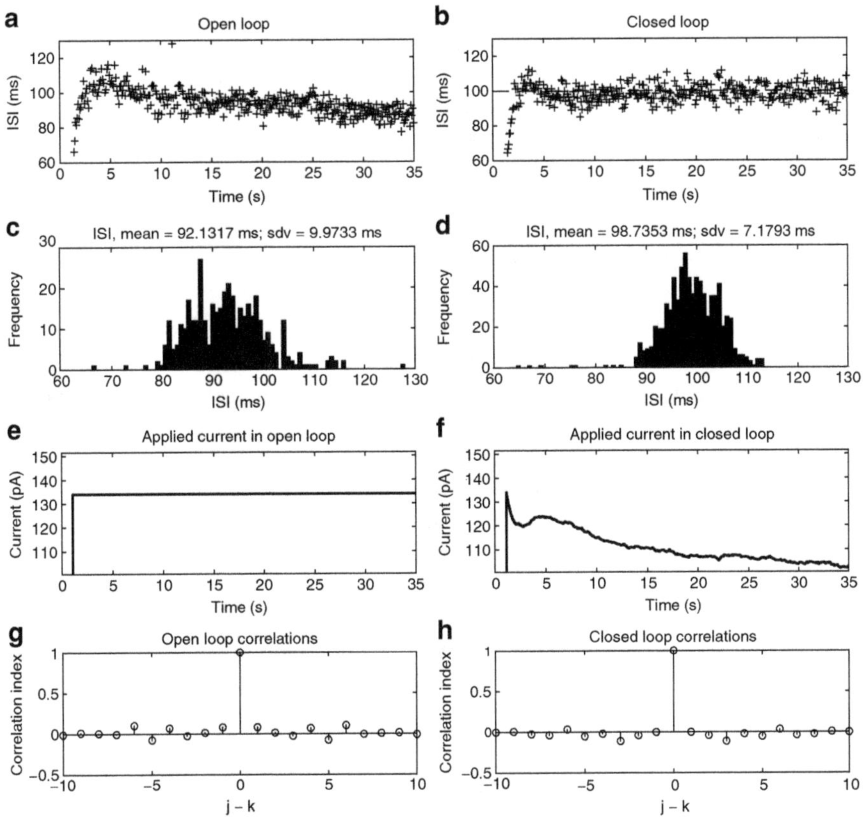

Fig. 5.11 Spike rate control using ePI controller. (**a**) In the open-loop configuration, inter-spike interval experiences significant drift over the ~30 second time interval in which the baseline applied current is applied, and the average inter-spike interval near the end of the trace (30 to 35 s) is over 10 ms away the target interval of 100 ms. Autocorrelations of first lag is nearly zero (see bottom row). This indicates that the error from one cycle is almost completely independent of the previous cycle. (**b**) With closed-loop control, the inter-spike interval converges quickly to the target rate of 100 ms. (**c** and **d**) The mean inter-spike interval, after the initial transient, is statistically indistinguishable from the target rate throughout the time interval in which the baseline applied current is applied. (**e** and **f**) During this time, the current injected into the cell is varying to maintain the neuron close to the target spike rate. Standard deviation of the error in open-loop and closed-loop are similar, indicating that the closed-loop is only reducing the drift in the inter-spike interval rate and not the variability from spike to spike. (**g** and **h**) The autocorrelation at the first lag is nearly zero for both the open and closed loop controller. If feedback gain (from the proportional feedback coefficient) is too high, the first lag of the autocorrelation will be negative, indicating a ringing of the controller

in which a current is adjusted to maintain the neuron at a desired firing rate. The mean ISI in the open-loop experiments undergoes a drift of $\sim$10–20 ms, whereas the mean ISI in the closed-loop experiments stays very close target period of 100 ms. The autocorrelation is also shown to show that the method does not introduce any significant correlation, which occurs if the feedback loop begins to oscillate.

4.3 Phase-Sampling Methods

When generating a PRC for a deterministic computational model of a neuron, it is easy to systematically sample the response to stimuli at various phases by simply stepping through phases of stimulation, while measuring the phase shift in response to each stimulus, and restarting the model neuron at a particular initial condition on the limit cycle after each measurement. In generating PRCs for real neurons, stimuli are delivered sequentially to a oscillating neuron. Experimentally, it is best to leave several unstimulated interspike intervals after each stimulus to minimize any interactions between the effects of stimuli. This can be achieved by periodically stimulating the neuron at intervals several times longer than the neuron's natural period. Assuming that the neuron has some variability in its period (i.e., jitter) or by choosing the ratio between the period of the neuron and the period of stimulation corresponds to an irrational number, this method should sample the period close to uniformly and in an unbiased fashion. The advantage of this somewhat haphazard sampling method is that it can be done open loop. The disadvantage is that, in practice, it may result in oversampling of some phases and undersampling of others.

With a closed-loop experimental system, the phase at which the stimuli are applied can be selected directly (i.e., by triggering the stimuli off of spike times). By randomly selecting the phases of stimulation, you can ensure that there are no biases introduced by the experimental protocol. However, even with randomly selecting phases it does not sample the phases most efficiently. Efficiency is paramount in experiments, because you are racing the slow death of the cell, and thus optimum sampling can improve your PRC estimates. Quasi-random selection of phases, using a "low-discrepancy" sequence such as a Sobol sequence, can cover the phases in $\frac{1}{\sqrt{N}}$ the time as it would take a random sequence, where N is the number of data points (Press 1992). The Sobol sequence as a function of stimulus cycle is illustrated at the bottom of Fig. 5.12.

5 Fitting Functions to PRCs

The interspike intervals measured after applying stimuli can be highly variable for real neurons, even if the stimuli are applied at the same phase. Because of this variability, PRC experiments yield scatter plots of phase shift *vs* the phase

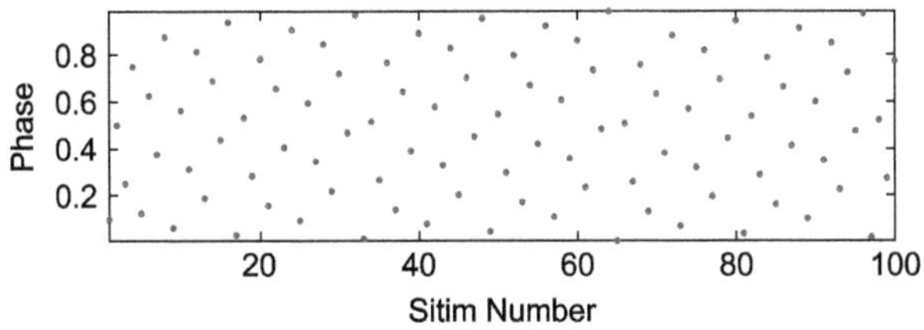

Fig. 5.12 Sobol sequence to sample the phase of stimulation. With closed-loop experiments stimuli can be applied at selected phases. The Sobol sequence is a quasi-random sequence, which efficiently samples phase and minimizes bias. The plot represents the selected stimulus phase plotted against the stimulus number. The intervals are not random, but not periodic

of stimulation. By appropriately fitting noisy PRC data, a functional relationship can be obtained to characterize the mean response of the neuron. This functional form of the PRC can then be used in conjunction with coupled oscillator theory to predict of the network behaviors. In this section, we discuss fitting polynomial functions (Sect. 5.1) and Fourier series to PRC data (Sect. 5.2) and address the issue of determining optimal number of fit coefficients in terms of the Aikake Information Criterion (Sect. 5.3). We also discuss statistical models of the variance in PRC data (Sect. 5.4).

5.1 *Polynomials*

Simple functions that are sufficiently flexible to accommodate the shapes of PRCs are polynomials (Netoff et al. 2005; Tateno and Robinson 2007). Fitting polynomials to PRC data is easy to implement: Matlab and many other data analysis programs have built-in functions that provide the coefficients of a kth degree polynomial to fit data in the least squares sense. A kth order polynomial fit to PRC data has the form

$$\mathrm{PRC}(\phi) = C_k\phi^k + C_{k-1}\phi^{k-1} + \cdots + C_2\phi^2 + C_1\phi + C_0,$$

where $PRC(\phi)$ is the change in phase as a function of the phase of the stimulus ϕ, C_x's are the coefficients that are determined by the fit to the data.

Often, spiking neurons are insensitive to perturbations during and immediately following spikes. This property is manifested in PRCs with noisy but flat portions at the early phases, which can sometimes cause spurious oscillations in polynomial fits. These oscillations in the fit can be reduced or eliminated by constraining the PRC to be zero at $\phi = 0$ by using the following constrained polynomial:

$$\mathrm{PRC}(\phi) = (C_k\phi^k + \cdots + C_2\phi^2 + C_1\phi + C_0)\phi$$

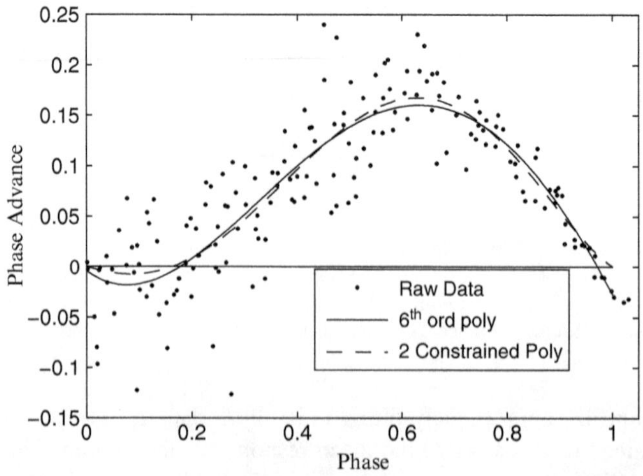

Fig. 5.13 Free and constrained polynomial fits to PRC data for excitatory input to a neuron. Phase advance as a function of stimulus phase is measured for a pyramidal neuron in hippocampus. The neuron was firing at 10 Hz (100 ms intervals). The *solid line* is an unconstrained 6th order polynomial fit (using 7 coefficients) to the points. Notice that the line does not meet the (0,0) point or the (1,0). The *dashed line* is a two-ended constrained polynomial fit (4 coefficients and 2 constraints) that forces the curve to start at (0,0) and end at (1,0)

Moreover, because excitatory inputs can only advance the phase of the next spike to the point that the neuron actually spikes, excitatory synaptic inputs to spiking neurons generally elicit a PRC with no phase shifts at $\phi = 1$. Thus, it is useful to constrain the fit of the PRC to be zero at both $\phi = 0$ and $\phi = 1$,

$$\text{PRC}(\phi) = (C_k\phi^k + \cdots + C_2\phi^2 + C_1\phi + C_0)\phi(1-\phi).$$

To obtain a constrained polynomial for the general period-1 polynomial case, a constant term $C_.$ must added to the above polynomial (Tateno & Robinson 2007).

Examples of a two-end constrained fit and a no-constraint fit to raw PRC data generated with excitatory stimuli are illustrated in Fig. 5.13. Figure 5.14 shows examples of a one-end constrained fit ($PRC(\phi) = 0$), a two-end constrained fit and a no-constraint fit for PRC data generated with inhibitory inputs. In the case of inhibitory input, there are almost zero phase shifts at early phases, but input causes considerable phase shifts at late phases.

5.2 *Fourier Series*

Due to the periodic nature of many PRCs, PRC data is often fit using Fourier series (e.g. (Galan, Ermentrout, & Urban 2005; Mancilla, Lewis, Pinto, Rinzel, & Connors 2007; Ota, Nomura, & Aoyagi 2009)). A kth order Fourier series fit to PRC data can be written as

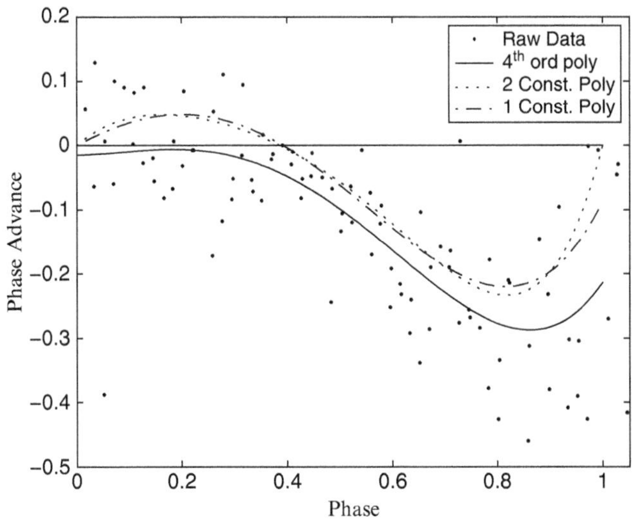

Fig. 5.14 Fits to PRC data generated with inhibitory input. PRCs generated with inhibitory inputs have different shapes than those generated with excitatory curves. This is predominantly because phase shifts are not limited by causality. The largest delays usually occur immediately prior to the neuron spiking. A 4th order polynomial (5 coefficients) fit is plotted with a solid line. A 6th order polynomial fit (4 coefficients and 2 constraints) with the beginning constrained to (0,0) and the end to (0,1) is plotted with a dotted line. This function does not fit the right hand side of the data well. A 5th order polynomial fit (4 coefficients and 1 constraint) constrained only at the beginning to (0,0) is plotted with a dot-dashed line. This curve provides the best fit to the data

$$\mathrm{PRC}(\phi) = a_0 + \sum_{j=1}^{k} \{a_j \cos(2\pi j\phi) + b_j \sin(2\pi j\phi)\},$$

where the Fourier coefficients are given by

$$a_0 = \frac{1}{N}\sum_{n=1}^{N} \Delta\phi_n, \quad a_j = \frac{2}{N}\sum_{n=1}^{N} \Delta\phi_n \cos(2\pi j\phi_n), \quad b_j = \frac{2}{N}\sum_{n=1}^{N} \Delta\phi_n \sin(2\pi j\phi_n),$$

where $\Delta\phi_n$ is the phase advance measured on stimulus number n that was delivered at phase ϕ_n, and N is the number of data samples. Because many PRCs are zero for $\phi = 0$ and $\phi = 1$, a better fit for fewer parameters can sometimes be obtained by using the Fourier sine series

$$\mathrm{PRC}(\phi) = \sum_{j=1}^{k} b_j \sin(\pi j\phi), \quad b_j = \frac{2}{N}\sum_{n=1}^{N} \Delta\phi_n \sin(\pi j\phi_n).$$

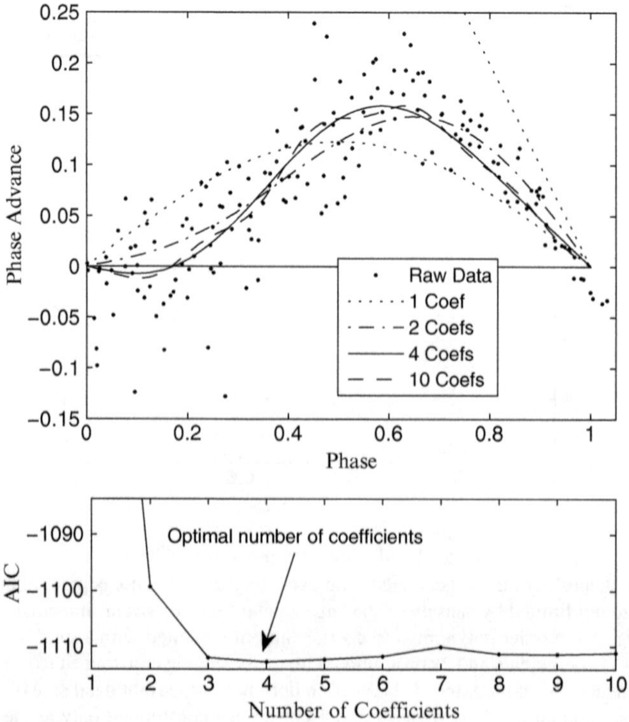

Fig. 5.15 Fourier sine series fit to PRC data for a hippocampal, CA1 pyramidal neuron. (*Top panel*) The same raw data as used in Fig. 5.13, but data is fit using a Fourier sine series. Curves for fits using different numbers of modes (coefficients) are indicated in the legend. The *dotted* (1 Coef), *dot-dashed* (2 Coefs) and *solid* (4 Coefs) show that the fit improves with more coefficients. However, while the *dashed line* (10 Coefs) technically has lower residual error, the curve exhibits spurious oscillate, indicating it is overfitting to the data. (*Bottom panel*) the Akaike Information Criterion (AIC) is used to determine the optimal number of coefficients. The minimum at 4 coefficients indicates that no more than 4 coefficients should be used to fit the PRC

Figure 5.15 illustrates PRC data that is fit using the Fourier sine series with $k = 1, 2, 3$, and 10. It can be seen that the PRC data set is fit well with only the first few modes. Seemingly spurious oscillations appear when the first 10 modes are used to fit the PRC data, suggesting the data are over fit.

One advantage that Fourier series has over polynomials is that one can get a reasonably good idea of the shape of the PRC by considering the values of the coefficients. Furthermore, the H-function, which is defined as $H(\Delta_i) = \Delta_{i+1} - \Delta_i \approx \text{PRC}_1(\Delta_1) + \text{PRC}_2(1 - \Delta_i)$, where $PRC_1([\Delta]_i)$ represents the phase advance of cell 1 given the synaptic input from cell 2 and $PRC_2(1 - \Delta_i)$ is phase advance of cell 2 given the approximate phase of cell 1's input (assuming the phase advance of cell 1 from cell 2's input is nearly zero). This is the difference between the two neuron's spikes on can simply be estimated by summing only the odd Fourier Coefficients (Galán, Ermentrout, & Urban 2006).

5.3 *Over- and Underfitting PRC Data: Akaike Information Criterion (AIC)*

Because Fourier modes are orthogonal to one another, each Fourier coefficient can be determined sequentially[5] and the fitting process can be stopped when the quality of the fit is satisfactory. As indicated above, when too few modes are included, the data will not be well fit, and as more modes are included, the residual error of the fit will decrease. However, while including additional modes can decrease the residual error, the decreased error may not be justified by the additional fitting parameters. To determine how many modes (i.e., number of fitting parameters) should be included, one can use the Akaike information criterion (AIC) (Burnham & Anderson 1998). The Akaike information criterion[6] is calculated using the function

$$\mathrm{AIC}(k) = 2k + n\left(\ln\left(\frac{1}{n}\sum_{j=1}^{n} \in_j (k)^2\right)\right),$$

where k is the number of fitting parameters (e.g., Fourier coefficients), n is the number of points in the data set, and $\in_j (k)$ is the residual error for the jth data point to the fitted PRC using the k fitting parameters. The optimal number of parameters is determined when $AIC(k)$ is at its minimum. In Fig. 5.15 (bottom), AIC is plotted as a function of the number of parameters used to fit neuronal PRC data. It can be seen that the minimum occurs at $k = 4$, thus using more than 4 Fourier modes is overfitting the PRC data.

The AIC can be used in a similar manner to select the optimal number of coefficients for polynomial fits. In fact, the AIC can be used to determine which model (i.e., a Fourier series, a constrained polynomial, etc.) yields the optimal fit. We note, however, that the AIC does not determine whether the fits are statistically significant.

There are alternative approaches to check for validity of fits to PRC data. Galan et al. (2005) fit raw PRC data with a Fourier series and then tested their fit by comparing it to smoothed data and to fits when the data had been shuffled along the phase axis (abscissa) of the PRC. There are also techniques that employ Bayesian methods to produce a maximum a posteriori (MAP) estimation of the iPRC ((Ota, Omori, & Aonishi 2009) see also Chap. 8).

---

[5]Note that this could be done for polynomial fits too by using orthogonal polynomials (e.g., Legrendre or Chebychev polynomials).

[6]This formula for the *AIC* assumes that errors are independently distributed and described by a Gaussian distribution.

5.4 Fitting Noise Around the PRC

PRC data for real neurons can be quite noisy. Models that use PRCs to predict phase-locking dynamics usually do not account for the variable phase-response properties of neurons. Accounting for the variance of PRC data into these models could provide insight into inherently stochastic behaviors such as random leader swapping and jitter around "stable" phase-locked states. Therefore, it could be very useful to obtain a good description of the variability of PRC data.

The variance in PRC data could be generated from several sources. One source of variability could be due to external synaptic noise, which will influence the neurons' spike times along with the simulated input applied through the electrode. We find that blocking synaptic inputs in slice experiments did not dramatically reduce the variability (Netoff, unpublished), indicating that synaptic noise may not be a major source of variability of *in vitro* PRC data. Another source of variability could be the stochastic fluctuation of the ion channels in the neurons themselves. It has not yet been identified how much of the variability can be attributed to this source. Identifying the source of the noise may be important in determining how the variability is related to the shape of the PRC, i.e., the variability in spike time may be phase dependent.

The variance around the PRC can be strongly phase dependent, as can be seen in Figs. 5.13 and 5.16. For moderate to large-sized inputs, the variability in response to excitatory inputs earlier in the cycle is usually greater than inputs arriving at the end of the cycle. There are two causes for the decreased variability at late phases. One is that, as the neuron approaches threshold toward the end of the cycle, synaptic inputs are more likely to directly elicit an action potential. A directly elicited action potential has significantly less variability than a spike whose time has been modulated by synaptic inputs early in the cycle. Inhibitory synaptic inputs generally do not elicit an action potential, and therefore generate PRCs with more uniform variability across phase, as shown in Fig. 5.14.

The simplest way to estimate the noise is to bin the data and estimate the standard deviation in each phase bin. The drawback to this method is that dividing the data into finer temporal bins results in fewer points in each bin and a less accurate estimate of the standard deviation. This also leads to a piecewise model of the variance.

Another approach is to fit a continuous function relating the variance to the phase. A simple function that can be fit to the standard deviation around the PRC data for excitatory stimuli is $\hat{\sigma}(\phi) = n_1 + n_2\sqrt{1-\phi}$. At the end of the cycle when $\phi = 1$, the second term is zero and the standard deviation is equal to n_1. As the phase of the input decreases, the variance increases as a square root of the phase. The motivation for this function is ad hoc, but is based on the premise that the noise is summed from the time of the synaptic input to the end of the period. Therefore, the variance increases linearly in time (and the standard deviation as a square root) as the synaptic input is applied earlier in the phase.

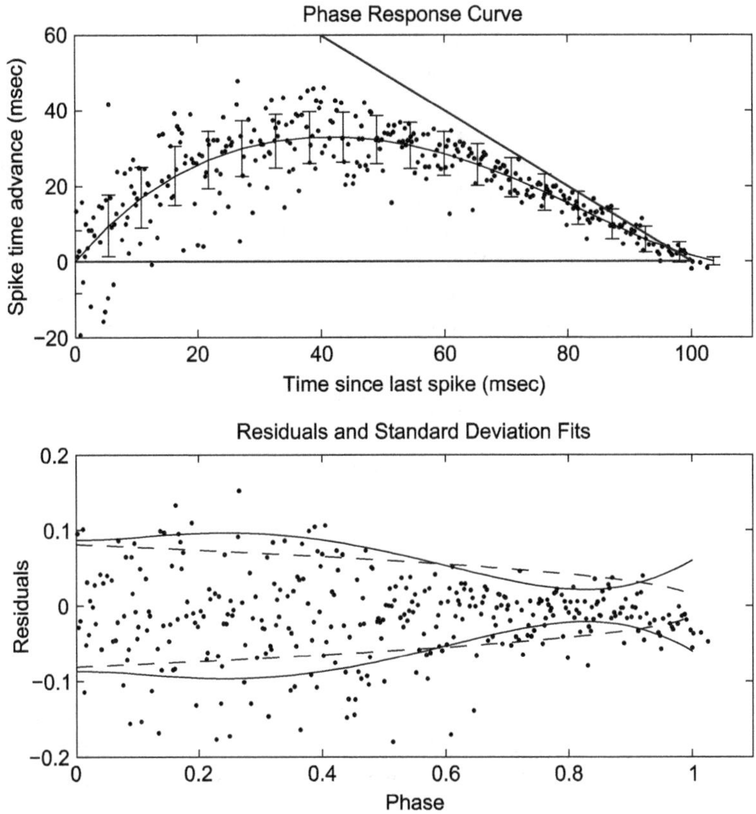

Fig. 5.16 Fitting a function to the phase-dependent noise. (*Top panel*) Raw PRC data fit with a function to estimate the mean PRC. The standard deviations are shown with error bars at each phase of the PRC. The slanted blue line represents the line of causality, the maximum phase advance that can occur (i.e., neuron spikes at time of stimulus). (*Bottom panel*) The estimated PRC is subtracted from the raw data leaving the residuals of the PRC. The *dashed line* represents the standard deviation of the PRC at each phase fit with a simple function using maximum likelihood. The *solid line* represents a fit function that makes use of the PRCs shape in predicting standard deviation of the noise

Fitting a function to the noise is not as easy as fitting a function to the mean. Rather than optimizing the least squares error from the fit function, we must find the maximum likelihood function instead. First, we start with removing our best estimate of the PRC from the raw data $r(i) = \text{DATA\_PRC}(\phi(i)) - \text{PRC\_FIT}(\phi(i))$ to get the residuals. The residuals of the PRC are plotted in lower panel of Fig. 5.16. Next, we need to estimate the probability of seeing the actual measured interspike intervals given an estimate of the variance at each phase. We can choose the initial conditions for the function $\hat{\sigma}(\phi)$ as: $n_1 = \sqrt{\frac{1}{n}\sum_i^n r(i)^2}$ and slope $N_2 = 0$. Assuming that the residuals are Gaussianly distributed and independent, the probability of observing each point given our function for the variance is $p(i) = \frac{1}{\delta(i)\sqrt{2x}} \exp\left(-\frac{r(i)^2}{2\delta(i)^2}\right)$. The total likelihood of all the points observed is the product of the probabilities at each point, $\mathcal{L} = \prod_i^n p(i)$. Because this probability can become very small very quickly and approach the limits of the machine precision, it is usually calculated as the log likelihood, $\log(\mathcal{L}) = -\sum_i^n \left[\frac{1}{2}\log(2\pi\hat{\sigma}(i)^2) + \log(p(i))\right]$. Optimizing the log likelihood by adjusting the parameters of $\hat{\sigma}$, we can fit the function to the variance of the data. Standard deviation as a function of phase fitting our function to the noise is shown in Fig. 5.16.

Recently, Ermentrout, Beverlin, 2nd, Troyer, & Netoff (2011) has shown that, when a neuron is subjected to additive white noise, the relationship between the variance in phase response of a neuron and the shape of the iPRC (Z) is

$$\mathrm{var}(\phi) = \in^2 \left([1 + \beta Z'(\phi)]^2 \int_0^{\phi} Z^2(s)\mathrm{d}s + \int_{\phi}^{T} Z^2(s + \beta Z(\phi))\mathrm{d}s \right),$$

where $\in$ is the magnitude of the (white) noise and β is the strength of the (delta function) stimulus. Note that, to leading order in β, this variance is phase independent for small β and is equivalent to the intrinsic jitter in the ISIs

$$\mathrm{var}(\phi) = \in^2 \left(\int_0^{T} Z^2(s)\mathrm{d}s \right).$$

The parameters $\in$ and β are usually unknown; therefore, they are used as free parameters to fit the function to the data optimizing the maximum likelihood. Fits to the residuals using this function is plotted in Fig. 5.16. This function gives slightly higher accuracy in fitting the variance over the simpler square-root function given the same number of free parameters.

6 Measuring iPRC with "White Noise" Stimuli

In this section, we outline an alternative method for measuring the infinitesimal phase response curve (iPRC). The method consists of continuously stimulating the neuron with a small-amplitude highly fluctuating input over many interspike intervals, measuring the phase shifts of all spikes due to the stimulus, and then deconvolving the stimulus and the phase-shifts to obtain the iPRC. The method is suggested in Izhikevich (Izhikevich 2007) and is related to work in Ermentrout et al. (Ermentrout, Galan, & Urban 2007) and Ota et al. (Ota et al. 2009).

As described in Sect. 2.3, we assume that the stimuli are sufficiently small so that stimulus has a linear effect on the phase of the neuron. Therefore, the phase shift $\Delta\psi_k$ of the kth spike during the stimulus is approximated by integral of the product of the iPRC $Z(t)$ and the stimulus $I_{\mathrm{stim},k}(t) = \sigma\xi_k(t)$ over the kth interspike interval

$$\Delta\psi_k = T - T_k \cong \int_0^{T_k} Z(\theta(t))\sigma\xi_k(t)\mathrm{d}t, \tag{5.12}$$

where T_k is the duration of the kth interspike interval, T is the intrinsic period of the neuron, and $\theta(t)$ is the absolute phase of the neuron. The stimulus $\sigma\xi_k(t)$ is chosen to be a piecewise constant function that is a realization of Gaussian white noise, i.e., time is broken up into very small intervals of width Δt and the amplitudes of $\sigma\xi_k(t)$ is each subinterval is drawn from a Gaussian distribution with zero mean and variance σ^2, where σ is assumed to be small. Note that this stimulus is composed of a wide range of Fourier modes that will typically form a basis for the iPRC Z.

The phase of the unstimulated neuron increases linearly with time, and therefore we approximate the phase of the weakly stimulated neuron as $\theta(t) = T/T_k t$. By changing variables so that the integration is in terms of phase, (5.12) becomes

$$\Delta\psi_k \cong \int_{\theta=0}^{T} Z(\theta)\sigma\xi_k(t(\theta))\frac{T_k}{T}\mathrm{d}\theta. \tag{5.13}$$

Note that, in this form, the upper limit of integration is independent of k, i.e., it is the same for all cycles. By discretizing phase into $M \sim 20$ equal bins of width $\Delta\theta = T/M$, (5.13) can be approximated using a middle Riemann sum

$$\Delta\psi_k \cong \sum_{j=1}^{M} Z(\theta_j)\langle\sigma\xi_k(t_j)\rangle\frac{T_k}{T}\Delta\theta, \tag{5.14}$$

where $\langle\sigma\xi_k(t_j)\rangle$ is the average of the stimulus in the jth bin during the kth cycle

$$\langle\sigma\xi_k(t_j)\rangle\frac{1}{\Delta}\int_{t_j-\Delta_k/2}^{t_j+\Delta_k/2}\sigma\xi_k(t)\mathrm{d}t.$$

and $\theta_j = \left(j - \frac{1}{2}\right)\Delta\theta$, $t_j = \frac{T_k}{T}\theta_j$, and $\Delta_k = \frac{T_k}{T}\Delta\theta$. Figure 5.17 shows an example of a fluctuating stimuli (second panel) and its binned and averaged version for a single cycle and $M = 20$.

If the stimulus is presented over N cycles, (5.14) with $k = 1\ldots N$ yields a system of N equations with M unknowns, i.e., the equally spaced points on the iPRC $Z(\theta_j)$. In matrix notation, this system of equations is

$$\overline{\Delta\psi} \cong \Gamma\bar{Z} \tag{5.15}$$

where $\overline{\Delta\psi}$ is an $N \times 1$ vector containing the phase shifts of each spike during the stimulus, Γ is an $N \times M$ matrix in which the j,k element is $\langle\sigma\xi_k(t_j)\rangle\frac{T_k}{T}\Delta\theta$ containing the binned and averaged stimuli for each spike, and $\bar{Z}$ is a $M \times 1$ vector containing the points on the iPRC. Thus, estimation of the iPRC is reduced to solving a "simple" linear algebra problem. Typically, there should be more phase shifts recorded (N) than points on the iPRC (M), so that system (5.15) is overdetermined and can be solved using a least squares approximation. Figure 5.18 shows an example of an iPRC estimated using this technique for the

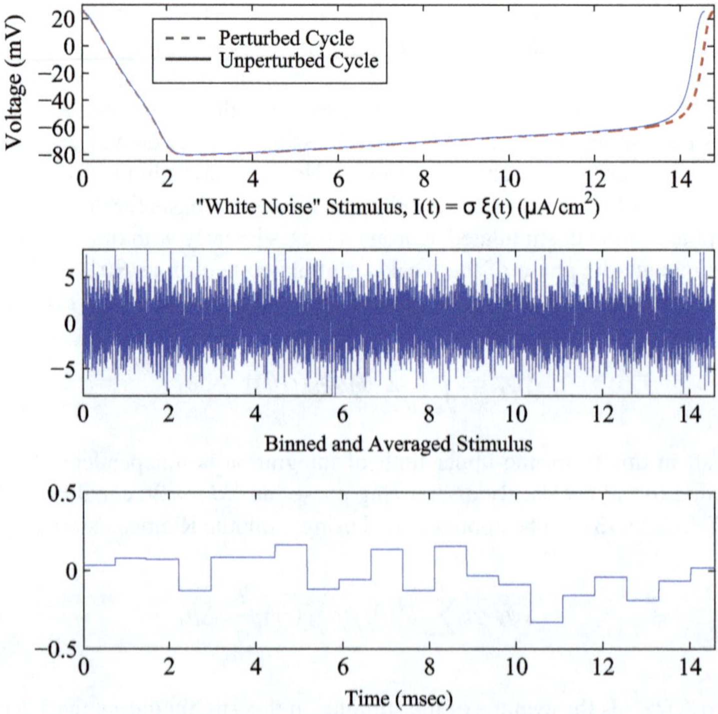

Fig. 5.17 [*Top panel*] The membrane potential of a Hodgkin–Huxley model neuron subject to an applied current of $10\,\mu A/cm^2$ for a single cycle. The *blue* trace represents the unperturbed oscillation, and the *red* trace represents the oscillation perturbed by the "white noise" stimulus in the middle panel. [*Middle panel*] A realization of the "white noise" stimuli $\sigma\xi_k(t)$ that is (sampled with a time step $\Delta t = 0.005$ ms and $\sigma = 1.5\,\mu A/cm^2$). [*Third panel*] The stimulus in the second panel after being divided into $M = 20$ bins and averaged for each bin. The amplitudes from this averaged signal make up one row of the matrix Γ

Hodgkin–Huxley (HH) model neuron with additive current noise. There are 20 points on the estimated iPRC ($M = 20$) and 40 spikes were sampled ($N = 40$). The error in this estimated iPRC is 0.30, where the error is computed as the normalized ℓ_2-norm $\|\bar{Z} - \overline{Z_a}\|_2 / \|\overline{Z_a}\|_2$, where $\overline{Z_a}$ is the iPRC calculated using the adjoint method evaluated at the appropriate phases.

Because Γ is a random matrix, it could sometimes have a high condition number, which could lead to significant error in the estimation of the discretized iPRC $\bar{Z}$. However, we can reduce the chance of this error by making the number of spikes considered N sufficiently larger than the number of points on the estimated iPRC M. Figure 5.19 shows the decrease in the error of the estimation of the iPRC for the HH model neuron as the number of spikes is increased. Note the steep initial decrease in the error. In practice, we find that about twice as many spikes (phase shifts) as number of points on the estimated iPRC yields a relatively low error. Typically, 20 points provides a good representation of an iPRC for a spiking neuron.

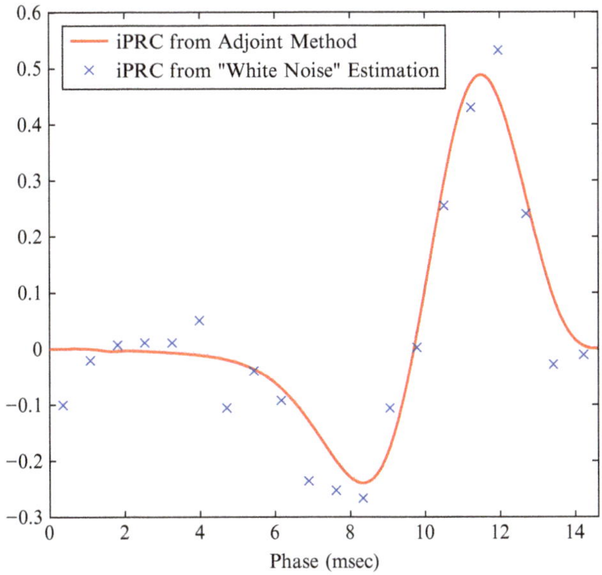

Fig. 5.18 Example of an estimated iPRC using the "white noise" method for the Hodgkin–Huxley neuron with "unknown" additive noise. The *red* trace is the iPRC calculated using the adjoint method (Ermentrout & Kopell 1991) and the crosses are the estimates of the iPRC found from solving (5.15) for $\overline{Z}$. There are $M = 20$ points on the estimated iPRC, and $N = 40$ cycles were used to calculate the iPRC. The stimulus has parameters $\Delta t = 0.005$ ms and $\sigma = 1.5\,\mu\text{A/cm}^2$. The signal (stimulus) to noise ratio was ~5.0. The error in the estimated iPRC is 0.30

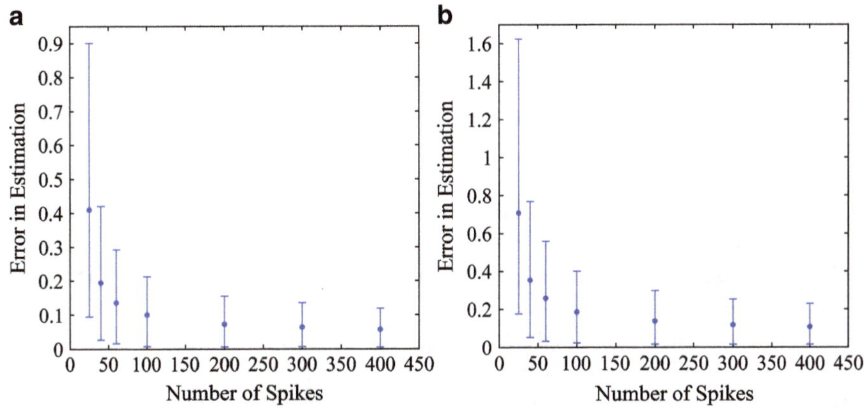

Fig. 5.19 Error in the estimated iPRC with $M = 20$ versus the number of interspike intervals. The error in the estimated iPRC is computed as $\|\bar{Z} - \overline{Z_\mathrm{a}}\|_2 / \|\overline{Z_\mathrm{a}}\|_2$, where $\overline{Z_\mathrm{a}}$ is the iPRC calculated using the adjoint method evaluated at the appropriate phases. (**a**) The system with no noise. (**b**) The system with "unknown" additive white noise with signal (stimulus) to noise ratio of ~5.0. Estimates were made for $M = 20$ points on the iPRCs. For both cases, the error decreases quickly as more trials are recorded. The stimulus has parameters $\Delta t = 0.005$ ms and $\sigma = 1.5\,\mu\text{A/cm}^2$ in (**a**) and $\sigma = 8\,\mu\text{A/cm}^2$ in (**b**). 700 trials were used to generate the statistics for every point on the graphs. Data points are mean values and error bars represent the limits that included ±30% of data

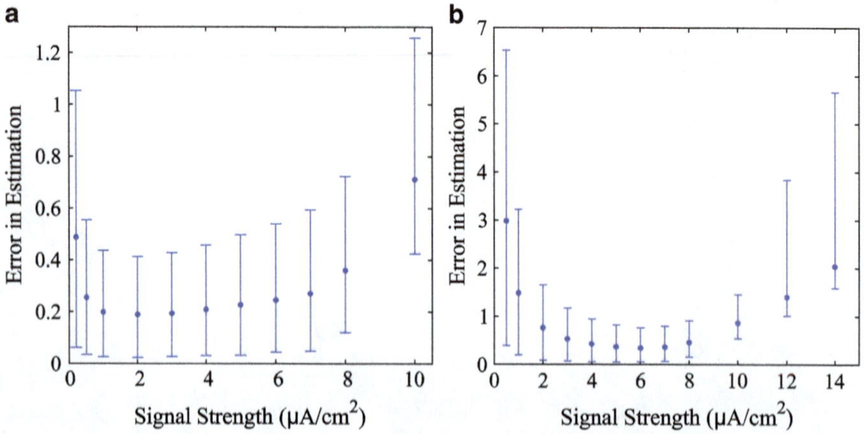

Fig. 5.20 Error in the estimated iPRC versus signal strength. The error is calculated as described in Fig. 5.19. The error is shown as a function of the strength of the random signal when (**a**) the system has no noise and (**b**) the system has "unknown" additive white noise. In both cases, there is an optimal value of the signal strength which minimizes the error in the estimation. Furthermore, both the mean and standard deviation of the error increase significantly as the signal strength becomes too large, i.e., the neuron no longer responds linearly. Estimates were made with $N = 40$ recorded spikes, $M = 20$ points on the iPRCs, and $\Delta t = 0.005$ ms. The unknown noise had a magnitude such that the signal (stimulus) to noise ratio was $\sim$5.0 when $\sigma = 8\,\mu\text{A/cm}^2$. 700 trials were used to generate the statistics for every point on the graphs. Data points are mean values and error bars represent the limits that included $\pm$30% of data

Therefore, if a neuron is firing at 10–20 Hz on average, and the phase shifts are measured over 40 spikes, it only takes 2–4 seconds to record the data needed to estimate the iPRC.

The strength of the random stimulus, σ, also affects the quality of the estimated iPRC. In practice, the stimulus amplitude must be small in order for the estimation to be theoretically valid, but it must also be large enough to overcome the intrinsic noise in the system. Figure 5.20 plots the error in the estimation as a function of σ when there is no unknown additive noise in the system (a), and when there is unknown additive noise in the system (b). In both cases, there is an optimal value of σ that minimizes the error in our estimation. This optimal value is larger when there is noise in the system.

While the method described above is perhaps the most straightforward "white noise" method, other methods that use white noise stimuli to measure the iPRC have also been proposed. Ermentrout et al. 2007 showed that, when an oscillating neuron is stimulated with small amplitude white noise, the spike triggered average (STA) is proportional to the derivative of its iPRC. As such, the iPRC can be calculated by integrating the STA. Ota (Ota et al. 2009) recently addressed several practical issues concerning the results of Ermentrout (Ermentrout et al. 2007) and outlined a procedure to estimate iPRCs for real neurons by using an appropriately weighted STA.

Izhikevich (Izhikevich 2007) comments that white noise methods for iPRC estimation should be more immune to noise than standard pulse methods because the stimulus fluctuations are spread over the entire cycle and not concentrated at the moments of pulses. However, to our knowledge, there has been no systematic comparison of the white noise methods and the standard pulse method. More work is needed to determine the optimal method for different situations (i.e., different noise levels, limitations on number of spikes, etc.). Furthermore, we expect that refinements could be made to improve most of these methods.

7 Summary

- The first step in generating a phase response curve for a neuron is choosing an appropriate stimulus waveform. When estimating the infinitesimal PRC (for use with theory of weakly coupled oscillators), a small brief delta-function-like stimulus pulse can be used. If synaptic inputs are not expected to sum linearly, then a realistic synaptic waveform should be used to measure the PRC to include the proper nonlinear responses of the neuron.
- The effects of a pulse stimulus on neuronal firing may last longer than a single cycle and give rise to measureable changes in ISIs in the cycles following the stimulated interval. These effects can be quantified with secondary and higher order PRCs and can be incorporated into models to increase their accuracy (Maran & Canavier 2008; Oprisan & Canavier 2001). Alternatively, the stimulus can be repeated at the same phase until the higher order effects accumulate and stabilize, and then the steady state response to the synaptic input at a phase can be measured. This results in measuring a "functional PRC" (Cui et al. 2009).
- Neurons exhibit considerable amounts of noise, making phase response data variable. There are two sources of noise: drift and jitter. Drift in the dynamics of the neuron occurs from slow timescale neuronal processes and "run down" (slow death) of the neuron during the experiment. This can be compensated to some degree by maintaining the firing rate of the neuron with a spike rate controller. While it is not a panacea, it keeps one aspect (the period) approximately constant over the duration of the experiment.
- To decrease the duration of the PRC experiment and thereby reduce the effects of drift on PRC estimation, the sampling of stimulus phase can be optimized. Using a Sobol sequence to sample the phases is much more efficient than random, or quasi-periodic sampling.
- The jitter in the phase response can be overcome by fitting a function to the data to estimate the deterministic portion of the neuron's PRC. Polynomials with constraints or Fourier series usually provide good fits to PRC data. The Akaike information criterion can be used to determine the appropriate number of coefficients when fitting either a polynomial or the Fourier series.
- The variability in a neuron's phase response can also be quantified and modeled. Ermentrout, Beverlin, 2nd, Troyer, & Netoff (2011) has recently shown that the

phase dependence of the variance is dependent on the shape of the PRC. We also present a simple function that can be fit to the variance by optimizing the maximum likelihood that does a reasonably good job.

- White noise stimulus approaches provide alternatives to pulse stimulation methods for measuring infinitesimal PRCs. This approach uses linear algebra to estimate the iPRC from neuronal response to white noise applied to several periods. More work must be done to optimize these methods and to systematically compare them to the standard pulse stimulation methods.

Acknowledgments TJL and MAS were supported by the National Science Foundation under grants DMS-09211039 (TJL), DMS-0518022 (TJL), TIN was supported by NSF CAREER Award CBET-0954797 (TIN).

References

Burnham, K. P., & Anderson, D. R. (1998). *Model selection and inference: A practical information-theoretic approach*. New York: Springer.

Canavier, C. C. (2005). The application of phase resetting curves to the analysis of pattern generating circuits containing bursting neurons. In S. Coombes, P. C. Bressloff & N. J. Hackensack (Eds.), *Bursting: The genesis of rhythm in the nervous system* (pp. 175–200) World Scientific.

Cruikshank, S. J., Lewis, T. J., & Connors, B. W. (2007). Synaptic basis for intense thalamocortical activation of feedforward inhibitory cells in neocortex. *Nature Neuroscience, 10*(4), 462–468. doi:10.1038/nn1861.

Cui, J., Canavier, C. C., & Butera, R. J. (2009). Functional phase response curves: A method for understanding synchronization of adapting neurons. *Journal of Neurophysiology, 102*(1), 387–398. doi:10.1152/jn.00037.2009.

Dorval, A. D., 2nd, Bettencourt, J., Netoff, T. I., & White, J. A. (2007). Hybrid neuronal network studies under dynamic clamp. *Methods in Molecular Biology (Clifton, N.J.), 403*, 219–231. doi:10.1007/978-1-59745-529-9_15.

Dorval, A. D., Bettencourt, J. C., Netoff, T. I., & White, J. A. (2008). Hybrid neuronal network studies under dynamic clamp. In *Applied patch clamp* Humana.

Dorval, A. D., Christini, D. J., & White, J. A. (2001). Real-time linux dynamic clamp: A fast and flexible way to construct virtual ion channels in living cells. *Ann Biomed Eng, 29*(10), 897–907.

Ermentrout, G. B., Beverlin, B., 2nd, Troyer, T., & Netoff, T. I. (2011). The variance of phase-resetting curves. *J Comput Neurosci*, [Epub ahead of print].

Ermentrout, G. B., & Chow, C. C. (2002). Modeling neural oscillations. *Physiol Behav, 77*(4–5), 629–633.

Ermentrout, G. B., Galan, R. F., & Urban, N. N. (2007). Relating neural dynamics to neural coding. *Physical Review Letters, 99*(24), 248103.

Ermentrout, G. B., & Kopell, N. (1991). Multiple pulse interactions and averaging in systems of coupled neural oscillators. *J Math Biol, 29*, 195–217.

Galan, R. F., Ermentrout, G. B., & Urban, N. N. (2005). Efficient estimation of phase-resetting curves in real neurons and its significance for neural-network modeling. *Physical Review Letters, 94*(15), 158101.

Galán, R. F., Bard Ermentrout, G., & Urban, N. N. (2006). Predicting synchronized neural assemblies from experimentally estimated phase-resetting curves. *Neurocomputing, 69*(10–12), 1112–1115. doi:DOI: 10.1016/j.neucom.2005.12.055.

Golomb, D., & Amitai, Y. (1997). Propagating neuronal discharges in neocortical slices: Computational and experimental study. *Journal of Neurophysiology, 78*(3), 1199–1211.

Hille, B. (1992). *Ionic channels of excitable membranes* (2nd ed.). Sunderland: Sinauer Associates.

Izhikevich, E. M. (2007). *Dynamical systems in neuroscience: The geometry of excitability and bursting*. Cambridge: MIT.

Kuramoto, Y. (1984). *Chemical oscillations, waves, and turbulence*. Berlin: Springer.

Lewis, T. J., & Rinzel, J. (2003). Dynamics of spiking neurons connected by both inhibitory and electrical coupling. *Journal of Computational Neuroscience, 14*(3), 283–309.

Mancilla, J. G., Lewis, T. J., Pinto, D. J., Rinzel, J., & Connors, B. W. (2007). Synchronization of electrically coupled pairs of inhibitory interneurons in neocortex. *The Journal of Neuroscience: The Official Journal of the Society for Neuroscience, 27*(8), 2058–2073. doi:10.1523/JNEUROSCI.2715–06.2007.

Maran, S. K., & Canavier, C. C. (2008). Using phase resetting to predict 1:1 and 2:2 locking in two neuron networks in which firing order is not always preserved. *Journal of Computational Neuroscience, 24*(1), 37–55. doi:10.1007/s10827–007–0040-z.

Miranda-Dominguez, O., Gonia, J., & Netoff, T. I. (2010). Firing rate control of a neuron using a linear proportional-integral controller. *Journal of Neural Engineering, 7*(6), 066004. doi:10.1088/1741–2560/7/6/066004.

Netoff, T. I., Acker, C. D., Bettencourt, J. C., & White, J. A. (2005). Beyond two-cell networks: Experimental measurement of neuronal responses to multiple synaptic inputs. *Journal of Computational Neuroscience, 18*(3), 287–295. doi:10.1007/s10827–005–0336–9.

Netoff, T. I., Banks, M. I., Dorval, A. D., Acker, C. D., Haas, J. S., Kopell, N., & White, J. A. (2005). Synchronization in hybrid neuronal networks of the hippocampal formation. *Journal of Neurophysiology, 93*(3), 1197–1208. doi:00982.2004 [pii]; 10.1152/jn.00982.2004.

Oprisan, S. A., & Canavier, C. C. (2001). Stability analysis of rings of pulse-coupled oscillators: The effect of phase-resetting in the second cycle after the pulse is important at synchrony and for long pulses. *Differential Equations and Dynamical Systems, 9*, 243–258.

Ota, K., Omori, T., & Aonishi, T. (2009). MAP estimation algorithm for phase response curves based on analysis of the observation process. *Journal of Computational Neuroscience, 26*(2), 185–202. doi:10.1007/s10827–008–0104–8.

Ota, K., Nomura, M., & Aoyagi, T. (2009). Weighted spike-triggered average of a fluctuating stimulus yielding the phase response curve. *Physical Review Letters, 103*(2), 024101. doi:10.1103/PhysRevLett.103.024101.

Press, W. H. (1992). *Numerical recipes in C: The art of scientific computing* (2nd ed.). Cambridge: Cambridge University Press.

Prinz, A. A., Abbott, L. F., & Marder, E. (2004). The dynamic clamp comes of age. *Trends in Neurosciences, 27*(4), 218–224. doi:10.1016/j.tins.2004.02.004.

Sharp, A. A., O'Neil, M. B., Abbott, L. F., & Marder, E. (1993). Dynamic clamp: Computer-generated conductances in real neurons. *J Neurophysiol, 69*(3), 992–995.

Tateno, T., & Robinson, H. P. (2007). Phase resetting curves and oscillatory stability in interneurons of rat somatosensory cortex. *Biophysical Journal, 92*(2), 683–695. doi:10.1529/biophysj.106.088021.

Van Vreeswijk, C., Abbott, L. F., & Ermentrout, G. B. (1994). When inhibition not excitation synchronizes neural firing. *J Comput Neurosci, 1*(4), 313–321.

Chapter 6
A Geometric Approach to Phase Resetting Estimation Based on Mapping Temporal to Geometric Phase

Sorinel Adrian Oprisan

Abstract The membrane potential is the most commonly traced quantity in both numerical simulations and electrophysiological experiments. One quantitative measure of neuronal activity that could be extracted from membrane potential is the firing period. The phase resetting curve (PRC) is a quantitative measure of the relative change in the firing period of a neuron due to external perturbations such as synaptic inputs. The experimentally recorded periodic oscillations of membrane potential represent a one-dimensional projection of a closed trajectory, or limit cycle, in neuron's multidimensional phase space. This chapter is entirely dedicated to the study of the relationship between the PRC and the geometry of the phase space trajectory. This chapter focuses on systematically deriving the mappings $\sigma = \sigma(\varphi, \mu)$ between the temporal phase φ and the geometric phase σ when some parameters μ are perturbed. For this purpose, both analytical approaches, based on the vector fields of a known theoretical models, and numerical approaches, based on experimentally recorded membrane potential, are discussed in the context of phase space reconstruction of limit cycle. The natural reference frame attached to neuron's unperturbed limit cycle, γ breaks the perturbation of control parameter μ into tangent and normal displacements relative to the unperturbed γ. Detailed derivations of PRC in response to weak tangent and normal, perturbations of γ are provided. According to the geometric approach to PRC prediction, a hard, external perturbation forces the figurative point to cross the excitability threshold, or separatrix, in the phase space. The geometric method for PRC prediction detailed in this chapter gives accurate predictions both for hard inhibitory and excitatory perturbations of γ. The geometric method was also successfully generalized to a more realistic case of a neuron receiving multiple inputs per cycle.

S.A. Oprisan (✉)
Department of Physics and Astronomy, College of Charleston, 66 George Street, Charleston, SC 29424, USA
e-mail: oprisans@cofc.edu

N.W. Schultheiss et al. (eds.), *Phase Response Curves in Neuroscience: Theory, Experiment, and Analysis*, Springer Series in Computational Neuroscience 6,
DOI 10.1007/978-1-4614-0739-3_6,

1 Introduction

Any bursting neuron can be modeled as a limit cycle oscillator, i.e., a periodic oscillator in which all variables, such as membrane potentials, repeat themselves exactly during each cycle. A convenient way to study periodic behavior is by using a phase representation. The term phase has been utilized in the literature in several different ways. This chapter adopted the definition of temporal phase introduced by Winfree (1980): the elapsed time measured from an arbitrary reference divided by the intrinsic period. A limit cycle oscillation requires a minimum of two variables (Guckenheimer and Holmes, 1983; Hoppensteadt and Izhikevich, 1997; Murray, 1993; Winfree, 1980). The limit cycle oscillator can exhibit a range of dynamic activity that depends upon the characteristic timescales of the two variables (Rinzel and Lee, 1986; Rinzel and Ermentrout, 1998). If they are comparable in magnitude, the qualitative dynamics of a phase oscillator arise, and the two variables tend to vary smoothly in unison. If, however, one variable varies much more slowly than the other, a relaxation oscillator can be produced, in which alternating rapid increases and decreases in the fast variable (jumps) are followed by slow relaxations in the slow variable. A phase resetting curve (PRC) tabulates the effects on the period of the current cycle as a result of brief perturbations applied to an uncoupled oscillator at various phases (Canavier et al. 1997, 1999; Ermentrout, 1985; Ermentrout and Kopell, 1991; Kopell and Ermentrout, 1988; Murray, 1993; Pavlides, 1973).

PRCs have been utilized to analyze certain simple circuits, specifically unidirectional rings comprised entirely of bursting neurons (Canavier et al. 1997, 1999). The unidirectional ring architecture arises from the assumption that each neuron receives input from exactly one other neuron during each cycle. The methods developed by Canavier et al. (1997, 1999) enable the design of circuits capable of generating multiple patterns of periodic behavior and offer efficient mechanisms for the control of multi stability. The firing pattern prediction of a fully-connected neural network based on the PRC assumes that the complete structure of the solution space of any unidirectional ring network can be predicted from the PRCs of the isolated neurons. The two basic hypotheses on which this method is built are: (1) the input received in the circuit by each component oscillator from its presynaptic oscillator has the same effect in the closed loop circuit condition as an identical input delivered in an open loop condition, such as that used for generating PRC curves, and (2) the effect of such an input effectively dies out by the time the next input is received. The ring geometry, which is widely used in biophysical modeling studies (Collins and Stewart, 1993; Collins and Richmond, 1994), was extended to include bidirectional coupling and arbitrary network geometries that allow for more than one input per neuron per cycle, and for the inclusion of neurons that do not burst endogenously when isolated from the circuit. The first step in generalizing circuit analysis is to understand the phenomenology of the phase resetting so that we can predict the effect of multiple and/or overlapping inputs within a single cycle.

For this purpose, this chapter presents, in a systematic way, the ideas behind the mapping of temporal resetting observed in neurons subject to external stimuli to a phase space perturbation of the corresponding limit cycle (Sect. 2). Any external perturbation (μ) impacts the phase space trajectory and the mapping from the geometric (σ) to temporal (φ) phase can be used to predict how a given external stimulus changes the temporal response of a neuron. While the phase space dynamics can be easily unveiled if the explicit mathematical model is known, in actual experiments only membrane potential recordings are accessible. Using the delay embedding method, it is possible to reconstruct the phase space from membrane potential recordings (Sect. 3). The anterior burster (AB) neuron sets the rhythm of the pyloric circuit in the stomatogastric ganglion (STG) of the lobster *Homarus americanus*. The actual dynamic clamp experiments were carried out on the pyloric dilator (PD) neuron, which is electrically coupled with AB neuron and experimentally more accessible. Once the phase space of a neural oscillator is accessible, either through explicit mathematical models or through reconstructed dynamics from membrane voltage recording using the delay embedding method, then a mapping between the geometric phase (σ) measured along the limit cycle and the temporal phase (φ) can be obtained (Sect. 4). Any external perturbation (μ) of the neuron moves the figurative point from the unperturbed limit cycle (γ) to a phase space point outside a new, or perturbed, limit cycle. A natural choice for the reference frame for a figurative point moving along a limit cycle (γ) has one axis tangent to the limit cycle and the other(s) normal to the limit cycle. For pedagogical reasons, we split the investigation of mapping between geometric and temporal phase into two classes: tangent, respectively, normal perturbations to the limit cycle trajectory. Moreover, the derivation of mapping splits also with respect to the strength of external perturbation in weak, respectively, strong perturbations. Mapping of the geometric phase into a temporal phase for the purpose of extracting the PRC was successfully applied to predicting 1:1 entrainment modes in small neural networks such as central pattern generators (CPGs). CPGs are networks of neurons that are capable of producing rhythmic output (Beer et al. 1999; Chiel et al. 1999; Golubitsky et al. 1998; Kopell and Ermentrout, 1988; Marder and Calabrese, 1996) in the absence of both input from higher centers and sensory feedback. CPGs are believed to mediate certain rhythmic, stereotyped behaviors, such as walking, flying (Robertson and Pearson 1985), swimming (Arshavsky et al. 1985; Satterlie, 1989), chewing, feeding, and scratching (Pearson 1993). The pyloric network of the STG is an example of a well-studied CPG that produces a three-phase rhythm that drives muscles in the stomach of lobsters and crabs (Selverston and Moulins 1987). Unlike STG neurons, bursters, which are part of most neural networks, receive more than one input per cycle. The geometric to temporal phase mapping was subsequently generalized to derive the PRCs for multiple inputs per cycle (Sect. 5).

2 Temporal and Geometric Phases for Limit Cycle Oscillators

2.1 *Temporal Phase*

Most neurons are intrinsic oscillators with a specific intrinsic period P_{i}. External stimuli (μ), such as presynaptic inputs, could perturb the periodic activity, or limit cycle trajectory (γ), of a neural oscillator by advancing or delaying the subsequent action potential(s). The transient change of the firing rate of a neural oscillator due to an external perturbation (Fig. 6.1a) is measured by the phase response curve, or PRC (Fig. 6.1b). Winfree defined the temporal phase φ as the elapsed time t measured from an arbitrary reference divided by the intrinsic period $\varphi = t/P_{\mathrm{i}}$ (Winfree 1987). The most significant change of the neuron's firing period takes place during the current cycle, i.e., the cycle that includes the perturbation. The first order PRC, $F^1(\varphi)$, measures the relative change in the firing period of a neuron due to an external perturbation received during the current cycle (Fig. 6.1):

$$F^1(\varphi) = F^1(t_{\mathrm{s}}/P_{\mathrm{i}}) = P_1/P_{\mathrm{i}} - 1, \tag{6.1}$$

where t_{s} is the stimulus time, which is the time elapsed between the most recent spike and the stimulus, P_{i} is the intrinsic period of the oscillator, and P_1 is the transiently modified firing period due to a stimulus applied during the current cycle (Oprisan and Canavier, 2002, 2005; Oprisan et al. 2003, 2004; Oprisan and Boutan, 2008; Oprisan 2009, 2010).

The PRC can be obtained numerically if the model equations are known or through electrophysiological measurements of membrane potential from biological

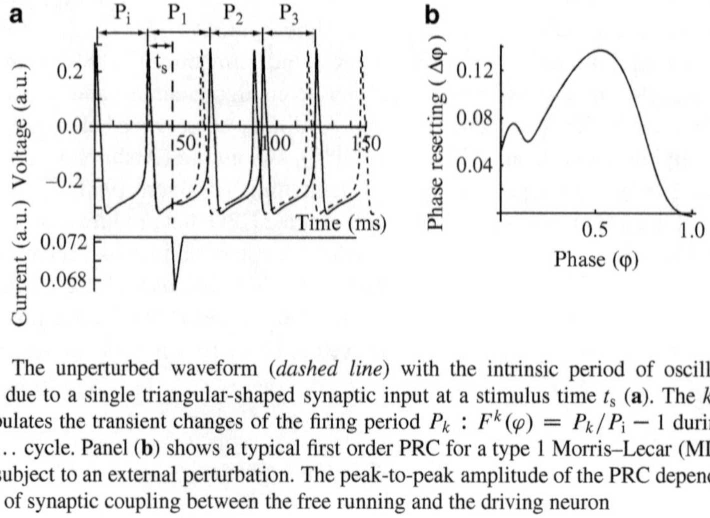

Fig. 6.1 The unperturbed waveform (*dashed line*) with the intrinsic period of oscillation P_{i} changes due to a single triangular-shaped synaptic input at a stimulus time t_{s} (**a**). The kth order PRC tabulates the transient changes of the firing period P_k : $F^k(\varphi) = P_k/P_{\mathrm{i}} - 1$ during kth = 1, 2, 3, ... cycle. Panel (**b**) shows a typical first order PRC for a type 1 Morris–Lecar (ML) model neuron subject to an external perturbation. The peak-to-peak amplitude of the PRC depends on the strength of synaptic coupling between the free running and the driving neuron

neurons. First, the intrinsic period P_{i} must be determined from the (filtered) membrane potential recording of the unperturbed neuron. The intrinsic period is the time between two successive crossings of an arbitrary voltage threshold with positive slope (Fig. 6.1a). Once the intrinsic period is determined, the membrane potential of a model or biological neuron in response to one current, or conductance, perturbation of specific shape (Fig. 6.1a) can be recorded. A stored activation profile of the current, or artificial synapse, is delivered to the isolated neuron at a controlled stimulus time t_{s} and its response is recorded. Individual stimuli are delivered a few cycles apart to ensure that the neuron returned to its unperturbed activity between stimuli. The experimental PRCs shown in this chapter were obtained using the above-described procedure by delivering 40 equally spaced stimuli at phases between 0 and 1 (Oprisan et al. 2004). The PRC is readily accessible through direct electrophysiological measurements from biological neurons and does not require any knowledge of the mechanisms involved in generating the action potential.

Although the most obvious candidate for electrophysiological recordings is the membrane potential, neural activity is in fact governed by complex and often unknown detailed kinetics of individual ionic channels spanning the neuron's membrane. The parameters defining the activation and inactivation of different ionic channels together with the membrane potential define the so-called phase space of the dynamic system that is the neuron. The theory of dynamic systems asserts that relevant measures of the complex, unknown, phase space dynamics can be recovered from the measurements of a single observable. Therefore, membrane potential recordings could be successfully used in order to reconstruct the multidimensional phase space dynamics of a neuron. Moreover, although not all mathematical models of neural oscillators are continuous, it is usually assumed that the neural model is described by a first order system of differential equations:

$$\mathrm{d}x/\mathrm{d}t = f(x;\mu), \tag{6.2}$$

where x is a biophysically relevant n-dimensional vector of independent variables such as membrane potential, gating variables, etc., and μ is a m-dimensional vector of control parameters such as external stimuli, synaptic currents, etc. (Hoppensteadt and Izhikevich, 1997; Izhikevich, 2007).

In the phases space, (6.2) defines a set of n remarkable curves called nullclines, which correspond to $f_i(x;\mu) = 0$, for all i from 1 to n (for a more formal introduction to the theory of dynamical systems see Baker and Gollub, 1996; Eckmann and Ruelle, 1985; Guckenheimer and Holmes, 1983; Hirsch et al. 2004; Hoppensteadt and Izhikevich, 1997; Izhikevich, 2007; Murray, 1993; Nayfeh and Balachandran, 1995; Ott, 1993). According to (6.2), a nullcline corresponds to a constant value of the state variable x_i. The intersection(s) of nullclines determine the steady-state(s) of the model (Hoppensteadt and Izhikevich 1997; Hirsch et al. 2004; Izhikevich, 2007). Stable steady-states attract all phase space trajectories. Unstable states can lead to complex neural activities such as spiking, bursting, and even chaos (Ruelle and Takens, 1971; Milnor, 1985; Izhikevich, 2007). A periodic solution of (6.2) is a close trajectory γ in the phase space x.

2.2 Geometric Phase

In contrast to Winfree's definition of temporal phase (Winfree 1987), other authors defined the phase as an angle that parameterizes the unit circle obtained through a homeomorphism of the actual phase space dynamics (Murray, 1993; Hoppensteadt and Izhikevich, 1997). For a constant angular velocity along a limit cycle γ, the temporal and angular phases are equivalent. The geometric phase is the normalized distance l measured from an arbitrary origin along the limit cycle $\sigma = l/L$, where L is the length of the limit cycle (Oprisan and Canavier, 2000, 2001, 2002; Oprisan and Boutan, 2008; Oprisan, 2009, 2010):

$$\sigma(t,\mu) = \frac{1}{L}\int_0^t \sqrt{\sum_{i=1}^{n} f_i^2(x(\tau);\mu)\mathrm{d}\tau}, \tag{6.3}$$

where n is the dimension of the vector field, $L = \int_0^{P_\mathrm{i}} \sqrt{\sum_{i=1}^{n} f_i^2(x(\tau);\mu)\mathrm{d}\tau}$ is the length of the limit cycle γ, and τ is a dummy variable. Since the elapsed time is $t = \varphi P_\mathrm{i}$, the above definition actually maps the geometric phase $\sigma = \sigma(\varphi;\mu)$ into the temporal phase φ.

For pedagogical reasons, and without reducing the generality of the results regarding the mapping between geometric (σ) and temporal phase (φ), a generic two-dimensional neural oscillator was considered. There are many biophysically relevant model neurons with a two-dimensional phase space (Morris and Lecar, 1981; FitzHugh, 1955; Nagumo et al. 1962). Throughout this chapter, we used a Morris–Lecar (ML) model oscillator, originally formulated to model action potentials in barnacle muscle (Morris and Lecar, 1981; Rinzel and Lee, 1986; Ermentrout, 1996). ML model neuron was preferred for two reasons. First, its limit cycle is only two-dimensional and can be captured by the membrane potential V and a negative feedback variable w that mimics the gating variable for a generic slow ionic channel (Hoppensteadt and Izhikevich, 1997; Morris and Lecar, 1981). More importantly, switching between type 1 and type 2 excitability in ML model neuron involves only three control parameters: V_3, V_4, and ϕ (see Appendix for a detailed mathematical model).

Neural excitability refers to the capability of a neuron to fire repetitively in response to an externally applied current. In their seminal work on giant squid axon, Hodgkin and Huxley (Hodgkin 1948) experimentally identified three classes, or types, of axonal excitability: type 1, where the repetitive firing is controlled by the intensity of an external stimulus; type 2, where the firing frequency is almost independent on stimulus intensity; and type 3, where there are no endogenous bursters regardless of stimulus intensity or duration. For type 1 excitability, the onset of the repetitive firing occurs at arbitrarily low frequency whereas for type 2 the repetitive firing occurs at a nonzero frequency (Hodgkin 1948). The type of membrane excitability reflects the geometry of the limit cycle through its bifurcation structure. Type 1 oscillations arise via an unstable saddle-node fix point, whereas type 2 oscillations are determined by a Hopf bifurcation of an unstable fix point (Hoppensteadt and Izhikevich, 1997; Rinzel and Ermentrout, 1998; Izhikevich 2007). The PRCs of a neuron depend upon the type of excitability it exhibits (Ermentrout 1996). In general, a type 1 neuron exhibits either only delays or only advances in response to an external perturbation (Fig. 6.1b). For the same perturbation, type 2 neurons exhibit both advances and delays, depending upon the phase of the stimulus. As a result of the topology of bifurcation leading to a limit cycle, two important characteristics of type 1 excitability class are worth mentioning: (1) the shape and phase space location of the limit cycle does not change significantly due to external stimuli, and (2) the period of the limit cycle motion is highly sensitive to perturbations (Oprisan and Canavier 2000, 2001, 2002, 2003, 2005, 2006; Oprisan et al. 2003, 2004; Oprisan and Boutan 2008; Oprisan 2009). For type 2 excitability, (1) the shape and phase space position of the limit cycle attractor is very sensitive to external stimuli, and (2) the period of the limit cycle motion does not change significantly with the external stimuli (Oprisan and Canavier 2000, 2001, 2002, 2003, 2005, 2006; Oprisan et al. 2003, 2004; Oprisan and Boutan 2008; Oprisan 2009, 2010).

Representative examples obtained by integrating the differential equation for ML model neuron (see Appendix) show steady oscillations near the lowest possible value for the bias current I_0 right at the bifurcation point, which is close to $I_0 = 35.0\,\mu\text{A/cm}^2$ for type 1, respectively, $I_0 = 77\,\mu\text{A/cm}^2$ for type 2 excitability. The voltage traces for type 1 excitability (Fig. 6.2a1) suggests spiking behavior and the shape of the corresponding limit cycle (Fig. 6.2a2) does not change with the injected current I_0. For type 1 excitability, a figurative point on the limit cycle spends a significant fraction of the intrinsic period around the saddle-node bifurcation point, which is at the crossing of the two nullclines shown with dashed lines in Fig. 6.2a2. Once it escapes that region of the phase space near the bifurcation point it quickly travels a full cycle (Hoppensteadt and Izhikevich 1997; Oprisan and Canavier 2000, 2001, 2002, 2003, 2005, 2006; Oprisan and Boutan 2008; Oprisan 2009, 2010). As a result, while the figurative point is near the saddle-node bifurcation it only travels a very small distance along the limit cycle (horizontal line in Fig. 6.2a3). As soon as it escapes the saddle-node region, quickly covers almost the entire limit cycle in about 20% of the firing period (Fig. 6.2a3). For type 2 excitability, the membrane potential trace could be assimilated to the envelope of the membrane voltage for a bursting neuron (Fig. 6.2b1) and the shape of the limit cycle (Fig. 6.2b2) is very sensitive to externally injected currents I_0 (Oprisan and Canavier 2000, 2001, 2002, 2003, 2005, 2006; Oprisan et al. 2003, 2004; Oprisan and Boutan 2008; Oprisan 2009, 2010). A type 2 ML model neuron has two regions of slow motions along the left and right branches of the voltage nullcline that determine two corresponding small slope regions in the geometric versus temporal phase plot (Fig. 6.2b3). At the same time, a type 2 ML oscillator has two steep-slope regions in the geometric versus temporal phase plot that correspond to fast motions of the figurative point when it jumps between the left and right branches of the voltage nullclines at the upper respectively, lower "knee" of the voltage nullcline (long-dashed curve in Fig. 6.2b2).

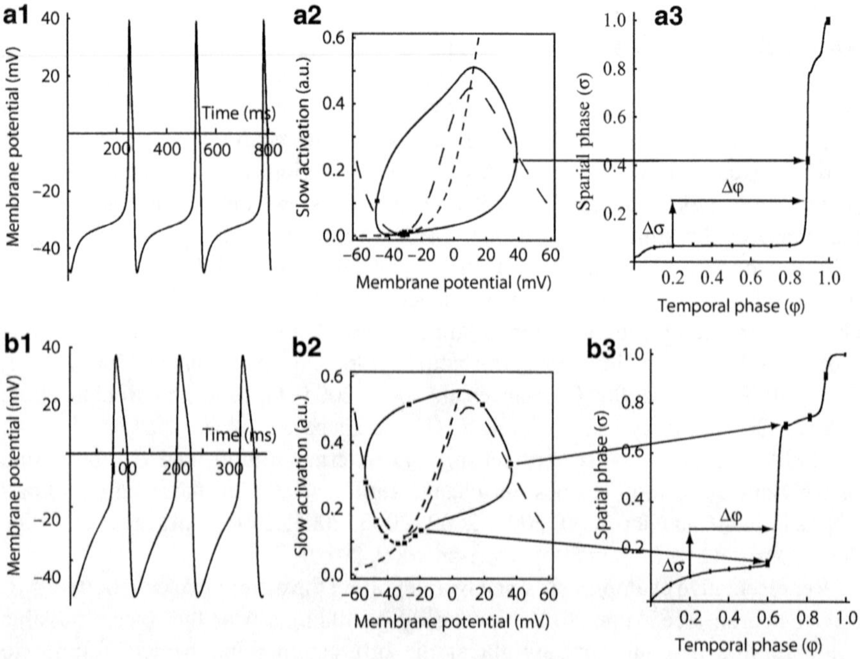

Fig. 6.2 Typical voltage traces for type 1 (**a**1) and type 2 (**b**1) neural excitability classes obtained by integrating the differential equations for a ML model neuron. Equally spaced markers (*filled squares*) are shown along each limit cycle together with the voltage (*long-dashed line*) and gating variable (*short-dashed line*) nullclines (**a**2 and **b**2). The hyperpolarized crossing of the nullclines indicates the existence of a saddle-node bifurcation for type 1 excitability (**b**1) and, respectively, a Hopf bifurcation for type 2 (**b**2). The geometric (σ) versus temporal (φ) phase plots show regions of significant fast motion of figurative point over some very short time intervals (**a**3 and **b**3). The temporal phase reference ($\varphi = 0$) was arbitrarily set to the most hyperpolarized point of the limit cycle

The voltage nullcline (long dashed lines in Fig. 2a2 and 2b2) has two branches with negative slope that attract nearby phase space trajectories. Their basin of attraction is delimited by a separatrix, which is the positive-slope segment of the voltage nullcline. Therefore, every phase space trajectory starting on the depolarized side of the separatrix will end up on the depolarized branch of the limit cycle and vice versa.

2.3 *Resetting induced by External Stimuli*

How can the geometry of the limit cycle attractor (Fig. 6.2) determine the temporal response of a neural oscillator to external stimuli? From a geometric viewpoint, an external perturbation (μ) forces the figurative point to switching between two

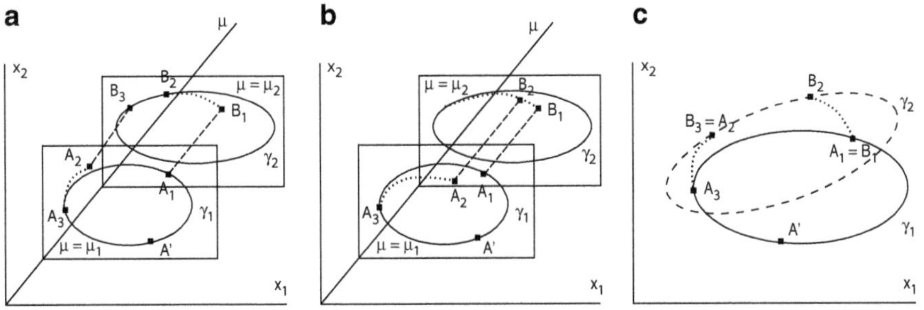

Fig. 6.3 A sudden change in one control parameter from μ_1 to μ_2 shifts the limit cycle from γ_1 to γ_2. For a long-lasting stimulus (**a**), a figurative point A_1 on the unperturbed limit cycle γ_1 projects outside γ_2 to B_1 and moves along the isochron until it reaches γ_2 at B_2 and continued along γ_2 until reaches B_3. At the moment the stimulus switches back to μ_1, the figurative point jumps back from B_2 to A_2 and then asymptotically approaches the unperturbed limit cycle γ_1 along another isochron. For a brief stimulus, the figurative point cannot reach the steady limit cycle γ_2 before it returns back on an isochron of the γ_1 (**b**). A two-dimensional projection of the unperturbed γ_1 and perturbed γ_2 limit cycles (**c**) similar to (**a**) shows some overlapping phase space points

limit cycles (Oprisan and Canavier 2000, 2001, 2002, 2003, 2005, 2006; Oprisan et al. 2003, 2004; Oprisan and Boutan 2008; Oprisan 2009, 2010). For example, a perturbation that suddenly switches the value of one control parameter μ in (6.2) between μ_1 and μ_2 forces the figurative point to jump from the unperturbed limit cycle (γ_1) to the perturbed limit cycle (γ_2) in Fig. 6.3. The control parameter μ could be a bias current, synaptic input, neuromodulator concentration, etc. Furthermore, let us assume that for each value of the control parameter μ a unique and stable limit cycles γ exists. Let A_1 be the figurative point on the limit cycle that corresponds to μ_1 at the time instant when the control parameter suddenly changes from μ_1 to μ_2. As a result, the figurative point no longer resides on the limit cycle γ_1 but it switched to B_1, in the attractive basin of the perturbed limit cycle γ_2. Therefore, the figurative point remains in the attractor basin of the limit cycle γ_2 for the entire duration τ of the perturbation. If the duration τ of the perturbation is longer than the time it takes the figurative point B_1 to reach the limit cycle γ_2, then the figurative point reaches the limit cycle γ_2 (path B_1B_2 in Fig. 6.3a) and continues along it for the time remaining (path B_2B_3 in Fig. 6.3a). Otherwise, if the duration τ of the perturbation is shorter than the time it takes the figurative point B_1 to reach the limit cycle γ_2, then it may only covers a small portion of the asymptotic path towards γ_2, called isochron (Coddington and Levinson, 1955; Ermentrout, 2002; Guckenheimer, 1975; Keener and Sneyd, 1998; Pavlides, 1973; Rabinovitch and Rogachevskii, 1999; Wiggins 1994), which leads to the limit cycle (path B_1B_2 in Fig. 6.3b) before returning back to γ_1. When the control parameter switches back from μ_2 to μ_1, the figurative point suddenly jumps back to the attractive basin of the unperturbed limit cycle γ_1 (point A_2 on Fig. 6.3a and b) and relaxes (approaches asymptotically) toward the unperturbed limit cycle trajectory along the isochron A_2A_3 (Fig. 6.3a and b). At the same time, in the absence of any perturbation, the figurative point A would have moved from A_1 to A'. The distance A_3A', measured along the unperturbed limit cycle γ_1, is the geometric (distance-based) phase difference $\Delta\sigma$ induced by a perturbation of amplitude $\Delta\mu = \mu_2 - \mu_1$ and duration τ. Therefore, a geometric to temporal phase mapping, similar to those shown in Fig. 6.2a3 and b3, convert the geometric phase difference $\Delta\sigma$ into a temporal phase $\Delta\varphi$ and, as a result, retrieve the PRC based on geometric considerations.

The nature of the bifurcation that generates the limit cycle significantly impacts its geometry and, therefore, the dynamic response of the neural oscillator captured by the PRC. For type 1 excitability, the two limit cycles shown in Fig. 6.3 have the same shape and phase space location. Therefore, the figurative point A_1 from γ_1 projects directly onto the second limit cycle γ_2 and, after the perturbation is over, back onto γ_1. As a result, the figurative point always travels along the same phase space path only with different speeds. For type 2 oscillators, the speed of the figurative point does not change much when switching between γ_1 and γ_2, but the two limit cycles do not overlap. As a result, there are significant contributions of the relaxation processes toward the limit cycles to the PRC (Oprisan and Canavier 2000, 2001, 2002, 2003, 2005, 2006; Oprisan et al. 2003, 2004; Oprisan and Boutan, 2008; Oprisan 2009, 2010).

3 Phase Space Reconstruction Using Action Potentials

A topological approach to phase resetting has the potential of connecting the geometry of the bifurcation leading to repetitive firing with transient temporal resetting in response to external stimuli. In order to fully exploit this potential benefit, we must be able to extract topological information directly from membrane potential recordings. How can the geometric phase σ be computed if the evolution equations are not explicitly known? How can the limit cycle be extracted directly from membrane potential recordings?

There are two methods that allow phase space reconstruction based on a time series: the delay embedding and the derivative embedding (Abarbanel et al. 1993; Mane 1981; Packard et al. 1980; Sauer et al. 1991; Takens, 1981). The delay coordinate embedding allows multidimensional phase space trajectory reconstruction based on one-dimensional recording. Using discrete membrane potential recordings V_i, a d-dimensional figurative point is defined as follows:

$$P(i) = [V_i, V_{i-\theta}, V_{i-2\theta}, \ldots, V_{i-(d-1)\theta}], \qquad (6.4)$$

where θ is called the embedding lag, and d is the embedding dimension. Takens' theorem gives the necessary conditions for a delay embedding reconstruction from an infinitely long series of noiseless data (Takens 1981). The derivative embedding works in a similar way, although it has the drawback of enhancing the biological noise present on recorded data. We used both the autocorrelation and the average mutual information methods to estimate the embedding lag θ (Abarbanel et al. 1993; Broomhead and King 1986; Eckmann and Ruelle 1985;

Grassberger and Procaccia 1983; Schouten et al. 1994; Wolf et al. 1985). The autocorrelation function measures the linear dependence of the time series values separated by a time lag of θ, and it may be used to detect deterministic components masked in a random-like background. The average mutual information function is the average information about V_i gained when observing $V_{i+\theta}$ and it estimates the reduction in uncertainty of V_i gained by observing $V_{i+\theta}$. The embedding lag, or time delay, θ must be large enough such that each reconstructed vector contains independent information about the system. At the same time, the embedding lag θ must not be too large in order to still capture the correlation between the components of the reconstructed vector (Fraser and Swinney 1986; Kennel et al. 1992).

To estimate the embedded dimension d, we used both the method of the false nearest neighbors and the spectrum of the fractal dimensions. The false nearest neighbors uses the geometric observation that, in a m-dimensional delayed reconstructed space, the reconstructed attractor must preserve the topological properties of the original attractor (Takens 1981; Abarbanel et al. 1993; Broomhead and King 1986; Eckmann and Ruelle 1985). Thus, the neighbors of a given point are mapped onto neighbors in the delay space. However, any delay embedding with a dimension k that is less than the optimum value m does not preserve the topological structure of the original m-dimensional attractor. As a result, figurative points in a suboptimal k-dimensional embedding are projected onto neighborhoods of other points to which they do not belong in the original m-dimensional phase space. These points are called false neighbors, and the embedding dimension should be increased until no further reduction in false nearest neighbors can be achieved (Broomhead and King 1986; Eckmann and Ruelle 1985; Schreiber 1995). The spectrum of the fractal dimensions estimates the number of variables necessary to define a dynamical system (Sauer and Yorke 1993). The most used measures in our phase space reconstructions were (Oprisan et al. 2003, 2004): (1) the capacity dimension implementing a simple box counting algorithm (Baker and Gollub 1996; Nayfeh and Balachandran 1995), (2) the information dimension that corrects the deficiency of capacity dimension by weighing the boxes proportionally to the number of data points they contain (Nayfeh and Balachandran 1995; Ott 1993), and (3) the correlation dimension that counts the total number of pairs of points that have a distance between them that is less than a given value r (Grassberger and Procaccia 1983). Among the most used algorithms we mention the brute-force pairwise distance counter, which gives the pairwise distribution over all distances, and the isotropic m-dimensional counter, which determines the number of points in each sphere of radius r (Grassberger 1983). The global embedding dimension d is the minimum number of time-delay coordinates needed in order to avoid self-intersections of the reconstructed trajectories. In suboptimal embeddings, trajectories intersect each other and violate the uniqueness requirement for any solution of a system of differential equations because they project down onto too few dimensions. On the other hand, if the delay-embedding dimension is too large, the noise corrupts other calculations because it fills any dimension (Casdagli et al. 1991; Sauer and Yorke 1993; Sauer et al. 1991; Wolf et al. 1985).

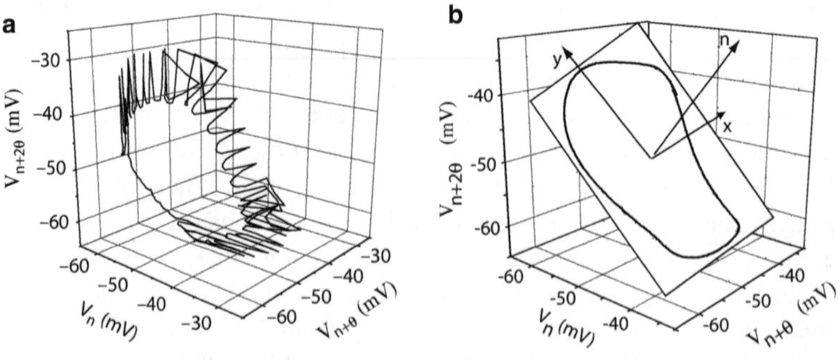

Fig. 6.4 Reconstructed tree-dimensional attractor using membrane potential recorded for AB/PD complex of STG. The embedding lag $\theta = 600$ time steps (300 ms) obtained with TISEAN package is optimal for untangling the phase space trajectory. The unfiltered phase space attractor (**a**) displays spurious spiking due to reconstruction artifacts. Filtered data removed the spikes and retained only the envelope of the burst revealing a "flat" two-dimensional attractor that can be represented in x–y phase space (**b**)

Experimental data recorded from a PD neuron of the STG of lobster (STG) with a sampling time of 0.5 ms and an average period of about 1500 ms were used (Oprisan et al. 2002, 2004). The PD neuron is a component of the pyloric circuit of the crustacean STG of *Homarus americanus*, which controls the movements of the pylorus (Bal et al. 1988; Bartos and Nusbaum 1997; Marder and Calabrese 1996; Marder et al. 1998; Mulloney 1977; Ijspeert 2008). Because of the electrical coupling between the AB pacemaker neuron and the two PD neurons of the STG, we considered them as a single functional unit called the AB/PD complex. External inputs were simulated by a synaptic current $I_{syn} = g_{syn}m \cdot (V_{postsyn} - E_{rev})$, where g_{syn} is the conductance of the synapse, m is the synaptic activation, $V_{postsyn}$ is the membrane potential of the postsynaptic neuron, and E_{rev} is the synaptic reversal potential, which was set to -90 mV for all synapses based on voltage clamp experiments. All synapses were instantaneously activating and deactivating. The synaptic activation function, $m = (1 + e^{(V_{1/2} - V_{filt})/s})^{-1}$, was chosen such that the synapse is fully activated during the burst and deactivated between bursts. Because the membrane potential between spikes reaches very hyperpolarized levels, and the activation of the synapses is instantaneous, a filtered version, V_{filt}, of the presynaptic membrane potential was used to compute the synapse activation. The phase space reconstruction was carried out using the free package TISEAN (Hegger et al. 1999) which gave us an embedding lag of $\theta = 600$ time steps ($\approx$ 300 ms) and an estimated embedding dimension of $d = 3$. A reconstructed phase space (see Fig. 6.4a) displays unrealistically fast oscillations of all state variables (Oprisan et al. 2002, 2004). This is an artifact of the delay embedding method that temporally shifts the same time series several times to unfold the attractor so that the rapid changes corresponding to an action potential are reflected in every reconstructed variable. As a result, each reconstructed state variable has the same timescale, which points to a major shortcoming of the delay embedding method that is its inability to capture the true timescales of the phase space dynamics of the original dynamic system. After filtering the fast spikes off the reconstructed attractor for the AB/PD complex (see Fig. 6.4b), the attractor looks relatively flat. This means that, despite the fact that the actual dynamics is described by at least three independent variables, only two "normal" coordinates in the plane of the limit cycle are relevant (marked x and y in Fig. 6.4b) and the third variable is the normal to the limit cycle plane.

4 Phase Resetting Induced by Soft Tangential and Normal Perturbations to Limit Cycles

For neural oscillators, perturbations generally occur in the form of an increase or decrease in the externally applied transmembrane current. Although there are no restrictions on the shapes of the transmembrane current perturbations, here we only refer to rectangular-shaped perturbations characterized by abruptly stepping up a control parameter (externally injected current, synaptic current, conductance pulse, etc.) from μ_1 to μ_2, holding it at μ_2 for a duration τ, and stepping it down to μ_1 afterward. Depending on its amplitude, or strength, $\Delta\mu_1 = \mu_2 - \mu_1$ and duration τ, a perturbation can induce a soft resetting, which results in small displacements of the figurative point around the unperturbed state, or a hard resetting, which forces the neural oscillator to switch from a depolarized to a hyperpolarized state or vice versa (Oprisan and Canavier 2000, 2001, 2002, 2003, 2005, 2006; Oprisan and Boutan 2008; Oprisan 2009, 2010). From a dynamic systems viewpoint, hard resetting is induced by perturbations that are able to move the figurative point outside the basin of attraction of the current state past the separatrix between the hyperpolarized and the depolarized branches of the limit cycle (see Fig. 6.2a2 and 2b2). The first derivative of the membrane potential is proportional to the externally applied transmembrane current, whereas the transmembrane current does not appear in the derivatives of the other state variables, which for neural oscillators are generally gating variables. In neural oscillators for which the frequency of the temporal waveform is simply proportional to the transmembrane current, such as type 1 or integrate and fire models, the direction (advance or delay) of a brief perturbation can be determined based on the sign of the current perturbation and the sign of the first derivative of membrane potential (the instantaneous transmembrane current) at the time of the perturbation (Oprisan and Canavier 2000, 2001, 2002, 2003, 2005, 2006; Oprisan and Boutan 2008; Oprisan 2009, 2010). By convention, a depolarizing (hyperpolarizing) externally applied current has a positive (negative) sign and produces a positive (negative) change, or increase (decrease) in the membrane potential. An advance of the next spike occurs if the membrane potential derivative and the external perturbation have the same sign, whereas if they have opposed signs a delay occurs. For an ionic current, such as a synaptic current, the sign convention is opposite. Furthermore, the magnitude of the instantaneous geometric phase shift due to a brief perturbation is proportional to the magnitude of the time integral of the perturbing current (Oprisan 2010).

4.1 Soft Tangent Perturbations to Limit Cycle

In Fig. 6.3c, a perturbation of amplitude $\Delta\mu = \mu_2 - \mu_1$ applied at t_0 finds the figurative point along the unperturbed limit cycle γ_1 at A_1, which projects onto B_1 located in the basin of attraction of the perturbed limit cycle γ_2. While the control parameter is held at $\mu = \mu_2$ for a duration τ, the figurative point approaches the perturbed limit cycle γ_2 along the path A_1B_2. For brief perturbations, B_2 is also the return point back to the unperturbed limit cycle γ_1. When the perturbation ends, the trajectory may be at some distance from the unperturbed limit cycle γ_1 requiring a time τ' to relax back to the unperturbed limit cycle (Oprisan and Canavier 2001, 2002, 2005, 2006; Oprisan and Boutan 2008; Oprisan 2009, 2010).

Physiological perturbations generally arise in the form of a (synaptic) current, which only affects the derivative of one variable in (6.2), namely the membrane potential, or $f_1(x;\mu)$ (see also Izhikevich 2007; Nishii 1999, 2006). The membrane potential has the general form (Hodgkin & Huxley 1952a,b, c, d):

$$V' = \left(-\sum I_{\text{ion}} + I_{\text{ext}}\right)/C, \tag{6.5}$$

where I_{ion} represent ionic currents, C is the membrane capacitance, and I_{ext} is a time-dependent externally applied current. The (normalized) infinitesimal displacement along the limit cycle γ_1 induced by a time-dependent current perturbation $i(t)$ is the dot product between the perturbation vector $(i(t)/C, 0, 0, \ldots)$ and the vector field given by (6.2):

$$d\sigma_1(t) = \frac{1}{L}\frac{i(t)}{C}\frac{f_1(t)}{||f(t)||}\mathrm{d}t. \tag{6.6}$$

For brief perturbations, the normalized vector filed $f_1/||f||$ is almost constant over the duration τ of the perturbation (Glass and Mackey 1988; Guckenheimer and Holmes 1983; Hoppensteadt and Izhikevich 1997; Murray 1993; Pavlides, 1973) and the finite geometric distance projected along the unperturbed limit cycle γ_1 gives the tangent displacement (Oprisan and Canavier 2000, 2001, 2002; Oprisan and Boutan 2008)

$$\Delta\sigma_{||}(t_0, i, \tau) = \frac{1}{L}\frac{i\tau}{C}\frac{f_1(t_0)}{||f(t_0)||}. \tag{6.7}$$

Based on (6.3), the temporal phase resetting induced by a perturbation of amplitude $\Delta\mu$ and duration τ is determined by inverting the geometric versus temporal phase map:

$$\Delta\varphi_{||} = \sigma^{-1}(\Delta\sigma_{||}). \tag{6.8}$$

The mapping between a geometric displacement $\Delta\sigma$ and a phase resetting $\Delta\varphi$ is exemplified in Figs. 6.2a3 and 6.2b3 (Oprisan and Canavier 2000, 2001, 2002).

A type 1 ML model neuron was used to test the geometric to temporal phase mapping method and predict the PRC based on geometric information regarding the limit cycle attractor (Oprisan and Boutan 2008). The bias current I_0 was set close to

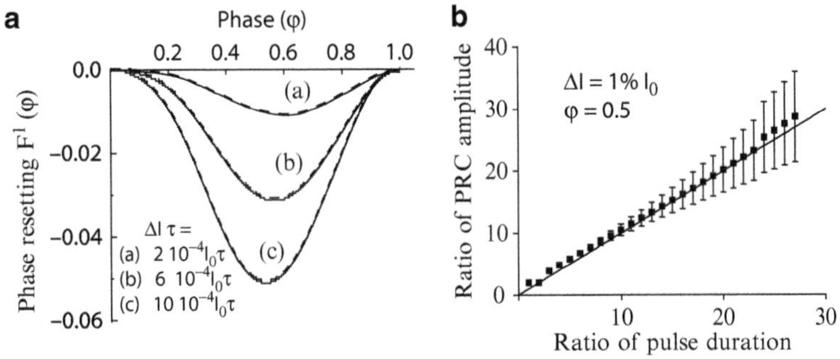

Fig. 6.5 Phase response curves obtained by numerical integration of differential equations (*continuous lines* in panel **a**). The geometric to temporal phase mapping method predicts the dashed PRCs (**a**). As predicted by the geometric method, the PRC is proportional to the product between the amplitude $\Delta\mu = \Delta I$ and the duration τ of the perturbation. As panel **a** suggests, a perturbation that is three times longer (curve b in panel **a**) than the reference perturbation (curve a in panel **a**) produces a PRC whose amplitude is three times larger at all phases. Panel **b** summarizes the results for a fixed phase ($\varphi = 0.5$) and amplitude of perturbation ($\Delta I = 1\% I_0$) with the perturbation duration sweeping the range from $0.1\% P_{\mathrm{i}}$ up to $30\% P_{\mathrm{i}}$ with a step of $0.1\% P_{\mathrm{i}}$. The amplitude of the PRC (at $\varphi = 0.5$) for $\tau_1 = 1\% P_{\mathrm{i}}$ was 10 times larger than for $\tau_2 = 0.1\% P_{\mathrm{i}}$, suggesting that the ratio of the PRC amplitudes might scale linearly with the ration of the pulse durations (τ_1/τ_2). Also, the PRC amplitude was approximately 10 times larger for a pulse duration of $\tau_1 = 2\% P_{\mathrm{i}}$ compared to the amplitude of the PRC for $\tau_2 = 0.2\% P_{\mathrm{i}}$. Panel **b** shows the average ratio of the PRC amplitudes obtained for the same fix ratio τ_1/τ_2 of pulse duration. For example, the point corresponding to a pulse ratio of $\tau_1/\tau_2 = 10$ in panel **b** is the average of the PRC amplitude ratios for the following pulse ratios: $1\% P_{\mathrm{i}}/0.1\% P_{\mathrm{i}}, 2\% P_{\mathrm{i}}/0.2\% P_{\mathrm{i}}, \ldots, 10\% P_{\mathrm{i}}/1\% P_{\mathrm{i}}, 20\% P_{\mathrm{i}}/2\% P_{\mathrm{i}}$, etc

the bifurcation point (the onset of oscillations). The control parameter used for the purpose of perturbing the limit cycle was a rectangular current pulse of amplitude ΔI, which was expressed as a fraction of the bias current I_0. The duration τ of the perturbation was also expressed as a fraction of the intrinsic period P_{i}. The PRC defined by (6.1) was determined by integrating the differential equations (see Appendix) with the additional current pulse of amplitude ΔI and duration τ applied at different temporal phases $\varphi = t_s/P_{\mathrm{i}}$ (solid line in Fig. 6.5).

For type 1 excitability, the topology of the limit cycle does not change significantly due to a brief perturbation. Therefore, only the tangent projection $\Delta\sigma_{||}$ of the current perturbation along the unperturbed limit cycle γ_1 has a significant contribution to the PRC and the normal displacement can be neglected (Oprisan and Canavier 2000, 2001, 2002, 2005, 2006; Oprisan et al. 2002, 2003, 2005, Oprisan and Boutan 2008). More quantitatively, the normal displacement from the unperturbed limit cycle is proportional to $f_2(t_0)/||f_1(t_0)|| = 2.8 \times 10^{-4}$ for type 1 ML, whereas for type 2 ML the average ratio was 2.8. Thus, the normal displacement induced by the same perturbation is four orders of magnitude smaller for type 1 oscillators compared to type 2 oscillators. This explains the excellent results obtained by using only tangent geometric displacement $\Delta\sigma_{||}$ for computing the PRC of type 1 model neurons (Oprisan and Canavier 2000, 2001, 2002, 2005, 2006;

Oprisan et al. 2002, 2003, 2005; Oprisan and Boutan 2008). The tangent geometric displacement $\Delta\sigma_{||}$ was computed using (6.7) where the norm of the vector field for ML model at time instant t_0 is $||f(t_0)|| = \sqrt{f_1^2(t_0) + f_2^2(t_0)}$. Subsequently, the geometric displacement $\Delta\sigma_{||}$ was mapped using (6.8) into a phase resetting $\Delta\varphi$ (see Fig. 6.2a3 and 2b3) and plotted with dashed lines in Fig. 6.5a. The agreement between the geometric prediction and the direct numerical integration of differential equations is excellent. The scaling of PRCs with the product of pulse amplitude $\Delta\mu = \Delta I$ and duration τ (Fig. 6.5b) was tested by integrating the differential equations for a fixed amplitude $\Delta I = 1\% I_0$ and different pulse durations from $\tau = 0.1\% P_{\mathrm{i}}$ up to $30\% P_{\mathrm{i}}$ with a step of $0.1\% P_{\mathrm{i}}$ for a total of 300 PRCs. Although the ratio of the PRC amplitudes somewhat depends on the particular phase at which it is evaluated, the linear relationship seen in Fig. 6.5b was generally preserved over a wide range of stimulus phases. Figure 6.5b summarizes all results regarding the 300(300–1)/2 = 44,850 pairs of PRCs that could be obtained out of the 300 PRCs generated. Each point on Fig. 6.5b represents the average value of the PRC ratio for the same ratio of perturbation duration. For example, the point on Fig. 6.5b corresponding to a pulse ratio of 10 (horizontal axis) is the average of PRC amplitude ratios for the following pulse ratios: $(1\% P_{\mathrm{i}})/(0.1\% P_{\mathrm{i}})$, $(2\% P_{\mathrm{i}})/(0.2\% P_{\mathrm{i}})$, $\ldots$, $(10\% P_{\mathrm{i}})/(1\% P_{\mathrm{i}})$, $(20\% P_{\mathrm{i}})/(2\% P_{\mathrm{i}})$, etc. (Oprisan and Canavier 2000, 2001, 2002, 2005, 2006; Oprisan et al. 2002, 2003, 2005; Oprisan and Boutan 2008). The error bars in Fig. 6.5b represent the standard deviation of the PRC amplitude ratios for the same fix ratio of the pulse durations. Similar linear relationship holds for the plot of the PRC amplitude ratios versus the perturbation amplitudes (data not shown). There is linear scaling of the PRC amplitudes at all phases with the product $\Delta I \tau$ below $10^{-4} I_0 P_{\mathrm{i}}$, which validates the derivation of the theoretical formula (6.6). As expected, for long pulse durations, the PRC scaling deviates from the linear relationship predicted by (6.7).

4.2 *Soft Normal Perturbations to Limit Cycle*

The normal displacement from the unperturbed limit cycle could be obtained by estimating the orthogonal projection of the perturbation to the unperturbed limit cycle. For a two-dimensional model neuron, the normal displacement Δh to the unperturbed limit cycle could be determined from similar triangles (Oprisan and Canavier 2000, 2001, 2002, 2005, 2006; Oprisan et al. 2002, 2003, 2005):

$$\Delta h = \Delta\sigma_{||} f_2(t_0)/f_1(t_0). \tag{6.9}$$

In order to estimate the effect of a normal displacement Δh in the phase plane on the phase resetting $\Delta\varphi$, we assumed that the corresponding normal geometric phase shift $\Delta\sigma_{\perp}$ induced by a normal displacement Δh is:

$$\Delta\sigma_{\perp} = k\Delta h = k\Delta\sigma_{||} f_2(t_0)/f_1(t_0), \tag{6.10}$$

where k is a yet to be determined constant. In order to find k, we reverse-engineered the geometric to temporal phase mapping: (1) computed the time required for a figurative point to relax to the unperturbed limit cycle γ_1 from a known distance Δh measured along the perpendicular to the limit cycle, (2) assumed that the temporal phase resetting $\Delta\varphi$ was proportional to the relaxation time, and (3) converted this temporal resetting $\Delta\varphi$ to a geometric resetting $\Delta\sigma_\perp$ in order to estimate k. First, we determined the maximum normal distance $\Delta h_{\max}$ between the unperturbed limit cycle and a perturbed limit cycle with a current perturbation of amplitude $\Delta\mu = i$ and duration τ. Next, we determined numerically the normalized time $\Delta\varphi_{\max}$ (with respect to the intrinsic period of the unperturbed limit cycle) necessary for the figurative point to relax from the maximum displacement sufficiently close to the unperturbed limit cycle. Finally, $\Delta\varphi_{\max}$ was converted into a geometric displacement $\Delta\sigma_{\max}$ (Fig. 6.2a3 and 2b3) and the constant k was found as the ratio $\Delta\sigma_{\max}/\Delta h_{\max}$. We used an ε-neighborhood condition to determine when a figurative point "reached" the limit cycle since the time required to relax to a limit cycle is, theoretically, infinite (Hoppensteadt and Izhikevich 1997; Izhikevich 2000, Rabinovitch and Rogachevskii 1999). The normalized distance ε was determined by the computational accuracy, which was set to 10^{-6}. Once the normal displacement to the limit cycle $\Delta\sigma_\perp$ was computed, the temporal phase resetting was recovered from the invertible map $\Delta\varphi_\perp = \sigma^{-1}(\Delta\sigma_\perp)$. Obviously, the total phase resetting is (Oprisan and Canavier 2000, 2001, 2002, 2005, 2006; Oprisan et al. 2002, 2003, 2005):

$$\Delta\varphi = \Delta\varphi_{||} + \Delta\varphi_\perp. \tag{6.11}$$

Since type 2 excitability arises via a Hopf bifurcation, the geometry of the associated limit cycle attractor is very sensitive to any changes of the control parameters, although the firing period is somewhat insensitive to the amplitude of the perturbation. As expected, predicting the PRC using only the tangent contribution to the geometric displacement $\Delta\sigma_{||}$ (dotted line, Fig. 6.6a) did not produce a good match for type 2 oscillators. The normal displacement using (6.10) was also computed and mapped into a phase difference $\Delta\varphi$. Although for type 2 oscillators the geometric to temporal phase prediction does not produce a perfect match with the actual PRC (solid line in Fig. 6.6), it proves that in principle the geometric interpretation of the PRC is valid (Oprisan and Canavier 2000, 2001, 2002, 2005, 2006; Oprisan et al. 2002, 2003, 2005).

4.3 *Hard Resetting Induced by Strong Perturbations of the Limit Cycle*

The examples given for soft resetting used explicit model equations to determine the geometric to temporal phase mapping (see Appendix). However, in most cases

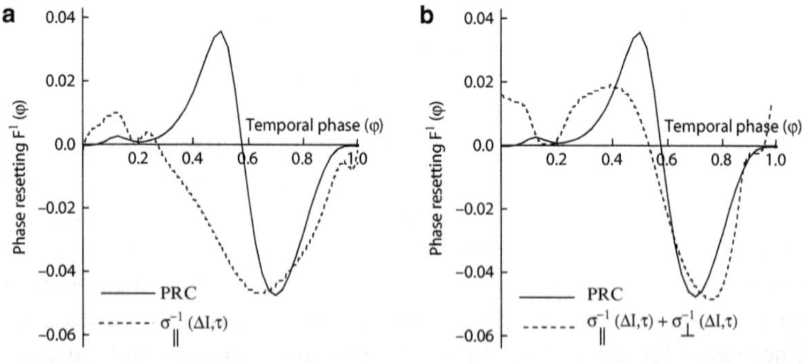

Fig. 6.6 Numerically generated PRC for a ML type 2 neuron (*continuous line*) has phase regions allowing both advance and delay of the next spike. Mapping tangential geometric displacement $\Delta\sigma_{||}$ only (*dashed line* in **a**) cannot accurately predict the PRC. A somewhat improved mapping that considered both the tangential $\Delta\sigma_{||}$ and the normal $\Delta\sigma_{\perp}$ displacements from the unperturbed limit cycle gives a better match to the actual PRC (*dashed line* in **b**)

of biological relevance the mathematical model is unknown. Therefore, using phase space reconstruction methods (see Sect. 2) to predict the PRC based on geometric considerations is the only alternative when only the membrane potential recordings are available. As already shown, the AB/PD complex of the STG could be well approximated by a two-dimensional limit cycle (Fig. 6.4). When using the phase space reconstruction, the limit cycle cannot capture the multiple time scales of the dynamic system since all reconstructed dimensions are simply time-shifted copies of the same measured time series. In order to mimic different timescales known to be involved in generating action potentials, the attractor was reconstructed (Fig. 6.7) by plotting the recorded time series (fast variable) versus a filtered and time-shifted version of it ("slow" variable).

4.3.1 Hard Inhibitory Resetting of Bursting Neurons

A hard resetting is induced by a large amplitude perturbation that has the ability to force a figurative point to switch from the depolarized (spiking) branch to the hyperpolarized (silent) branch (Fig. 6.7b). In fact, both the amplitude and the duration of the stimulus must be strong enough only to move the figurative point from the depolarized, or the hyperpolarized, branch of the limit cycle past the separatrix represented by the positive slope part of the voltage nullcline (Fig. 6.2a2 and 2b2). For example, a brief and strong inhibitory perturbation applied near phase $\varphi_{\text{old}} = 0.25$ (the horizontal arrow in Fig. 6.7b) shifts the figurative point from the depolarized to the hyperpolarized branch of the limit cycle. The new phase of the neural oscillator is about $\varphi_{\text{new}} = 0.6$ (Fig. 6.8a). The neuron skipped ahead a significant portion of the limit cycle due to the above inhibition and spikes again after $1 - \varphi_{\text{new}} = 0.4$ instead of $1 - \varphi_{\text{old}} = 0.75$ of its intrinsic period. This gives

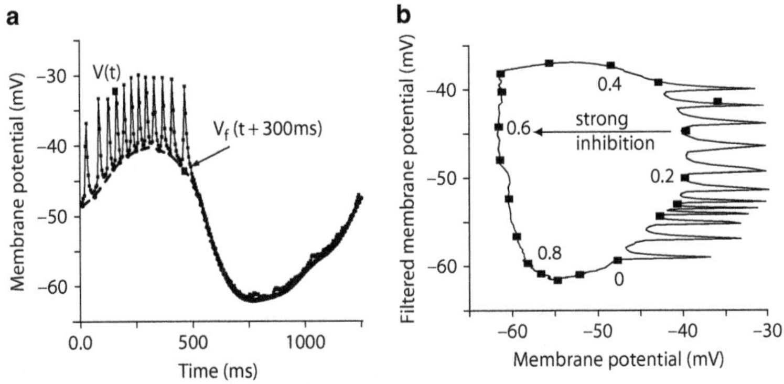

Fig. 6.7 The time-delayed envelope of the membrane potential provides the "slow" (filtered) variable (*dashed line*) for the reduced two-dimensional attractor reconstruction (**a**). The attractor retains the spikes of the "fast" variable (membrane potential) and is smooth with respect to the "slow" variable (**b**). The mapping has the same drawback as any phase space reconstruction using delay embedding technique – the attractor is always a phase oscillator

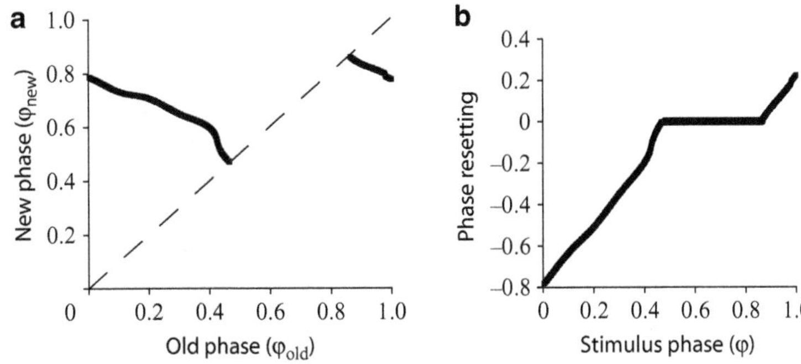

Fig. 6.8 Instantaneous PRC (**b**) induced by a hard stimulus derived from the mapping between the old and new phase (**a**). Any strong inhibition during the burst (or the depolarized branch) forces the figurative point to jump to the hyperpolarized branch and induces an advance of the next burst (negative phase resetting in panel **b**). Any inhibition during the hyperpolarized phase has no resetting effect (the horizontal line corresponding to zero resetting in panel **b**)

a phase advance of $\varphi_{\text{old}} - \varphi_{\text{new}} = 0.35$ as indicated at phase $\varphi = \varphi_{\text{old}} = 0.25$ in Fig. 6.8b. In general, the infinitesimal first order PRC induced by a brief and strong stimulus is given by (Oprisan and Canavier 2000, 2001, 2002; Oprisan et al. 2003, 2004; Oprisan and Boutan 2008):

$$F^1(\varphi) = \begin{cases} \varphi_{\text{old}} - \varphi_{\text{new}}, & \varphi \text{ on depolarized branch} \\ 0, & \text{otherwise} \end{cases}, \tag{6.12}$$

where φ_{old} is the (old) phase on the depolarized branch and φ_{new} is the (new) phase after the inhibitory stimulus was applied.

A very good agreement was found between the above theoretically predicted phase resetting using the phase space reconstructed limit cycle and the normal displacement hypothesis and the experimentally measured PRC from the AB/PD complex (Oprisan et al. 2003, 2004). Synaptic pulses of 100 ms and strengths in the range from 100 nS to 1000 nS were tested with very good agreement between the geometric mapping method and the experimentally measured PRC (Fig. 6.8b) (Oprisan et al. 2003, 2004). It must also be noted that the strength of inhibition must be above a certain threshold in order to force the figurative point to switch between branches of the limit cycle. If the strength of the perturbation is below threshold, then the perturbation slightly displaces the figurative point and the resetting must be computed according to the geometric method for soft perturbations (Sects. 3.1 and 3.2). The strong resetting threshold is determined by the distance to the separatrix of the phase space, which for two-dimensional neural oscillators is represented by the positive-slope voltage nullcline (Fig. 6.2a2 and 2b2). Both negative-slope depolarized and hyperpolarized branches of the voltage nullcline attract phase space trajectories, whereas the positive-slope segment of voltage nullcline repels all phase space trajectories (Fig. 6.2a2 and 2b2). Therefore, any hyperpolarization that is strong enough to move the figurative point outside the attraction basin of the depolarized branch past the separatrix advances the subsequent burst.

These PRCs also suggest that the effective control parameter that determines the shape of the PRC is not only the pulse amplitude but the pulse duration. The stimulus strength must be sufficient to move the figurative point outside the attraction basin of the current state past the separatrix. How does the duration of the inhibition affect the PRC? The finite duration τ of the perturbation induces additional delay to the total phase resetting. Therefore, the total phase resetting is given by

$$\Delta\varphi = \Delta\varphi_{\perp} + \Delta\varphi_{\tau}, \tag{6.13}$$

where $\Delta\varphi_{\tau}$ is the component of the phase resetting due to the time spent by the figurative point waiting for the suprathreshold inhibitory pulse to be over before a new burst can be initiated (Oprisan et al. 2003, 2004). On the most hyperpolarized branch there is no additional correction $\Delta\varphi_{\tau} = 0$ if the pulse duration τ is shorter than the remaining time interval until the neuron switches to the depolarized branch by itself, which is $(1 - \varphi_{\text{new}})P_{\text{i}}$. For example, a hard inhibitory pulse of duration $\tau = 0.3\ P_{\text{i}}$ applied at phase $\varphi_{\text{old}} = 0.25$ (Fig. 6.9) does not induce additional phase resetting $\Delta\varphi_{\tau}$ due to the finite pulse duration τ because the inhibition will be removed by the time the neuron switches to the depolarized branch of the limit cycle by itself. However, a correction of $\Delta\varphi_{\tau} = \varphi_{\text{new}} - 1 + \tau/P_{\text{i}}$ is otherwise required. For example, the same inhibitory perturbation applied at a phase $\varphi_{\text{old}} = 0.05$ maps into a new phase $\varphi_{\text{new}} = 0.725$, which gives an advance of $\Delta\varphi_{\perp} = 0.725$ (Fig. 6.9). However, since the time left until the neuron fires again, i.e., $(1 - \varphi_{\text{new}})P_{\text{i}} = 0.275\ P_{\text{i}}$, is shorter than the duration of the perturbation $\tau = 0.3\ P_{\text{i}}$, the neuron will be held hyperpolarized by the applied perturbation for an additional delay $\Delta\varphi_{\tau} = \varphi_{\text{new}} - 1 + \tau/P_{\text{i}} = 0.025$ (Oprisan et al. 2003, 2004).

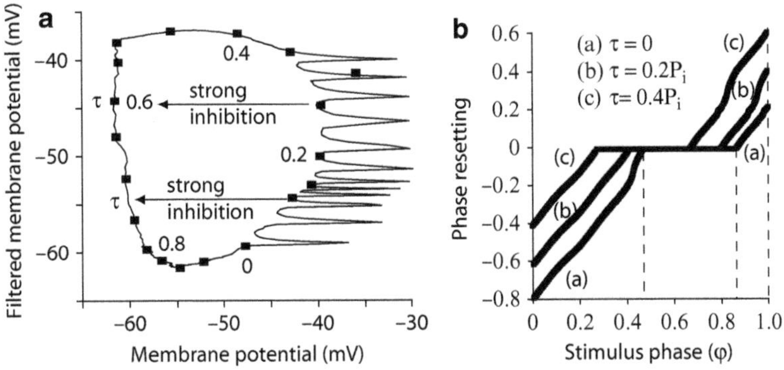

Fig. 6.9 Two independent strong and long-lasting ($\tau = 0.3\ P_i$) inhibitions map the old phase $\varphi_{old} = 0.05$ onto the new phase $\varphi_{new} = 0.755$ and, respectively $\varphi_{old} = 0.25$ onto $\varphi_{new} = 0.60$ (**a**). The figurative point is held on the hyperpolarized branch for the duration $\tau = 0.3\ P_i$ of the inhibition and released immediately afterwards. Holding the point that was mapped onto $\varphi_{new} = 0.755$ for $\tau = 0.3\ P_i$ induces an effective delay of $\Delta\varphi_\tau = \tau/P_i - \varphi_{new}$, which reduces the net advance (negative resetting) by $\Delta\varphi_\tau$ (**b**)

How is the instantaneous PRC in Fig. 6.8b affected by the finite duration τ of the perturbation? First, according to Figs. 6.7a and 6.8a, the maximum possible advance corresponds to an inhibition applied at $\varphi_{old} = 0$ that is mapped into $\varphi_{new} = 0.8$ (Figs. 6.7b and 6.8b). Any perturbation of duration $\tau > (1 - \varphi_{new})P_i = 0.2\ P_i$ will induce an additional delay of $\Delta\varphi_\tau = \varphi_{new} - 1 + \tau/P_i$. For example, if $\tau = 0.3\ P_i$ then $\Delta\varphi_\tau = \varphi_{new} - 1 + \tau/P_i = 0.1\ P_i$ and the total resetting will be $\Delta\varphi = \Delta\varphi_\perp + \Delta\varphi_\tau = -0.8 + 0.1 = -0.7$. Geometrically, the correction due to the finite duration of the perturbation shifts the entire PRC in Fig. 6.8b with $\Delta\varphi_\tau = 0.1\ P_i$ upward. As long as the stimulus duration τ is shorter than the time it takes to cover the entire hyperpolarized branch, which is $\Delta\varphi_{hyperpol} P_i = 0.85\ P_i - 0.45\ P_i = 0.4\ P_i$ (Fig. 6.9a), the extent of zero phase region in the PRC is constant. The above two situations are represented in the predicted PRC shown in Fig. 6.9b. The experimental PRCs were in very good agreement with the geometric-based predictions (Oprisan et al. 2004).

4.3.2 Hard Excitatory Resetting of Bursting Neurons

The geometric-based prediction of the PRC for rectangular and strong excitatory perturbations works the same way as for inhibitory pulses. Assuming that the neuron is already on the depolarized branch when it receives the excitatory input, it remains on that branch as long as the perturbation is active. As explained in Sect. 4.3.1, if the pulse duration τ is shorter than the time left until the neuron switches spontaneously from the depolarized branch to the hyperpolarized branch then the perturbation has no effect on the PRC. Otherwise, a delay equal to the difference between the pulse duration τ and the time left until the neuron spontaneously switches the branches

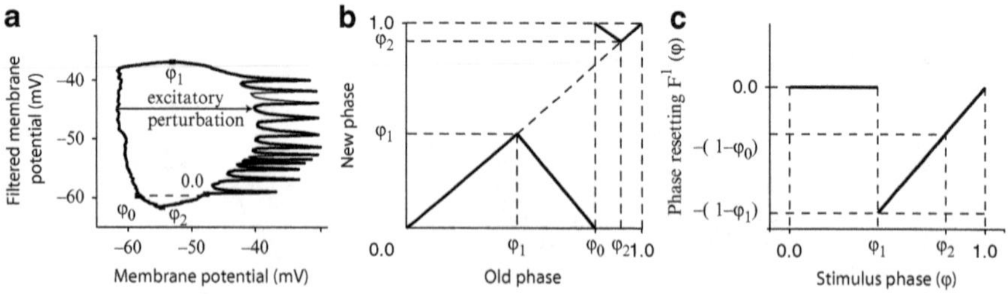

Fig. 6.10 A brief and strong excitatory perturbation (horizontal arrow in panel **a**) moves the figurative point from the attraction basin of the hyperpolarized branch past the separatrix into the attractive basin of the depolarized branch of the limit cycle. Any infinitesimal and strong excitatory perturbation has no resetting effect if applied during the burst (**a**). A sketch of the mapping between the two branches of the limit cycle shows no resetting (45° line in panel **b** and horizontal line of no resetting in panel **c**) during the burst. The perturbation always advances the next burst if it was received during the interburst interval (**b** and **c**)

of the limit cycle must be added to the PRC. The mapping of the phases along the depolarized branch φ_{new} versus the phases on the hyperpolarized branch φ_{old} using the reconstructed two-dimensional limit cycle is shown on Fig. 6.10 along with the corresponding infinitesimal PRC.

The two-dimensional reconstructed limit cycle (Fig. 6.10a) has a few remarkable points: (a) the hyperpolarized figurative point at phase φ_0 corresponding to zero phase on the depolarized branch, (b) the lowest filtered membrane potential point at phase φ_2, and (c) the highest filtered membrane potential at phase φ_1. Any instantaneous and hard excitation delivered during the burst should heave no effect and, as a result, determines the line of no resetting $\varphi_{\text{new}} = \varphi_{\text{old}}$ in Fig. 6.10b. Strong excitations applied during the interburst interval elicit immediate and truncated bursting activity with the old phase φ_{old} mapped to a smaller phase φ_{new}. The phase resetting is computed as:

$$\Delta\varphi = \varphi_{\text{new}} - \varphi_{\text{old}}, \tag{6.14}$$

and leads to a region of no resetting during the depolarized phase (Fig. 6.10c) and advances of the next burst otherwise.

As in Sect. 4.3.1, for a finite duration of the excitatory perturbation, an additional resetting $\Delta\varphi_\tau$ must be considered. If the stimulus duration τ is shorter than the intrinsic burst duration $\tau < \varphi_1 P_{\text{i}}$ then a strong excitation applied during the depolarized phase elongates the burst at phase larger than $\varphi_1 - \tau/P_{\text{i}}$ and induces a corresponding delay. The maximum phase delay of τ/P_{i} corresponds to the phase φ_1 that terminates the burst (Fig. 6.10a). For longer perturbations, the region of no resetting is completely replaced by a large delay equal to the difference between the pulse duration τ and the intrinsic burst duration $\varphi_1 P_{\text{i}}$ (Fig. 6.11b).

A striking characteristic of the PRC induced by a rectangular and finite duration excitatory perturbation is that the phase resetting during the interburst interval has the same value as for an instantaneous perturbation. The explanation is that a strong, finite duration excitatory perturbation always succeeds in moving the figurative

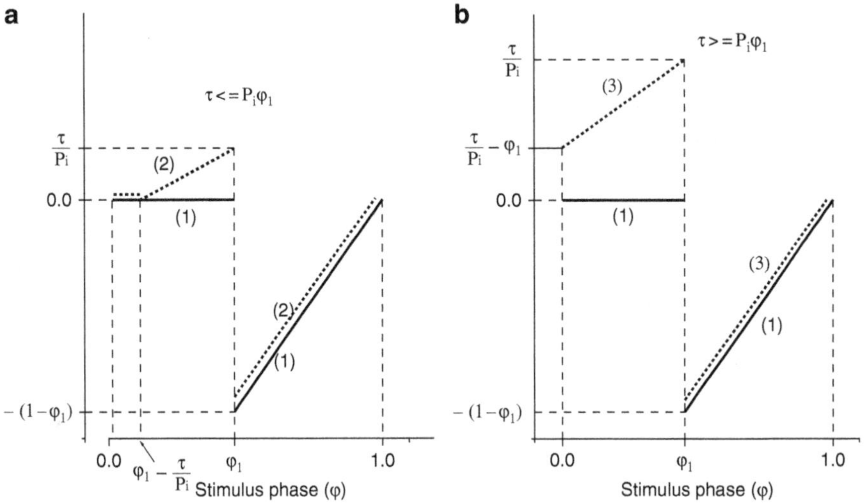

Fig. 6.11 Infinitesimal PRC in response to strong excitatory perturbations (*solid line* in panels **a** and **b**). The PRC determined by a strong and finite duration perturbation (*dotted line* in panels **a** and **b**) is sensitive to stimulus duration τ. Stimuli shorter then the burst duration $\tau < \varphi_1 P_i$ applied during the burst prolonged it. The additional delay is equal to the difference between the duration of the stimulus and the duration of the intrinsic burst (**a**). For infinitesimal and strong excitatory perturbations there is no predicted effect on the PRC if the stimulus is applied during the burst (**a** and **b**)

point from the attracting basin of the hyperpolarized branch past the separatrix into the attracting basin of the depolarized branch. Therefore, a new burst is generated and the induced phase resetting is the same as for any infinitesimal perturbation strong enough to force the neuron switch from the hyperpolarized to the depolarized branch of the limit cycle. The predict PRCs for strong excitatory perturbations (Fig. 6.11) are in good agreement with the experimental results (Prinz et al. 2003).

4.4 Applications to Circuit Analysis and Limitations of Geometric Mapping

Section 4 presented the theoretical framework of a general method for adding hard perturbations. If the perturbations were above the threshold required to move the figurative point over the separatrix, the amplitude of the synaptic conductance was not a determinant of the amount of phase resetting. Thus, the effect of an overlapping second perturbation with the same timing and duration and of arbitrary amplitude, would be independent of the effect produced by other perturbations (Prinz et al. 2003). If the pulses are of different durations, the longer duration alone would determine the phase resetting of the combined pulse. Similarly, the effect of two pulses whose durations did not overlap would be a simple sum of the effects of the two single pulses. This chapter showed how the PRC could be predicted both from known model equations and from a time series recording of the membrane potential. For type 1 excitability, the neural oscillators fire at arbitrarily low frequencies due to an underlying saddle-node bifurcation, the tangential geometric resetting predicts the temporal resetting with high accuracy (Oprisan and Canavier 2001). Such type 1 neural oscillators function as integrators (Izhikevich 2000), so the effect of a perturbation is converted to a change in the velocity of the figurative point moving around the limit cycle. On the other hand, the tangential approach does not work very well (Oprisan and Canavier 2001) for type 2 excitability, i.e., neural oscillators which cannot oscillate below a minimum firing frequency due to an underlying Hopf bifurcation. Such neural oscillators function as resonators, so a current perturbation changes the shape of the limit cycle with minimal effect on the phase space velocity. The major source of errors in predicting phase resetting from a time series is the trial-to-trial variability inherent in the experimental determination of the PRC. Other sources of noise are background activities of other neurons, channel noise, shot noise, and noisy release of transmitter. Another potential source of error is the delay embedding reconstruction method. Correct mapping between the points on the limit cycle with respect to the assumed second variable were obtained whether the slow variable is a low-pass filtered version of the membrane potential or the envelope of the membrane potential. Only the values of the time lag used for delay embedding is slightly different in each case. Finally, hard resetting was assumed to induce instantaneous normal switch to the other side of the limit cycle. In reality, activation and inactivation processes associated with voltage-gated channels all have kinetics associated with them, so the actual path may be curved and/or multi-dimensional. We expect that this will cause more problems with subthreshold pulses.

5 Generalization to Multiple Inputs per Cycle

5.1 *Soft Tangential and Normal Resetting due to Multiple Inputs per Cycle*

The above-described geometric approach to PRC prediction was generalized to the case when neural oscillators receive more than one input during a cycle (Oprisan and Canavier 2000, 2001, 2002, 2003, 2005, 2006; Oprisan et al. 2003, 2004; Oprisan and Boutan 2008; Oprisan 2009, 2010). The following derivation refers to a compound current pulse consisting of two rectangular current pulses of equal amplitudes and durations τ_1, respectively τ_2, separated by a gap of duration τ delivered during the same firing period. The geometric to temporal phase mapping method is general and could be applied to an arbitrary number of pulses with arbitrary shapes. As a consequence of the first current pulse of amplitude ΔI and duration τ_1, the geometric phase shift $\Delta\sigma_{||1}$ induced at the temporal phase

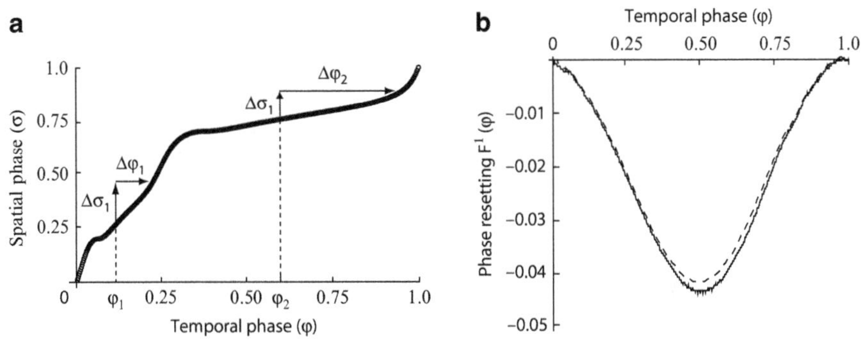

Fig. 6.12 A mapping between the temporal and the geometric phase (**a**) with two different perturbations applied at phases φ_1, respectively, φ_2. For each pulse, the geometric displacement $\Delta\sigma$ was mapped onto a phase resetting $\Delta\varphi$ that determined the cumulative PRC for a two-pulse per cycle perturbation. The predicted (*dashed line*) and the actual PRCs (*solid line*) are in very good agreement (**b**)

$\varphi_1 = t_1/P_\mathrm{i}$ is given by (6.5). Another current pulse of amplitude ΔI and duration τ_2, applied τ seconds later, induces an additional geometric phase shift $\Delta\sigma_{||2}$ given by (6.5). However, the temporal phase when the second pulse is delivered is not simply $(t_1 + \tau)/P_\mathrm{i}$ since the first pulse already reset the phase of the oscillator to a new value $\varphi_2 = \varphi_1 + \Delta\varphi_1$, where $\Delta\varphi_1$ is the temporal phase shift corresponding to the geometric phase shift $\Delta\sigma_{||1}$ computed using (6.6). This additional resetting was taken into account by defining a phase resetting for two inputs per cycle $F(\varphi)$ based entirely on the PRC for a single input per cycle (Oprisan and Canavier 2000, 2001, 2002, 2003, 2005, 2006):

$$F(\varphi) = F^1(\varphi) + F^1(F^1(\varphi) + \tau/P_i), \tag{6.15}$$

where $F^1(\tau_1/P_\mathrm{i}) + (\tau + \tau_2)/P_\mathrm{i} < 1$ in order to deliver both current pulses during the same cycle (Ermentrout and Kopell 1991). Numerical simulations using ML model neuron subject to two inputs per cycle are in very good agreement with the predictions based on the geometric to temporal phase mapping given by (6.15) as seen in Fig. 6.12. We found that (6.15) is also valid for different pulse shapes (Oprisan and Canavier 2000, 2001, 2002, 2003, 2005, 2006).

5.2 *Hard Resetting due to Multiple Inputs per Cycle*

As discussed in Sect. 4.3, for the inhibitory or the excitatory hard perturbations, the amplitude of stimulus does not affect the amount of phase resetting as long as it is strong enough to force the figurative point to switch limit cycle branches. This is a threshold effect, in that a pulse is either sufficient to cause a jump between branches or it is not. As a result, if two overlapping stimuli are both subthreshold, then neither will have an appreciable effect, and the method highlighted in Sects. 4.1 and 4.2 and (6.8) for soft resetting must be used to compute the total resetting. On the other hand, if each stimulus or their combined sum is above the threshold, a large phase resetting will be produced and computed with the method highlighted in Sect. 4.3 for hard resetting.

6 Conclusions

Neurons are excitable cells that can rapidly change their membrane potential from very hyperpolarized (below −65 mV) to highly depolarized (above +40 mV) states that could be sustained over a wide range of durations from milliseconds to a few seconds. Neurons are also information processing units. Neural information, viewed as an ordered sequence of symbols, could be assimilated to a sequence of hyperpolarized ("0" or "silent") and depolarized ("1," or "active") states. The simplest quantitative measure of information encoding the state of a neuron is its firing period, which is the time a neuron needs to return to the same state. The PRC is a quantitative measure of the relative change in the firing period of an excitable cell due to external perturbations, such as presynaptic inputs, externally injected currents, etc. Since the PRC contains all the necessary information regarding the response of an isolated cell to a particular stimulus, it could be used to predict the activity of neural networks (see Part 4 of this book). The current chapter is entirely dedicated to the study of the relationship between the PRC, which is accessible experimentally through direct membrane potential measurements, and the complex topology of the phase space associated with an excitable cell.

We started investigating this relationship by establishing a firm connection between the temporal φ and the geometric phase σ (see Fig. 6.2a3 and b3). There are at least two reasons why such a mapping $\sigma = \sigma(\varphi, \mu)$ is relevant for our understanding of the PRC: (1) Any external perturbation directly changes the phase space dynamics of a (few) parameter(s) μ, which in turn affect all dynamic variables of the system via nonlinear couplings. As a result, although we only perturb one, or a very limited number of, control parameter(s) μ (synaptic conductance, etc.), the response of the neuron translates the perturbation, among other things, into a change in the firing period. Therefore, the main goal of this chapter was reviewing the methods used to derive quantitative relationships between the firing period and the characteristics of the perturbation (shape, amplitude, duration, etc.) via the PRC. We derived theoretically and checked numerically the validity of such quantitative relationships by using a biophysically realistic computer model (Sect. 1). (2) In contrast to mathematical models, actual electrophysiological measurements of the membrane potential give direct access to only one variable, or dimension, of the multidimensional phase space. In order to extract the PRC from electrophysiological data, the original phase space dynamics must be reconstructed first. A review of methods used for delay embedding of a biological neuron was covered in Sect. 2.

The phase space trajectory of a periodically firing neuron is a close trajectory, or limit cycle, γ. The most obvious way of characterizing a limit cycle is by using a moving reference frame that travels with the figurative point though the phase space. The natural axes of such a reference frame are (1) the tangent line to γ oriented along the direction of travel and, (2) the normal axes oriented perpendicular to the tangent line to γ. As a result, an external perturbation naturally splits into tangent and normal perturbations with respect to γ (Sect. 3). Although the phase space dynamics is pretty dull, there are some remarkable points, curves, and planes that make it interesting. For example, fix points, which are at the intersection of nullclines, determine the existence, stability, and period of the limit cycle(s) that orbit them. Another example, the separatrices are curves, or hyperplanes, that follow nullclines and separate qualitatively different regions of the limit cycle such as the depolarized from the hyperpolarized branch. In the theory of dynamic systems, the terms soft (hard) perturbation refers to a perturbation that is too weak (strong enough) to force the system to cross a specific separatrix. Naturally, the easiest to model are weak and tangent perturbations of the phase space dynamics since they simply speed up or slow down the motion of the figurative point along the limit cycle. The amount of acceleration, respectively, deceleration induced by a soft tangential perturbation readily translates into a change in the neuron's firing period that can be tabulated by the PRC (Sect. 4.1). On the other hand, a soft perturbation that acts normal to the limit cycle involves estimations of the relaxation times needed for the figurative point to return to the unperturbed limit cycle (Sect. 4.2). Although the estimation of the relaxation time from detailed computational models is straightforward, there is no established method for computing the relaxation time from a phase space reconstruction based on experimental recordings of the membrane potential. It is easier and more accurate to predict the phase space dynamics in response to hard, or strong, external perturbations than for soft perturbations (Sect. 4.3). The reason is that, if an external perturbation is strong enough to force the figurative point to cross a separatrix, the process is usually very fast and both the initial and final phase space locations are very well defined because they are conjugated points on or near the limit cycle. The experimentally measured membrane potentials from biological neurons were used in this chapter to reconstruct the phase space limit cycle. The mapping of the geometric phase onto a temporal phase was successfully used to predict the PRC due to hard resetting based on pure topological considerations.

In general, a neuron receives multiple synaptic inputs simultaneously. Therefore, we generalized the topological method to complex perturbation patterns. The geometric mapping approach to the PRC prediction gave very good results both for hard inhibitory and excitatory perturbations. In this entire chapter, detailed, step-by-step explanations of the numerical procedures and the theoretical derivations were provided such that the reader can connect phase space and temporal phase through the PRC.

Acknowledgments I would like to thank my former postdoc advisor, Dr. Carmen C. Canavier, for introducing me to the fascinating field of phase resetting, and for the freedom she has allowed me in pursuing my own ideas and interests. I deeply appreciate the guidance, support, and inspiration she has given me.

I would especially like to thank my wife, Dr. Ana Oprisan, for her patience, good humor, and to my children Andrei and Andra for their understanding and patience. I dedicate this work to the memory of my parents.

I gratefully acknowledge the helpful comments and feedback I received from reviewers while preparing this chapter.

This work was partly supported by the National Science Foundation CAREER grant IOS – 1054914 to SAO.

Appendix

The numerical simulations were preformed using a single-compartment neural oscillator described by a standard conductance-based, or Hodgkin–Huxley (HH), mathematical model (Hodgkin and Huxley 1952d; Bean 2007; Clay 1998; Connor et al. 1977; Hille 2001). The rate of change of the membrane potential is proportional to the sum of the ionic currents plus any externally injected current flowing through the cell's membrane:

$$\begin{aligned} \mathrm{d}V/\mathrm{d}t &= -I_{\mathrm{Ca}} - I_{\mathrm{K}} - I_{\mathrm{Leak}} + I_0 \\ &= -\bar{g}_{\mathrm{Ca}} m_\infty(V)(V - V_{\mathrm{Ca}}) - \bar{g}_{\mathrm{K}} w(V - V_{\mathrm{K}}) - \bar{g}_{\mathrm{L}}(V - V_{\mathrm{L}}) + I_0, \end{aligned}$$

where V is the membrane potential, $\bar{g}_{\mathrm{ch}}$ and V_{ch} are the maximum conductance and, respectively, the reversal potential for the ionic channel *ch* (only calcium, potassium, and leak were considered), w is the instantaneous probability that a potassium channel is open, and I_0 is an external bias current. Each ionic current is the product of a voltage-dependent conductance and a driving force $I_{\mathrm{ch}} = g_{\mathrm{ch}}(V)(V - V_{\mathrm{ch}})$, where $g_{\mathrm{ch}}(V)$ is the product of the maximum conductance for that channel and a voltage-dependent gating variables. Morris and Lecar (ML) model neuron is described by two noninactivating voltage-sensitive gating variables: one instantaneous, voltage-dependent, calcium activation $m_\infty(V)$ and a delayed voltage-dependent potassium w given by a first order differential equation (Morris and Lecar 1981):

$$\mathrm{d}w/\mathrm{d}t = \phi(w_\infty(V) - w)/\tau(V),$$

where ϕ is a temperature-dependent parameter and a voltage-dependent relaxation time constant is defined by $\tau(V) = \mathrm{Cos}h^{-1}((V - V_{w0.5})/(2V_{w\,\mathrm{slope}}))$. All open-state probability functions, or steady-state gating variables x_∞, have a sigmoidal form (Morris and Lecar 1981):

$$x_\infty(V) = (1 + Tanh(V - V_{x\,0.5})/V_{x\,\mathrm{slope}})/2,$$

where $V_{x\,0.5}$ is the half-activation voltage and $V_{x\,\text{slope}}$ is the slope factor for the gating variable x. The ML model is widely used in computational neuroscience because it captures relevant biological processes and, at the same time, by changing only a small subset of parameters it can behave either as a type 1 or a type 2 neural oscillator. The dimensionless parameters for a type 1 ML neuron are: $V_{m\,0.5} = -0.01$; $V_{m\,\text{slope}} = 0.15$; $V_{w\,0.5} = 0.1$; $V_{w\,\text{slope}} = 0.145$; $V_{\text{K}} = -0.7$; $V_{\text{L}} = -0.5$; $V_{\text{Ca}} = 1.0$; $g_{\text{Ca}} = 1.33$; $g_{\text{K}} = 2.0$; $g_{\text{L}} = 0.5$; $I_0 = 0.070$; $\phi = 0.6$ (Ermentrout 1996). The model's dimensionless parameters were obtained by dividing all voltages by the value of calcium reversal potential ${V_{\text{Ca}}}^0 = 120\,\text{mV}$, all conductances were normalized by ${g_{\text{Ca}}}^0 = 4\,\text{mS/cm}^2$, and all currents were normalized by ${V_{\text{Ca}}}^0 {g_{\text{Ca}}}^0 = 480\,\mu\text{A/cm}^2$ (Ermentrout 1996). For example, the dimensionless reversal potential for the leak current of $V_{\text{L}} = -0.5$ means $V_{\text{L}} = -0.5\,{V_{\text{Ca}}}^0 = -0.5 \times 120\,\text{mV} = -60\,\text{mV}$.

References

Arshavsky, Y. I., Beloozerova, I. N., Orlovsky, G. N., Panchin, Y. V., & Pavlova, G. A. (1985). Control of locomotion in marine mollusc Clione limacina. I. Efferent activity during actual and fictitious swimming. *Exp. Brain Res.* 58, 255–262.

Abarbanel, H.D.I., Brown, R., Sidorowich, J.J., & Tsimring, L.Sh. (1993). The analysis of observed chaotic data in physical systems. *Rev. Mod. Phys.*, 65(4), 1331–1392.

Baker, G.L., & Gollub, J.B. (1996). *Chaotic Dynamics: An Introduction*. Cambridge, England: Cambridge University Press.

Bal, T., Nagy, F., & Moulins, M. (1988). The pyloric central pattern generator in crustacean: a set of conditional neuronal oscillators. *J. Comp. Physiol.*, 163, 715–727.

Bartos, M., & Nusbaum M.P. (1997). Intercircuit control of motor pattern modulation by presynaptic inhibition. *J. Neurosci.*, 17, 2247–2256.

Bean, B.P. (2007). The action potential in mammalian central neurons. *Nat. Rev. Neurosci.*, 8, 451–465.

Beer, R. D., Chiel, H. J., & Gallagher, J. C. (1999). Evolution and analysis of model CPGs for walking. II. General principles and individual variability. *J. Comp. Neurosci.*, 7, 119–147.

Broomhead, D.S., & King, G.P. (1986). Extracting qualitative dynamics from experimental data. *Physica D*, 20, 217–236.

Canavier, C. C., Butera, R. J., Dror, R. O., Baxter, D. A., Clark, J. W., & Byrne, J. H. (1997). Phase response characteristics of model neurons determine which patterns are expressed in a ring circuit model of gait generator. *Biol. Cybern.*, 77, 367–380.

Canavier, C. C., Baxter, D. A., Clark, J. W., & Byrne, J. H. (1999). Control of multistability in ring circuits of oscillators. *Biol. Cybern.*, 80, 87–102.

Casdagli, M., Stephen, E., Farmer, J.D., Gibson, J. (1991). State space reconstruction in the presence of noise. *Physica D*, 51, 52–98.

Chiel, H. J., Beer, R. D., & Gallager J. C. (1999). Evaluation and analysis of model CPGs for walking. I. Dynamical models. *J. Comp. Neurosci.*, 7, 1–20.

Clay, J.R. (1998). Excitability of the squid giant axon revisited. *J. Neurophysiol.*, 80, 903–913.

Coddington, E.A., & Levinson, N. (1955). *Theory of Ordinary Differential Equations*. McGraw-Hill, New York.

Collins, J.J., & Stewart, I.N. (1993). Coupled nonlinear oscillators and the symmetries of animal gaits. *J. Nonlin. Sci.*, 3, 349–392.

Collins, J.J., & Richmond, S.A. (1994). Hard-wired central pattern generators for quadrupedal locomotion. *Biol. Cybern.*, 71, 375–385.

Connor, J.A., Walter, D., & McKown, R. (1977). Modifications of the Hodgkin–Huxley axon suggested by experimental results from crustacean axons. *Biophys. J.*, 18, 81–102.

Eckmann, J.P., & Ruelle, D. (1985). Ergodic theory of chaos and strange attractors. *Rev. Mod. Phys.*, 57, 617–620.

Ermentrout, G.B. (1985). The behavior of rings of coupled oscillators. *J. Math. Biol.*, 23, 55–74.

Ermentrout, G.B., & Kopell, N. (1991). Oscillator death in systems of coupled neural oscillators. *SIAM J. Appl. Math.*, 29, 195–217.

Ermentrout, G.B. (1996). Type I membranes, phase resetting curves, and synchrony. *Neural Computation*, 8, 979–1001.

Ermentrout, G.B. (2002). *Simulating, Analyzing, and Animating Dynamical Systems: A Guide to XPPAUT for Researchers and Students*. SIAM, Philadelphia.

FitzHugh, R. (1955). Mathematical models of threshold phenomena in the nerve membrane. *Bull. Math. Biophys.*, 17, 257–278.

Fraser, A.M., & Swinney, H.L. (1986). Independent coordinates for strange attractors. *Phys. Rev.*, 33A, 1134–1140.

Golubitsky, M., Stewart, I., Buono, P.-L., & Collins, J.J. (1998). A modular network for legged locomotion. *Physica D.*, 115, 56–72.

Grassberger, P. (1983). Generalized Dimensions of Strange Attractors. *Physics Letters A*, 97(6), 227–230.

Grassberger, P., & Procaccia, I. (1983). Estimation of Kolmogorov entropy from a chaotic signal. *Phys. Rev. A*, 28, 2591–2599.

Grassberger, P., & Procaccia, I. (1983). Characterization of Strange Attractors. *Physical Review Letters*, 50 (5), 346–349.

Guckenheimer, J. (1975). Isochrons and phaseless sets. *J Math Biol.*, 1, 259–273.

Guckenheimer, J., & Holmes, P. (1983). *Nonlinear oscillations, dynamical systems, and bifurcation of the vector fields*. Berlin: Springer-Verlag

Hille, B. (2001). *Ion channels of excitable membranes*. Sunderland, MA: Sinauer.

Hirsch, M.W., Smale, S., & Devaney, R.L. (2004). *Differential Equations, Dynamical Systems and an Introduction to Chaos*. Academic Press, Elsevier, San Diego.

Hegger, R., Kantz, H., & Schreiber, T. (1999). Practical implementation of nonlinear time series methods: the TISEAN package. *Chaos*, 9,413–435.

Hodgkin, A.L. (1948). The local electric changes associated with repetitive action in a non-medullated axon. *J. Physiol.*, 107, 165–181.

Hodgkin, A.L., & Huxley, A.F. (1952a). Currents carried by sodium and potassium ions through the membrane of the giant axon of Loligo. *J. Physiol.*, 116, 449–472.

Hodgkin, A.L., & Huxley, A.F. (1952b). The components of membrane conductance in the giant axon of Loligo. *J. Physiol.*, 116, 473–496.

Hodgkin, A.L., & Huxley, A.F. (1952c). The dual effect of membrane potential on sodium conductance in the giant axon of Loligo. *J. Physiol.*, 116, 497–506.

Hodgkin, A.L., & Huxley, A.F. (1952d). A quantitative description of membrane current and its application to conduction and excitation in nerve. *J. Physiol.*, 117, 500–544.

Hoppensteadt, F.C., & Izhikevich, E.M. (1997). *Weakly connected neural networks*. New York: Springer-Verlag.

Ijspeert, A.J. (2008). Central pattern generators for locomotion control in animals and robots: A review. *Neural Networks*, 21(4), 642–653.

Izhikevich, E.M. (2007). *Dynamical Systems in Neuroscience: The Geometry of Excitability and Bursting*. MIT Press.

Kennel, M.B., Brown, R., & Abarbanel, H.D.I. (1992). Determining embedding dimension for phase-space reconstruction using a geometric construction, *Phys. Rev. A.*, 45, 3403.

Keener, J., Sneyd, J. (1998). *Mathematical Physiology*. Springer-Verlag, New York.

Kopell, N., & Ermentrout, G.B. (1988). Coupled oscillators and the design of central pattern generators. *Math. Biol.*, 90, 87–109.

Mane, R. (1981). On the dimension of the compact invariant sets of certain nonlinear maps. in D. A. Rand and L.-S. Young. Dynamical Systems and Turbulence, Lecture Notes in Mathematics, vol 898. Springer-Verlag. pp. 230–242.

Marder, E., & Calabrese, R.L. (1996). Principles of rhythmic motor patter generation. *Physiol Rev.*, 76, 687–717.

Marder, E., Manor, Y., Nadim, F., Bartos, M., & Nusbaum, M.P. (1998). Frequency control of a slow oscillatory network by a fast rhythmic input: pyloric to gastric mill interactions in the crab stomatogastric nervous system. *Ann. NY Acad. Sci.*, 860, 226–238.

Milnor, J. (1985). On the concept of attractor. *Commun. Math. Phys.*, 99, 177–195.

Morris, C., & Lecar, H. (1981). Voltage oscillations in the barnacle giant muscle fiber. *Biophys. J.*, 35, 193–213.

Mulloney, B. (1977). Organization of the stomatogastric ganglion of the spiny lobster. V. Coordination of the gastric and pyloric systems. *J. Comp. Physiol.*, 122, 227–240.

Murray, J.D. (1993). *Mathematical biology*. New York: Springer-Verlag.

Nagumo, J., Arimoto, S., & Yoshizawa, S. (1962) An active pulse transmission line simulating nerve axon. *Proc. IRE*, 50, 2061–2070.

Nayfeh, A.H., & Balachandran, B. (1995). *Applied Nonlinear Dynamics: Analytical, Computational, and Experimental Methods*. New York: Wiley.

Nishii, J. (1999). A learning model of a periodic locomotor pattern by the central pattern generator, *Adaptive Behavior*, 7(2), 137–149.

Nishii, J. (2006). An analytical estimation of the energy cost for legged locomotion, *Journal of Theoretical Biology*, 238, 636–645.

Oprisan, S.A., & Canavier, C.C. (2000). Phase response curve via multiple time scale analysis of limit cycle behavior of type I and type II excitability. *Biophys J.*, 78(1), 218A.

Oprisan, S.A., & Canavier, C.C. (2001). Stability Analysis of Rings of Pulse-Coupled Oscillators: The Effect of Phase Resetting in the Second Cycle After the Pulse Is Important at Synchrony and For Long Pulses. *Differential Equations and Dynamical Systems*, 9(3–4), 243–258.

Oprisan, S.A., & Canavier, C.C. (2002). The influence of limit cycle topology on the phase resetting curve. *Neural Computation*, 14, 1027–1057.

Oprisan, S.A., Thirumulai, V., Marder, E., & Canavier, C.C. (2002). Phase resetting in neural oscillators as a component of circuit analysis. 24th Annual Conference and the Annual Fall Meeting of the Biomedical Engineering Society. Engineering in Medicine and Biology 3:1960–1961.

Oprisan, S.A., Thirumalai, V., & Canavier, C.C. (2003). Dynamics from a time series: Can we extract the phase resetting curve from a time series? *Biophysical Journal*, 84, 2919–2928.

Oprisan, S.A., & Canavier, C.C. (2003). Stability analysis of entrainment by two periodic inputs with a fixed delay. *Neurocomputing*, 52–54, 59–63.

Oprisan, S.A., Prinz, A.A., & Canavier, C.C. (2004). Phase resetting and phase locking in hybrid circuits of one model and one biological neuron. *Biophys. J.*, 87, 2283–2298.

Oprisan, S.A., & Canavier, C.C. (2005). Stability criterion for a two-neuron reciprocally coupled network based on the phase and burst resetting curves. *Neurocomputing*, 65, 733–739.

Oprisan, S.A., & Canavier, C.C. (2006). Technique for eliminating nonessential components in the refinement of a model of dopamine neurons. *Neurocomputing*, 69, 1030–1034.

Oprisan, S.A., & Boutan, C, (2008), Prediction of Entrainmnet and 1:1 Phase-Locked Modes in Two-Neuron Networks Based on the Phase Resetting Curve Method. *International Journal of Neuroscience* 118, 867–890.

Oprisan, S.A. (2009). Stability of Synchronous Oscillations in a Periodic Network. *International Journal of Neuroscience*, 119(4), 482–491.

Ott, E. (1993). *Chaos in Dynamical Systems*. New York: Cambridge University Press.

Packard, N., Crutchfield, J., Farmer, D., & Shaw, R. (1980). Geometry from a time series. *Physical Review Letters*, 45, 712–716.

Pavlides, T. (1973). *Biological Oscillators: Their Mathematical Analysis*. Academic Press, New York.

Pearson, K. (1993). Common principles of motor control in vertebrates and invertebrates. *Annu. Rev. Neurosci.*, 16, 256–297.
Prinz, A.A., Thirumalai, V., Marder, E. (2003). The functional consequences of changes in the strength and duration of synaptic inputs to oscillatory neurons. *J. Neurosci.*, 23, 943–954.
Rabinovitch, A., & Rogachevskii, I. (1999). Threshold, excitability and isochrones in the Bonhoeffer–van der Pol system. *Chaos*, 9, 880.
Rinzel, J., & Lee, Y.S. (1986). On different mechanisms for membrane potential bursting. In H. G. Othmer (Ed.), *Nonlinear oscillations in biology and chemistry*. New York: Springer-Verlag.
Rinzel, J., Ermentrout, G.B. (1998). Analysis of neural excitability and oscillations. In C. Koch and I. Segev (Eds.), *Methods in neuronal modeling: From ions to networks* (pp. 251–292). Cambridge, MA:MIT Press.
Robertson, R.M., & Pearson, K.G. (1985). Neural circuits in the flight system of locusts. *J. Neurophysiol.*, 53, 110–128.
Ruelle, D., & Takens, F. (1971). On the nature of turbulence. *Commun. Math. Phys.*, 20, 167–192.
Sauer, T., Yorke, J.A., & Casdagli, M. (1991). Embedology. *J. Stat. Phys.*, 65, 579–616.
Sauer, T., & Yorke, J. (1993). How many delay coordinates do you need? *Int. J. Bifurcation and Chaos*, 3, 737.
Schouten, J.C., Takens, F., & van den Bleek, G.M. (1994). Estimating the dimension of a noisy attractor. *Phys. Rev. E*, 50, 1851–1861.
Schreiber, T. (1995). Efficient neighbor searching in nonlinear time series analysis, *Int. J. Bifurcation and Chaos*, 5, 349.
Selverston, A.I., & Moulins, M. (1987). *The crustacean stomatogastric system*. Springer-Verlag, Berlin.
Takens, F. (1981). Detecting strange attractors in turbulence. in D. A. Rand and L.-S. Young. *Dynamical Systems and Turbulence*, Lecture Notes in Mathematics, vol. 898. Springer-Verlag. pp. 366–381.
Wiggins, S. (1994). *Normally Hyperbolic Invariant Manifolds in Dynamical Systems*. Springer-Verlag, New York.
Winfree, A.T. (1980). *The geometry of biological time*. Springer-Verlag, New York.
Winfree, A.T. (1987). *The Timing of Biological Clocks*. Scientific American Books. New York.
Wolf, A., Swift, J.B., Swinney, H.L., & Vastano, J.A. (1985). Determining Lyapunov exponents from a time series. *Physica D*, 16, 285–317.

Chapter 7
PRC Estimation with Varying Width Intervals

Daniel G. Polhamus, Charles J. Wilson, and Carlos A. Paladini

Abstract The definition of the infinitesimal phase-resetting curve implies that the period of oscillation is known and constant for all values of the perturbation phase. Experimental estimation of the infinitesimal phase-resetting curve in neurons requires estimation of the unperturbed period of oscillation and the changes in the period in response to perturbations along the phase. Action potentials provide well-defined, yet stochastic endpoints of the cycle of oscillation. Experimental estimation of the phase-resetting curve is substantially complicated as a result. Here, we discuss a common problem with experimental PRC estimation caused by using the mean interspike interval (ISI) to describe the resting period. We propose a solution through truncated estimators of the firing period, conditioned upon observed phase: i.e., an estimate of the period given that we have experienced up to phase τ.

1 Introduction

The definition of the infinitesimal phase-resetting curve ($Z(\psi) = \frac{T-T'(\psi)}{T}$) implies that the period of oscillation (T, such that $g(t+T) = g(t)$ for all $t \geq 0$) is known and constant for all values of the perturbation phase $\phi = \frac{t \bmod T}{T}$. Experimental estimation of the phase-resetting curve in neurons requires knowledge of the unperturbed period of oscillation and the changes in the period in response to perturbations along the phase, $T'(\psi)$. In practice, action potentials provide well defined endpoints of the cycle of oscillation. If there is no variability in the difference in these endpoints [the interspike interval (ISI)], the dynamic system's period will be the interval length. As ISIs are stochastic, the interval differences are

D.G. Polhamus (✉) • C.J. Wilson • C.A. Paladini
UTSA Neurosciences Institute, University of Texas at San Antonio, One UTSA Circle, San Antonio, TX 78249, USA
e-mail: danp@metrumrg.com; charles.wilson@utsa.edu; carlos.paladini@utsa.edu

N.W. Schultheiss et al. (eds.), *Phase Response Curves in Neuroscience: Theory, Experiment, and Analysis*, Springer Series in Computational Neuroscience 6,
DOI 10.1007/978-1-4614-0739-3_7,

described by a probability distribution on the range of possible periods rather than by a single value. Experimental estimation of the infinitesimal phase-resetting curve is substantially complicated as a result.

Minimizing the error in both the periodic estimate and the perturbation interval would improve our estimate of the PRC as its variability is a complicated mixture of the ISI histogram variance and the variance of the periodic estimate. We can do little about the variation in the perturbation interval estimate – it is just the observed interval after perturbation at τ – so instead we focus on the periodic estimation. When there are no trends in firing rate over the course of the trials, a single ISI histogram approximates the probabilistic distribution of periods. From this, we wish to choose a statistic that best estimates the period. If the estimate is wrong, then the PRC will be systematically wrong.

The period can be misestimated through nonstationarity. In nonstationary data, the ISI histogram will begin to change (drift, expand or contract, change skew) over the course of observation. Common approaches to dealing with nonstationarity are either to estimate the period of the unperturbed oscillation using one or a few ISIs preceding the perturbation (e.g., Reyes and Fetz 1993; Izhikevich 2007) or to break the nonstationary ISI histogram into several stationary ISI histograms (e.g., Netoff 2005; Tateno and Robinson 2007). Each of these time continuous ISI histograms (denoted sessions) are then summarized by the session means. The former estimate controls deviations from stationarity by predicting the period as its immediate past, yet it is severely limited by use of unadjusted samples drawn randomly from the ISI distribution to predict the next randomly drawn sample from the ISI distribution. This error will propagate into the phase-resetting curve, making it necessary to greatly increase the number of trials required to extract a reliable mean PRC. For example, in a series of trials collected using a perturbation of size zero (no effect), the relative distribution of values at each point in the PRC would be greater than that of the ISI distribution, as the variability of the ratio of the ISI distribution and the periodic estimate is approximately (e.g., through the variance of an expansion of the pRC transformation) a scaled sum of the ISI variance and the variance of the draw or mean of the few draws acting as the periodic estimate. On the contrary, when using session means the variance of the periodic estimate will approach zero as the session sample size increases.

The ISI for each trial of the PRC (the interval perturbed to $T'(\psi)$) is a draw from the ISI distribution. As the phase of the perturbation approaches the mean period of the cell's oscillation, there will be trials in which the cell fires before the perturbation is presented. These trials must be rejected, and so for those values of perturbation phase, one is sampling from a truncated version of the ISI distribution. As the perturbation is presented later in the cycle, it is applied to an increasingly small subset of intervals from the interval histogram. Not only does this distort the variance of the PRC at large values of phase but also the estimate of the average period calculated from the sample of ISIs becomes increasingly invalid. The mean value of the PRC generated by averaging the results for a sample of trials is systemically distorted.

Phase dependency of the latent perturbation phase, resulting perturbation interval, and implied truncated sampling suggests an improved estimate of the period over

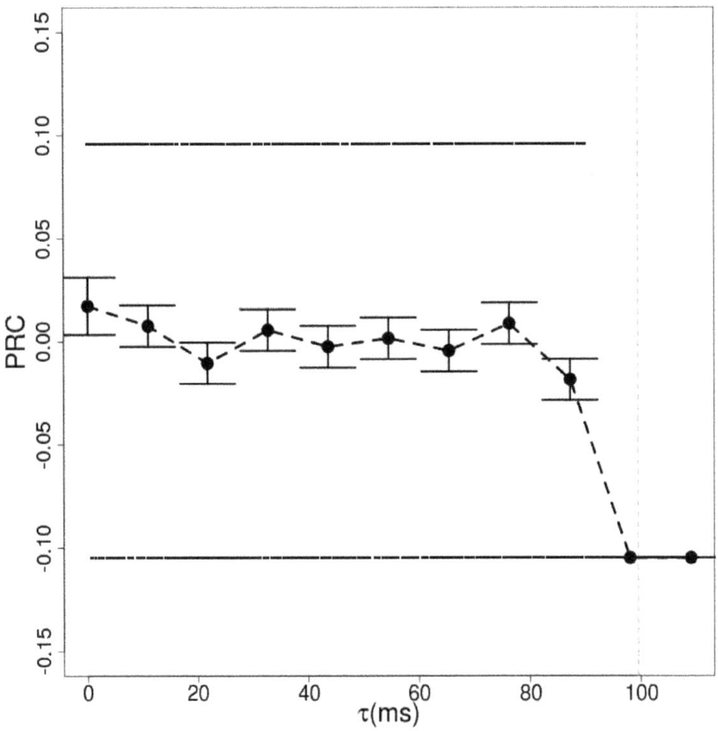

Fig. 7.1 Here we show the PRC of the degenerate neuron displaying no input. Shown are the PRC observations from a neuron firing only at 90 ms and 110 ms (with phase-resetting points of −0.1 and 0.1, respectively) as well as the means and standard error. Bin width corresponds to the error bar width. The period is estimated using the spontaneous mean (100 ms). Responses from the bimodal neuron were simulated in the R statistical package: 1,000 Bernoulli observations were generated, all 0s were assigned an interval of 90 ms and all 1s assigned intervals of 110 ms

sample means – namely, the mean interval conditional on the latent phase of perturbation. We estimate the period based upon only periods with probability greater than zero; these are intervals under the ISI histogram left truncated at τ. It is this observation that motivates our approach to noisy PRC estimation, an approach to minimize the variability in the PRC transformation by utilizing latent phase information to provide better periodic estimates.

For example, consider an oscillator that repeats with only two discrete interval lengths (e.g., 90 ms and 110 ms). Its bimodal ISI histogram will have an estimated period of 100 ms using the mean. By choosing the mean, we will never correctly predict the true period of oscillation. Rather, we will consistently misestimate the period by 10 ms, yet no other estimate will do as well on average. For instance, estimating the current underlying period with the mode (equal probability of 90 ms and 110 ms) means we correctly estimate the period half the time, and misestimate by 20 ms the other half of the time. This is illustrated in Fig. 7.1, where a Bernoulli sample (converted so that $0 \rightarrow 90$ and $1 \rightarrow 110$) was used to simulate a PRC experiment.

The standard statistical method to compare different estimates is mean squared error (MSE). For this example, we know that when using the mean every squared component of the MSE will be 10^2, and thus the MSE of the mean will be 100. Using either mode, we estimate half of the intervals being 90 ms and half being 110 ms. Regardless of which mode is selected, the error will be 0 on half of the trials, and 20 ms on the other half. The total MSE is 200 when using either mode. Despite never correctly guessing the underlying period, using the average provides a better (least squared error) estimate of the underlying period. A PRC calculated for this cell will have a bimodal distribution with zero mean at all perturbation times less than 90 ms. For perturbation times greater than 90 ms, however, the trials will be divided into two groups. One group will be rejected, because the cell fired before the perturbation was presented. The other set of trials will be used in calculating the PRC. In those, the variance will be zero, but because the true period is 110, not the value of 100 used to calculate the PRC, there will appear to be a 10 ms phase lag, which is entirely an artifact. Using 100 to the estimate a 90 period causes phases from 0.9 to 1.0 to be undersampled (the maximum calculated phase in a cycle would be 90/100), and using 100 to estimate the 110 period leads to fictive phases greater than 1. In this chapter, we demonstrate the origin of this artifact and suggest a method for correcting experimental PRCs to remove its effects.

2 Methods

2.1 Models

To demonstrate the improved estimation, we use three simulated examples and experimental data recorded from neurons of the subthalamic nucleus. The simulated examples include a repetitively firing Hodgkin–Huxley model (Hodgkin and Huxley 1952) subjected to Gaussian noise and with parameters adjusted to produce Type I and Type II phase resetting. Type I and Type II phase resetting (Hansel et al. 1995; Ermentrout 1996; Izhikevich 2007) show only advanced phases or the delayed and advanced phases due to stimulation, respectively. The Hodgkin–Huxley neuron models were made noisy by the addition at each time sample of a current drawn from a zero mean normal distribution. Histograms of the spontaneous activity of the model neurons are depicted in Fig. 7.2. The equations for the model neurons were:

$$\frac{\mathrm{d}V}{\mathrm{d}t} = -I_L - I_K - I_{Na} + I_{\mathrm{app}} - I_{\mathrm{noise}}$$

$$x_\infty(V) = \left[1 + \exp\left(-\frac{-(V - V_{Hx})}{V_{Sx}}\right)\right]^{-1}$$

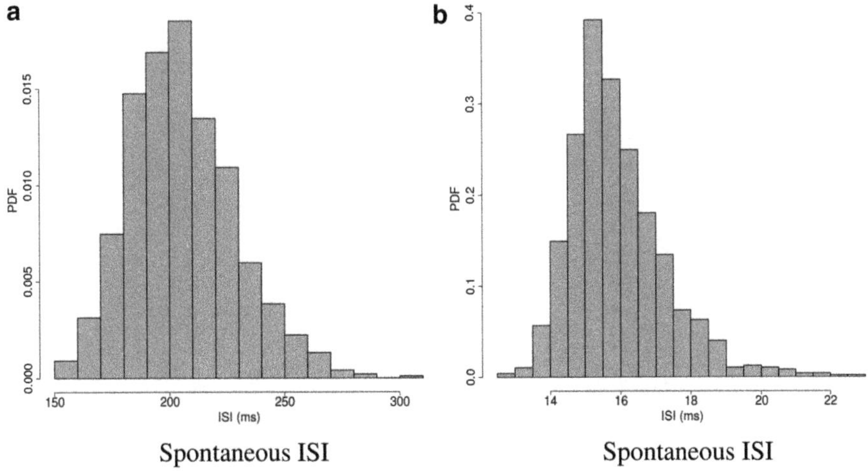

Fig. 7.2 Interspike interval histograms for the Hodgkin-Huxley model under Type I and Type II phase-resetting: $\mu_I = 205.67$, $\sigma_{II} = 22.93$, $n = 1{,}000$; and $\mu_{II} = 15.98$, $\sigma_{II} = 0.065$, $n = 1{,}000$ (**a**) Type I Spontaneous ISI (**b**) Type II Spontaneous ISI

$$\tau_x(V) = \tau_{x_0} + \tau_{x_1}\left(1 + \exp\{(V - th_x)/\sigma_x\}\right)^{-1}$$
$$I_L = g_L(V - V_L);\ I_K = \bar{g}_K n^4(V - V_K);\ I_{Na} = \bar{g}_K m^3 h(V - V_K)$$
$$I_{\text{noise}} = w_t \sigma_{\text{noise}};\ w_t - w_{t-1} \sim N(0, 1)$$

in which V is the membrane potential, I_y is the current due to ionic species $y \in \{Na^+, K^+, Leak\}$, g_y and $\bar{g}_y$ are the conductances and maximal conductances due to ionic species y, and $\tau_x(V)$ are the fitted time constant for activation/inactivation variable $x \in \{m, n, h\}$. The values of parameters for the two neurons are shown in Table 7.1.

The model was integrated using Xppaut (Ermentrout 2002) with a time step of 0.05 ms for the Type I model and a time step of 0.01 ms under the Type II model. The total variability in these two models scales with the time step as the noise current was added at each iteration.

2.2 Experimental

The phase resetting curves used in Fig. 7.7 were obtained using perforated patch recordings from neurons of the rat subthalamic nucleus in tissue slices. Slices were prepared from rats 17 to 25 days of age, and were cut in the sagittal plane using techniques described elsewhere (Teagarden et al. 2008) and superfused with ACSF containing 125 mM NaCl, 3.0 mM KCl, 1.25 mM NaH2PO4, 2 mM CaCl2, 1.5 mM

Table 7.1 Parameters used for the Type I and Type II H-H neuron models

Parameter	Type I	Type II	Parameter	Type I	Type II
V_L	−60	−60	V_{Sm}	10	10
V_K	−90	−90	V_{Hh}	−67	−67
V_{Hn}	−56	−56	V_{Sh}	7	7
V_{Sn}	9	18	τ_{m0}	0.09	0.09
g_L	0.1	0.1	τ_{m1}	0.6	0.6
τ_{n1}	6.5	6.5	σ_m	21	21
th_n	−53.5	−53.5	τ_{h0}	1	1
σ_n	22	22	τ_{h1}	9	9
$\bar{g}_K$	4	4	th_h	−57	−57
V_{Na}	60	60	σ_h	8	8
V_{Hm}	−45	−45	$\bar{g}_{Na}$	43	43
I_{app}	−1.9	7	V_{noise}	0	0
σ_{noise}	0.075	1.5			
Pulse	0.05 pA	5 pA			

MgCl, 26 mM NaHC03, and 10 mM glucose, bubbled with 95% O2, 5% CO2, and maintained at 32–35. Perforated patch recordings were obtained using electrodes filled with 140.5 mM KMeSO4, 7.5 mM NaCl, 10 mM HEPES buffer, to which had been added 100 g/ml gramicidin-D. Gramicidin-D stock solutions of 10 mg/ml gramicidin in DMSO were prepared daily. 5 l of this solution was added to a 500 l aliquot of electrode filling solution and mixed by sonication within 20 min of use. Phase-resetting curves were obtained after equilibration of the access resistance (30 min after sealing). Small (1–50 pA) current pulses were applied 1/second during ongoing spontaneous activity, so that the phase of the stimulus was asynchronous with firing. This randomized stimulus phases across trials. For calculation of the mean phase-resetting curve, phase was divided into 16 bins, and the mean and standard error of the phase shift was plotted for each bin. Each stimulus was preceded and followed by several unperturbed ISIs, and these were used to calculate the mean unperturbed interval.

3 Results

3.1 Conditional Means

The estimated phase-resetting curve is calculated as,

$$PRC(\tau) = \frac{\hat{T}(\tau) - T'(\tau)}{\hat{T}(\tau)}$$

using $\hat{T}(\tau)$ to denote the experimental estimation of the period.

We compare two different estimates of the period: the historical estimate – the average spontaneous ISI independent of the stimulus τ, $\hat{T}(\tau) = \frac{1}{n}\sum_{i=1}^{n} t_i = \bar{t}$, and the mean interval *given that the mean must be larger than* τ. We suggest two ways to estimate this conditional mean, the simplest being the left truncated sample average, $\hat{T}(\tau) = \sum_{i=1}^{n_\tau} t_i = \bar{t}|t_i > \tau$. Here, n_τ is the number of spontaneous ISIs greater than the latent perturbation time, τ (i.e., $n_\tau = \sum_{i=1}^{n} \mathrm{I}_{(>\tau)}(t_i)$). This formula describes the average of only the observed spontaneous intervals greater than the latent perturbation time, τ. The truncated mean is motivated by the fact that when applying a perturbation at time τ, it is impossible for the perturbation interval to be less than τ. That is, we have more information about the period of the current interval than simply its average spontaneous activity, the average needs to be conditioned on the fact that $\hat{T}(\tau)$ must be greater than τ.

While the left-truncated sample average approaches the conditional mean with large enough samples, we often are limited to noisy samples that have yet to converge. As the latent perturbation time enters the range of ISI variability, the number of spontaneous intervals greater than this perturbation time will progressively decrease. As a consequence, the variability in the periodic estimate increases. By approximating some of the probability components, we can alleviate the dependence on sample size of the estimated conditional mean. To do so, we must introduce some distributional probability results.

The conditional probability of observing an ISI of length t is $P(T=t|T>\tau) = \frac{P(T>\tau)\cap P(T=t)}{P(T>\tau)}$. For ISI's at least τ (i.e., $t > \tau$), $P(T = t|T > \tau) = \frac{P(T=t)}{1-F(\tau)}$ where F denotes the cumulative distribution of T (the CDF is $F(\tau) = F(T \le \tau) = \int_0^\tau f(u)\mathrm{d}u$). If the derivative ($f = F'$) exists, the left truncated distribution follows:

$$f_T(\tau) = \frac{f(t)}{1 - F(\tau)}; \quad \tau \le t < \infty. \tag{7.1}$$

The way to think about this is that the probability density function f retains its shape, but as it must still integrate to 1 (every probability distribution must integrate to 1) it must be multiplied by a proportionality constant. As τ is fixed, $1-F(\tau)$ is fixed as well and represents this proportionality constant which increases monotonically in τ.

Using the conditional mean, $\mathrm{E}[T|T > \tau] = \int_\tau^\infty [u|u > \tau]\, f(u)\mathrm{d}u$, improves the estimate of the period by utilizing the extra information contained in the latent spike time. The conditional probability density function (pdf) at τ and the conditional mean given τ are illustrated in Fig. 7.3, along with the concept of multiplying by a proportionality constant of the integral. Overlaid is the conditional mean.

3.2 *Estimation*

The truncated mean can be estimated in several ways: (a) One may use the truncated sample average, as earlier mentioned. Alternatively, (b) the parametric form derived from (7.1) in terms of the spontaneous ISI histogram will yield estimates paralleled

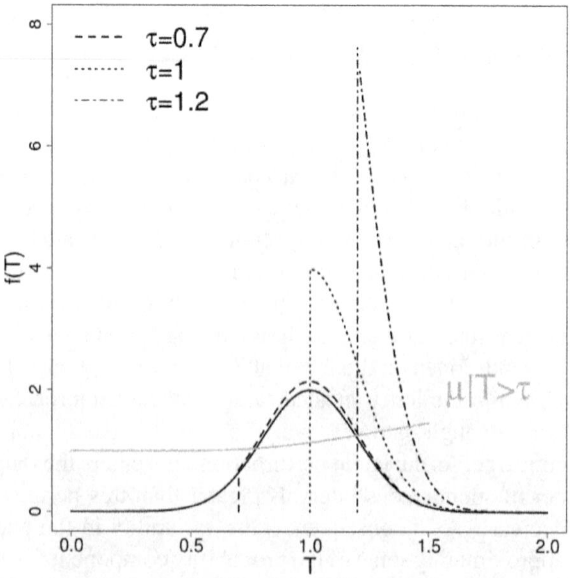

Fig. 7.3 The probability density function of a normal random variable ($\mu = 1, \sigma = 0.2$) random variable under left truncation. The truncated mean is shown in *gray*

in quality by the goodness of fit to the ISI histogram. Another approach for ISI histograms that yield poor parametric fits is to (c) approximate the ISI histogram distribution discretely and then use a numerical integration to find the conditional mean. While (a) is simple conceptually and practically, it suffers from relying on degrees of freedom generated from the right tail of the distribution, which consequently occur with lower probability as τ increases. The second method (b) yields conditional mean estimates that are as accurate as the parametric fit to the ISI histogram, but finding a solution to the expectation over the truncated distribution is typically a burdensome task. The third method is simple and robust, assuming the ISI distribution has been sampled densely. To summarize, the following estimates and notation will be used:

- The historical estimate of the period, the sample mean of the spontaneous activity,

$$\hat{T}(\tau) = \frac{1}{n}\sum_{i=1}^{n} t_i = \bar{t}.$$

- The truncated sample average of the spontaneous activity,

$$\hat{T}(\tau) = \frac{1}{n_\tau}\sum_{i=1}^{n_\tau} t_i = \bar{t}|T > \tau.$$

- The conditional mean in terms of a parametric distributional fit to the spontaneous ISI histogram. An example of this for the normal distribution is provided in the appendix, but the basic idea is to solve for the parametric conditional mean in terms of an appropriate parametric distribution. For example, if we assume a normal ISI histogram, then we solve for analytic value of the conditional mean of a normal distribution. Once the analytical form is found, computational time and algorithmic ease is trivial. As the conditional mean is solved in terms of the distributional parameters of the assumed distribution for the ISI histogram, it is necessary that we use invariant estimates of the parameters to estimate the conditional mean. Fortunately, the maximum likelihood estimates (MLE) are simple to find and meet the invariance criteria. With f denoting the parametric pdf of the spontaneous ISI histogram and F denoting the cdf, we will denote this estimate of the conditional mean the "ML plug-in" written with MLEs on the parameter vector $\hat{\theta}$,

$$\hat{T}(\tau) = \int_{\tau}^{\infty} t \frac{f(\hat{\theta}, t)}{1 - F(\hat{\theta}, \tau)} \mathrm{d}t = \hat{T}_{ML} | T > \tau.$$

- Finally, we illustrate the most robust of the group of estimators (due to their nonparametric nature). After the ISI histogram is approximated (N total draws), we have found a bin-width (Δt), bin number ($b_j,\ j = 1, 2, ..., J$), and a frequency associated with a bin number ($\{n_j : \sum_{j=1}^{j} n_j = N\}$). The probability mass function (discrete pdf) is estimated as $\hat{f}(t \in b_j) = \frac{n_j}{N}$ and the cumulative distribution function as $\hat{F}(t \in b_j) = \sum_{i=1}^{j} \hat{f}(b_i)$. We approximate the interval in the conditional mean giving

$$\hat{T}(t|T > \tau) = \sum_{j=1}^{J} \left(j \times \Delta t - \frac{\Delta t}{2} \right) \times \left(\frac{\hat{f}(b_j)}{1 - \hat{F}(b_j)} \right) = \hat{T}_{DA} | T > \tau.$$

It is important to note that the monotonic nature of the conditional mean is preserved in all three methods. We recommend inspection of the conditional mean as a function of τ to check that it is a relatively smooth monotonically increasing function when approximating the expectation. To this end, adjustment of the bin-width will help as will smoothing or interpolation on the actual conditional mean function. Additionally, smoothing will help to alleviate sampling error introduced by using a histogram to estimate the pdf.

3.3 *Simulations*

The estimated PRC is misestimated unless it is estimated using truncated means. To demonstrate this, we show a neuron following a Gaussian spontaneous ISI histogram with no perturbation, so that $\mathrm{E}[PRC] = 0$. After using this nonresponsive example to clearly demonstrate the artifact presented by using the historical estimate of the period, we demonstrate with a more realistic example, a Hodgkin–Huxley model under Type I and Type II phase resetting.

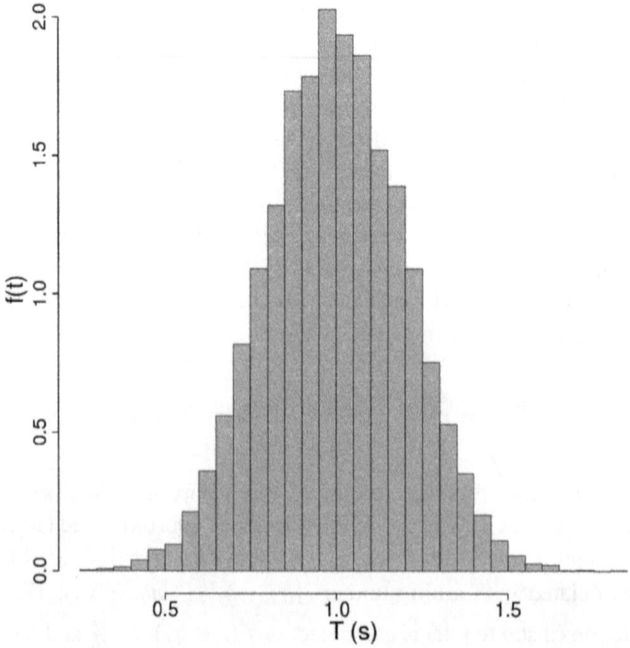

Fig. 7.4 ISI histogram for the neuron firing normally around $\mu = 1$ and $\sigma = 0.2$

3.3.1 Example 1: No Response, Normal

First, we consider a more realistic neuron generating spontaneous ISIs described by a Normal distribution function, parameterized by $\mu = 1$ s and $\sigma = 0.2$ ms (Fig. 7.4). This neuron also shows no response to stimuli, and its PRC should be centered around 0. Figure 7.5 clearly demonstrates the benefit of using the truncated estimate as the latent phase, τ, begins to overlap into the uncertainty of the period (1 ± 0.6). Again, the dashed vertical line shows $\tau = \bar{t}$, the traditional estimate of the period. The historical estimate of the PRC reflects the uncertainty in spontaneous firing, showing negative bias as $\bar{t}$ becomes a less likely periodic estimate for the underlying perturbation interval. There is a noticeable decrease in the variability for both *a* and *b* of Fig. 7.5 but we point out that the error bars are reflecting bin variability. There is little difference between the variability around the PRC in the two graphs, but the presence of a linear effect at the end of the phase in Fig. 7.6a inflates the estimated variability of the PRC (by the binned average) considerably.

3.3.2 Example 2: Type I and Type II Phase Resetting, Hodgkin–Huxley

We now consider the previously described Hodgkin–Huxley model under Type I and Type II phase resetting, shown in Fig. 7.2. Figures 7.6 a,b display the expected

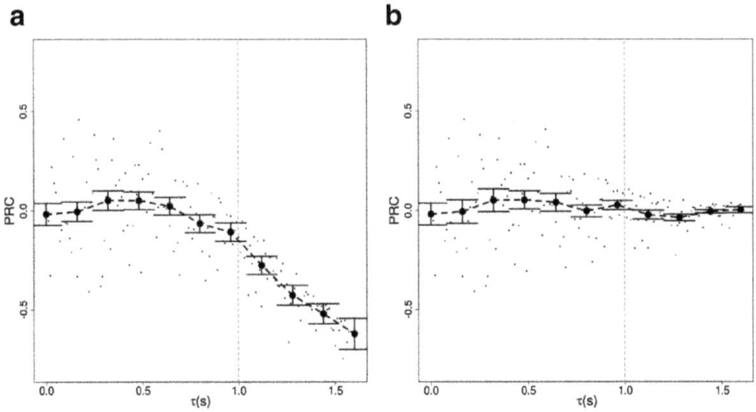

Normal PRC, $\hat{T}(\tau) = \bar{t}$ Normal PRC, $\hat{T}(\tau) = \hat{T}_{ML}/T > \tau$

Fig. 7.5 The phase-resetting curve for a neuron firing normally, under the assumption that it shows no response to stimulus. The phase-resetting transformation was calculated using the spontaneous average ISI and the fully parametric ML plug-in method to estimate the period. The error bars are again standard errors of the bins. The *vertical gray line* shows the conventional estimate of the period (**a**) Normal PRC, $\hat{T}(\tau) = \bar{t}$ (**b**) Normal PRC, $\hat{T}(\tau) = \hat{T}_{ML}|T > \tau$

PRC behavior of a Type I model. The perturbation intervals only decrease as a function of the latent perturbation time. With Type II excitability (Figs. 7.6 c,d), the functional resonator shape is apparent (Izhikevich 2000), in which the perturbation interval is increased in response to an early phase perturbation and decreased in response to one occurring later in the phase. The behavior of the curves under different estimations of the period is similar to the first two examples; as the phase of perturbation ranges through the area of firing uncertainty both the previous interval and mean interval begin to underestimate the underlying period. This decreases the estimated PRC from the expected limit of 0 (or at least PRC(0)) as $\tau \rightarrow T$. The error bars represent standard deviation to illustrate the effects on variance, as well as mean of the PRC. Also shown are the theoretical phase-resetting curves from the noiseless models. These were calculated using the adjoint method (Ermentrout and Kopell 1991), and then scaled to match the maximum of the estimated PRC (a binned average). As the adjoint solution comes from the noiseless systems ($\sigma_{\text{noise}} = 0$), we do not anticipate to replicate them in the noisy systems. Rather, we are attempting to recover the general shape, and they are recovered well in both cases, ignoring the spontaneous excitability (seen as compression along phase) added by increasing the noise.

3.3.3 Example 3: Experimental

We have shown that the truncated mean corrects noisy phase-resetting plots for theoretical situations where we have control over the dynamics of the oscillating

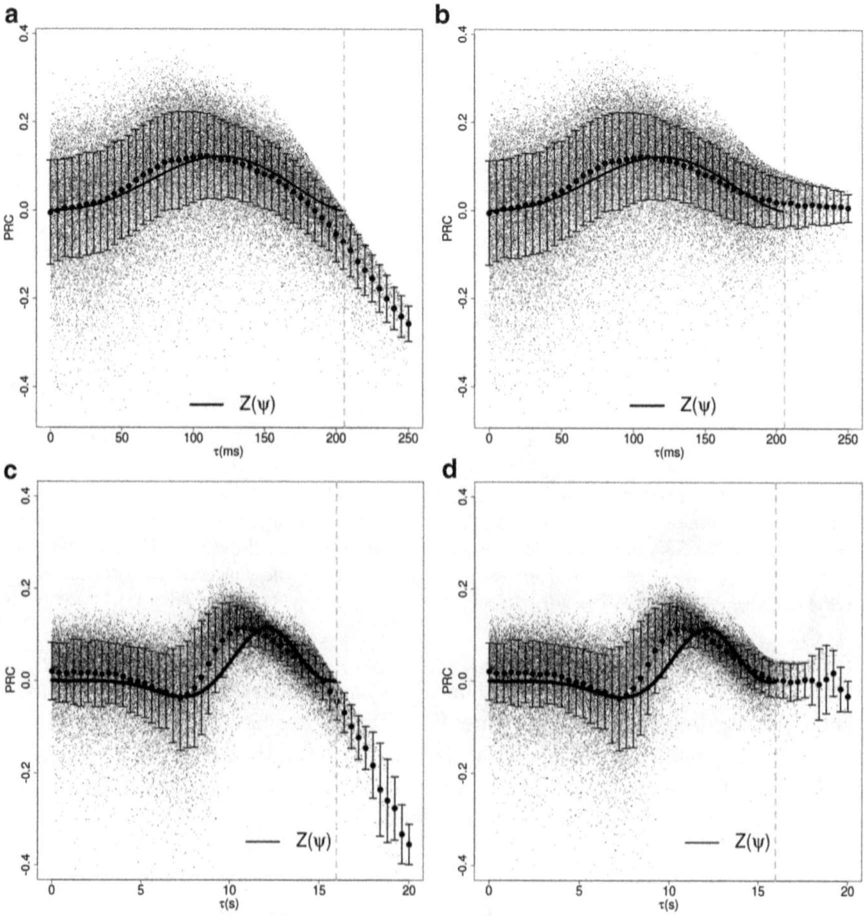

Fig. 7.6 Estimated phase-resetting for our noisy Hodgkin-Huxley model using the mean spontaneous ISI and the discretely approximated conditional mean to transform the perturbation interval. Observations were generated at 50,000 values of τ over the entire period. The *black trace* is the adjoint solution to the noiseless model ($\sigma_{\text{noise}} = 0$), scaled for comparison to the estimated phase-resetting observations (**a**) Type I Hodgkin-Huxley, $\hat{T} = \bar{t}$ (**b**) Type I Hodgkin-Huxley, $\hat{T} = \hat{T}_{DA}|T > \tau$ (**c**) Type II Hodgkin-Huxley, $\hat{T} = \bar{t}$ (**d**) Type II Hodgkin-Huxley, $\hat{T} = \hat{T}_{DA}|T > \tau$

system. We now apply the simplest of the methods, the truncated sample mean, to neurons of the subthalamic nucleus. This basal ganglia nucleus consists of a single population of neurons that fire autonomously at rates ranging from 3 to 20/s. Their oscillation is maintained by persistent sodium current, and periodic firing can be slowed by applied current to arbitrarily low rates (Bevan and Wilson 1999). We controlled for small rate changes by using episode averages, each episode

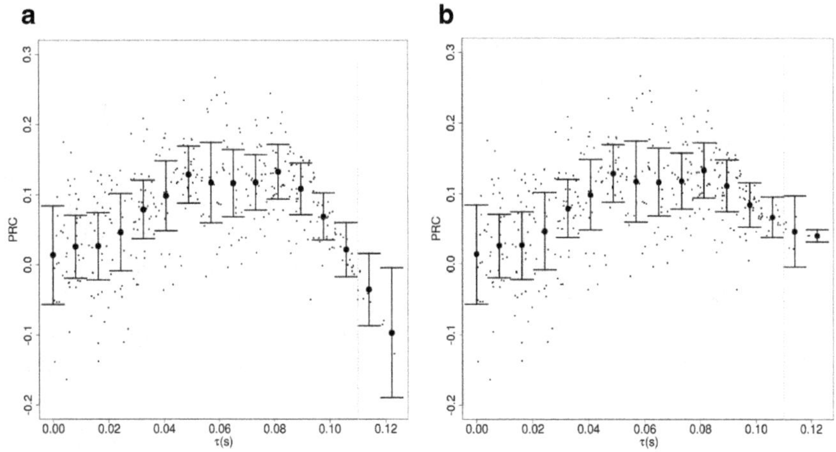

STN phase resetting, $\hat{T} = \bar{t}$ STN phase resetting, $\hat{T} = \bar{t} \mid T > \tau$

Fig. 7.7 Estimated phase-resetting for an STN cell. Using eight baseline trials to control for non-stationarity, n=399 perturbation intervals were taken. Error bars are standard error with bar width reflecting bin size (**a**) STN phase-resetting, $\hat{T} = \bar{t}$ (**b**) STN phase-resetting, $\hat{T} = \bar{t}|T > \tau$

consisting of spontaneous firing and perturbation intervals. Each episodes spontaneous ISIs are used to calculate the corresponding conditional means for the phase-resetting plot, rather than the collection of all spontaneous ISIs. We applied the truncated sample mean correction, $\bar{t}|T > \tau$, to find each of the PRC observations. Figure 7.7 is similar to our HH model under Class I excitability.

4 Discussion

We have demonstrated that using truncated means to estimate the period preserves the periodic shape of PRCs when variability of ISIs would otherwise cause them to appear nonperiodic. Three methods of estimating the truncated mean were suggested: conditional sample means, using maximum likelihood estimation from the spontaneous ISI histogram fit, and an approximation of the expectation integral. The conditional sample mean will approximate the period well and is simple to implement, but when the end of the phase is under consideration, the accuracy of the estimate truly becomes a matter of sample size. The ML estimate does not, but accuracy will be affected by the quality of the fit to the spontaneous activity. Solving for the statistical moments of the truncated distribution can also be unwieldy. Using numerical analysis to discretely approximate the integral in the conditioning avoids poorly fitting the spontaneous distribution and is easily implemented. It is generally more accurate and the resulting conditional mean estimation is more correctly generalized to the population level (as opposed to sample level) with smoothing on either $\hat{f}$ or the conditional mean itself. Because of this smoothing, better estimates will be made than the simple conditional sample average – a step function in observed spontaneous activity.

We stated that using the truncated mean preserves the periodic nature of the infinitesimal PRC, and while this statement itself is correct it is not necessarily true for the left truncated mean. The left truncated estimate places no upper bound on the period. Determination of where to decide what the maximum period could be is ambiguous. When using the left truncated sample mean or discretely approximating the conditioning we essentially bound the estimated mean at the maximum observed ISI. Noisy PRC behavior at large stimulus phases will be quite variable, however, as there are few opportunities to apply a perturbation this close to the impending action potential. As can be seen in the STN example provided, there are few perturbation intervals contributing to the PRC at the latent phases so large that the right boundary selection problem would begin. Experimental phase-resetting curves derived from estimating the period through the mean (or any statistic independent of the lag perturbation time) lead to an irregularity (Fig. 7.1) apparent to even the scientist unfamiliar with PRCs. Rather than using an incorrect PRC due to ignoring this irregularity, or discarding experimental information in the form of discontinuing the phase-resetting curve where the perturbation phase enters the region of periodic uncertainty, we suggest using latent phase based estimates of the period to correct the irregularity. The estimated phase-resetting curve using the truncated mean yields a more accurate description of the oscillator response, removing the subjectivity inherent to a priori expectation of an artifact by the experimenter.

Appendix

ML-Estimation: Parametric Mean of Truncated Normal

We illustrate the alternative to using sampling averages with an ISI histogram that can be approximated by a normal distribution using the definition of the conditional expectation. The conditional pdf is,

$$f_T(t) = \frac{\frac{1}{\sigma}\phi\left(\frac{t-\mu}{\sigma}\right)}{1-\Phi\left(\frac{\tau-\mu}{\sigma}\right)}; \quad \tau \le t,$$

where ϕ and Φ are the standard normal probability density function and cumulative density function. The moment generating function may be used to find the moments of the truncated distribution $\mathrm{E}[T|T > \tau] = \frac{\mathrm{d}}{\mathrm{d}s} M_T(s = 0)$,

$$
\begin{aligned}
M_T(s) &= \mathrm{E}_\tau\left[\mathrm{e}^{st}\right] = \int_\tau^\infty \mathrm{e}^{st} f_T(t)\mathrm{d}t \\
&= \mathrm{e}^{\frac{(s\sigma^2+\mu)^2}{2\sigma^2} - \frac{\mu^2}{2\sigma^2}} \left(\frac{\Phi(\infty) - \Phi\left(\frac{\tau-(\mu+s\sigma^2)}{\sigma}\right)}{1-\Phi\left(\frac{\tau-\mu}{\sigma}\right)} \right) \\
&= \mathrm{e}^{\frac{(s\sigma^2+\mu)^2}{2\sigma^2} - \frac{\mu^2}{2\sigma^2}} \left(\frac{1 - \Phi\left(\frac{\tau-(\mu+s\sigma^2)}{\sigma}\right)}{1-\Phi\left(\frac{\tau-\mu}{\sigma}\right)} \right), \\
M_T'(0) &= \mu + \sigma \frac{\phi\left(\frac{\tau-\mu}{\sigma}\right)}{1-\Phi\left(\frac{\tau-\mu}{\sigma}\right)}.
\end{aligned}
$$

The parameters and can be estimated from the spontaneous ISI histogram, the invariance of the maximum likelihood estimators imply the resulting truncated mean estimate will remain ML.

References

Mark D. Bevan and Charles J. Wilson. Mechanisms underlying spontaneous oscillation and rhythmic firing in rat subthalamic neurons. *Journal of Neuroscience*, 19(17):7617–7628, 1999.

G. Bard Ermentrout. *Simulating, Analyzing, and Animating Dynamical Systems: A Guide To Xppaut for Researchers and Students*. Society for Industrial and Applied Mathematics, Philadelphia, PA, 2002.

G. Bard Ermentrout. Type i membranes, phase resetting curves, and synchrony. *Neural Computation*, pages 979–1001, 1996.

G. Bard Ermentrout and Nancy Kopell. Multiple pulse interactions and averaging in systems of coupled neural oscillators. *Journal of Mathematical Biology*, pages 195–217, 1991.

D. Hansel, G. Mato, and C. Meunier. Synchrony in excitatory neural networks. *Neural Computation*, pages 307–337, 1995.

A. L. Hodgkin and A. F. Huxley. A quantitative description of membrane current and its application to conduction and excitation in nerve. *Journal of Physiology*, pages 500–544, 1952.

Eugene M. Izhikevich. Neural excitability, spiking, and bursting. *International Journal of Bifurcation and Chaos*, pages 1171–1266, 2000.

Eugene M. Izhikevich. *Dynamical Systems in Neuroscience*. MIT Press, Cambridge, MA, 2007.

T. I. Netoff. Synchronization in hybrid neuronal networks of the hippocampal formation. *Journal of Neurophysiology*, pages 1197–1208, 2005.

Alex D. Reyes and Eberhard E. Fetz. Two modes of interspike interval shortening by brief transient depolarizations in cat neocortical neurons. *Journal of Neurophysiology*, pages 1661–1672, 1993.

T. Tateno and Hugh Robinson. Phase resetting curves and oscillatory stability in interneurons of rat somatosensory cortex. *Biophysical Journal*, pages 683–695, 2007.

Mark Teagarden, Jeremy E. Atherton, and Charles J. Wilson. Accumulation of cytoplasmic calcium, but not apamin-sensitive. *Journal of physiology - London*, pages 817–833, 2008.

Chapter 8
Bayesian Approach to Estimating Phase Response Curves

Keisuke Ota and Toru Aonishi

Abstract Many research groups have sought to measure phase response curves (PRCs) from real neurons. In contrast to the numerical calculations of the PRCs for the mathematical neuron models, electrophysiological experiments on real neurons face serious problems whereby PRCs have to be retrieved from noisy data. However, methods for estimating PRCs from noisy spike response data have yet to be established. In this chapter, we explain our Bayesian approach to estimating the PRCs and its application to physiological data. In the first half of this chapter, we describe a Bayesian algorithm for estimating PRCs from noisy spike response data. This algorithm is based on a likelihood function derived from a detailed model of the spike response in PRC measurements that is formulated as a Langevin phase equation. We construct a maximum a posteriori (MAP) estimation algorithm based on the analytically obtained likelihood function. This algorithm gives estimates of not only the PRC but also the Langevin force intensity. In the last half of this chapter, we apply the MAP estimation algorithm to physiological data measured from a hippocampal CA1 pyramidal neuron. We explain the protocol of the PRC measurement in a dynamic clamp, which maintains the baseline firing frequencies as close to a target value for as long as the perturbation experiment lasts. Finally, we verify the reliability of the estimated PRC by testing whether the Fokker–Planck equation based on the estimated PRC and Langevin force intensity captures the stochastic oscillatory behavior of the same neuron disturbed by periodic perturbations.

K. Ota
Brain Science Institute, RIKEN, Wako, Japan
e-mail: k-ota@brain.riken.jp

T. Aonishi (✉)
Interdisciplinary Graduate School of Science and Engineering, Tokyo Institute of Technology, Yokohama, Japan
e-mail: aonishi@dis.titech.ac.jp

N.W. Schultheiss et al. (eds.), *Phase Response Curves in Neuroscience: Theory, Experiment, and Analysis*, Springer Series in Computational Neuroscience 6,
DOI 10.1007/978-1-4614-0739-3_8,

1 Background

Synchrony of neural populations has been reported in various regions of the brain. Phase reduction (Kuramoto 1984; Winfree 1967) is an efficient way to gain an understanding of such cooperative behavior in neural populations. For a quarter century, many researchers have conducted phase-reduction-based theoretical analyses of neural synchronies, for instance, synchronization of two coupled oscillatory neurons (Netoff et al. 2005; Oprisan et al. 2004), mutual entrainment of globally coupled oscillatory neurons (Neltner et al. 2000; Tsubo et al. 2007), and synchronization of neural populations induced by a common noise (Galán et al. 2007; Marella and Ermentrout 2008; Nakao et al. 2007; Teramae and Tanaka 2004). Furthermore, several groups have provided insight into the nature of neural coding with phase-reduction-based theoretical analyses (Ermentrout et al. 2007; Lengyel et al. 2005).

To apply phase-reduction-based theoretical analyses to real nervous systems, many research groups, including those responsible for the theoretical studies mentioned above, have sought to measure phase response curves (PRCs) from real neurons, e.g., from mitral cells of mouse olfactory bulb (Ermentrout et al. 2007; Galán et al. 2005), stellate cells of rat entorhinal cortex (Netoff et al. 2004, 2005), pyramidal cells of rat neocortex layer 5 (Goldberg et al. 2007; Stoop et al. 2000), pyramidal cells of rat hippocampal CA3 (Lengyel et al. 2005), interneurons of rat and mouse barrel cortex (Mancilla et al. 2007) and so on (Oprisan et al. 2004; Preyer and Butera 2005; Reyes and Fez 1993; Tateno and Robinson 2007; Tsubo et al. 2007).

The PRC of a real neuron can be measured with the perturbation method shown in Fig. 8.1. A precisely timed perturbation is given to an oscillating neuron, and the transient change in the interspike interval (ISI) is measured. For instance, the point at which the membrane potential crosses 0 mV in the positive direction can be chosen to have a reference phase of $\phi = 0$. A short pulse perturbation current is applied at $\phi = t_0$, where t_0 is the time elapsed since the reference phase. The cycle containing the perturbation has an ISI T_p. Accordingly, we obtain the sensitivity of the phase to perturbation current as

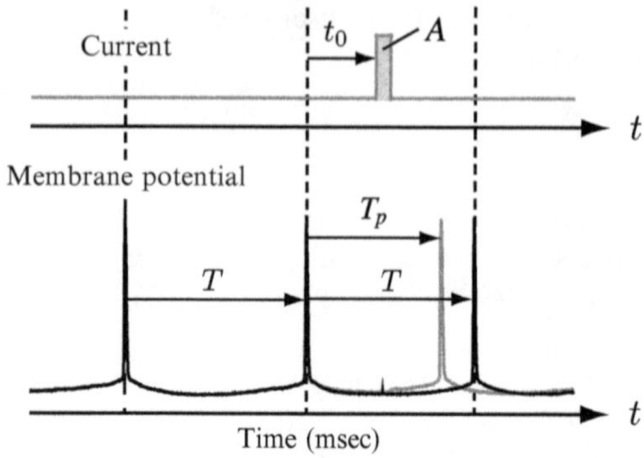

Fig. 8.1 Perturbation method for measuring the PRC from real neurons. A short pulse perturbation current is applied at phase $\phi = t_0$ and ISI T_p in the cycle containing the perturbation. The PRC is obtained as $Z(t_0) = (T - T_p)/A$. T is the baseline ISI, and A is the area of the pulse perturbation

$$Z(t_0) = \frac{T - T_\mathrm{P}}{A}, \tag{8.1}$$

where T is the baseline ISI and A is the area of the pulse perturbation.[1] The PRC $Z(t_0)$, which is sensitivity as a function of t_0, can be obtained by repeating this procedure for various perturbation timings t_0.

According to the definition of PRC, the baseline ISI must be constant for as long as the perturbation-response experiment lasts. However, it is difficult to keep the baseline ISI constant for a long time, because of vicissitudes of intracellular ionic concentrations. For instance, the accumulation of cytoplasmic calcium induces spike frequency adaptation. In much of the previous work, to cancel the baseline variations, the spike response was normalized with a calibration equation $(T_\mathrm{B} - T_\mathrm{P})/T_\mathrm{B}$, where T_P is the perturbed ISI mentioned above and T_B is the baseline ISI calculated by the mean value of consecutive unperturbed ISIs just before measuring T_P. Such a calibration can cause artifacts in experimentally obtained PRCs. Moreover, spike responses and baseline ISIs randomly fluctuate because of intrinsic noise. To address the random variation problem in measuring PRCs, many studies have resorted to basic statistical methods for estimating PRCs; these methods are classifiable into smoothing with a Gaussian filter (Lengyel et al. 2005; Reyes and Fez 1993) and regression with nonlinear functions (Galán et al. 2005; Netoff et al. 2005; Oprisan et al. 2004; Stoop et al. 2000; Tateno and Robinson 2007).

On the other hand, to address the baseline variation problem in measuring PRCs, our research group has made a protocol for PRC measurements executed on a dynamic clamp (Robinson and Kawai 1993; Sharp et al. 1993), which maintains the mean unperturbed ISI as close to the target value for as long as the perturbation experiment lasts. Moreover, to address the random variation problem in measuring PRCs, our research group has recently developed a statistical algorithm based on a Bayesian framework for estimating PRCs from noisy spike response data (Ota et al. 2009).

In this chapter, we shall outline our Bayesian algorithm for estimating PRCs and its application to physiological data measured with a dynamic clamp. Section 2 describes the Bayesian algorithm. This algorithm is based on a likelihood function derived from a detailed model of the spike response in a PRC measurement that is formulated as a Langevin phase equation. We then construct a maximum a posteriori (MAP) estimation algorithm based on the analytically obtained likelihood function. This algorithm estimates the PRC and the Langevin force intensity. We evaluate its efficacy in numerical simulations. In Sect. 3, we apply the Bayesian algorithm to physiological data measured from a hippocampal CA1 pyramidal neuron. We briefly explain the protocol of the PRC measurement with the dynamic clamp. In Sect. 4, we verify the reliability of the PRC and Langevin force intensity estimated from physiological data. By using the dynamic clamp, we inject a periodic perturbation consisting successive pulses into the same neuron from which the PRC

[1]Sensitivity is defined as the phase response normalized by the strength of the pulse perturbation.

was measured. We obtain histograms of phase differences between spikes and pulses and ascertain whether the Fokker Planck equation based on the estimated PRC and Langevin force intensity can predict this histogram. Consistency between the histogram and the distribution predicted by the Fokker–Planck equation is taken to be verification that the protocol of the phase response measurement and the Bayesian algorithm is valid. At the last of this chapter, Appendix 1 describes the necessary mathematical background, and Appendix 2 describes the mathematical detail of the Bayesian algorithm.

2 Bayesian Approach for Estimating PRCs from Noisy Spike Response Data

In Sect. 2.1, we explain the outline of our Bayesian algorithm and show how to execute it. Although our algorithm is mathematical, the guideline is helpful to use it. The mathematical detail is represented in Appendix 2. In Sect. 2.2, we evaluate the efficacy of this approach in numerical simulations.

2.1 *Outline of Bayesian Algorithm*

Figure 8.2 schematically shows the outline of the Bayesian approach for estimating PRCs from noisy spike response data. The observation of spike responses for real neurons can be regarded as a signal transmission through a noisy band limited channel. The PRC (original signal) is deteriorated by the perturbation and the inherent noise during the perturbation response experiment. The observed data, which we can actually obtain, is the noisy spike response data (received signal). Our aim is to reconstruct the PRC from the noisy spike response data. For this, we firstly derive the observation model of PRC $P(\psi | Z(t), \sigma, G_o(t - t_0))$ from a detailed model of the spike response in a PRC measurement that is formulated as a Langevin phase equation (see Appendix 2). This observation model is equivalent to the transmission of a signal $Z(t)$ through the band-limited Gaussian channel with impulse response $G_o(t)$. Then, we introduce a prior $P(Z(t)|D, n)$ which represent the smoothness of PRC. According to Bayes' theorem (see Appendix 1), we obtain the probability density of PRC as the posterior $P(Z(t)|\{\psi_i(\Delta tj)\}, \sigma, D, n, \{G_o(t - \Delta tj)\})$ representing the reverse of the signal transmission. We can estimate the PRC from the noisy received signal by maximizing the posterior.

As shown in Fig. 8.1, the PRC of a real neuron can be measured in the perturbation-response experiment (Fig. 8.1). When our Bayesian algorithm is carried out, there is an experimental condition we should follow. We should give perturbations to an oscillating neuron at equidistant points, because we introduce Fourier transform in order to obtain an executable algorithm as described below. Here, we consider the situation where spike response is observed at

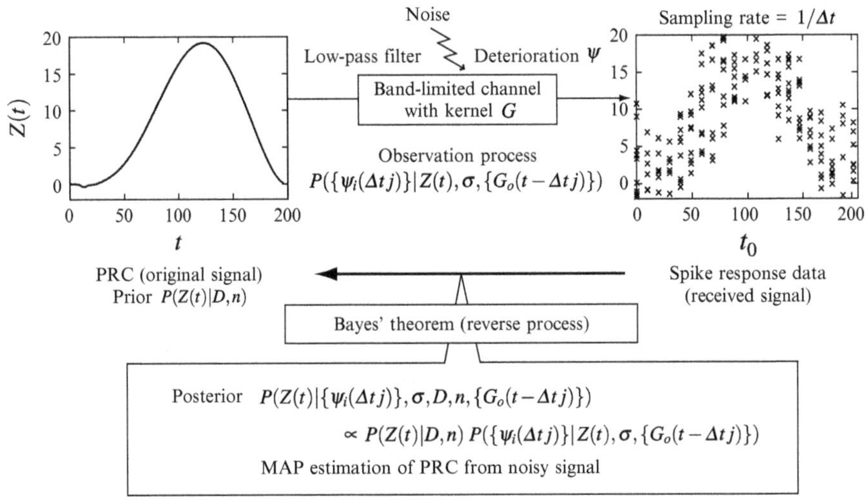

Fig. 8.2 Outline of Bayesian algorithm

$t_0 = \Delta tj (j = 0, 1, \cdots, 2L-1)$ whose sampling period satisfies $\Delta t = T/2L$, and N samples $\{\psi_1(\Delta tj), \cdots, \psi_N(\Delta tj)\}$ are measured at each sampling point, $t_0 = \Delta tj$. The total number of experimental data is $2LN$.

We can obtain the MAP estimate of PRC by maximizing the following a posterior probability density

$$P(Z(t)|\{\psi_i(\Delta tj)\}, \sigma, D, n, \{G_o(t-\Delta tj)\})$$
$$= \frac{\mathscr{Z}_{\mathrm{pr}}\mathscr{Z}_{\mathrm{like}}^{2LN}}{\mathscr{Z}_{\mathrm{ps}}} \exp\left(-\frac{ND^{2n}}{2}\int_0^T \left(\frac{\partial^n Z(t')}{\partial t'^n}\right)^2 \mathrm{d}t' \right.$$
$$\left. -\sum_{i=1}^{N}\sum_{j=0}^{2L-1} \frac{\left(\psi - \int_0^T \mathrm{d}t\, Z(t) G_o(t-t_0)\right)^2}{4\sigma^2 \int_0^T \mathrm{d}t\, Z(t)^2}\right), \quad (8.2)$$

where $\mathscr{Z}_{\mathrm{pr}}$, $\mathscr{Z}_{\mathrm{like}}$, and $\mathscr{Z}_{\mathrm{ps}}$ is a partition function for normalizing the prior probability density, the probability density of the spike response and the posterior probability density, respectively. Z is PRC, G_o is a perturbation input current, σ is the intensity of the inherent noise in the neuron, $\{\psi_i(\Delta tj)\}$ is the set of spike response data, and D and n are hyperparameters to control the smoothness of Z.

We transform the posterior (8.2) described in the time domain to (8.34) described in the frequency domain by using the Fourier series of $Z(t)$, $G_o(t)$ and $\psi_i(\Delta tj)$ defined in (8.31)–(8.33) (see Appendix 2), and we obtain an executable algorithm in the frequency domain. As a result, we can find that the minimum point of the Hamiltonian $\mathscr{H}'$ (8.36) in Appendix 2 approximates the MAP solution of the posterior density if λ is sufficiently large.

The estimated PRC $\hat{Z}$ depends on the hyperparameters D and n and the noise intensity σ. In our approach, we can estimate these parameters by maximizing the following log marginal likelihood (see Appendix 1):

$$\mathscr{F}(\sigma, D, n|\{\psi_i(\Delta tj)\}) = \log \mathscr{Z}_{\text{ps}} - \log \mathscr{Z}_{\text{like}}^{2LN} - \log \mathscr{Z}_{\text{pr}}. \tag{8.3}$$

Equations (8.38)–(8.40) in Appendix 2 show $\log \mathscr{Z}_{\text{like}}$, $\log \mathscr{Z}_{\text{pr}}$, and $\mathscr{Z}_{\text{ps}}$, respectively, which can be analytically derived in the frequency domain. As a result, we can find maximum point of the log marginal likelihood by numerically calculating $\mathscr{Z}_{\text{ps}}$ with frequency domain spike response data.

Note that before performing MAP estimation and maximization of the log marginal likelihood, to reduce the parameter redundancy we should initially decide the value of β, which is the variance of spike responses given by $\frac{1}{\beta} = 2\sigma^2 \int_0^T \mathrm{d}t\, Z(t)^2$ as defined in (8.30). Here, we decide β as the variance of sampled spike response data.

2.2 Numerical Simulation

To verify the algorithms by using numerical simulations, we used the Morris–Lecar model (Morris and Lecar 1981) in the form of

$$C\frac{\mathrm{d}V}{\mathrm{d}t} = -g_{Ca}m_\infty(V - V_{Ca}) - g_K w(V - V_K) - g_{\text{leak}}(V - V_{\text{leak}}) + I_{\text{app}}, \tag{8.4}$$

where each ion channel has the following activation profile:

$$\begin{aligned}
\frac{\mathrm{d}w}{\mathrm{d}t} &= \phi \cdot \frac{w_\infty - w}{\tau},\\
m_\infty &= 0.5[1 + \tanh((V - v_1)/v_2)],\\
w_\infty &= 0.5[1 + \tanh((V - v_3)/v_4)],\\
\tau &= 1/\cosh((V - v_3)/(2 \cdot v_4)).
\end{aligned}$$

Depending on the parameters, the Morris–Lecar model has different bifurcation structures, classified as Type I and Type II (Ermentrout 1996). The parameters used in the numerical simulations are as follows. Type I: $C = 20\,\mu\text{F}$, $V_K = -84\,\text{mV}$, $g_K = 8\,\text{mS}$, $V_{Ca} = 120\,\text{mV}$, $g_{Ca} = 4\,\text{mS}$, $V_{\text{leak}} = -60\,\text{mV}$, $g_{\text{leak}} = 2\,\text{mS}$, $v_1 = -1.2\,\text{mV}$, $v_2 = 18\,\text{mV}$, $v_3 = 12\,\text{mV}$, $v_4 = 17.4\,\text{mV}$ and $\phi = 0.066$. Type II: $C = 20\,\mu\text{F}$, $V_K = -84\,\text{mV}$, $g_K = 8\,\text{mS}$, $V_{Ca} = 120\,\text{mV}$, $g_{Ca} = 4.4\,\text{mS}$, $V_{\text{leak}} = -60\,\text{mV}$, $g_{\text{leak}} = 2\,\text{mS}$, $v_1 = -1.2\,\text{mV}$, $v_2 = 18\,\text{mV}$, $v_3 = 2\,\text{mV}$, $v_4 = 30\,\text{mV}$, and $\phi = 0.04$. It has been proved that the PRC of the Type I neuron only has positive values, whereas the PRC of the Type II neuron has both positive and negative values (Ermentrout 1996).

The injection current into the cell has the form:

$$I_{\mathrm{app}} = I_{\mathrm{c}} + G_{\mathrm{o}}(t - t_0) + \overline{\sigma} s(t).$$

Here, I_{c} is the depolarizing constant current for inducing rhythmical firing with an average period T. $G_{\mathrm{o}}(t - t_0)$ represents a rectangular perturbation current to induce the transient change in the cycle ISI. $s(t)$ is the Langevin force satisfying $\langle s(t) \rangle = 0$ and $\langle s(t)s(t') \rangle = 2\delta(t - t')$. $\overline{\sigma}$ is the intensity of the Langevin force used in the numerical simulations. We measured stochastic fluctuating spike responses of the cycle containing the perturbation.

We tried to estimate PRCs and three parameters (D, n, σ), which are not directly measurable,[2] from artificial data generated by Type I and Type II of the Morris–Lecar model (8.4) with the Langevin force intensity of $\overline{\sigma} = 0.45$. Our algorithm can be said to be effective if the estimated $\hat{\sigma}$ is close to the true value $\overline{\sigma}$.

Figure 8.3 shows the resulting estimates of PRC, D, n and noise intensity σ. As revealed in Figs. 8.3a2,b2, the estimate $\hat{\sigma}$ that maximized the marginal likelihoods was close to the true value $\overline{\sigma} = 0.45$. Thus, our algorithm to estimate the intensity of the Langevin force works for Type I and Type II models. Figures 8.3a3,b3 are the MAP solutions of PRCs with the optimal $(\hat{D}, \hat{n}, \textit{and } \hat{\sigma})$ with respect to the log marginal likelihood. The algorithm with the optimal parameters retrieved the detailed structure of the original PRC. Note that the MAP estimation algorithm with nonoptimal parameters (small D and σ) gave oscillatory solutions (see Ota et al. 2009).

3 Measurement of PRC with Dynamic Clamp

As mentioned in the Introduction, the baseline ISI must be constant during the perturbation-response experiment for measuring the PRC. In the case of the mathematical neuron model used in Sect. 2, its firing frequency is easily kept constant. However, in the case of real neurons, the baseline ISI continuously changes during a depolarizing stimulation with a certain amount of DC current, because of variations in intracellular ionic concentrations.

One of the best ways to address this problem is to introduce the dynamic clamp system (Robinson and Kawai 1993; Sharp et al. 1993), which can maintain the baseline ISI as close to a target value by adjusting the DC injection current depending on the recorded firing period in real-time. Figure 8.4 shows the protocol of the perturbation-response experiment using the dynamic clamp on a rat hippocampal CA1 pyramidal neuron. In the first step, the mean ISI is adjusted to 100 ms by

[2] β defined in (8.30) (see Appendix 1) is directly measurable, because $1/\beta$ is the variance of the inter-spike interval. In the following numerical calculations, we determined β by calculating it from the sampling data.

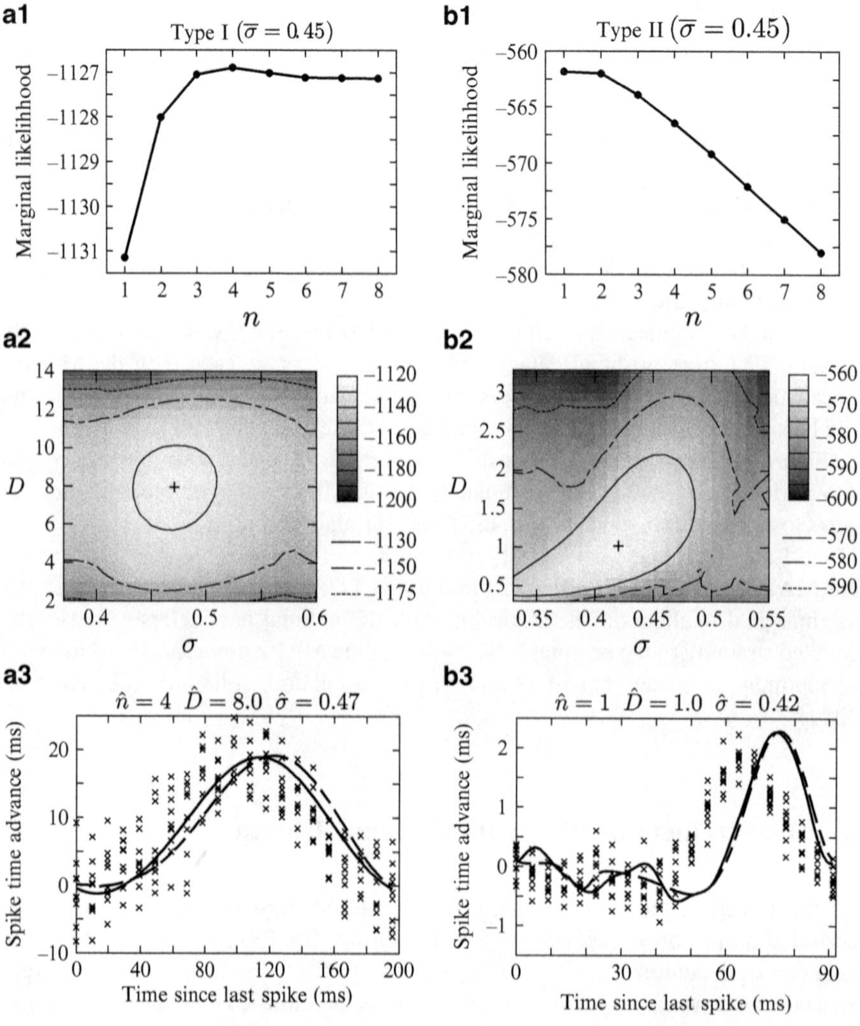

Fig. 8.3 Estimated PRCs and other parameters. (**a1**)–(**a3**) Type I. (**b1**)–(**b3**) Type II of Morris–Lecar model. (**a1**) (**b1**) The log marginal likelihood as a function of the hyperparameter n. The values in this graph are given by maximizing the log marginal likelihood on two parameters D and σ. (**a1**) Type I. $\hat{n} = 4$. (**b1**) Type II. $\hat{n} = 1$. (**a2**) (**b2**) Landscapes of the log marginal likelihood $\mathscr{F}$ in (D, σ) space when n is optimal. (**a2**) Type I. The maximum of $\mathscr{F}$ is $-1,126.89$ at $(\hat{D}, \hat{\sigma}) = (8.0, 0.47)$. (**b2**) Type II. The maximum of $\mathscr{F}$ is -561.82 at $(\hat{D}, \hat{\sigma}) = (1.0, 0.42)$. (**a3**) (**b3**). In each panel, the solid curve denotes the MAP solution with the optimal parameter $(\hat{D}, \hat{n}, \hat{\sigma})$. The dashed curve indicates the infinitesimal PRC obtained with the adjoint method (Ermentrout 1996; Kuramoto 1984). Each data set (plotted as *crosses* in each panel) contains 200 samples ($2L = 20$ and $N = 10$), and each set was numerically generated using Type I or Type II of the Morris–Lecar model (8.4) with a $\overline{\sigma} of 0.45$ and a rectangular pulse with a width of 20 ms and height of 1.0 pA

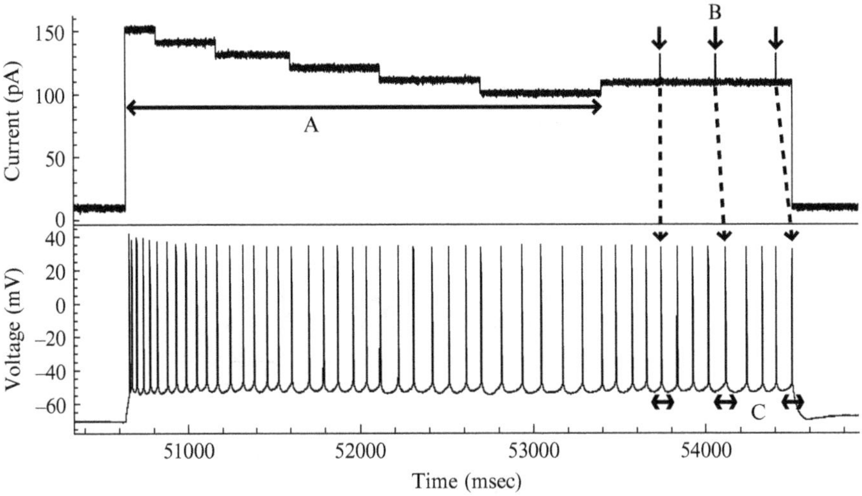

Fig. 8.4 Protocol of the perturbation-response experiment for the PRC measurement using the dynamic clamp. *Top and bottom* traces are examples of the injected current and membrane potential of a rat hippocampal CA1 pyramidal neuron, respectively. In (A), the mean unperturbed ISI is adjusted to 100 ms by tuning the DC depolarizing current. In (B), a one-shot pulse perturbation is evoked using various timings relative to the last spike, and in (C), we measure how the perturbation current disturbs the timing of the next spike firing. Here, the neuron was perturbed every five periods in order to reduce the effect of the last perturbation. The dynamic clamp automatically repeats this protocol for 200+ samples

tuning the DC depolarizing current. In the next step, every five periods, one-shot pulse perturbation is evoked using various timings relative to the last spike, and we measure how the perturbation current disturbs the timing of the next spike firing. The dynamic clamp automatically repeats this protocol consisting of short-term perturbation-response measurements for 200+ samples.

To assess performance of the protocol, before and after the PRC measurement trials, we carried out two sets of extra 20 trails without the pulse perturbation to measure the unperturbed ISIs and the coefficient of variation (CV) after the first step of the protocol. As shown in Fig. 8.5, the mean unperturbed ISI could be maintained as close to 100 ms, and the CV could be maintained at almost 10% of the mean unperturbed ISI, except for the 8th trial in the extra trials after the PRC measurement trials. Note that there was no such outlier during thc PRC measurement trials. This result suggests that the proposed protocol implemented in the dynamic clamp system was successful in controlling the baseline firing frequency except for one trial.

Figure 8.6a shows the spike response data measured from a rat hippocampal CA1 pyramidal neuron. We applied the Bayesian algorithm to the physiological data. Figures 8.6b,c show the log marginal likelihood as a function of n, D, and σ. The estimated intensity of the Langevin force $\hat{\sigma}$ was 9.0. This result suggests that the noise inherent in the neuron is 9.0 pA. A solid curve in Fig. 8.6a shows the estimated PRC based on optimal values of $\hat{n}$, $\hat{D}$, and $\hat{\sigma}$.

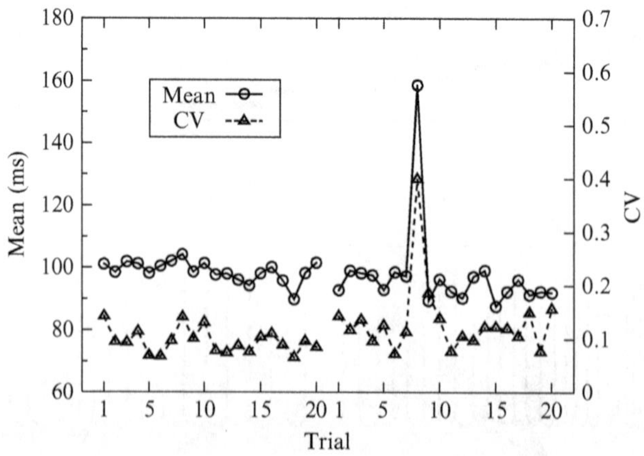

Fig. 8.5 Mean and CV of unperturbed ISI in two sets of 20 extra trials without the pulse perturbation before and after the PRC measurement trails

4 Reliability of Estimated PRC and Noise Intensity

Next, we verified the reliability of the estimated result in Sect. 3. The numerical simulation in Sect. 2.2 showed that our Bayesian approach accurately estimates the PRCs and noise intensity for the Morris–Lecar model. However, it is still unclear whether our method can be applied to physiological data obtained from a real neuron. In Sect. 2.1, the PRC measurement process was formulated as a Langevin phase equation (8.15) in Appendix 1, but it is still unclear whether the stochastic dynamics of a real neuron can be modeled by it. Here, we demonstrate that the Langevin phase equation with the estimated PRC and Langevin force intensity can predict the stochastic behavior of a real neuron in response to arbitrary small perturbations. If the prediction is consistent with the experimental result, we can say that the Langevin phase equation is good for modeling a real neuron, and at the same time, we can say that the PRC and noise intensity estimated by our Bayesian algorithm are reliable.

We shall focus on the stochastic phase motion of the oscillating neurons whose spike are entrained by a periodic perturbation $G_p(t)$. $G_p(t)$ consists of successive pulses and its period is the same as the unperturbed ISI, $G_p(t) = G_p(t + T)$. As Fig. 8.7 (left) shows, we define the relative phase of each spike to the onset time of the each pulse as the phase difference ψ for each spike. The positive value of ψ indicates that the spike comes after the pulse. Figure 8.7 (right) shows the membrane potential in response to the periodic perturbation $G_p(t)$ in the perturbation-response experiment for the phase difference measurement, and this experiment was conducted with the same neuron whose PRC and noise intensity had been estimated.

We attempted to predict the stochastic fluctuation of the phase difference in the stationary state by using the Langevin phase equation. As Appendix 1 shows, the

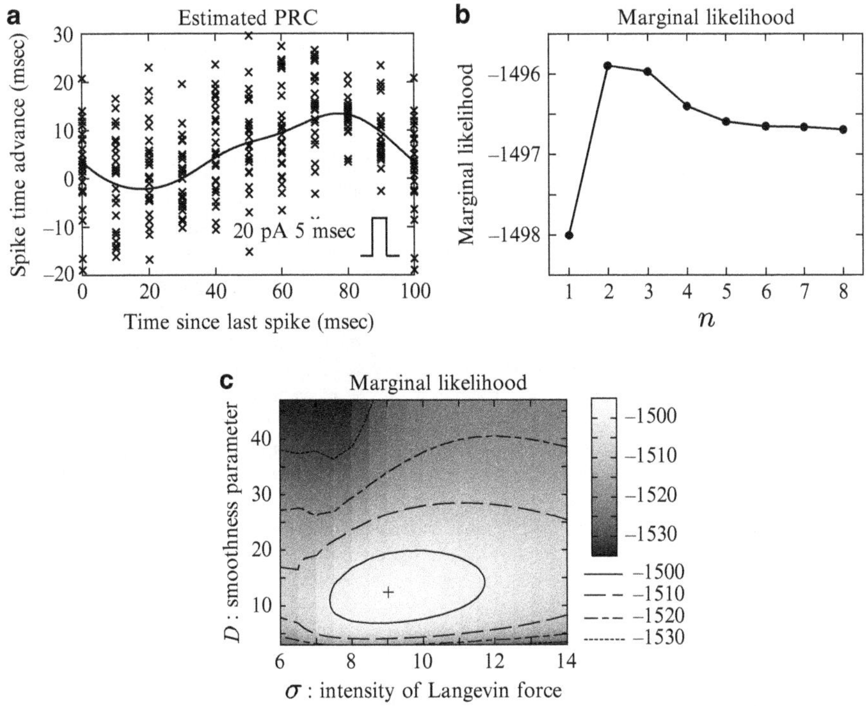

Fig. 8.6 Estimated PRC and other parameters of a pyramidal neuron in a rat hippocampal CA1. (**a**) The plotted *crosses* indicate the phase responses at various timings (*horizontal axis*) measured in the electrophysiological experiment (Fig. 8.4). A positive spike time advance (*vertical axis*) means that the inter spike interval (ISI) becomes shorter than the unperturbed period (100 ms), and a negative value means a longer ISI. The regular firing with an average period 100 ms was disturbed by rectangular pulses with a height of 20.0 pA and width of 5.0 ms (displayed at the *bottom right*), and a total of 260 samples were obtained. The *solid line* denotes the MAP solution with the optimal parameters $(\hat{D}, \hat{n}, \hat{\sigma})$ estimated by maximizing the log marginal likelihood. (**b**) The log marginal likelihood as a function of the hyperparameter n. The values in this graph are given by maximizing the log marginal likelihood on two parameters D and σ. In this case, $\hat{n} = 2$. (**c**) The landscape of the log marginal likelihood in (D, σ) space when $n = 2$. It indicates that the optimal values of these parameters are $(\hat{D}, \hat{\sigma}) = (12.0, 9.0)$ (*cross*)

Fokker–Planck equation describing the evolution of the probability density of the phase difference ψ is obtained from the Langevin phase equation with the estimated PRC and Langevin force intensity, as

$$\frac{\partial P(\psi, t)}{\partial t} = -\frac{\partial}{\partial \psi} D^{(1)}(\psi) P(\psi, t) + D^{(2)} \frac{\partial^2}{\partial \psi^2} P(\psi, t), \tag{8.5}$$

$$D^{(1)}(\psi) = \frac{1}{T} \int_0^T \mathrm{d}t' \hat{Z}(t' + \psi) G_{\mathrm{p}}(t'), \; D^{(2)} = \frac{\hat{\sigma}^2}{T} \int_0^T \mathrm{d}t' \hat{Z}(t')^2, \tag{8.6}$$

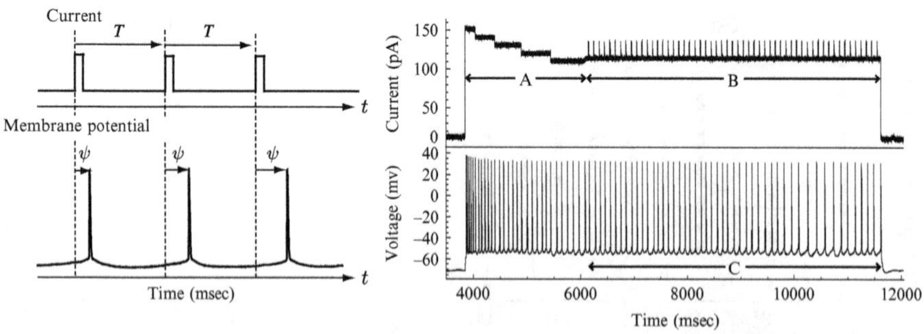

Fig. 8.7 (*Left*) Phase difference between spikes and periodic perturbation. A periodic perturbation $G_p(t)$ consists of successive pulses whose interval is the same as the target ISI of spikes, $100\,ms$. The phase difference ψ is determined as the relative phase of each spike to the onset time of the pulse. (*right*) Protocol for measuring phase differences between periodic spikes and perturbations using the dynamic clamp. The trace is an example of the injected current and membrane potential of the same neuron used for the PRC measurement in response to the injected current. In (A), the mean ISI is adjusted to 100 ms by tuning the DC depolarizing current as in Fig. 8.4. In (B), a periodic perturbation with a period of 100 ms is superimposed on the DC depolarizing current. In (C), phase differences between the periodic perturbation and spikes are measured. The dynamic clamp automatically repeats this protocol for about 1,000 samples

where $\hat{Z}$ and $\hat{\sigma}$ represent the estimated PRC and the Langevin force intensity, respectively. The stationary solution of the phase difference density $P(\psi, \infty)$ can be approximated by performing numerical calculations with sufficiently long time steps.

Figure 8.8 compares the theoretically predicted probability density (solid lines) and the histogram obtained from the experimental data (dashed lines). The solid lines match the dashed lines, even when the shape of the pulse is not the same as the one used in the PRC measurement (Fig. 8.4). This result indicates that the Langevin phase equation (8.15) can predict the stochastic behavior of neurons driven by periodic perturbations, and at the same time, that the PRC and noise intensity estimated by our Bayesian algorithm are reliable.

5 Discussion

We proposed a Bayesian algorithm for estimating the PRC and intensity of the Langevin force. First, we formulated the spike response in a PRC measurement as a Langevin phase equation and analytically derived the likelihood function for the PRC and the Langevin force intensity. Next, we introduced a smoothness prior for the PRC and applied Bayes' theorem to obtain the posterior probability density. Finally, by maximizing the posterior probability density, we obtained a MAP estimate of the PRC. The MAP estimate strongly depends on the hyperparameters

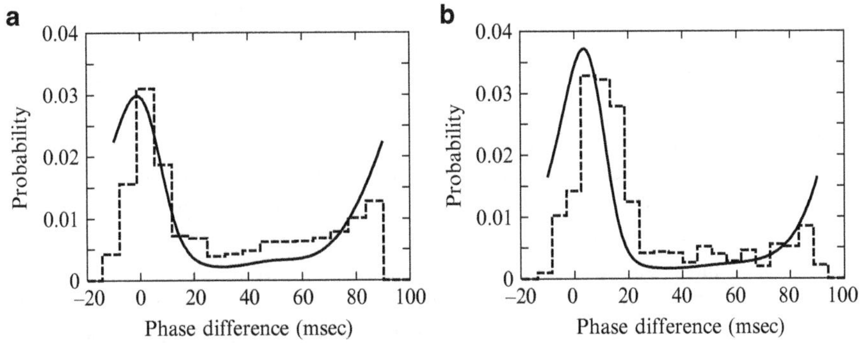

Fig. 8.8 Comparison of the stationary solution of the Fokker–Planck equation (8.5) (*solid lines*) and the histogram of phase differences calculated from the physiological data measured in the perturbation-response experiment illustrated in Fig. 8.7 (*dashed lines*). In each graph, the *horizontal axis* indicates the phase difference between the spikes and pulses, and the *vertical axis* denotes the probability density. A positive horizontal axis value means that the spike timing was delayed relative to the periodic perturbation, and a negative value represents advanced spike timing. The shape of each periodic rectangular perturbation was (**a**) height: 20.0 pA, width: 5.0 ms (the same shape used in the PRC estimation), and interval: 95.0 ms and (**b**) height: 10.0 pA, width: 20.0 ms (different shape), and interval: 80.0 ms. The histograms were made from (**a**) 950 samples and (**b**) 1,200 samples

D and n and the Langevin force intensity σ. Thus, at the same time, D, n, and σ were estimated by maximizing the log marginal likelihood.

We verified the efficacy of our algorithm through a numerical simulation using the Morris–Lecar model and a physiological experiment on rat hippocampal CA1 pyramidal neurons. In numerical simulation, the estimated PRC matched the infinitesimal PRC obtained with the adjoint method and the estimated Langevin force intensity was close to the true value. In the case of real neurons, however, it is still unclear whether the stochastic spike response in the PRC measurement can be formulated as a Langevin phase equation (8.15). For this reason, we ascertained whether the Langevin phase equation with the estimated PRC and noise intensity could predict the stochastic behavior of oscillating neurons in response to a periodic perturbation. We showed that the result predicted by the Fokker–Planck equation based on the Langevin phase equation coincided with the experimental result. These results indicate that the modeling of the stochastic spike response of the real neuron with the Langevin phase equation is effective and that the PRC and noise intensity estimated by our Bayesian algorithm are reliable. Moreover, the consistency between the experimental result and the theoretical prediction indicates a verification that the protocol of the phase response measurement with the dynamic clamp works well.

Our algorithm has the following advantages over those of previous studies.

- Estimation of the Langevin force intensity – Our algorithm is based on a likelihood function derived from a detailed model of the spike response in PRC measurements that is formulated as a Langevin phase equation. In the likelihood function (8.26), the variance of the spike responses is given by $2\sigma^2 \int_0^T \mathrm{d}t\, Z(t)^2$, where σ is the Langevin force intensity and $Z(t)$ is the PRC. This relational expression enables us to estimate the Langevin force intensity, which cannot be directly measured from the spike response data, by maximizing the log marginal likelihood. Therefore, we can determine the Langevin phase equation describing continuous-time stochastic dynamical processes of the neuron. This stochastic differential equation is more mathematically tractable and versatile than other stochastic phase equations (Ermentrout and Saunders 2006).

- Deconvolution of the perturbation from the spike response data – The wide rectangle pulse perturbation works as a low-pass filter in a PRC measurement and the detailed structure of the original PRC deteriorates as Fig. 8.3b3 shows. Note that in the likelihood function (8.26), the mean of the spike response is represented by convolving the PRC Z with the perturbation G. The MAP estimation algorithm deconvolves the wide rectangle pulse perturbation from the spike response data and retrieves the detailed structure of the original PRC.
- Hyperparameter estimation. – As mentioned in Ota et al. (2009), the advantage of our Bayesian algorithm is that it can give optimal truncation of the Fourier components. Our algorithm encompasses basic statistical methods used in previous studies, such as smoothing with a Gaussian filter (Lengyel et al. 2005; Reyes and Fez 1993) and regression with a Fourier series (Galán et al. 2005; Netoff et al. 2005; Oprisan et al. 2004; Stoop et al. 2000; Tateno and Robinson 2007). The second-order differential smoothness prior is closely related to the Gaussian filter because the Green function of the diffusion equation is a Gaussian. The MAP estimation algorithm derived here is described in the frequency domain, and it can also be regarded as regression with a Fourier series. Therefore, the estimation of the hyperparameter D corresponds to selecting the optimal variance of the Gaussian filter and optimal truncation of the Fourier components. These parameters were selected ad-hoc in the previous studies. In our algorithm, D is estimated by maximizing the marginal likelihood.

Appendix 1: Mathematical Preparations

A1.1 Phase Reduction Theory

Let us consider the following limit cycle oscillator driven by a perturbation $\mathbf{G}(t)$:

$$\frac{\mathrm{d}\mathbf{X}}{\mathrm{d}t} = \mathbf{F}(\mathbf{X}) + \mathbf{G}(t). \tag{8.7}$$

Here, we assume that the unperturbed oscillator, $\mathrm{d}\mathbf{X}/\mathrm{d}t = \mathbf{F}(\mathbf{X})$, has a stable limit cycle orbit $\mathbf{X}_s(t)$ parametrized by a phase ϕ with a period T. In a neural context, the unperturbed ISI of the neuron is T, and the point having phase ϕ in a high-dimensional state space is the value of state variables representing the membrane potential, calcium concentration, and gate variables of ion channels on the limit cycle orbit $\mathbf{X}_s(t)$. According to phase reduction methods (Kuramoto 1984; Winfree 1967), if $\mathbf{G}(t)$ is sufficiently small, a high-dimensional system (8.7) can be reduced to a one-dimensional system expressing the motion of ϕ as

$$\frac{\mathrm{d}\phi}{\mathrm{d}t} = 1 + \mathrm{grad}_{\mathbf{X}}\phi|_{\mathbf{X}=\mathbf{X}_s(t)} \cdot \mathbf{G}(t), \tag{8.8}$$

where $\cdot$ represents the inner product, and the gradient vector $\mathrm{grad}_{\mathbf{X}}\phi$ of an equi-phase plane[3] on the limit cycle orbit $\mathbf{X}_s(t)$ is called the PRC. PRC measures how sensitively the oscillator responds to external perturbations. Instead of $\mathrm{grad}_{\mathbf{X}}\phi|_{\mathbf{X}=\mathbf{X}_s(t)}$, $\mathbf{Z}(t)$ is often used to express the PRC. The PRC can be measured by using the perturbation method explained in the Introduction.

A1.2 Langevin Equation

We explain the general Langevin equation for one variable and the Fokker–Planck equation for calculating the probability density of the stochastic variable in the Langevin equation.

Let us consider the following one-dimensional differential equation containing a stochastic fluctuating force:

$$\frac{\mathrm{d}x}{\mathrm{d}t} = h(x,t) + g(x,t)s(t), \tag{8.9}$$

where $s(t)$ is white noise satisfying $\langle s(t)\rangle = 0$ and $\langle s(t)s(t')\rangle = 2\delta(t-t')$. Note that $\langle\ \rangle$ denotes ensemble averaging. Equation (8.9) is called the Langevin equation and $s(t)$ is called the Langevin force. The Langevin equation is a powerful tool for describing many continuous-time stochastic processes in stochastic neural dynamics (Risken 1989).

Next, we introduce the Kramers–Moyal expansion coefficient $D^{(n)}(x,t)$, which is defined as

$$D^{(n)}(x,t) = \frac{1}{n!}\lim_{\tau\to 0}\frac{1}{\tau}\langle (x(t+\tau)-\xi)^n\rangle\Big|_{\xi=x(t)}. \tag{8.10}$$

This coefficient represents the n-th moment for measuring the time variation of the stochastic variable x from t to $t+\tau$. $D^{(1)}(x,t)$ describes the systematic drift,

[3]The equi-phase plane is defined as a set of initial points which converges to the same point on the limit cycle orbit after an infinite number of periods.

and $D^{(2)}(x,t)$ corresponds to the diffusion coefficient. In the case of (8.9), these coefficients all vanish for $n \geq 3$, and we obtain

$$D^{(1)}(x,t) = h(x,t) + \frac{\partial g(x,t)}{\partial x} g(x,t), \quad D^{(2)}(x,t) = g(x,t)^2. \tag{8.11}$$

The second term in $D^{(1)}(x,t)$ is called the noise-induced drift (Risken 1989).

The Langevin equation (8.9) is classified as a Markov process. The probability density of the stochastic variable x obeys the following equation:

$$P(x,t+\tau) = \int \mathrm{d}x' P(x,t+\tau|x',t)P(x',t), \tag{8.12}$$

where $P(x,t+\tau|x',t)$ is called a the transition probability density and satisfies

$$P(x,t+\tau|x',t) = \frac{1}{\sqrt{4\pi\tau D^{(2)}(x',t)}} \exp\left(-\frac{1}{4\tau D^{(2)}(x',t)}\left(x - x' - \tau D^{(1)}(x',t)\right)^2\right).$$

In the limit of $\tau \to 0$, (8.12) becomes a Fokker–Planck equation in the form of

$$\frac{\partial P(x,t)}{\partial t} = -\frac{\partial}{\partial x} D^{(1)}(x,t)P(x,t) + \frac{\partial^2}{\partial x^2} D^{(2)}(x,t)P(x,t). \tag{8.13}$$

A1.3 Langevin Phase Equation and Fokker–Planck Equation

We outlines the special Langevin equation for the stochastic phase variable, i.e., the Langevin phase equation, and an averaging expression for the Fokker–Planck equation recently proposed by Kuramoto (unpublished) and Teramae and Fukai (2008).

We consider a limit-cycle neural oscillator with a period T driven by a perturbation input current $G(t)$ and noise current $\sigma s(t)$, which can be formulated in the general form:

$$\frac{\mathrm{d}\mathbf{X}}{\mathrm{d}t} = \mathbf{F}(\mathbf{X}) + (G(t) + \sigma s(t))\mathbf{U}. \tag{8.14}$$

The term $(G(t)+\sigma s(t))\mathbf{U}$ represents the driving current, in which the vector $\mathbf{U}$ ($\mathbf{U} = [1, 0, \cdots, 0]^T$) results in perturbing one degree of freedom of the neural oscillator. $s(t)$ is a one-dimensional Langevin force satisfying $\langle s(t)\rangle = 0$ and $\langle s(t)s(t')\rangle = 2\delta(t-t')$, and σ is the intensity of the Langevin force.

As explained in A1.1, we assume that the unperturbed oscillator $\mathrm{d}\mathbf{X}/\mathrm{d}t = \mathbf{F}(\mathbf{X})$ has a stable limit cycle orbit. When σ and $G(t)$ are sufficiently small, a high-dimensional system (8.14) can be reduced to a one-dimensional one expressing the motion of the phase $\phi(t)$ on the limit-cycle orbit:

$$\frac{d\phi}{dt} = 1 + Z(\phi)\,(G(t) + \sigma s(t))\,. \tag{8.15}$$

where $Z(\phi)$ is the PRC denoting the sensitivity to the perturbation current. This is called the Langevin phase equation.

We apply the variable transformation $\phi(t) = 1 + \psi(t)$ to (8.15). $\psi(t)$ represents a slow-moving initial phase driven by a small perturbation and noise. Then, we can write the slow dynamics of the initial phase ψ as

$$\frac{d\psi(t)}{dt} = Z(t + \psi(t))(G(t) + \sigma s(t)). \tag{8.16}$$

In what follows, $G(t)$ is restricted to a periodic function with a period T for deriving the averaging expression of Fokker–Planck equation. The theory outlined below can be extended to the case of a perturbation with slightly different periods. We derive the Kramers–Moyal expansion coefficients to set up the Fokker–Planck equation. By integrating both sides of (8.16) from 0 to T, we get

$$\psi(T) - \psi(0) = \int_0^T dt' Z(t' + \psi(t'))G(t') + \sigma \int_0^T dt' Z(t' + \psi(t'))s(t'). \tag{8.17}$$

Assuming the slow dynamics of the initial phase ψ, by integrating both sides of (8.16) from 0 to t', $\psi(t')$ in (8.17) can be approximated as

$$\psi(t') = \psi(0) + \int_0^{t'} dt'' Z(t'' + \psi(0))G(t'') + \sigma \int_0^{t'} dt'' Z(t'' + \psi(0))s(t''). \tag{8.18}$$

Then, inserting (8.18) in (8.17) leads to

$$\begin{aligned}\psi(T) - \psi(0) &= \sigma \int_0^T dt' Z(t' + \psi(0))s(t') + \int_0^T dt\, Z(t' + \psi(0))G(t') \\ &\quad + \int_0^T dt' \int_0^{t'} dt'' Z'(t' + \psi(0))Z(t'' + \psi(0))s(t')s(t'') + \cdots .\end{aligned}$$

Thus, the first coefficient of the Kramers–Moyal expansion is approximately

$$\begin{aligned}D^{(1)}(\psi) &\sim \frac{\langle \psi(T) - \psi(0)\rangle}{T} \\ &= \frac{1}{T}\int_0^T dt' Z(t' + \psi(0))G(t') \\ &\quad + \frac{\sigma^2}{T}\int_0^T dt' \int_0^{t'} dt'' Z'(t' + \psi(0))Z(t'' + \psi(0))\langle s(t')s(t'')\rangle. \end{aligned} \tag{8.19}$$

Note that the second term on the right side, which corresponds to the noise-induced drift, vanishes because of the periodicity of the PRC. Similarly, the second coefficient of the Kramers–Moyal expansion is approximately

$$D^{(2)}(\psi) \sim \frac{\langle(\psi(T) - \psi(0))^2\rangle}{T}$$
$$= \frac{\sigma^2}{2!T} \int_0^T \mathrm{d}t' \int_0^T \mathrm{d}t'' Z(t' + \psi(0)) Z(t'' + \psi(0)) \langle s(t') s(t'') \rangle. \quad (8.20)$$

Note that we can safely replace $\psi(T)$ and $\psi(0)$ in (8.19) and (8.20) by $\psi(T + t)$ and $\psi(t)$ because of the periodicity of $Z(t)$ and $G(t)$. Consequently, we can obtain the averaging expression of the Fokker–Planck equation in the form (Teramae and Fukai 2008):

$$\frac{\partial P(\psi, t)}{\partial t} = -\frac{\partial}{\partial \psi} D^{(1)}(\psi) P(\psi, t) + D^{(2)} \frac{\partial^2}{\partial \psi^2} P(\psi, t), \quad (8.21)$$

$$D^{(1)}(\psi) = \frac{1}{T} \int_0^T \mathrm{d}t' Z(t' + \psi) G(t'),\ D^{(2)} = \frac{\sigma^2}{T} \int_0^T \mathrm{d}t' Z(t')^2, \quad (8.22)$$

where $D^{(1)}(\psi)$ is the convolution of Z and G, and $D^{(2)}$ is the power of PRC times the variance of the Langevin force σ^2.

Due to the averaging of the first and second coefficients for T, the smallest time resolution of this Fokker–Planck equation is T. Thus, as explained in the previous subsection, we can derive an expression for the transition probability density for the smallest time resolution T in the form:

$$P(\psi, t + T | \psi', t) = \frac{1}{\sqrt{4\pi T D^{(2)}}} \exp\left(-\frac{1}{4TD^{(2)}} \left(\psi - TD^{(1)}(\psi')\right)^2 \right). \quad (8.23)$$

A1.4 Bayes' Theorem and Parameter Estimation

We briefly explains Bayes' theorem and the parameter estimation method based on Bayes' theorem. Let us consider an observation x randomly generated according to a probability density

$$P(x|\Theta),$$

where Θ is the parameter of the probability density. The probability density allows us to predict x on the basis of the known parameter Θ. Letting X_o be the observed value of x, the function $P(X_o|\Theta)$ is called a likelihood function. The likelihood allows us to estimate the unknown parameter Θ based on the observed X_o. The method of maximum likelihood estimation gives $\hat{\Theta}$ as the mode of the likelihood $P\,(X_o|\Theta)$:

$$\hat{\Theta} = \mathrm{argmax}_{\Theta} P(X_o|\Theta).$$

On the other hand, by introducing a prior representing the probability density of the parameter Θ, Bayes' theorem gives the conditional probability density of the parameter Θ, given X_o, as follows:

$$P(\Theta|X_o,\eta) = \frac{P\,(\Theta|\eta)\,P\,(X_o|\Theta)}{P(X_o|\eta)},$$

$$P(X_o|\eta) = \int \mathrm{d}\Theta P\,(\Theta|\eta)\,P\,(X_o|\Theta).$$

Here, $P\,(\Theta|\eta)$ is the prior probability density we introduced. η in $P\,(\Theta|\eta)$ is called the hyperparameter. The conditional probability density $P(\Theta|X_o,\eta)$ given by Bayes' theorem is called the posterior probability density. $P\,(X_o|\Theta)$ is the likelihood function mentioned above. $P(X_o|\eta)$ acts as a normalizing constant, which is called the marginal likelihood, and $\ln P(X_o|\eta)$ is called the log marginal likelihood. Note that if the prior $P(\Theta|\eta)$ is a uniform distribution, the posterior $P(\Theta|X_o,\eta)$ is proportional to the likelihood $P\,(X_o|\Theta)$.

The MAP estimation gives $\hat{\Theta}$ as the mode of the posterior $P(\Theta|X_o,\eta)$:

$$\hat{\Theta} = \mathrm{argmax}_{\Theta} P(\Theta|X_o,\eta).$$

Since $\hat{\Theta}$ depends on the hyperparameter η, we should also estimate η when estimating the optimal $\hat{\Theta}$. We can estimate η by maximizing the log marginal likelihood $\ln P(X_o|\eta)$. Note that if the prior $P(\Theta|\eta)$ is a uniform distribution, this method gives $\hat{\Theta}$ as the mode of the likelihood $P\,(X_o|\Theta)$. This special case corresponds to the maximum likelihood estimate.

Appendix 2: Mathematical Detail for Bayesian Algorithm

A2.1 Likelihood Function

The PRC measurement experiment is shown in Fig. 8.1. A precisely timed perturbation is given to an oscillating neuron, and the transient change in the ISI is measured. A short pulse perturbation $G_o(t)$ is applied at the phase $\phi = t_0$, where t_0 is the time passed since the reference phase. The cycle containing the perturbation has an ISI

T_p. We obtain the spike response as $\psi = T - T_\mathrm{p}$. The spike response of real neurons stochastically fluctuates because of intrinsic noise. This process can be described in terms of the Langevin phase equation:

$$\frac{\mathrm{d}\psi(t)}{\mathrm{d}t} = Z(t + \psi(t))(G_0(t - t_0) + \sigma s(t)). \tag{8.24}$$

Here, the reference phase is at $t = 0$, and $\psi(T)$ represents the spike response of the cycle containing the perturbation. In this formulation, we can calculate the probability density of the spike response with the transition probability density (8.23)

$$P(\psi, T) = \int_{-\infty}^{\infty} \mathrm{d}\psi' P(\psi, T|\psi', 0)\delta(\psi'), \tag{8.25}$$

where $\delta(\psi)$ denotes the initial probability density of ψ at $t = 0$. The probability density of the spike response is thus

$$P(\psi|\, Z(t), \sigma, G_\mathrm{o}(t - t_0)) = \frac{1}{\mathscr{Z}_\mathrm{like}} \exp\left(-\frac{\left(\psi - \int_0^T \mathrm{d}t\, Z(t) G_\mathrm{o}(t - t_0)\right)^2}{4\sigma^2 \int_0^T Z(t')2\mathrm{d}t} \right), \tag{8.26}$$

where $\mathscr{Z}_\mathrm{like}$ is a normalization factor referred to as the partition function in physics. The probability density allows us to predict unknown spike-timing responses from known $Z(t)$, σ, and $G_\mathrm{o}(t - t_0)$.

If $\{\psi_i(\Delta tj)\}$ is the set of observed spike response data defined below, we obtain the likelihood function

$$L(Z(t), \sigma|\{\psi_i(\Delta tj)\}) = \prod_{i=1}^{N} \prod_{j=0}^{2L-1} P(\{\psi_i(\Delta tj)\}|Z(t), \sigma, G_\mathrm{o}(t - \Delta tj)), \tag{8.27}$$

where the spike response data are sampled as a sequence of equidistant points $\Delta tj (j = 0, 1, \ldots, 2L - 1)$ whose sampling period satisfies $\Delta t = T/2L$, and N samples $\{\psi_1(\Delta tj), \ldots, \psi_N(\Delta tj)\}$ are measured at each sampling point, $t_0 = \Delta tj$. The total number of experimental data is $2LN$.

A2.2 Maximum a Posteriori Estimation Algorithm

In (8.26), the mean values of the spike responses are governed by the convolution of PRC $Z(t)$ and perturbation $G_\mathrm{o}(t)$. Therefore, as shown in Fig. 8.2, a measurement of $Z(t)$ can be regarded as a signal transformation through a noisy band-limited channel with impulse response $G_\mathrm{o}(t)$. In this channel, the input is PRC and the output is the spike-timing response. We estimate the input PRC from the output spike response.

We introduce the following prior probability density to represent the smoothness of the PRC:

$$P(Z(t)|D,n) = \frac{1}{\mathscr{Z}_{\mathrm{pr}}} \exp\left(-\frac{ND^{2n}}{2} \int_0^T \left(\frac{\partial^n Z(t')}{\partial t'^n} \right)^2 \mathrm{d}t' \right). \tag{8.28}$$

This prior probability density has a $2n$-th order differential smoothness constraint. N is the number of samples at each sampling point mentioned above, and D and n are hyperparameters to control the smoothness of $Z(t)$. Here, D is scaled to ND^{2n} in order to obtain a consistent estimate and prevent D from diverging with increasing n. $\mathscr{Z}_{\mathrm{pr}}$ is the partition function for normalizing this probability density.

According to Bayes' theorem, we obtain a posterior probability density from (8.27) and (8.28)

$$P(Z(t)|\{\psi_i(\Delta tj)\}, \sigma, D, n, \{G_{\mathrm{o}}(t-\Delta tj)\})$$
$$= \frac{\mathscr{Z}_{\mathrm{pr}} \mathscr{Z}_{\mathrm{like}}^{2LN}}{\mathscr{Z}_{\mathrm{ps}}} P(Z(t)|D,n) \prod_{i=1}^{N} \prod_{j=0}^{2L-1} P(\{\psi_i(\Delta tj)\}|Z(t), \sigma, G_{\mathrm{o}}(t-\Delta tj)), \tag{8.29}$$

where $\mathscr{Z}_{\mathrm{ps}}$ is a partition function for normalizing the prior density.

Here, we introduce an auxiliary variable β and the δ function term in order to deal with the second coefficient of the Kramers–Moyal expansion $D^{(2)}$ depending on $Z(t)$ in a rigorous manner. We can write (8.29) as

$$P(Z(t)|\{\psi_i(\Delta tj)\}, \sigma, D, n, \{G_{\mathrm{o}}(t-\Delta tj)\})$$
$$= \frac{\mathscr{Z}_{\mathrm{pr}} \mathscr{Z}_{\mathrm{like}}^{2LN}}{\mathscr{Z}_{\mathrm{ps}}} \exp\left(-\frac{ND^{2n}}{2} \int_0^T \left(\frac{\partial^n Z(t')}{\partial t'^n} \right)^2 \mathrm{d}t' - \sum_{i=1}^{N} \sum_{j=0}^{2L-1} \frac{\beta}{2} \left(\psi_i(\Delta tj) \right.\right.$$
$$\left.\left. - \int_0^T Z(t') G_{\mathrm{o}}(t'-\Delta tj) \mathrm{d}t' \right)^2 \right)$$
$$\times\, \delta\left(\frac{1}{\beta} - 2\sigma^2 \int_0^T Z(t')^2 \mathrm{d}t' \right). \tag{8.30}$$

In this case, the δ function works as a constraint for keeping the power of $Z(t)$ constant. Because $Z(t)$ is restricted to a high-dimensional spherical space of radius $1/\sqrt{2\sigma^2\beta}$, the posterior density (8.30) is equivalent to the spherical spin model (Berlin and Kac 1952).

In the algorithm for estimating the PRC by maximizing the posterior density, the Fourier series of $Z(t)$, $G_{\mathrm{o}}(t)$, and $\psi(t_0)$ are used as follows to derive an executable algorithm to maximize the posterior density.

$$Z(t) = \sum_{l=-M}^{M} (Z(l)_c - iZ(l)_s) \exp(il\omega t),$$

$$Z(l)_c = \frac{1}{T}\int_0^T Z(t)\cos(l\omega t)\,dt,\ Z(l)_s = \frac{1}{T}\int_0^T Z(t)\sin(l\omega t)\,dt, \tag{8.31}$$

$$G_o(t) = \sum_{l=-M}^{M} (G_o(l)_c - iG_o(l)_s) \exp(il\omega t),$$

$$G_o(l)_c = \frac{1}{T}\int_0^T G_o(t)\cos(l\omega t)\,dt,\ G_o(l)_s = \frac{1}{T}\int_0^T G_o(t)\sin(l\omega t)\,dt, \tag{8.32}$$

$$\psi_i(\Delta tj) = \sum_{l=-L}^{L} (\Psi_i(l)_c - i\Psi_i(l)_s) \exp(il\omega\Delta tj),\quad L \le M,$$

$$\Psi_i(l)_c = \frac{1}{2L}\sum_{j=0}^{2L-1} \psi_i(\Delta tj)\cos(l\omega\Delta tj),\ \Psi_i(l)_s$$

$$= \frac{1}{2L}\sum_{j=0}^{2L-1} \psi_i(\Delta tj)\sin(l\omega\Delta tj). \tag{8.33}$$

To obtain an executable algorithm, we truncate the Fourier series after the M-th term.

Inserting these series in (8.30), the posterior density can be transformed into the following Fourier form:

$$P(\{Z(l)_c\},\{Z(l)_s\}|\{\Phi_i(l)_c\},\{\Phi_i(l)_s\},\sigma,D,n,\{G_o(l)_c\},\{G_o(l)_s\})$$
$$\propto \exp(-\mathscr{H})\,\delta\left(\frac{1}{\beta} - 2\sigma^2 T\left(Z(0)_c^2 + 2\sum_{l=1}^{M}(Z(l)_c^2 + Z(l)_s^2)\right)\right), \tag{8.34}$$

$$\mathscr{H} = \beta L\sum_{i=1}^{N}\Psi_i(0)_c^2 + 2\beta L\sum_{l=1}^{M}\sum_{i=1}^{N}(\Psi_i(l)_c^2 + \Psi_i(l)_s^2) + \frac{1}{2}a(0)Z(0)_c^2 - b_c(0)Z(0)_c$$
$$+ 2\sum_{l=1}^{M}\left(\frac{1}{2}a(l)(Z(l)_c^2 + Z(l)_s^2) - b_c(l)Z(l)_c - b_s(l)Z(l)_s\right), \tag{8.35}$$

where

$$a(l) = NTD^{2n}\omega^{2n}l^{2n} + 2\beta LT^2 N(G_o(l)_c^2 + G_o(l)_s^2),$$

$$b_c(l) = 2\beta LT\sum_{i=1}^{N}(\Psi_i(l)_c G_o(l)_c - \Psi_i(l)_s G_o(l)_s),$$

$$b_s(l) = 2\beta LT\sum_{i=1}^{N}(\Psi_i(l)_c G_o(l)_s + \Psi_i(l)_s G_o(l)_c).$$

Here, to avoid aliasing, we truncate the terms of $\Phi_i(l)_c$ and $\Phi_i(l)_s$ higher than the L-th term as $\Phi_i(L+1)_c = \Phi_i(L+2)_c = \cdots = \Phi_i(M)_c = 0$ and $\Phi_i(L+1)_s = \Phi_i(L+2)_s = \cdots = \Phi_i(M)_s = 0$.

By minimizing the following Hamiltonian $\mathscr{H}'$, we can approximately calculate the MAP value of $\hat{Z}$ as

$$\mathscr{H}' = \mathscr{H} + \lambda \left(\frac{1}{\beta} - 2\sigma^2 T \left(Z(0)_c^2 + 2 \sum_{l=1}^{M} \left(Z(l)_c^2 + Z(l)_s^2 \right) \right) \right)^2, \tag{8.36}$$

where $\mathscr{H}$ is defined in (8.35). The minimum point of the Hamiltonian $\mathscr{H}'$ approximates the MAP solution of the posterior density if λ is sufficiently large.

A2.3 Marginal Likelihood

$\hat{Z}(t)$ depends on the hyperparameters D and n and the intensity of the Langevin force σ. Therefore, we should also estimate these parameters. For this, we can maximize the following log marginal likelihood:

$$\mathscr{F}(\sigma, D, n | \{\psi_i(\Delta t j)\}) = \log \mathscr{Z}_{\text{ps}} - \log \mathscr{Z}_{\text{like}}^{2LN} - \log \mathscr{Z}_{\text{pr}}, \tag{8.37}$$

which is referred to as the free energy in physics. We can obtain an analytical solution of $\mathscr{F}$ and estimate the optimal $\hat{D}$, $\hat{n}$, and $\hat{\sigma}$ that maximize the log marginal likelihood $\mathscr{F}$.

The logarithms of the partition functions of the prior and the likelihood function are

$$\log \mathscr{Z}_{\text{like}}^{2LN} \propto NL \log 2\pi - NL \log \beta, \tag{8.38}$$

$$\log \mathscr{Z}_{\text{pr}} \propto M \log \pi - M \log NTD^{2n}\omega^{2n} - 2n \sum_{l=1}^{M} \log l. \tag{8.39}$$

The partition function of the posterior density can be solved by introducing an auxiliary variable x as follows:

$$\mathscr{Z}_{\text{ps}} \propto \exp\left(-\beta L \sum_{i=1}^{N} \Phi_i(0)_c^2 - 2\beta L \sum_{l=1}^{M} \sum_{i=1}^{N} \left(\Phi_i(l)_c^2 + \Phi_i(l)_s^2 \right) \right)$$
$$\times 2 \int_0^\infty \mathrm{d}x \sqrt{\frac{2\pi}{\sqrt{a(0)^2 + 16x^2\sigma^4 T^2}}} \prod_{l=1}^{M} \left(\frac{1}{\sqrt{a(l)^2 + 16x^2\sigma^4 T^2}} \right)$$

$$\times \exp\left(\frac{b_c(0)^2 a(0)}{2a(0)^2 + 32x^2\sigma^4 T^2} + \sum_{l=1}^{M}\left(\frac{b(l)^2 a(l)}{a(l)^2 + 16x^2\sigma^4 T^2}\right)\right)$$

$$\times \cos\left(\left(\frac{x}{\beta} - \frac{1}{2}\tan^{-1}\left(\frac{4x\sigma^2 T}{a(0)}\right) - \frac{b_c(0)^2 2x\sigma^2 T}{a(0)^2 + 16x^2\sigma^4 T^2}\right)\right.$$

$$\left. + \sum_{l=1}^{M}\left(-\tan^{-1}\left(\frac{4x\sigma^2 T}{a(l)}\right) - \frac{b(l)^2 4x\sigma^2 T}{a(l)^2 + 16x^2\sigma^4 T^2}\right)\right), \qquad (8.40)$$

where

$$b(l)^2 = b_c(l)^2 + b_s(l)^2$$

$$= 4\beta^2 L^2 T^2 \left(G(l)_c^2 + G(l)_s^2\right)\left(\left(\sum_{i=1}^{N} \Phi_i(l)_c\right)^2 + \left(\sum_{i=1}^{N} \Phi_i(l)_s\right)^2\right).$$

The detailed derivation is explained in Ota et al. (2009).

References

Berlin TH, Kac M (1952) The spherical model of a ferromagnet. Phys Rev 86:821.

Ermentrout GB, (1996) Type I membranes, phase resetting curves, and synchrony. Neural Comput, 8:979-1001.

Ermentrout GB, Saunders D (2006) Phase resetting and coupling of noisy neural oscillators. J Comput Neurosci 20:179-190.

Ermentrout GB, Galán RF, Urban NN (2007) Relating neural dynamics to neural coding. Phys Rev Lett 99:248103.

Galán RF, Ermentrout GB, Urban NN (2005) Efficient estimation of phase-resetting curves in real neurons and its significance for neural-network modeling. Phys Rev Lett 94:158101.

Galán RF, Ermentrout GB, Urban NN (2007) Stochastic dynamics of uncoupled neural oscillators: Fokker-Planck studies with the finite element method. Phys Rev E 76:056110.

Goldberg JA, Deister CA, Wilson CJ (2007) Response properties and synchronization of rhythmically firing dendritic neurons. J Neurophysiol 97:208-219.

Kuramoto Y (1984) Chemical oscillations, waves, and turbulence. Springer-Verlag.

Kuramoto Y (unpublished).

Lengyel M, Kwag J, Paulsen O, Dayan P (2005) Matching storage and recall: hippocampal spike timing-dependent plasticity and phase response curves. Nat Neurosci 8:1677-1683.

Mancilla JG, Lewis TJ, Pinto DJ, Rinzel J, Connors BW (2007) Synchronization of electrically coupled pairs of inhibitory interneurons in neocortex. J Neurosci 27:2058-2073.

Marella S, Ermentrout GB (2008) Class-II neurons display a higher degree of stochastic synchronization than class-II neurons. Phys Rev E 77:041918

Morris C, Lecar H (1981) Voltage oscillations in the barnacle giant muscle fiber. Biophys J, 35: 193-213.

Nakao H, Arai K, Kawamura Y (2007) Noise-induced synchronization and clustering in ensembles of uncoupled limit-cycle oscillators. Phys Rev Lett 98:184101.

Neltner L, Hansel D, Mato G, Meunier C (2000) Synchrony in heterogeneous networks of spiking neurons. Neural Comput 12:1607-1641

Netoff TI, Acker CD, Bettencourt JC (2004) Beyond two-cell networks: experimental measurement of neuronal responses to multiple synaptic inputs. J Comput Neurosci 18:287-295

Netoff TI, Banks MI, Dorval AD, Acker CD, Haas JS, Kopell N, White JA (2005) Synchronization in hybrid neuronal networks of the hippocampal formation. J Neurophysiol 93:1197-1208.

Oprisan SA, Prinz AA, Canavier CC (2004) Phase resetting and phase locking in hybrid circuits of one model and one biological neuron. Biophysical J 87:2283-2298.

Ota K, Omori T, Aonishi T (2009) MAP estimation algorithm for phase response curves based on analysis of the observation process. J. Comput. Neurosci. 26:185-202.

Preyer AJ, Butera RJ (2005) Neuronal oscillators in *aplysia californica* that demonstrate weak coupling *in vitro* Phys Rev Lett 95:138103.

Reyes AD, Fez EE (1993) Two modes of interspike interval shortening by brief transient depolarizations in cat neocortical neurons. J Neurophysiol 69:1661-1672.

Risken H (1989) The fokker-planck equation. Methods of solution and applications. Springer, Berlin.

Robinson HP, Kawai N (1993) Injection of digitally synthesized synaptic conductance transients to measure the integrative properties of neurons. J Neurosci Methods 49:157-165.

Sharp AA, O'Neil MB, Abbott LF, Marder E (1993) Dynamic clamp: computer-generated conductances in real neurons. J Neurophysiol 69:992-995.

Stoop R, Schindler K, Bunimovich LA (2000) Neocortical networks of pyramidal neurons: from local locking and chaos to macroscopic chaos and synchronization. Nonlinearity 13:1515-1529.

Tateno T, Robinson HP (2007) Phase resetting curves and oscillatory stability in interneurons of rat somatosensory cortex. Biophysical J 92:683-695.

Teramae J, Tanaka D (2004) Robustness of the noise-induced phase synchronization in a general class of limit cycle oscillators. Phys Rev Lett 93:204103.

Teramae J, Fukai T (2008) Temporal precision of spike response to fluctuating input in pulse-coupled networks of oscillating neurons. Phys Rev Lett 101:248105.

Tsubo Y, Takada M, Reyes AD, Fukai T (2007a) Layer and frequency dependencies of phase response properties of pyramidal neurons in rat motor cortex. Eur J Neurosci 25:3429-3441.

Tsubo Y, Teramae J, Fukai T (2007b) Synchronization of excitatory neurons with strongly heterogeneous phase responses. Phys Rev Lett 99:228101.

Winfree AT (1967) Biological rhythms and the behavior of populations of coupled oscillators. J Theor Biol 16:15-42.

Part III
Cellular Mechanisms of Neuronal Phase Response Properties

Introduction

A neuron's spiking mechanism is a product of the contingent of membrane conductances it possesses, and the phase response curve reflects the mechanism of spike generation or class of excitability exhibited by a neuron. Membrane conductances can also sculpt the shape of neuronal PRCs without qualitatively altering a neuron's spiking mechanism. Part III of this book focuses on the cellular mechanisms underlying not only neuronal excitability but also the shaping of PRCs by specific membrane conductances when neurons spike intrinsically or are driven to different spike frequencies by current injection or synaptic input. Additionally, chapters in this part of the book address how cellular mechanisms can shape the way a neuron spikes in response to stimulus features, and how those stimuli can be recovered given an output spike train and insight into the neuron's phase response properties. A neuron's cellular mechanisms define the manner in which it processes inputs, and, therefore, embody the input–output characteristics of the neuron which is of direct consequence for network interactions addressed in Part IV.

Chapter 9 by Ermentrout and colleagues surveys the effects of many common membrane conductances on the underlying bifurcation structures and excitability mechanisms. This is related to the PRCs of conductance-based models and the activation profile of those conductances during the spike cycle. In Chap. 10, Galan introduces the concepts of spike-time reliability and stochastic synchronization whereby neurons generate consistent spike patterns in response to repeated or shared fluctuating input. Changes in these phenomena are related to effects of membrane conductances, particularly the A-type potassium current, on the PRC. As the PRC can be used to determine the reliability with which a neuron will spike in response to a fluctuating stimulus, it can also be used to recover the fluctuating stimulus from an output spike pattern. In Chap. 11, Kim and Lazar introduce the use of conditional PRCs for stimulus recovery. The last two chapters in Part III delve directly into specific mechanisms by which the qualitative form of neuronal PRCs, i.e., type I or type II, can depend on cellular mechanisms. Chapter 12 by Stiefel and Gutkin details how cholinergic modulation of cortical pyramidal neurons can switch the neuron's PRC between type II and type I by adjusting the strength of slow potassium currents including the M-type potassium current. In Chap. 13, Schultheiss identifies how another slow potassium current, the small conductance calcium-activated potassium current (SK) in the dendrite of a morphologically reconstructed neuron model can underlie type II dendritic PRCs whereas the somatic PRC is type I.

Chapter 9
Phase Response Curves to Measure Ion Channel Effects on Neurons

G. Bard Ermentrout, Bryce Beverlin II, and Theoden Netoff

Abstract In this chapter, we explore how different common ionic currents change the shape of the PRC using several well-known conductance-based models. By also studying the simplest possible network – two pulse coupled neural oscillators – we can use the shape of the PRC to quantify the degree of synchrony in such a pair. We find a number of qualitative effects that include changing the sign of the PRC during certain phases and altering the skew of the PRC. Using bifurcation theory, we trace these effects back to the underlying dynamics that govern the onset of repetitive firing as the neuron is depolarized by a constant current. The key bifurcation involves the so-called Takens-Bogdanov bifurcation. We show that many of the shape changes of the PRC can be understood by examining the dynamics of the neuron near this bifurcation.

1 Introduction

The phase response curve (PRC) provides a useful tool for analyzing the dynamics of forced and coupled neural oscillators. The relative ease of experimentally determining the PRC makes it a good way to study small networks of coupled neurons without requiring knowledge of the biophysical details. Since the PRC is itself dependent on the underlying dynamics that govern the repetitive spiking of a

G.B. Ermentrout (✉)
Department of Mathematics, University of Pittsburgh, Pittsburgh, PA, USA
e-mail: bard@pitt.edu

B. Beverlin II
Department of Physics, University of Minnesota, Minneapolis, MN, USA
e-mail: beverlin@physics.umn.edu

T. Netoff
Department of Biomedical Engineering, University of Minnesota, Minneapolis, MN, USA
e-mail: tnetoff@umn.edu

N.W. Schultheiss et al. (eds.), *Phase Response Curves in Neuroscience: Theory, Experiment, and Analysis*, Springer Series in Computational Neuroscience 6,
DOI 10.1007/978-1-4614-0739-3_9, © Springer Science+Business Media, LLC 2012

given neuron, its shape and form can be very sensitive to the relative strengths of the currents which comprise the spiking of the neuron. The shape of the PRC, in turn, plays a major role in determining whether networks of coupled rhythmically firing neurons can synchronize. Thus, it is useful to better understand how different ionic currents alter the shape of the PRC and thus how easily networks synchronize.

In this chapter, we explore how different common ionic currents change the shape of the PRC using several well-known conductance-based models. By also studying the simplest possible network – two pulse coupled neural oscillators – we can use the shape of the PRC to quantify the degree of synchrony in such a pair. We find a number of qualitative effects that include changing the sign of the PRC during certain phases and altering the skew of the PRC. Using bifurcation theory, we trace these effects back to the underlying dynamics that govern the onset of repetitive firing as the neuron is depolarized by a constant current. The key bifurcation involves the so-called Takens–Bogdanov bifurcation. We show that many of the shape changes of the PRC can be understood by examining the dynamics of the neuron near this bifurcation.

2 Neuronal Oscillations

The role of synchronous neural activity has been an intense focus of brain research in the past decade. Several phenomena have been associated with synchronous activity from normal brain functions such as motor coordination, active working memory, and emotion to abnormal brain function observed in diseases such as epilepsy, Parkinson's, and schizophrenia. Measurement of synchronous behavior in the brain is achieved through various techniques ranging from multielectrode recording to electroencephalography (EEG). Ion channels, whose opening and closing kinetics are responsible for generating action potentials, function at molecular scales; at much larger scales, millions of neurons generate population oscillations as large ensemble phenomenon, detectable outside the skull with EEG. Considering a growing knowledge and discovery across the size continuum, there remains a large intellectual gap between ion channel and neuronal population behaviors. The importance in bridging molecular to population scales increases as more evidence amasses that mutations and modulations of ion channels are major factors in many population driven diseases. Even with multiple scale methods of study, the exact role of synchrony in neural assemblies has not yet been determined. To complicate matters, networks of modeled neurons function in different ways based on factors including ion channel kinetics, network connectivity, and neuron cell type. Therefore, it is necessary to study synchronous brain dynamics from various size scales and conditions to solve some of the most interesting and pressing problems in neuroscience.

Synchronous oscillations may combine sensory signals from several brain areas or even through different sensory modalities into a single perception (Gray 1994). For example, the phase of a place cell firing in the theta rhythm of a neuronal population may indicate whether an animal is in the center of the place field (Skaggs et al. 1996). Also, cholinergic stimulation of the hippocampus and entorhinal cortex produces strong oscillations (Chrobak and Buzsaki 1996). Disruption in normal population oscillations is an indication, or possibly even a cause, of many neuropathological diseases such as Parkinson's disease (Uhlhaas and Singer 2006; Hutchison et al. 2004), schizophrenia and epilepsy (Netoff and Schiff 2002; Lehnertz et al. 2009; McCormick and Contreras 2001; Huguenard and McCormick 2007). These population oscillations, which can be measured at the scalp in EEG measurements, are generated by the synchronization of millions of neurons. We posit that changes in the dynamics of individual neurons and synapses are responsible for generating the emergent oscillations at the network level. Oscillatory changes in the neuron dynamics are caused by the modulation of ion channels by neurotransmitters, neuropeptides and phosphorylation states. One feature of neuronal oscillations is that they can disappear as quickly as they are formed; for example, gamma oscillations produced in the rat hippocampus during sniffing and exploring their environment may only last a few cycles and disappear. Gamma oscillations are also found to wax and wane with the phase of theta oscillations (Chrobak and Buzsaki 1998). Therefore, a model of population oscillations must describe how neuronal oscillations are generated and destroyed. The PRC function applied to neuronal oscillators is an excellent method to explain how synchrony arises in populations of neuronal oscillators and thus produces *population oscillations*.

3 Oscillator Theory

The interplay of ionic currents and the PRC is subtle and complex to address with analytical methods. Nevertheless, there are some basic principles which apply and allow us to infer how the shape of the PRC is altered as ionic currents are added. Before describing this work, we will first motivate the study of PRCs from the viewpoint of synchronization of neural oscillators. In order to model neuronal oscillators, we need to describe an oscillator theory which will be relevant to our neuronal system.

3.1 Weak Pulse Coupling

Suppose that two identical weakly coupled oscillators with period T and phases θ_1 and θ_2 are coupled in such a way that each time oscillator 1 spikes, it corresponds to crossing $\theta_1 = T$. Then the phase of oscillator 2, θ_2, is advanced by an amount that is proportional to its PRC, that is

$$\theta_2 = F(\theta_2) := \theta_2 + a\Delta(\theta_2)$$

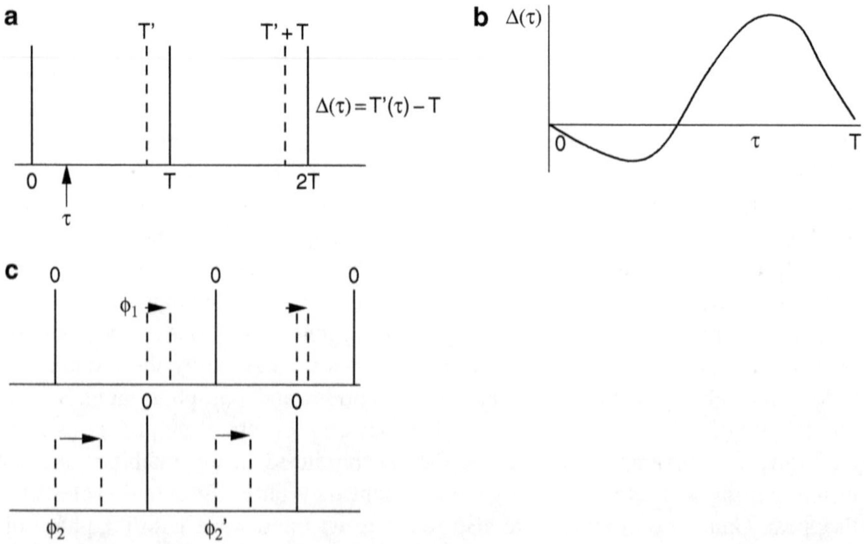

Fig. 9.1 The PRC and pulse coupling. (**a**) Construction of the PRC. An oscillator with period T is perturbed at a point τ after its last spike. This causes it to fire at a time T' that is generally not the same as the unperturbed cycle time. The $T - T'$ difference is the magnitude of the PRC at τ, $\Delta(\tau)$. By performing a perturbation at many values of τ between 0 and T, a continuous PRC function is constructed. (**b**) A PRC showing delays in the early part of the cycle with negative values of $\Delta(\tau)$ and advances in the second half of the period with positive values of $\Delta(\tau)$. (**c**) Here oscillator 1 spiking (*solid line* with 0 on *top row)* influences the spike timing of oscillator 2 at ϕ_2 with magnitude $\Delta_2(\phi_2)$. Oscillator 2 then spikes (*solid line* with 0 on *bottom row*), influencing the spike timing of oscillator 1 at ϕ_1 with magnitude $\Delta_1(\phi_1)$. Thus, a map of pulse coupling may be constructed for two reciprocally connected oscillators

where a (assumed to be small) is the amplitude of the coupling and $\Delta(\theta_2)$ is the PRC of oscillator 2. For simplicity, each oscillator has the same frequency corresponding to period T. Figure 9.1a shows how a perturbation at time τ advances the next spike time from T to T'. Conversely, each time oscillator 2 spikes, oscillator 1 is similarly advanced. This effect of reciprocal pulse coupling on oscillator phase is demonstrated in (c) of Fig. 9.1. The result of this coupling is a map which takes the unperturbed phase of oscillator 2 at the time that oscillator 1 fires to its new phase at the next cycle. Let ϕ_n be this phase, thus the map (Goel and Ermentrout 2002) satisfies:

$$\phi_{n+1} = T - F(T - F(\phi_n)) := G(\phi).$$

We refer to $G(\phi)$ as the *spike phase difference map.* Thus, we can see that the phase-locked states of a pair of pulse coupled oscillators are approximately the roots of this map, $G(\phi)$, which is the odd part of the PRC. In general, a fixed point x_0 for a map $x \to Q(x)$ is stable if $|Q'(x_0)| < 1$. For small a, this means $G'(\phi) < 0$. Thus, for two oscillators with identical PRCs, by looking at the shape of

$$G(\phi) = \Delta(\phi) - \Delta(-\phi), \tag{9.1}$$

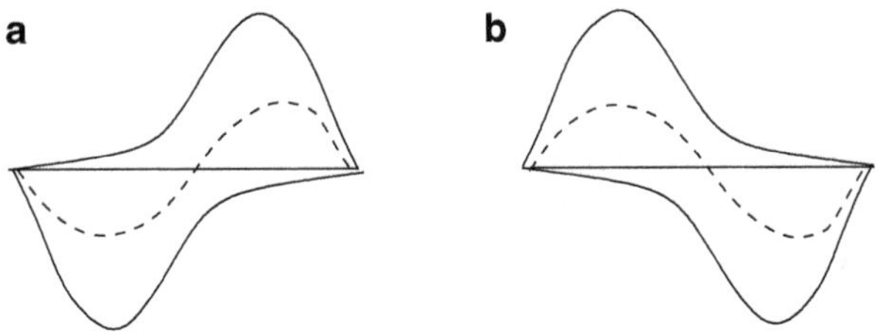

Fig. 9.2 Skew of the PRC determines stability of synchrony and antiphase states in two reciprocally connected identical oscillators. (**a**) PRC $\Delta(\phi)$ is skewed toward the right (*solid line* above axis) and $-\Delta(-\phi)$ (*solid line* below axis) is displayed to visualize $G(\phi) = \Delta(\phi) - \Delta(-\phi)$ (*dashed line*), demonstrating stable synchronous state due to $G'(0) < 0$. and unstable anti-phase state due to $G'(\frac{T}{2}) > 0$. (**b**) PRC is now skewed toward the left with $-\Delta(-\phi)$ below the axis, indicating the system has an unstable synchronous solution because $G'(0) > 0$ and has a stable anti-phase state because $G'(\frac{T}{2}) < 0$

we can find both the phase-locked points and their stability. Equation (9.1) may be generalized to two oscillators with different PRCs as $G(\phi) = \Delta_1(\phi) - \Delta_2(-\phi)$.

A phase-locked solution is a fixed point of this map satisfying $\phi = G(\phi)$. The fixed points or roots are stable if $\Delta'(\phi) + \Delta'(-\phi) < 0$. We can rewrite this last condition as $\Delta'(\phi) + \Delta'(T - \phi) < 0$ since $\Delta(\phi)$ is $T - periodic$. For most PRCs measured in neurons, $\Delta(0) = \Delta(T) = 0$. Thus for identical neurons, the root $\phi = 0$ is a fixed point; this is the *synchronous* solution. Synchrony in this system is stable if and only if $-1 < G'(0) < 1$. Using the definition of G, this condition implies that for synchrony to be stable, we must have

$$-1 < F'(0)F'(T) := [1 + a\Delta'(0)][1 + a\Delta'(T)] < 1.$$

Thus, we reduce a pair of coupled oscillators to a single scalar equation for the phase difference, ϕ. Zeros of $G(\phi)$ correspond to phase-locked states and their stability is determined by the slope near zero, where positive slopes are unstable and negative slopes are stable. For example, $G(0) = 0$ so that synchrony is a solution for identical reciprocally coupled oscillators. Figure 9.2 shows two different PRCs and the resulting functions $G(\phi)$ that are obtained from them. In panel (a), the skew of the PRC is to the right and in panel (b), it is to the left. Synchrony is stable in the former while the antiphase state (oscillators are half a cycle out of phase) is stable in the latter.

4 Measuring PRCs

The response of periodic oscillators to external perturbation can be described by the PRC. As previously discussed, a PRC is a continuous function which describes how an oscillator's period will temporarily change depending on its phase at the moment of external perturbation (we remark that biologically measured PRCs are not necessarily continuous, especially for strong stimuli or for stimuli that are long lasting, e.g., Reyes and Fetz 1993). The PRC function when positive (negative) will decrease (increase) the oscillator's period. A PRC may contain both positive and negative values, meaning that in certain parts of the oscillator's phase, a perturbation will decrease the period (positive PRC magnitude) while in other parts of the oscillator's phase a perturbation will increase the period (negative PRC magnitude) as shown in Fig. 9.1b.

The PRC may be used to bridge ion channel modulation and synapse variability at the cellular level to changes in network synchrony at larger population scales. One way to model population oscillations is to assume that the neurons have been activated through synaptic input and in the activated state they are firing periodically. We put forth the hypothesis that activation of the population may result in periodically firing neurons, while modulation of ion channels can transition a network from a disorganized state to a network that oscillates. PRC models help explain how these networks may slip in and out of oscillations as facile as oscillations in the brain seem to come and go. First, we introduce a method of measuring PRCs and explain the mathematical basics of bifurcation in order to describe how neuronal oscillations may be modeled.

4.1 The Adjoint Method

The PRC can be measured with the *adjoint method*, which uses differential equations rather than perturbations (the *direct method* of PRC measurement). Let us denote the directly measured PRC as $\Delta(a,t)$ where a is the width of the perturbation. The *infinitesimal PRC* is defined as

$$\Delta_0(t) := \lim_{a \to 0} \frac{\Delta(a,t)}{a}.$$

For small widths of perturbation, $\Delta(a,t) \approx a\Delta_0(t)$, so that computing $\Delta_0(t)$ can be useful for understanding the effects of finite (nonzero) perturbations. Butera showed this kind of linearity holds in certain neural preparations (Butera and Preyer 2005). The reason to study $\Delta_0(t)$ is that it is readily computed for any system of differential equations that has a limit cycle without having to perform perturbations. Specifically, if

$$X' = F(X)$$

has a stable $T-$periodic solution, $Y(t)$, then to compute $\Delta_0(t)$ we must solve the linear equation:

$$Z' = -D_X(Y(t))^T Z$$

along with the condition that $Y(t)^T Z(t) = 1$, where D_X is the derivative of F(X) with respect to X and $Z(t)$ is a vector of components corresponding to the n variables in the differential equation defining the limit cycle. Each component represents the phase shift that would result from a perturbation of the corresponding variable. However, for membrane models where perturbations are applied currents (using the *direct method* of measurement), $\Delta_0(t)$ is the voltage component of the vector $Z(t)$. Therefore, the PRC of a neuron due to a current injection is proportional to the component of Z corresponding to the voltage variable. For example, in the Hodgkin and Huxley equations the vector of states is (V,m,n,h) with (zV,zm,zn,zh) as the four components of Z (the specifics of these equations will be explained in the next section on bifurcations). We see that zV(t) is proportional to the PRC $\Delta(t)$ obtained by perturbing the voltage with a current pulse. We can efficiently explore the roles of ionic currents on the PRC by exploiting this mathematical connection of the adjoint method.

4.2 *Bifurcations*

Many neurons will fire repetitively when sufficiently depolarized. The transition from a stable quiescent state to active firing is characterized by a bifurcation (Izhikevich 2007); that is, there is a qualitative change in the dynamics of the neuron. The onset of repetitive firing for many neurons takes place via two dynamical mechanisms: (1) saddle-node on a limit cycle (SNIC) and (2) Hopf bifurcation (HB) (Rinzel and Ermentrout 1989). Associated with each of these bifurcations is a so-called *normal form*, which describes the local dynamics near the onset of oscillation. Numerous authors have computed the PRC for these normal forms and thus, there is a connection between the way that a neuron makes the transition from rest to repetitive firing and its PRC through this reduction of the full model to its normal form. For the SNIC, the PRC is given as (Fig. 9.3)

$$\Delta_{\mathrm{SN}}(t) = 1 - \cos(2\pi t/T) \tag{9.2}$$

while for the HB, the PRC is

$$\Delta_{\mathrm{HB}}(t) = \sin(2\pi t/T + \xi) \tag{9.3}$$

where $-\pi/2 < \xi < \pi/2$ is a parameter that depends on the model parameters (Brown et al. 2004). Observe that in the case of the SNIC, the PRC is nonnegative while for the HB, the PRC has both positive and negative components. What this says is that near the bifurcation depolarizing current pulses can only advance the phase for the SNIC and can both advance and delay the phase for the HB. Thus the SNIC accumulates perturbations in a monotonic fashion and is thus called an integrator. The HB neuron will more easily synchronize both to common input and when coupled with excitation since oscillators which lead can be slowed down and those that follow can be sped up. In the SNIC, spikes are always advanced.

These facts imply that changes in currents which result in changes of bifurcations will lead to qualitative differences in the PRCs. Consider a neuron that has a SNIC bifurcation. The SNIC bifurcation always involves the coalescence of a stable node

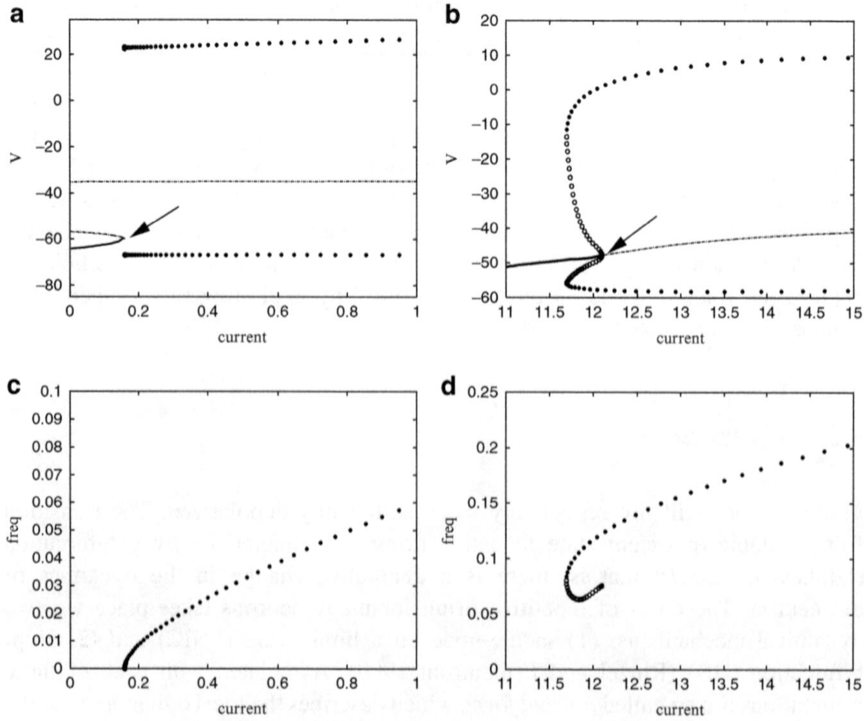

Fig. 9.3 Bifurcation diagrams for a Wang–Buzsaki neuron for leakage conductance $g_l = 0.1$ (**a**,**c**) and $g_l = 1$ (**b**,**d**). *Top panels* show voltage as a function of current where the *thick solid line* represents the stable fixed points and the *thin solid lines* are the unstable fixed points. The *filled circles* represent the stable periodic solution and the *open circles* are the unstable periodic solution. Each *arrow* shows the point of bifurcation for the separate neurons. (**a**) Saddle node on invariant circle (SNIC) bifurcation. (**b**) Subcritical Hopf bifurcation (HB). (**c**) Frequency vs current for neuron in (A) where the frequency goes to 0 (may be made arbitrarily low) at the current value of SNIC bifurcation. (**d**) Frequency vs current for neuron in (**b**) where the frequency remains bounded above 0 for all values of current, including those near the HB

with a saddle point as the current changes so that for a range of currents, there are at least three equilibrium points. We can explore the number of equilibrium points for conductance-based models by examining the steady-state current voltage curve (Rinzel and Ermentrout 1989). Specifically, suppose the membrane model has the form:

$$C\frac{\mathrm{d}V}{\mathrm{d}t} = I - \sum_j g_j m_j^{p_j} h_j^{q_j}(V - E_j), \tag{9.4}$$

$$\frac{\mathrm{d}m_j}{\mathrm{d}t} = \frac{m_j^{\infty}(V) - m_j}{\tau_j^m(V)}, \tag{9.5}$$

$$\frac{\mathrm{d}h_j}{\mathrm{d}t} = \frac{h_j^{\infty}(V) - h_j}{\tau_j^h(V)}, \tag{9.6}$$

where p_j, q_j are nonnegative integers. The steady state current-voltage curve is found by setting $m_j = m_j^{\infty}(V)$ and $h_j = h_j^{\infty}(V)$ so that we can express the equilibria as

$$I_{ss} = F(V) = \sum_j g_j [m_j^{\infty}(V)]^{p_j} [h_j^{\infty}(V)]^{q_j} (V - E_j).$$

A SNIC occurs only if $F(V)$ is nonmonotonic, so one way to change the bifurcation structure of a neural model is the add or remove currents such that I_{ss} becomes monotonic. Similarly, adding or removing currents to a model that has an HB can also result in a gain of nonmonotonicity. We should note that this is a very crude manipulation since it is possible that removing or adding such currents can destroy the ability of the neuron to spike.

Currents that work near rest can also serve to destabilize the resting state without destroying the multiplicity of equilibria. We will see that this leads to an interesting "co-dimension 2" bifurcation; two parameters (rather than just the current) need to be manipulated to find it. As noted above, altering the skew of the PRC can also change the way in which a neuron interacts with other neurons and with the external environment. Currents which come on only when the neuron spikes cannot affect the local bifurcation structure but can have profound effects on the shape of the PRC.

5 Ionic Currents

Here, we employ the basic conductance-based membrane model and explain the role of bifurcations in our neuronal system. Currents are formed when ions such as sodium and potassium flow through neuronal membranes via ion-specific channels. The opening and closing of these channels are dependent on the voltage difference between intracellular and extracellular space. While open, ions will flow through their respective channels until an ion-specific equilibrium potential is achieved. The probability of each channel being open is described with differential equations; these probabilities are examined as a function of time through one complete oscillation to study when the channels are likely to be open or closed. Hodgkin and Huxley first developed equations that characterize the gating of the ion channels and we use similar methods here to model neurons.

In this section, we will explore both the bifurcation structure and PRCs of several models and show how individual channels change the shape of the PRC. Modulating each ionic current in the cell changes the functional shape of the PRC and using the spike phase difference map, $G(\phi)$, one may infer the modulation's effect on network synchrony. First, we start with a spatially compact model with a large complement of ionic currents and explore empirically how modulating the amplitudes of the currents alter the PRC. Next, using a very simple example of modeling a fast spike unit (inhibitory interneuron) we change the leak current with the consequence that the PRC changes drastically. Then, we turn to a model for pyramidal cell and add a variety of potassium currents and a sag current to observe the resultant PRC. In some cases, we will simplify the model to a skeletonized version and explain through dynamical systems theory how and why these currents affect the PRC.

In general, we use conductance-based differential equation models of ionic voltage gated channels to describe an individual neuron, which is then driven to periodically oscillate and is subsequently studied with input perturbation. By altering the parameters of these ionic channels, we are able to alter the response of the individual cells to this perturbation. Measuring the PRC as an input/output function is achieved by perturbing the oscillator and observing the resultant dynamics, which include a temporary alteration in oscillation frequency. This process of applying a perturbation at several points in the oscillator's phase with current stimulus is known as the direct method and is discussed in detail in Chap. 5.

5.1 Probabilities of Channel Opening

Ion channel gating can be controlled by the voltage across the membrane, neurotransmitters, or second messenger systems. Here, we focus on voltage gated ion channels and their effect on the PRC. Figure 9.4 contains the ion channel gating probabilities as a function of phase, displayed in time over one complete cycle. The top plot is the voltage of a model neuron cell from one spike (maximum positive voltage) to the next spike, which defines the period of oscillation, T. The remaining plots demonstrate how the conductance of a given channel is dependent on the cell's voltage. Some ion channels such as Na_{fast} closely follow the voltage while others like K_{slow} have a longer time course of opening and closing. From these dynamics, we are able to infer some changes on the cell's behavior depending on the level of individual ion channel conductance.

6 Golomb and Amitai Model

We use the Golomb and Amitai conductance-based model (GA model) (Golomb and Amitai 1997) to examine the effect of certain ion channel mutations on the PRC. The GA model is a sophisticated neuron model containing sodium, persistent sodium, potassium-delayed rectifier, slow potassium, potassium A, and leak currents. It is relatively easy to add additional conductance channels such as M and TASKρ to this complex model. By varying the maximum conductance of individual channels, we measure the PRC from the model neuron oscillator. From each PRC, $\Delta(\phi)$, we calculate the spike phase difference map $G(\phi) = \Delta(\phi) - \Delta(-\phi)$ and the slope of G at zero, $G'(0)$, to predict the stability of synchrony of a mutually connected identical two cell network.

Each PRC is measured using the direct method of perturbative stimulation with an AMPA-like input to the GA model. The baseline model transitions through

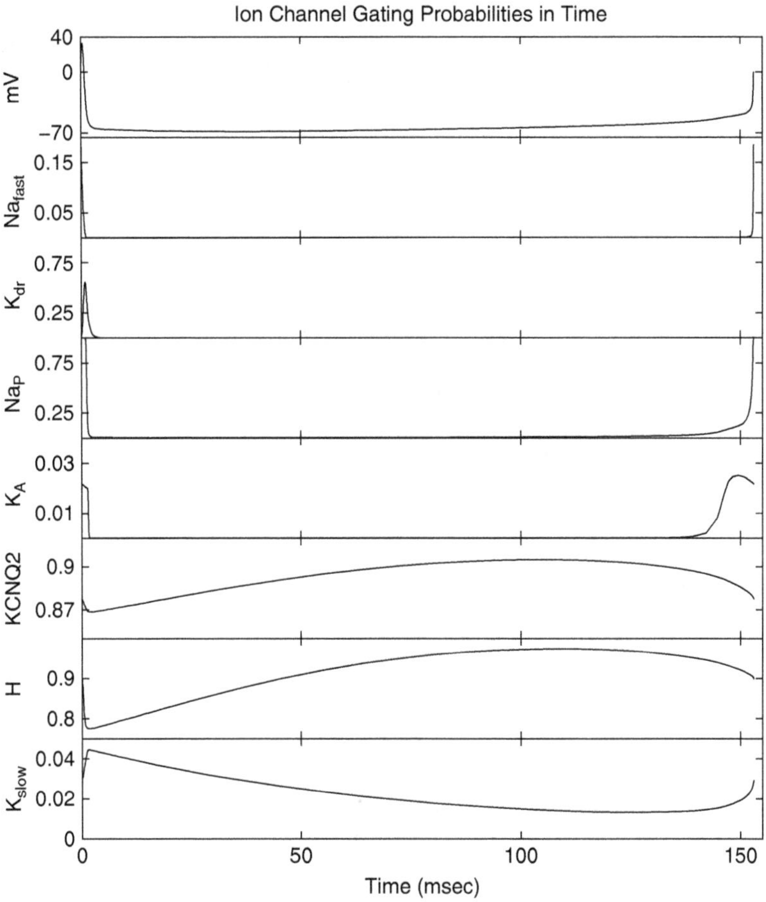

Fig. 9.4 Probabilities of individual ion channels being open as a function of time over one complete cycle. The oscillation's voltage trace (as the top plot) is displayed for reference. Channels Na_{fast}, K_{dr}, Na_P, and K_A show a quick response which closely follows the voltage while $K_M(KCNQ2)$, $H(sag)$, K_{slow} channels display slower change over the oscillation period

a SNIC bifurcation from rest to periodic oscillation and thus will produce a Type I PRC, which contains only positive values. We examine several PRCs with increasing values of maximum conductance from various ion channel mutations implicated in synchronous diseases. For each channel, we include the voltage trace of one complete cycle and a plot of the PRC as well as the spike phase difference map, $G(\phi)$; in addition, $G'(0)$ is plotted as a function of maximum conductance, $g_{\max}$, where positive (negative) values of $G'(0)$ indicate instability (stability) of the synchronous solution in a reciprocally connected two cell network.

While some channels studied are at zero conductance value during baseline measurements, others are examined for lower conductance values than baseline. In both cases, the maximum conductance values, $g_{\max}$, are chosen such that the oscillator is within limit cycle constraints, meaning that the oscillator continues to oscillate. In all cases, the baseline values are included as a dashed line in the phase-dependent plots and a solid star in the $G'(0)$ plot. A voltage trace for each measurement is included, which provides insight on the effect of an altered channel on repolarization following a spike generation and on the natural period of oscillation. However, input current was varied for each simulation to constrain the oscillator to a period of about 150 ms for control purposes.

6.1 *Fast Sodium*

Sodium channels are responsible for generating the spike of an action potential by allowing Na ions to flow into the cell, thus depolarizing from the rest potential of about −70 mV. The Na_{fast} channel is particularly important because several mutations in sodium channels have been linked to epilepsy. These fast gated voltage channels open very rapidly with small depolarization which causes a positive feedback loop, opening even more channels. When the Na_{fast} conductance is increased, the voltage peak during an action potential is increased, leading to a larger overshoot during depolarization as evidenced in the voltage trace in Fig. 9.5. After crossing threshold for spiking, the subsequent fall of the action potential is dependent on the inactivation of the sodium channels as well as the onset of the hyperpolarizing potassium current. Figure 9.4 shows how the Na_{fast} is a fast acting channel, which is open when the voltage is high, both at the beginning and end of the oscillation period. While changes in sodium channel density affect the shape of the action potential, the predominant effect on the functional PRC is the dependence of action potential threshold. By increasing the sodium conductance, we observe an increase in the PRC and a shift earlier in phase leading to an unstable synchronous solution as indicated by $G(\phi)$ displayed in Fig. 9.5. The $G'(0)$ clarifies that an increase in Na_{fast} conductance at first, decreases the stability of synchrony in a mutually coupled 2-cell network and with enough conductance, changes the synchronous solution from stable to unstable.

6.2 *Persistent Sodium*

Persistent sodium channels are noninactivating sodium channels, which make a small inward current, and have extremely slow kinetics, on the order of 30 s. Na_P can sum very slowly resulting in slow accommodation of spiking in rapidly firing neurons. This current can enhance excitability of the cells near threshold, increasing the frequency vs. current relationship range resulting in firing rates that can approach zero at low currents (Hansel and Sompolinsky 1996). Figure 9.6 indicates that increasing the Na_P maximum conductance tends to increase the PRC and influence earlier phase values. The slope of $G(\phi)$ at zero phase plotted as

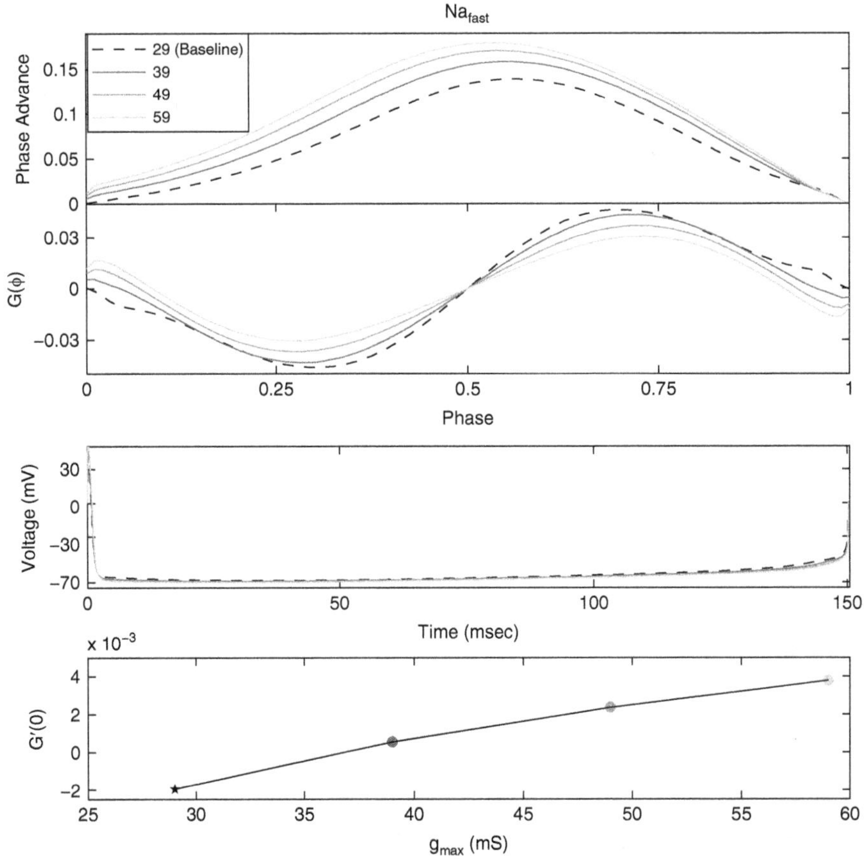

Fig. 9.5 Fast sodium channel shows a shift in the PRC toward the earlier phase with greater amplitude. The large difference between baseline and $g_{\max} = 59\,\text{mS}$ shows how increasing the maximum conductance of this channel strongly affects the PRC. $G(\phi)$ demonstrates the shift in stability of the in-phase solution as the slope switches from negative (stable) to positive (unstable). $G'(0)$ as a function of $g_{\max}$ shows how this slope increases with increasing conductance. Although the voltage trace remains fairly consistent across conductance values, the maximum voltage during spiking is increased for a higher conductance

$G'(0)$ in Fig. 9.6 increases as Na_P increases, thus leading to a predictive decrease in network synchrony. The voltage trace is very consistent across conductance values examined here.

6.3 Delayed Rectifier Potassium

While the depolarization phase of an action potential is caused by the voltage-gated sodium channel, the repolarization is due to the outward delayed rectifier potassium

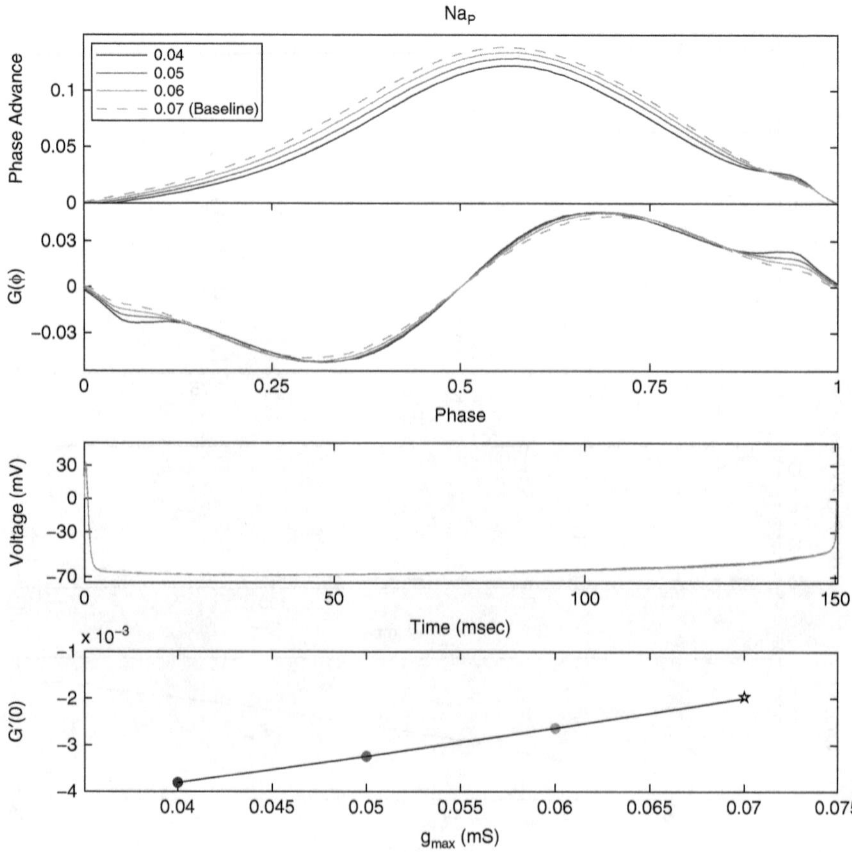

Fig. 9.6 Persistent sodium channels increase the PRC and slightly skew the peak to earlier phase values. Notice that values less than and up to baseline ($g_{\max} = 0.07\,\text{mS}$) are examined. Within the values tested, the network synchrony remains stable for the in-phase solution as evidenced in the $G(\phi)$. However, this channel tends to decrease the stability of synchrony with increasing maximum conductance as the values of $G'(0)$ increase with increasing $g_{\max}$. The slow kinetics of persistent sodium are evidenced in the voltage trace by the nearly identical values across variable conductance

current, K_{DR}. Figure 9.4 shows how the K_{DR} is quickly active immediately after a spike generation. Although these are fast activating channels, they are slower than the fast activating sodium channels. The inactivation of the fast sodium channels along with K_{DR} returns the cell back to its resting potential, even with a little overshoot. This phenomena is evidenced in the voltage trace of Fig. 9.7 which increasingly overshoots the resting potential with increasing conductance. Because the K_{DR} channels do not inactivate, they are simply voltage dependent. Following recovery from an action potential, the channels cause a small amount of current, which is phase independent in their activation. It is interesting to note

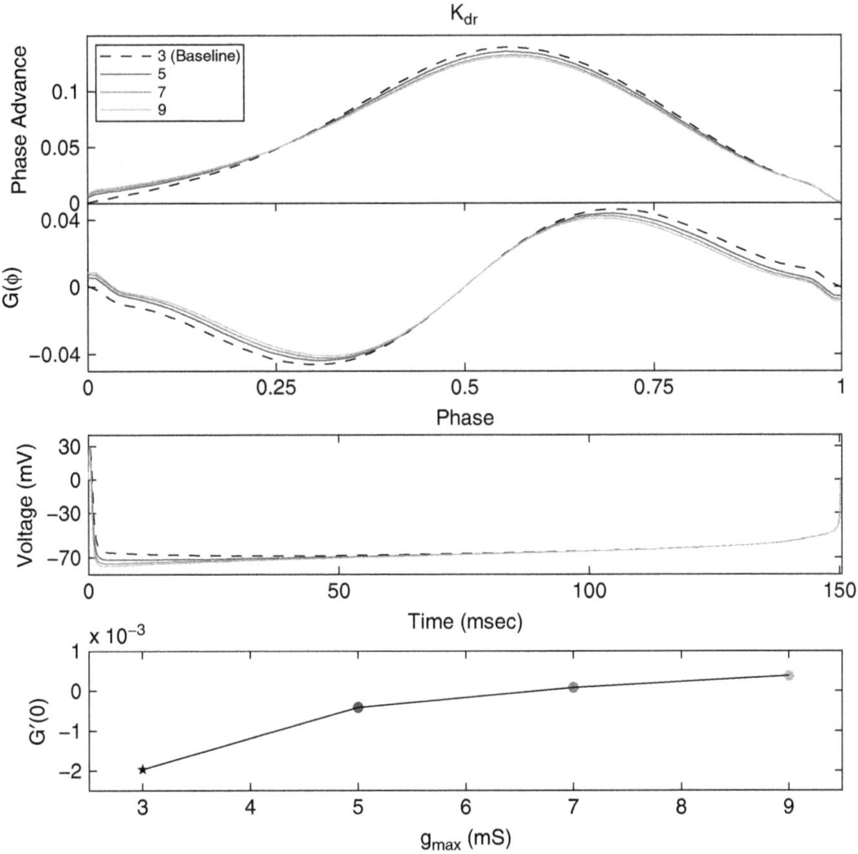

Fig. 9.7 The delayed rectifier potassium current acts to repolarize the cell to resting potential after a generated action potential. This current tends to generally reduce the advance amplitude and slightly decreases the stability of network synchrony. The PRC is increased in the first quarter of the phase and is decreased thereafter with increasing conductance. The peak is also slightly skewed to the left. Perhaps the most interesting effect due to the presence of this current is the change in repolarization which overshoots the $-70\,\mathrm{mV}$ resting potential with increasing maximum conductance as shown in the voltage trace. Although this current is outward, it decreases the synchronous solution stability as shown in the increasing values of $G'(0)$ with increasing conductance, eventually destabilizing the synchronous solution as $G'(0) > 0$

that Tetraethylammonium (TEA) is a general potassium channel blocker that when applied, widens the action potential by removing the K_{DR} current. Figure 9.7 shows a shallowing of the PRC as K_{DR} increases. The plot of $G'(0)$ indicates that as $g_{\max}$ increases, the stability of the in-phase solution decreases and as $G'(0) > 0$, the synchronous solution is no longer stable. However, we must keep in mind that values of $g_{\max}$ studied here are chosen within limit cycle constraints to ensure periodic oscillation, yet the conductance values may be outside the range of naturally occurring mutations. In addition, we are measuring PRCs at the cell soma, but

Goldberg et al. show that measuring PRCs from synaptic inputs on the dendrites can be significantly different from measuring PRCs with synaptic inputs at the soma (Goldberg et al. 2007).

As we previously observed, the increase of Na_{fast} channels (causing an inward current) increase the excitability of these model neurons, which skews the PRC. The skewing of the PRC toward the left predicts a less synchronous solution within a network. One may think that K_{DR} being an outward current would have the opposite effect, that increasing K_{DR} would cause a skewing of the PRC toward the right and lead to a more synchronous solution. However, this is not the case. When K_{DR} is increased, there is a hyperpolarization early in the phase which can be seen in the voltage trace in Fig. 9.7. This effect causes an increase in the sensitivity of Na_{fast} channels and they participate in the skewing of the PRC toward the left. Therefore, the system is highly complicated; simply increasing a given ion channel may have effects on other ion channels, which in turn, shape the cellular dynamics.

6.4 Potassium A

The K_{A} channels are activated as a cell depolarizes, but then inactivate. Maximum conductance occurs when the cell is hyperpolarized before an action potential. Following an action potential, the voltage overshoot caused by the other potassium currents deinactivates these channels which turns on the potassium current at a late phase. Figure 9.4 shows how this current turns on late in the phase and decreases rapidly after a spike is generated. The result is that this current extends the period by passing a negative current just as the voltage of the cell begins to reach the sodium threshold again, prolonging the onset of the spike (Hille 2001). The K_{A} current is blocked by 4-aminopyridine, a drug used to induce seizures in brain slice models of epilepsy (Perreault and Avoli 1991). Figure 9.8 demonstrates how K_{A} decreases the PRC with increasing conductance. This increase of K_{A} also subtly predicts a decrease in network synchrony as $G'(0)$ increases with increasing g_{max} conductance. It has also been noticed that this current reduces the gain of the neurons, limiting their maximum firing rate (Hansel and Sompolinsky 1996).

6.5 Potassium M

The potassium M channel, K_{M}, is a muscarinic acetylcholine (ACh) receptor. ACh is a neurotransmitter that modulates the behavior of neurons and is often involved in the generation of population oscillations. Drugs such as Carbachol block K_{M} and K_{AHP} channels. In addition, mutations in the KCNQ2 gene, which makes the M-channel, have been linked to certain forms of epilepsy (Meldrum and Rogawski 2007; Soldovieri et al. 2007). It is different from the other channels studied here

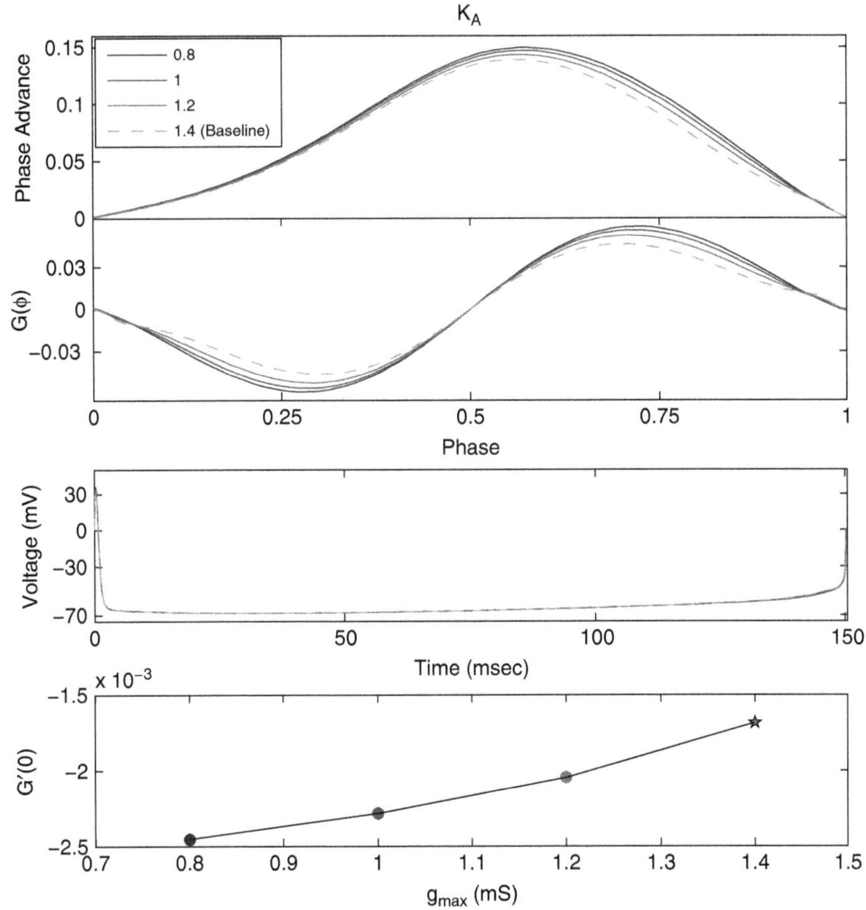

Fig. 9.8 Potassium A current has a subtle effect of decreasing the PRC amplitude and the slight increase in $G'(0)$ with increasing conductance predicts a decrease in synchrony. However, as implicated in slice models of epilepsy, K_A may be important in the generation of seizure like activity in vivo. Notice that here we examine values of K_A less than and including baseline, which is represented as a *dashed line*

because the receptor and the ion channel are coupled through a second messenger system. As a second messenger dependent channel, it is odd because it seems to be directly activated by the G-protein, which makes it very fast for a G-protein dependent channel, and limits the amplification of the signal. In Fig. 9.4, we find the K_M channel has a high probability of being open through most of the oscillation. This opening probability increases while the voltage is near resting potential during the period of oscillation until sufficient depolarization begins to decrease the number of open K_M channels. Figure 9.9 shows a smooth decrease of PRC values in the first half of the phase with increasing conductance, which become negative very early in

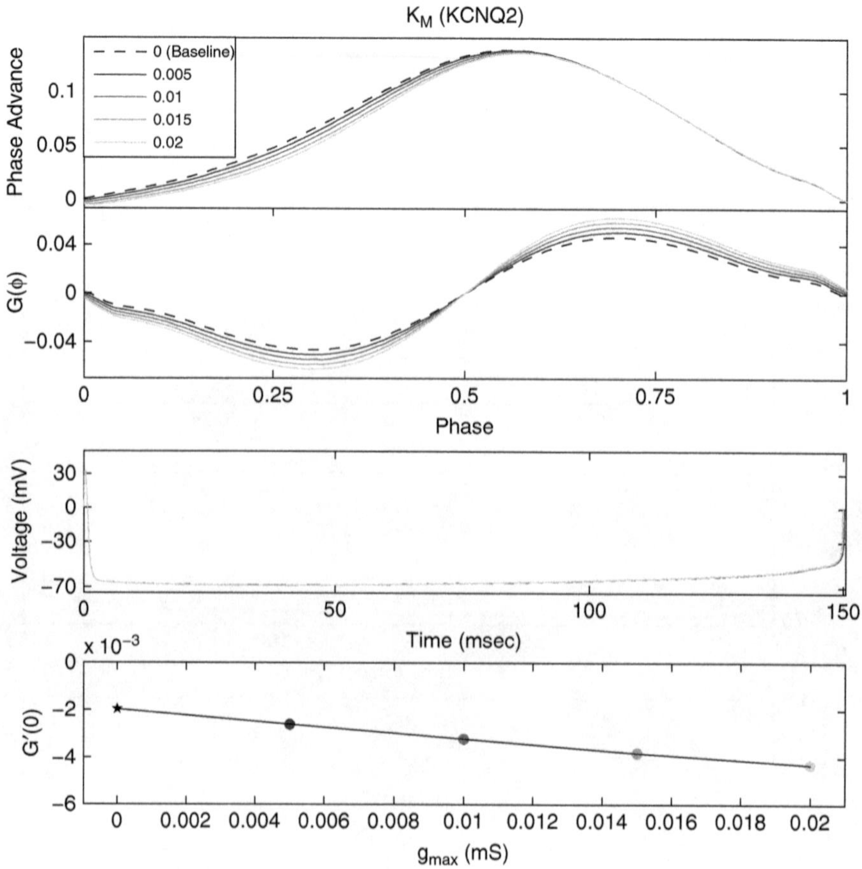

Fig. 9.9 In the case of increasing the conductance of the Potassium M channel (KCNQ2), the PRC is skewed toward the right and attenuated in the first half of the phase. This shift predicts an increase in network synchrony, demonstrated by $G'(0)$ becoming more negative with increasing maximum K_M conductance. It is also important to note how the increase in conductance tends to lengthen the period, but we have altered the input current to keep the oscillator near a period of 50 ms. Here, the baseline value of zero maximum conductance is represented as a *dashed line*

the phase. As conductance increases, $G'(0)$ becomes more negative, thus increasing the stability of a synchronous solution. This phenomena coupled with a change in oscillator period may play a major role in the generation of network oscillations.

6.6 Cationic H

The hyperpolarizing activated cation channel I_H, also known as *sag-current*, is very different from the potassium and sodium channels discussed so far. This current, as the description suggests, turns on when the cell is hyperpolarized as seen in the increased probability upon hyperpolarization and slow relaxation of opening probability in Fig. 9.4. Therefore, when current flows through this channel, the depolarization serves to turn the channel off, resulting in a sagging voltage. Conversely, when a cell is hyperpolarized, the current turns on, causing the cell to depolarize. It affects the slow depolarization between spikes and affects the interspike interval (Rush and Rinzel 1994). During the AHP current, this channel is opened resulting in a decrease in membrane resistance causing a neuron to be less sensitive to synaptic inputs following an action potential than later in the phase.

It has also been observed that I_H channels are modulated following seizures in animal models (Chen et al. 2001; Shah et al. 2004). This cationic channel with reversal potential around −40 mV alters the dynamics of the cell more than any other cell examined so far because of its high control of resting potential. The intrinsic period of the cell can be significantly changed with the addition and increase of I_H current, although here we alter the input current to keep the oscillator period near 150 ms. As the cell repolarizes following an action potential, the I_H channels are opened and a resulting depolarization current flows. At high enough levels of cationic H conductance, a cell will almost immediately depolarize following a hyperpolarization from action potential generation. Later in this chapter, we will see how the Traub model skews the PRC peak to earlier phase and it also pushes down the early phase of the PRC to have negative values. Similarly, in Fig. 9.10 the PRC transitions from Type I to Type II with a significant part of the early phase becoming more negative with increasing I_H conductance. The $G'(0)$ values decrease with increasing conductance, predicting an increase in network synchrony. From the range of values studied here, it is clear that a small amount of I_H current significantly alters the PRC in the first half of the oscillator's phase.

6.7 *TASK and TREK*

TASK and TREK channels are two pore potassium channels. These channels are sensitive to pH and open in acidic conditions. They are thought to be activated by volatile general anesthetics (Patel et al. 1999) and mutations in the TASK channel have been linked to Childhood Absence Epilepsy (Holter et al. 2005). Even though the current is small, these channels are located in the dendrites. Small changes in these currents can dramatically change the input efficacy by shunting the synaptic input as a postsynaptic potential travels down the dendritic tree. Unlike several of those examined so far, these channels are not voltage dependent. Figure 9.11 shows how an increase in conductance increases the PRC in the first half of the phase, making the neuron more affected by input early in its oscillation. This channel is of interest because it may be able to shift the stability of the synchronous solution from stable to unstable, as $G'(0)$ increases from negative to positive values with increasing conductance.

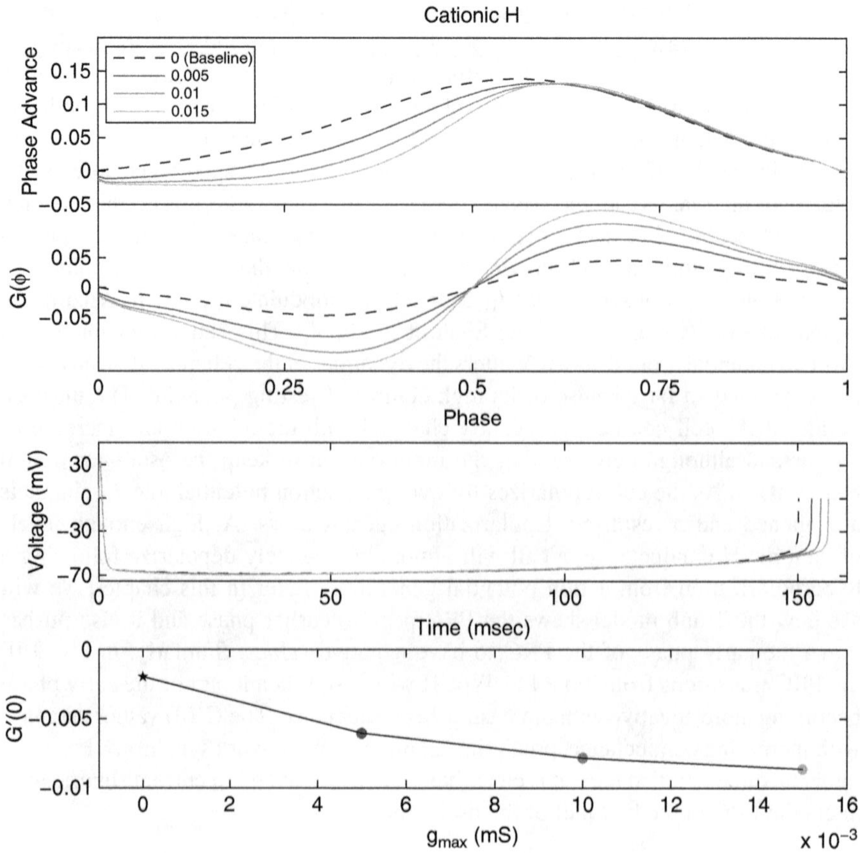

Fig. 9.10 Cationic H current (sag current) drastically alters the PRC towards synchrony with increasing maximum conductance. I_{H} is open at resting potential while subtle changes in this current cause a change in the resting potential of the cell. $G(\phi)$ shows how an increase in maximum conductance leads to an alteration in network dynamics by changing the in-phase synchronous solution. That is, $G'(0)$ shifts to more negative values as the channel conductance is increased. The period of oscillation is also greatly increased with increasing cationic H current, but we alter the input current to keep the oscillator period near 150 ms. Baseline value is represented as a *dashed line*

7 Erisir Model

So far, we have examined how changing maximum conductance while constraining the period affects the PRC. In this section, we will look at how the PRC changes by varying the input current across various leak currents. In a recent paper, Erisir et al. propose a simple membrane model for a fast spiking interneuron and we will look at the full version of this model in the next section. For now, we consider their model without the slow Kv3 potassium current. The model has only transient

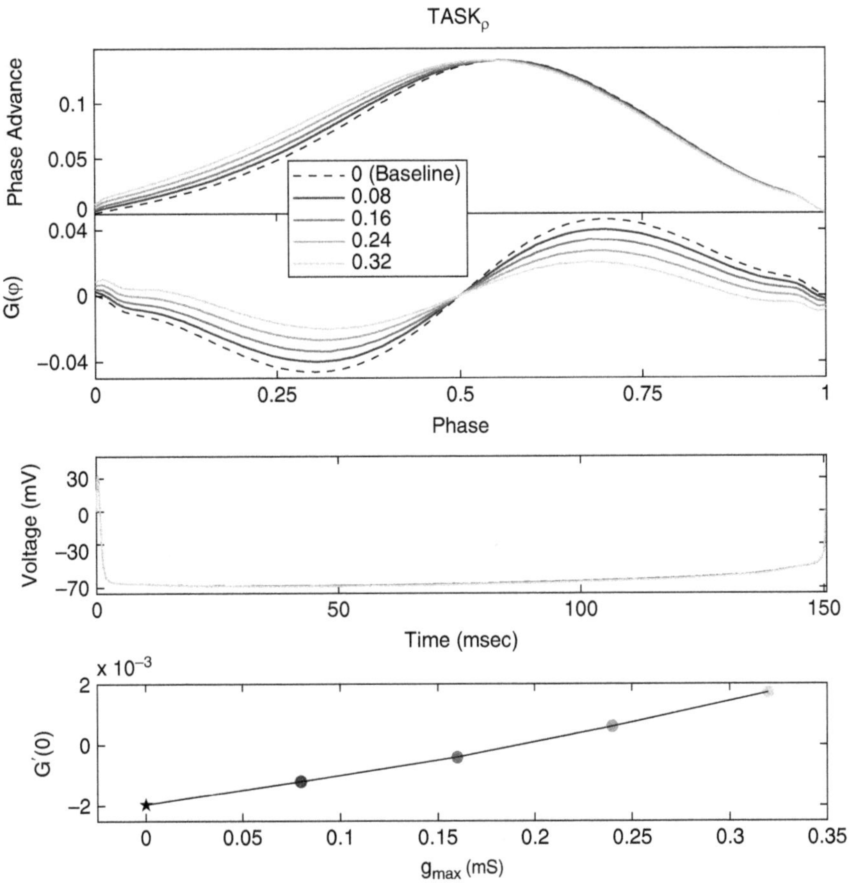

Fig. 9.11 TASKρ current alters the PRC by increasing its magnitude in the first half of the phase with increasing conductance. The $G(\phi)$ shows how this increase leads to an alteration in network dynamics by changing the in-phase solution. This effect is evidenced as $G'(0)$ moves from negative (synchronous) to positive (asynchronous) as the channel conductance is increased. Conversely, the $G(\frac{T}{2})$ solution is asynchronous with baseline value of zero and remains unstable as conductance increases to $g_{\max} = 0.32$ mS

sodium, delayed rectifier, and leak currents. The leak is mainly responsible for the dynamics at rest, so in this example, we will change the leak to see dramatic changes in the PRC and in the bifurcation structure. Figure 9.12 summarizes the mechanisms for the onset of rhythmic firing for this FSU model; here we show the bifurcation diagram of the membrane model as the current increases. In such diagrams, the voltage is often shown. However, presenting the norm along the y-axis let us see the SNIC bifurcation. Starting with $I = 0$, we increase the current. The steady state current voltage curve has a cubic shape (reflected in the cubic shape of the norm)

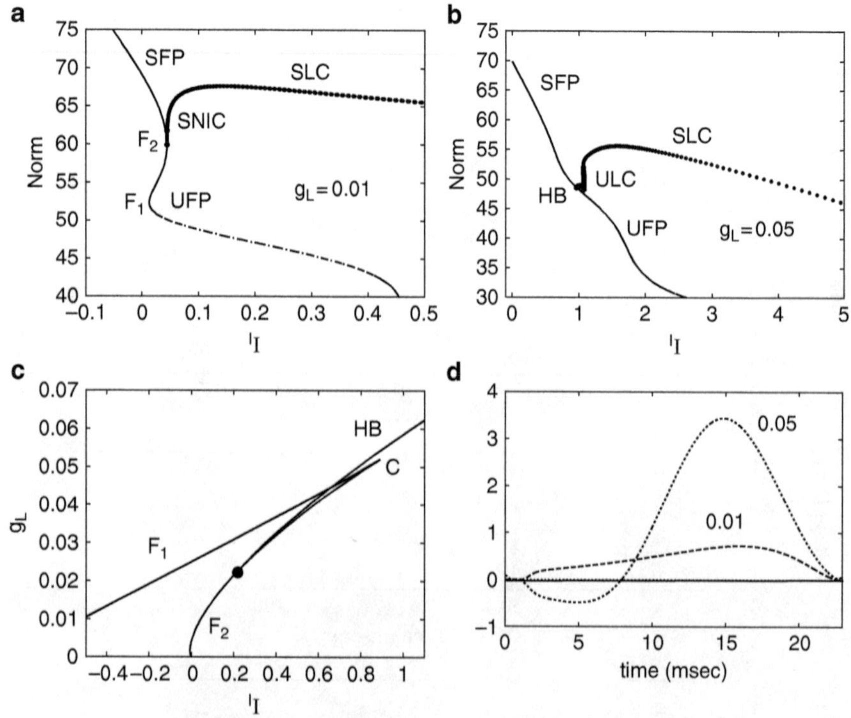

Fig. 9.12 Effects of changing g_l on a simple membrane model. (**a**) Bifurcation diagram as the current changes with $g_l = 0.01$. The y-axis shows the norm of the solution and not the voltage. SLC: stable limit cycle; SNIC: saddle-node on an invariant circle; SFP: stable fixed point; UFP: unstable fixed point; $F_{1,2}$: fold bifurcations. (**b**) Same as (**a**), but $g_l = 0.05$. HB: Hopf bifurcation; ULC: unstable limit cycle. (**c**) Two parameter diagram summarizing (**a,b**). There are two curves of fold points and a cuve of Hopf points. The fold points coalesce at a cusp (**c**) and the Hopf and fold point F_2 coalesce at a Takens-Bogdanov point (*filled circle*). (**d**) The adjoint or infinitesimal PRC for $g_l = 0.01, 0.05$ and the currents matched so that the period of oscillation is the same for both cases

and for a small range of currents, there are three fixed points. Only one of them is stable (labeled SFP). When the current reaches F_2 the fold bifurcation, there are no stable rest states. But, we can see that a branch of stable limit cycles meets the curve of equilibria exactly at this fold and results in the SNIC bifurcation. As the current approaches F_2 from the right, the frequency of the limit cycle goes to zero. In Fig. 9.12b, we have increased the leak to $g_l = 0.05$ with dramatic consequences for the bifurcation diagram. Instead of a cubic steady state curve, it is monotonic. It is stable until the current reaches the Hopf bifurcation (HB). A branch of unstable limit cycles (ULC) emerges from the HB and then reverses direction and becomes stable (SLC). Thus the onset of rhythmicity is via an HB rather than a SNIC. Figure 9.12c summarizes the local equilibrium dynamics as the current and leak conductance vary. For low values of leak, as the current increases, it passes through twofolds.

At F_2 there is a SNIC. For high leak, as the current increases, there is an HB. The twofold bifurcations meet at a *cusp* point. For g_l below the cusp point, there will always be a cubic steady state IV relation and above the cusp point, the curve is monotonic. The HB meets the lower fold at another special point, the Takens-Bogdanov point (TB). In the linearized system, there are two zero eigenvalues at this point. For g_l above the TB point, there will be a Hopf bifurcation on the branch of stable equilibria that makes this branch lose stability for currents below the fold point. Thus, there can be no SNIC above the TB point. Analysis of the dynamics near bifurcations that we alluded to in the previous section would suggest that the PRC would be nonnegative for the system in (a) and have a negative lobe for the system in (b). Figure 9.12d shows the PRC for each of the two cases. Clearly the HB case has a big negative lobe while the SNIC case is nonnegative. The HB case also has a considerably larger amplitude. This large amplitude is not due to the presence of the HB. Rather, it is because we picked the current to be near the transition between a SLC and ULC in Fig. 9.12b. As we approach this transition, the amplitude of the PRC tends to infinity. Thus, the higher amplitude is due to the presence of another bifurcation.

8 Traub Model

The "Traub model" is a three conductance model consisting of a leak, delayed rectifier, and transient sodium current with parameters taken from hippocampal pyramidal cells. As with the Erisir model, the three currents are the minimum needed for spiking. As such, the model provides an ideal platform on which to add other types of currents to study their effects on the PRC. We first note that the baseline model undergoes repetitive firing via a SNIC bifurcation so that the PRC is expected to be positive.

Figure 9.13 shows the effects of several model currents on the PRC of the Traub model. Panel (0) represents the baseline PRC. We have added enough current so that the frequency is between 40 and 50 Hz, so this value of current is far from the bifurcation to limit cycles. Nevertheless, the PRC is still qualitatively like the limiting PRC given by (9.2). While not symmetric, the baseline PRC is more symmetric than any of the other PRCs in Fig. 9.13. Two of the currents added to the model (A,KCa) have an effect on the skew and neither of them are sufficiently active at rest to affect the bifurcation diagram of the model. The M, Ks, and Sag currents affect both the skew and also push the PRC to have negative values. (This is barely evident in the sag current.) These latter currents act at threshold and *change the local bifurcation.* In particular, at sufficiently large values of these currents, a Hopf bifurcation destabilizes the rest state for currents smaller than those that reach the saddle-node bifurcation much like the effects of leak seen in Fig. 9.12. All five of these currents represent a form of negative feedback; this is clear for the potassium currents. For the sag current, while it is depolarizing, the gating variable *decreases* with increased voltage; thus, it is in fact a negative feedback current.

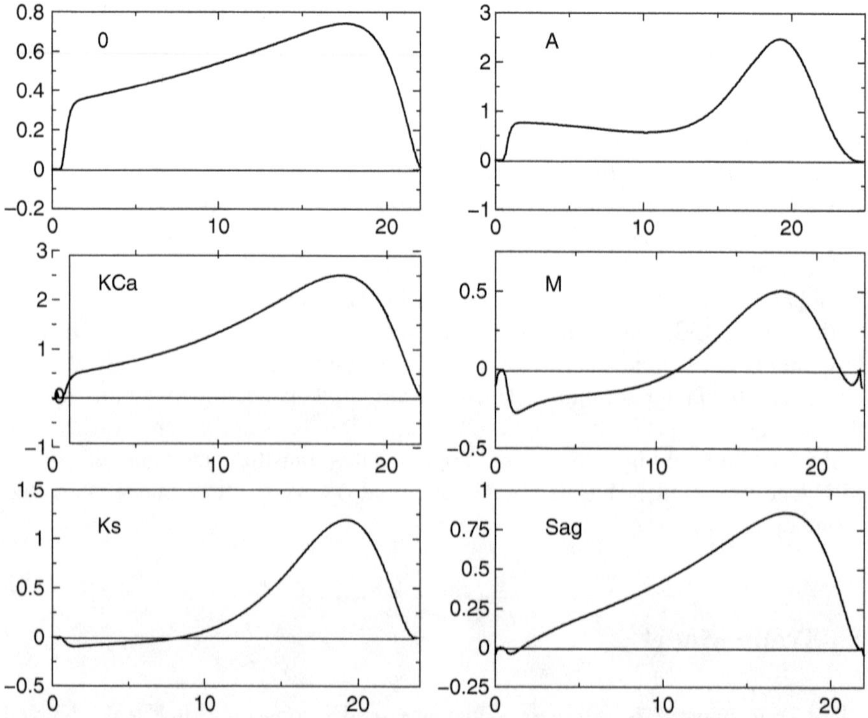

Fig. 9.13 Traub model PRCs with different currents. In each case enough current is added to make the cell spike at a frequency of 40 to 50 Hz. (0) Baseline currents; (**a**) $g_A = 10\,\mathrm{mS/cm^2}$; ($K_{CA}$) $g_{KCa} = 0.5\,\mathrm{mS/cm^2}$; (M) $g_M = 1.5\,\mathrm{mS/cm^2}$; (Ks) $g_{Ks} = 0.5\,\mathrm{mS/cm^2}$; (sag) $g_{sag} = 2\,\mathrm{mS/cm^2}$

Figure 9.14 shows a bifurcation diagram for the local dynamics of the Traub model in the presence of the M-type potassium current. Note in panel A that as the current increases, a Hopf bifurcation (HB) occurs on the stable lower rest-state branch at an applied current below the current required for the saddle node (or fold, F). This is due to the M-current; without this current, the stability of the resting state is lost at the fold and the neuron makes its transition from rest to firing via the SNIC. Figure 9.14b shows the two-parameter diagram. For low values of the M-current (e.g., $g_m = 0.1$) as he current increases, only the fold is encountered and the neuron undergoes a SNIC bifurcation and the PRC remains positive. For large enough M-current, the HB occurs for lower values of current and we expect a PRC that has a negative lobe. The transition occurs when the HB curve meets the F curve at the Takens–Bogdanov (TB) point, just as in Fig. 9.12. Once again, the crucial dynamics that underlie the transition from a type I to a type II PRC appear to converge at the Takens–Bogdanov point. The negative lobe in the PRC in the presence of M-current can be understood intuitively. After the spike, sodium currents are largely inactivated. A depolarizing pulse increases the potential and since the M-current is active at low potentials, a depolarizing pulse increases the M current and since it

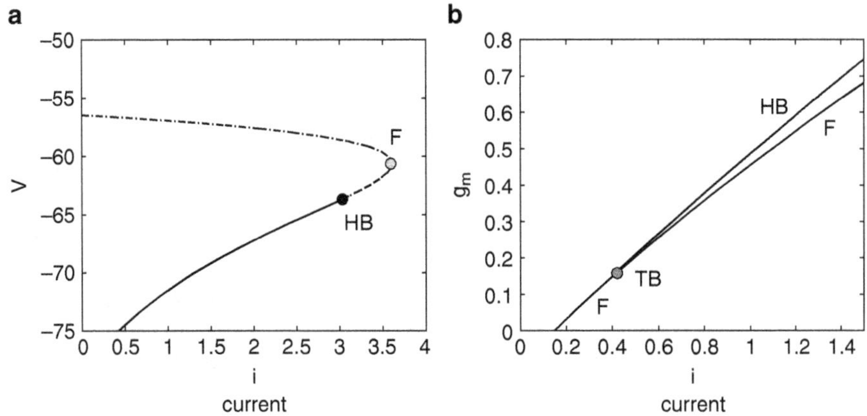

Fig. 9.14 Bifurcation diagram for the Traub model with an M-type potassium current. (**a**) Local subthreshold dynamics show destabilization of the lower branch at a Hopf bifurcation (HB) at currents below the fold (F). (**b**) Two parameter diagram (zoomed in view) showing the curve of HB and F points. The HB curve terminates on the curve of fold points at a Takens–Bogdanov (TB) point

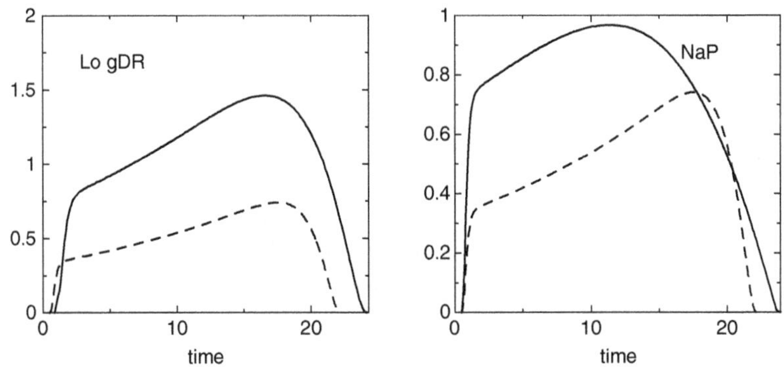

Fig. 9.15 Effects of reduction of the delayed rectifier (*left*) and addition of a persistent sodium current (*right*) on the shape of the PRC

lasts much longer than the other currents, it results in a delay to spiking and thus a negative lobe in the PRC. At later times, the inward currents have recovered and the pulse advances the next spike.

As we saw above, currents acting during the spike (such as the AHP and to some extent the A current) affect the skew of the PRC. What about inward currents? Figure 9.15 shows two effects. If the delayed rectifier in the Traub model is considerably reduced, there is a slight change in the skew, but mainly the amplitude of the PRC increases. Pfeuty et al. find a greater change in the skew of the PRC when lowering the delayed rectifier, but they used different kinetics for their channels

(Pfeuty et al. 2003). Addition of a persistent sodium current (instantaneous onset at a high potential and no inactivation) results in a leftward skew of the PRC, an effect also observed by Pfeuty et al.

Pfeuty et al. were among the first to notice that inward currents and the reduction of outward currents resulted in a shift of the skew or the PRC to the left (Pfeuty et al. 2003). They also briefly considered how outward currents could skew the PRC rightward. Gutkin et al. (2005) and Ermentrout et al. (2001) discussed how high-threshold outward currents skewed the PRC to the right and in the latter, used a very simple analytic approximation to show why this was the case. Prescott et al. have noted that there is a difference between the effects of an M-type adaptation versus that of KCa in terms of spike rate and spike time coding. A PRC which is strictly positive is much better at coding rates since it is relatively insensitive to the timing of inputs while a PRC with both positive and negative parts is better at synchronizing at short time windows (Marella and Ermentrout 2008; Abouzeid and Ermentrout 2009). Goldberg et al. noted how a sag current in dendrites resulted in a prominent negative lobe in the PRC (Goldberg et al. 2007). Stiefel et al. showed that blocking the M-current removed the negative lobe in the PRC for cortical neurons (Stiefel et al. 2008, 2009).

9 Toward a Normal Form

9.1 The Takens–Bogdanov Point

Bifurcations such as the HB and the SNIC have associated reduced dynamics and the PRCs for these reduced systems are readily computed, cf, (9.2) and (9.3). Like the HB and the SNIC, there are simplified dynamical models for systems that undergo the TB bifurcation. There are two such "normal forms" (as they are referred to by mathematicians); one is due to Bogdanov and the other to Takens. They are equivalent although the Bogdanov is more commonly encountered. However, the Takens form is more conveniently related to neural systems as we will see. The Takens normal form is

$$\frac{dx}{dt} = y + \beta x + x^2, \qquad \frac{dy}{dt} = \alpha + x^2.$$

There are two parameters since we need two different bifurcations to coalesce (the HP and the F). We let $V = x + (\beta + 1)/2$ and $w = x - y + \alpha + (\beta + 1)^2/2$ and obtain:

$$V' = I + V^2 - w, \qquad w' = bV - w$$

where $b = \beta + 1$ and $I = \alpha + \beta/2 + \beta^2/4$. For added flexibility, we incorporate a parameter for the temporal dynamics of w, and obtain the model

$$\frac{dV}{dt} = V^2 + I - w, \tag{9.7}$$

$$\frac{dw}{dt} = a(bV - w). \tag{9.8}$$

This is none other than the quadratic integrate and fire (QIF) model ($V' = I + V^2$) along with a feedback term w. The feedback is negative if $b > 0$ and positive if $b < 0$. Like the QIF model, V can "blow up" in finite time, so to make this into a neural model (like the integrate and fire), we introduce a reset condition. If $V(t) = V_{\text{spike}}$ then $V(t) \to c$ and $w(t) = w(t) + d$ where c, d are two more parameters. Equations (9.7) and (9.8) along with the reset conditions constitute the so-called Izhikevich model. We will abbreviate the model as quadratic integrate and fire with adaptation (QIFA). As Izhikevich has shown, the QIFA model can mimic dozens of neural spiking patterns with appropriate rescaling of V and choices of the parameters $a - d$. Here, the current, I has its usual meaning and represents what is injected into the neuron. We can associate the parameters with different current in the full Traub model. The decay rate of all outward currents is given by the parameter a. Outward currents that act in the subthreshold regime are implemented by $b > 0$. Spike dependent outward currents are implemented by d and fast repolarization by c. Because of the relationship between the QIFA and the TB as well as the importance of the TB in determining the transition from Type I to Type II, we spend the remainder of this section exploring the PRCs that arise from the QIFA which is equivalent to the TB normal form. We can regard this work as the next step in the evolution of our understanding of how bifurcations and PRCs are related.

9.2 *Adjoints for the QIFA*

Equations (9.7) and (9.8) are discontinuous. Thus, packages like AUTO can not be directly used for computing periodic solutions and solutions to the adjoint equation. However, we can solve these equations and their adjoints in a systematic manner by treating the equations as a boundary value problem. Specifically, we must solve (9.7) and (9.8) subject to the following conditions: $V(0) = c$, $V(P) = V_{\text{spike}}$, and $w(0) = w(P) + d$. There are two ODEs and three conditions. However, the period, P is unknown, so we can write $dP/dt = 0$ so that there are now three ODEs and three conditions. The adjoint equation is

$$\frac{du}{dt} = -2V(t)u - abz,$$

$$\frac{dz}{dt} = u + az.$$

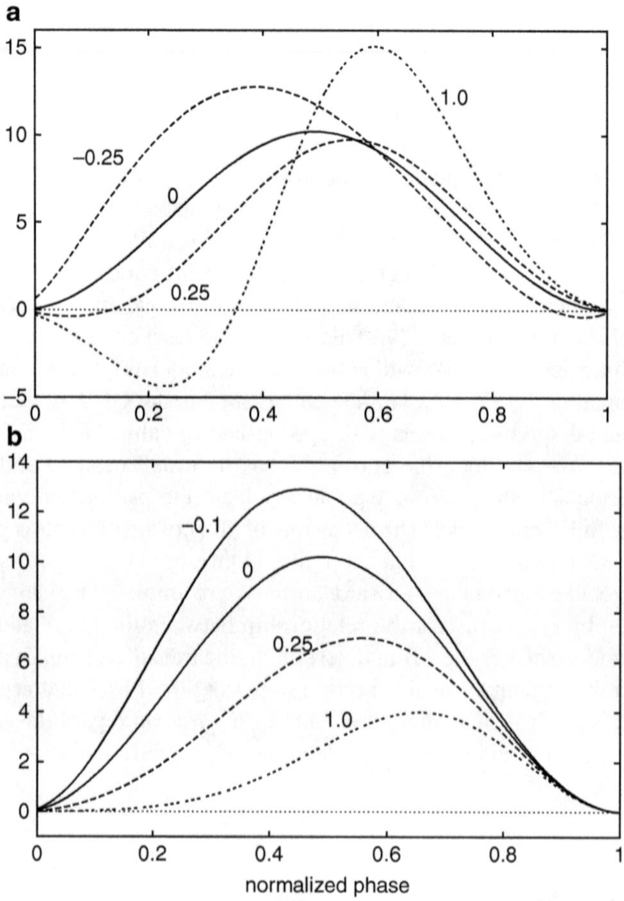

Fig. 9.16 PRCs for the QIFA model as (**a**) b subthreshold adaptation varies (note negative values mean augmentation) and (**b**) d, spike-based adaptation varies. Numbers correspond to the value of the parameter. Period is 9.74, I varies to keep the period fixed, $a = 0.3$, $V_{\text{spike}} = 20$, $c = -4$

We want solutions to this linear equation satisfying $z(0) = z(P)$ and $u(0)V'(0) + z(0)w'(0) = 1$. The former condition ensures periodicity and the latter is a normalization condition. Numerically, we rescale time $t = Ps$ so that $0 \le s < 1$ with the period as an unknown parameter. In the examples shown in the previous section, we tried to keep the period roughly the same as we altered the conductances of different currents. We did this by adjusting the current. Here, we can do the same thing. Fix P and choose I so that there is a solution.

With these numerical preliminaries, we present results on the PRCs for the QIFA. Figure 9.16 shows the PRC for the QIFA model for different types and degrees of feedback. When $d = b = 0$, the model is the QIF with finite reset and spiking. Here, we have chosen $V_{\text{spike}} = 20$ and $c = -4$. The PRC for infinite spike and reset is proportional to $1 - cos(2\pi s)$, a very close match to the truncated version shown as the black curves in Fig. 9.16a,b. In panel (a), we change the subthreshold adaptation parameter, b. Clearly as b increases (this would be like the M- or Ks-current in Fig. 9.13) the PRC develops a negative lobe right after the spike. For $b < 0$, corresponding to augmentation, the PRC is shifted leftward. In Fig. 9.16b, we change the parameter d which is the spike-related adaptation and corresponds to the KCa-current. No negative lobe occurs, but the PRC is very strongly skewed to the right. For $d < 0$, corresponding to augmentation, the PRC is larger and skewed to the left.

10 Conclusion

In this chapter, we have tried to show through a number of conductance-based models how the contributions of different ions affect the shape of the PRC. In general, we have found that inward currents tend to skew the PRC toward the left and as a consequence can make synchrony less stable. Outward currents have two major effects depending on whether or not they operate near rest or during the spike. The latter tend to skew the PRC to the right and encourage synchronization. Outward currents operating at rest can change the nature of bifurcation leading to spiking and so can give the PRC a large negative region at phases shortly after the spike. Since the sag current operates near rest and provides negative feedback, it is similar in its effect to an outward current operating at rest. Finally, we showed that many of the effects of different currents can be understood by mapping them onto a simple two variable model: the QIF model with adaptation (Izhikevich model) which itself is related to the local behavior of any biophysical neuron model near the Takens–Bogdanov bifurcation. This bifurcation, thus, provides a partial mathematical answer to the question of how different currents contribute to the shape of the PRC.

Parameters for models discussed in this chapter can be found at http://neuralnetoff.umn.edu/public/ionPRC.

References

C.M. Gray, Journal of computational neuroscience **1**(1-2), 11 (1994)

W.E. Skaggs, B.L. McNaughton, M.A. Wilson, C.A. Barnes, Hippocampus **6**(2), 149 (1996)

J.J. Chrobak, G. Buzsaki, The Journal of neuroscience : the official journal of the Society for Neuroscience **16**(9), 3056 (1996)

P.J. Uhlhaas, W. Singer, Neuron **52**(1), 155 (2006)

W.D. Hutchison, J.O. Dostrovsky, J.R. Walters, R. Courtemanche, T. Boraud, J. Goldberg, P. Brown, The Journal of neuroscience : the official journal of the Society for Neuroscience **24**(42), 9240 (2004)

T.I. Netoff, S.J. Schiff, The Journal of neuroscience : the official journal of the Society for Neuroscience **22**(16), 7297 (2002)

P. Goel, B. Ermentrout, Physica D **163**(3), 191 (2002)

E. Brown, J. Moehlis, P. Holmes, Neural Computation **16**, 673 (2004)
K. Lehnertz, S. Bialonski, M.T. Horstmann, D. Krug, A. Rothkegel, M. Staniek, T. Wagner, Journal of neuroscience methods **183**(1), 42 (2009)
D.A. McCormick, D. Contreras, Annual Review of Physiology 63, 815 (2001)
J.R. Huguenard, D.A. McCormick, Trends in neurosciences **30**(7), 350 (2007)
J.J. Chrobak, G. Buzsaki, The Journal of neuroscience : the official journal of the Society for Neuroscience **18**(1), 388 (1998)
A.D. Reyes, E.E. Fetz, Journal of Neurophysiology 69, 1673 (1993)
R. Butera, A. Preyer, Conference proceedings : ...Annual International Conference of the IEEE Engineering in Medicine and Biology Society. IEEE Engineering in Medicine and Biology Society. Conference 4, 4324 (2005)
E.M. Izhikevich, Dynamical systems in neuroscience: the geometry of excitability and bursting. (Cambridge, Mass.:MIT Press) (2007)
J. Rinzel, B. Ermentrout, Analysis of neural excitability and oscillations (MIT Press, Cambridge, Mass., 1989)
D. Golomb, Y. Amitai, Journal of neurophysiology 78, 1199 (1997)
D. Hansel, H. Sompolinsky, Journal of computational neuroscience **3**(1), 7 (1996)
J.A. Goldberg, C.A. Deister, C.J. Wilson, Journal of neurophysiology **97**(1), 208 (2007)
B. Hille, Ion channels of excitable membranes (Sinauer, Sunderland, Mass., 2001)
P. Perreault, M. Avoli, Journal of neurophysiology **65**(4), 771 (1991)
B.S. Meldrum, M.A. Rogawski, Neurotherapeutics : the journal of the American Society for Experimental NeuroTherapeutics **4**(1), 18 (2007)
M.V. Soldovieri, M.R. Cilio, F. Miceli, G. Bellini, E.M. del Giudice, P. Castaldo, C.C. Hernandez, M.S. Shapiro, A. Pascotto, L. Annunziato, M. Taglialatela, The Journal of neuroscience : the official journal of the Society for Neuroscience **27**(18), 4919 (2007)
M.E. Rush, J. Rinzel, Biological cybernetics **71**(4), 281 (1994)
K. Chen, I. Aradi, N. Thon, M. Eghbal-Ahmadi, T.Z. Baram, I. Soltesz, Nature medicine **7**(3), 331 (2001)
M.M. Shah, A.E. Anderson, V. Leung, X. Lin, D. Johnston, Neuron **44**(3), 495 (2004)
A.J. Patel, E. Honore, F. Lesage, M. Fink, G. Romey, M. Lazdunski, Nature neuroscience **2**(5), 422 (1999)
J. Holter, D. Carter, N. Leresche, V. Crunelli, P. Vincent, Journal of molecular neuroscience : MN **25**(1), 37 (2005)
B. Pfeuty, G. Mato, D. Golomb, D. Hansel, The Journal of neuroscience : the official journal of the Society for Neuroscience **23**(15), 6280 (2003)
B.S. Gutkin, G.B. Ermentrout, A.D. Reyes, Journal of neurophysiology **94**(2), 1623 (2005)
B. Ermentrout, M. Pascal, B. Gutkin, Neural Computation **13**(6), 1285 (2001)
S. Marella, G.B. Ermentrout, Physical review E **77**(4 Pt 1), 041918 (2008)
A. Abouzeid, B. Ermentrout, Physical Review E 80, 011911 (2009)
K.M. Stiefel, B.S. Gutkin, T.J. Sejnowski, PloS one **3**(12), e3947 (2008)
K.M. Stiefel, B.S. Gutkin, T.J. Sejnowski, Journal of computational neuroscience **26**(2), 289 (2009)

Chapter 10
Cellular Mechanisms Underlying Spike-Time Reliability and Stochastic Synchronization: Insights and Predictions from the Phase-Response Curve

Roberto F. Galán

Abstract Neurons can respond with highly consistent spike patterns to repetitions of the same stimulus. Analogously, similar neurons receiving a common stimulus can fire highly consistent spike patterns. The former phenomenon is referred to as spike-time reliability, whereas the latter is an example of stochastic synchronization. Both phenomena are quite general and in fact, they also manifest in simplified models of single neuron dynamics, like phase oscillator models, in which the activity of the neuron is determined by its phase-response curve, which in turn is determined by the membrane conductances. Here, we use two measures of spike-time reliability and stochastic synchronization for real neurons and conductance-based models that have been recently introduced in the theory of phase oscillators: the Lyapunov exponent of the oscillator dynamics and the variance of the phase difference between two identical oscillators. Analyzing data from simulations and experiments, we show that, in response to manipulations of membrane conductances, a change of the phase-response curve leading to lower variance of the relative phase is a good predictor of increased spike-time reliability and stochastic synchronization in real and simulated neurons. We also explain why the Lyapunov exponent is not sufficient by itself. Our approach is then exemplified by investigating the effect of certain potassium currents, A and A-like currents, on spike-time precision. Finally, we discuss the biological relevance of our results.

R.F. Galán (✉)
Case Western Reserve University, School of Medicine, Department of Neurosciences, 10900 Euclid Avenue, Cleveland, OH, 44106–4975, USA
e-mail: rfgalan@case.edu

N.W. Schultheiss et al. (eds.), *Phase Response Curves in Neuroscience: Theory, Experiment, and Analysis*, Springer Series in Computational Neuroscience 6,
DOI 10.1007/978-1-4614-0739-3_10,

1 Introduction

Spike-time reliability, or the ability of neurons to fire highly reproducible action potentials in response to a fluctuating stimulus, every time the stimulus is presented (Bryant & Segundo 1976; Mainen & Sejnowski 1995) enables neurons to encode sensory information with millisecond precision (Fig. 10.1), i.e., on a much faster time scale than firing-rate codes (Rieke, Warland, de Ryuter van Stevenick, & Bialek 1997). Not surprisingly, spike-time reliability has been widely investigated over the last few years (Galán, Ermentrout, & Urban 2008; Fellous, Tiesinga, Thomas, & Sejnowski 2004; Ermentrout, Urban, & Galán 2009; Hunter & Milton 2003; Hunter, Milton, Thomas, & Cowan 1998; Mainen & Sejnowski 1996; Movshon 2000; Schreiber, Fellous, Tiesinga, & Sejnowski 2004; Tiesinga, Fellous, & Sejnowski 2008). An important conclusion of those studies is that certain membrane conductances may play a major role controlling the timing of action potentials (Schreiber et al. 2004). It is also well-known that the slope of the membrane potential ($\mathrm{d}V/\mathrm{d}t$) as it approaches the firing threshold is inversely proportional to the width of the inter-spike interval distribution (Goldberg, Smith, & Fernandez 1984; Hunter et al. 1998; Stein 1965), or equivalently, to the coefficient of variation of the firing frequency. An effect of a fluctuating current is that threshold crossings occur with a steeper slope than they would with a constant current and thus, generate more reliably timed spikes in the presence of background noise (Hunter et al. 1998). An equivalent slope condition exists for the synchronization of neurons receiving common inputs (Gerstner, van Hemmen, & Cowan 1996). Altogether these findings suggest that the interplay of synaptic and intrinsic membrane currents that are active around the firing threshold controls spike-time reliability and neuronal synchronization. In this regard, phase-oscillator models are a useful tool to investigate the role of membrane conductances in these two phenomena. Indeed, the dynamics of a phase oscillator are characterized by its phase-response curve, which in the case of a neuronal oscillator (neuron firing periodically) is fully determined by its membrane conductances. The philosophy of this approach is sketched in Fig. 10.2.

The intrinsic membrane currents most likely to have the tightest control of spike-time reliability are those which modulate the slope of the membrane potential ($\mathrm{d}V/\mathrm{d}t$) around the firing threshold. In particular, outward currents activated near but below threshold will make the slope shallower, thereby decreasing the likelihood of firing in response to upstrokes of the stimulus and hence, affecting spike-time reliability. In agreement with this observation, a potassium A-like current has been recently shown to regulate the timing of spike discharges in mitral cells (Balu, Larimer, & Strowbridge 2004; Balu & Strowbridge 2007). In addition, the interplay between the I_A and other currents, especially the I_h is known to control the timing and rhythm of neuronal activity in certain central pattern generators (Marder & Bucher 2007; Marder & Goaillard 2006). Here, we exemplify an approach to elucidating the effect of membrane conductances on spike timing from the phase-response curve, by manipulating A-like currents in simulations and experiments.

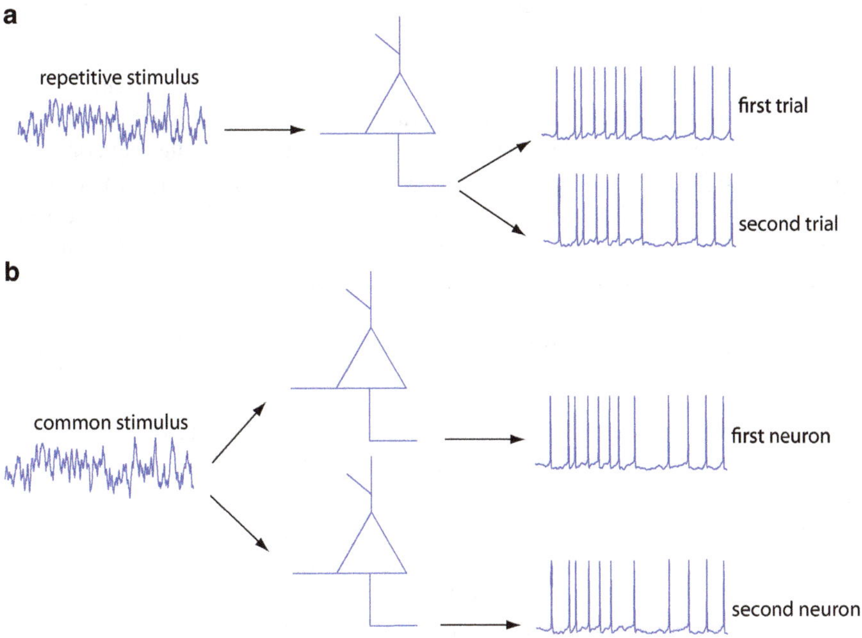

Fig. 10.1 Conceptual equivalence between spike-time reliability and stochastic synchronization. (**a**) A fluctuating stimulus to a single neuron evokes almost identical responses in successive repetitions. This phenomenon is known as spike-time reliability. (**b**) Analogously, two similar neurons receiving common fluctuating inputs will evoke almost identical responses, thereby synchronizing most of their spikes

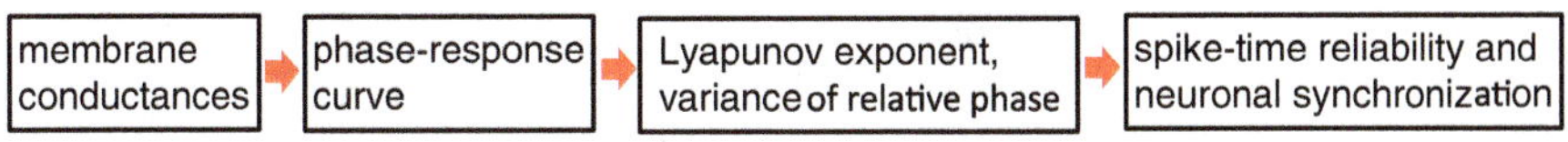

Fig. 10.2 Relating membrane conductances to spike-time reliability and neuronal synchronization. Membrane conductances determine the neuron's phase-response curve, which informs us about the phase shift of the membrane potential in response to a transient perturbation at any time between successive spikes. The phase-response curve allows us to calculate the Lyapunov exponent, which quantifies the stability of the neuron's dynamics with respect to external perturbations, as well as the variance of the phase difference between two identical neuronal oscillators driven by a common fluctuating stimulus. These parameters permit us to predict the reliability of the neuron's response to repetitions of the same fluctuating stimulus and its ability to synchronize with other neurons receiving similar inputs

2 Theoretical Framework

Since Hodgkin's and Huxley's celebrated work (Hodgkin & Huxley 1952), conductance-based models are the canonical computational approach to studying the interplay between membrane currents, synaptic inputs and the generation of action potentials. Consider a set of variables, $\vec{X} = (V, m, h, n, \ldots)$ denoting respectively, the membrane potential, and the activation or inactivation of various membrane currents. The conductance-based model of the neuron dynamics is expressed as a system of coupled differential equations

$$\frac{\mathrm{d}\vec{X}}{\mathrm{d}t} = \vec{F}(\vec{X}) + \vec{I}_{\text{total}}(t), \tag{10.1}$$

where $\vec{F}$ is a vector-valued function describing the change rate of the variables $\vec{X}$, and $\vec{I}_{\text{total}}(t) = \vec{I}_{\text{DC}} + \vec{I}(t) = (I_{\text{DC}} + I(t), 0, 0\ldots, 0)$ is the total input current to the neuron that only affects the equation of the membrane potential (first vector component), which consists of two parts: a tonic current leading to a mean excitability level of the neuron, I_{DC}; and a transient/fluctuating current due to the barrages of synaptic events, $I(t)$. The mean of these synaptic currents can be absorbed in I_{DC}, so we assume that $I(t)$ has zero mean. In the presence of the tonic current I_{DC} only, the neuron will fire with a roughly constant frequency, which corresponds to a closed trajectory (limit cycle, γ) of dynamical system (10.1). Under these circumstances, the dynamics of the whole cell are described by a single parameter: the phase, φ along the limit cycle. In the presence of an additional and fluctuating current, $\vec{I}(t)$ the phase will vary according to the equation: $\mathrm{d}\varphi/\mathrm{d}t = \omega + \vec{Z}(\varphi) \cdot \vec{I}(t)$ (Kuramoto 1984), which, since all the components of $\vec{I}(t)$ are zero except the first one, reduces to the scalar equation:

$$\frac{\mathrm{d}\varphi}{\mathrm{d}t} = \omega + Z(\varphi)I(t),$$

where ω is the angular frequency (2π over the mean interspike interval) and $Z(\varphi)$ is the neuron's phase sensitivity (Kuramoto 1984), or phase-response curve with respect to perturbations of the membrane potential, i.e., the first component of $\vec{Z}(\varphi)$. This property can be calculated directly from system (10.1) in computational models (Ermentrout 1996; Galán 2009b; Rinzel, Ermentrout, Koch, & Segev 1998), or estimated experimentally for real neurons (Galán, Ermentrout, & Urban 2005; Tateno & Robinson 2007; Tsubo, Takada, Reyes, & Fukai 2007). In particular, linearization of system (10.1) around the limit cycle γ, yields:

$$\mathrm{d}\vec{Z}/\mathrm{d}t = -J_{\gamma}^{T}\vec{Z},$$

with J_{γ}^{T} being the transpose of the Jacobian evaluated along γ. Thus, the phase response is fully determined by the membrane conductances via the Jacobian, J_{γ}, plus the normalization condition $\vec{Z} \cdot \vec{F} = \omega$ along the limit cycle.

Once the phase-response curve has been determined, several parameters of the neuron's dynamics can be calculated. Two of those parameters with special relevance for our studies are the Lyapunov exponent, λ and the variance of the phase difference for two identical neurons. The Lyapunov exponent of a dynamical system is a parameter characterizing the convergence of the system to the same state regardless of the initial conditions. For a general dynamical system, $\dot{\vec{X}} = \vec{F}(\vec{X})$, with N degrees of freedom, the Lyapunov exponent is defined as

$$\lambda = \lim_{t\to\infty} \sup \frac{1}{t} \ln \frac{|\delta \vec{X}(t)|}{|\delta \vec{X}(0)|}, \tag{10.2}$$

where "lim sup" stands for *limit superior* (the largest of the N limits of the expression on the right), and $\delta\vec{X}(t)$ represents the temporal evolution of a "perturbed" trajectory resulting from a slight change of the initial conditions, $\delta\vec{X}(0)$. More specifically, if λ is negative, $-1/\lambda$ is the characteristic time to converge to the same state, or equivalently, to forget the difference of the initial conditions. It therefore quantifies the robustness of the dynamics with respect to internal or external perturbations. Chaotic systems are characterized by a positive Lyapunov exponent (Schuster 1988), indicating the irreproducibility of the system's trajectory after a perturbation, or equivalently, after starting from different initial conditions. The unpredictable consequence that a small perturbation may have on a chaotic system is often referred to as "butterfly effect." On the contrary, a dynamical system with negative Lyapunov exponent will converge to the same state after being perturbed or after starting from different initial conditions. A negative Lyapunov exponent is therefore a necessary condition for a neural system to respond reliably to a stimulus in the presence of background noise. Indeed, neurons with chaotic dynamics, i.e., positive Lyapunov exponent, cannot display spike-time reliability (Goldobin & Pikovsky 2006). However, most neurons do not exhibit chaotic behavior, but rather behave as oscillators, i.e., they tend to fire regularly when depolarized with a constant current. In these circumstances, the Lyapunov exponent of the perturbed neuron is determined by the phase-response curve as follows (Ermentrout, Urban, & Galán 2009; Galán et al. 2008). We begin with the equations for two identical and uncoupled neuronal oscillators driven by a common fluctuating stimulus, $I(t)$, that start with different but very close initial conditions, $\varphi_1(0) \approx \varphi_2(0)$, i.e., they are almost synchronized. We thus have

$$\begin{cases} \dot{\varphi}_1 = \omega + \sigma Z(\varphi_1) I \\ \dot{\varphi}_2 = \omega + \sigma Z(\varphi_2) I \end{cases} \tag{10.3}$$

where σ is the mean amplitude (standard deviation) of the fluctuating stimulus, $I(t)$. We then consider the phase difference, $\Delta = \varphi_2 - \varphi_1$ and investigate its dynamics. From (10.3), we have

$$\dot{\Delta} = \sigma(Z(\Delta + \varphi_1) - Z(\varphi_1))I$$

Since the oscillators are close to synchrony, i.e., $|\Delta| \approx 0$, one can approximate the right hand side by its Taylor expansion up to first order in Δ: $\dot{\Delta} \approx \sigma Z'(\varphi_1) I \Delta$, whose solution is

$$\Delta(t) = \Delta(0) \exp\left(\sigma \int_0^t Z'(\varphi_1(t)) I(t) \mathrm{d}t\right). \tag{10.4}$$

From (10.2) and (10.4) the Lyapunov exponent is

$$\lambda = \lim_{T\to\infty} \frac{1}{T} \ln \frac{|\Delta(T)|}{|\Delta(0)|} = \lim_{T\to\infty} \frac{\sigma}{T} \int_0^T Z'(\varphi_1(t)) I(t) \mathrm{d}t = \sigma \langle Z'(\varphi_1(t)) I(t) \rangle, \tag{10.5}$$

where $\langle \ldots \rangle$ denotes temporal average. If $\lambda < 0$, the phase difference tends to decrease on average which ensures reliability: the neuron tends to correct perturbations of the phase encoding fluctuating stimuli. In other words, the neuron will be more reliable, the more negative λ is. If $\lambda > 0$, the background perturbation will destroy the relation between the phase of the neuron and the stimulus it tries to encode. The case $\lambda = 0$ corresponds to the movement along a stable limit cycle. Expression (10.5) can be calculated analytically by assuming that the stimulus is a weak perturbation of the oscillatory dynamics of the neuron. In this case, there is a straightforward mapping of the phase onto time to the lowest order in σ (we drop the phase subscript "1" for the remaining calculations)

$$\varphi(t) \approx \omega t + \sigma \int_0^t Z(\omega s) I(s) \mathrm{d}s,$$

which upon substitution in the integral of (10.5) leads to

$$\lim_{T\to\infty} \frac{\sigma}{T} \int_0^T Z'\left(\omega t + \sigma \int_0^t Z(\omega s) I(s) \mathrm{d}s\right) I(t) \mathrm{d}t.$$

Using the approximation, $f(x+\varepsilon) \approx f(x) + f'(x)\varepsilon$, we obtain

$$\lim_{T\to\infty} \frac{\sigma}{T} \left[\int_0^T Z'(\omega t) I(t) \mathrm{d}t + \int_0^T Z''(\omega t) \left(\sigma \int_0^t Z(\omega s) I(s) \mathrm{d}s\right) I(t) \mathrm{d}t \right].$$

The first term vanishes, as $Z'(\omega t)$ and $I(t)$ are uncorrelated, and the second can be rewritten as

$$\lim_{T\to\infty} \frac{\sigma^2}{T} \int_0^T Z''(\omega t) \int_0^t Z(\omega s) I(s) I(t) \mathrm{d}s \mathrm{d}t. \tag{10.6}$$

The temporal average in (10.5) is equivalent to an ensemble average over the phase and the values or the input signal, I. As a result, the product $I(s)I(t)$ in (10.6) can be replaced with its ensemble average. In turn, this ensemble average is equivalent to its temporal average, $\langle I(t)I(s)\rangle = C(t-s)$, where $C(s)$ is the autocorrelation function of the fluctuating input, defined as $C(s) = \langle I(t)I(t-s)\rangle$. We thus have

$$\lambda = \lim_{T\to\infty} \frac{\sigma^2}{T\omega^2} \int_0^T \frac{\mathrm{d}^2}{\mathrm{d}t^2} Z(\omega t) \int_0^t Z(\omega s)C(s-t)\mathrm{d}s\mathrm{d}t. \tag{10.7}$$

Typically, the autocorrelation function decays exponentially with time constant (autocorrelation time), τ, $C(s) = \exp(-|s|/\tau)$. Also, since the phase-response curve is a periodic function of its argument, one can expand it as a Fourier series

$$Z(\varphi) = \sum_n A_n \cos(n\varphi) + B_n \sin(n\varphi).$$

Finally, one can integrate the expression for the Lyapunov exponent (10.7), thereby obtaining

$$\lambda = -\frac{\sigma^2}{2} \sum_n \frac{\left(A_n^2 + B_n^2\right) n^2\tau}{1+n^2\tau^2\omega^2}. \tag{10.8}$$

Incidentally, similar calculations are carried out to compute the mean frequency change $\langle\Delta\omega\rangle = \langle\dot{\varphi}-\omega\rangle$ of an oscillator driven by colored noise (Galán 2009a). The mean frequency change reads

$$\langle\Delta\omega\rangle = \sigma\langle Z(\varphi(t))I(t)\rangle = -\frac{\sigma^2}{2} \sum_n \frac{\left(A_n^2 + B_n^2\right) n^2\tau^2\omega}{1+n^2\tau^2\omega^2},$$

which implies that a fluctuating input with finite time constant τ, decreases the average frequency of the oscillator. The relationship between the Lyapunov exponent and the frequency shift thus reads $\langle\Delta\omega\rangle = \tau\omega\lambda$, which allows one to estimate λ from the mean frequency shift observed experimentally. This is an important result, as Lyapunov exponents are in general quite difficult to estimate from experimental data (Eckmann, Kamphorst, Ruelle, & Ciliberto 1986). Note that $\langle\Delta\omega\rangle$ is of order σ^2, compared to ω, which is of order σ^0, and hence, it may be ignored in most applications to neuroscience. However, this small frequency shift may have important consequences in the design of oscillators for radiofrequency devices (Demir 2002; Galán 2009a; Gleeson 2006; Maffezzoni 2007; O'Doherty & Gleeson 2007).

The Lyapunov exponent (10.7) is always negative, provided that the autocorrelation function of the inputs decays exponentially and that the phase-response curve $Z(\varphi)$ is twice differentiable. In the typical case, when $Z(\varphi)$ is a "smooth," sinusoidal-like curve, it will always be differentiable. Hence, λ will be negative and the neuron will display reproducible responses to a repetitive fluctuating stimulus. Note, however, that expression (10.8) is a nonmonotonic function of τ. In fact, for a phase-response curve that is a pure sinusoidal ($n = 1$ only), λ is minimized at $\tau_{\text{opt}} = 1/\omega$. For example, if the neuron is firing at ~40 Hz (gamma frequency band), then the Lyapunov exponent is minimal at $\tau_{\text{opt}} = 1/(2\pi \times 40) \approx 4$ ms. This implies that input signals fluctuating with this time constant will be optimally encoded by the neuron, in the sense that the reliability of the responses is maximal. Experiments with real neurons corroborate this theoretical prediction based on the Lyapunov exponent (Ermentrout et al. 2009; Galán et al. 2008). From a physiological perspective, the fact that fast synaptic transmission occurs on a timescale of ca. 5 ms suggests that synapses filter (modulate) input barrages, so that they can be encoded/decoded by neurons with highest fidelity in the gamma frequency band.

Whereas the Lyapunov exponent successfully predicts neuronal reliability as a function of τ, we will see below that it does not accurately predict changes of reliability due to changes in the shape of the phase-response curve. Note that λ depends on the shape of the phase-response curve via $A_n^2 + B_n^2$. Since the weights of the sines and cosines are squared and summed, λ is not sensitive to the sign or symmetry of the Fourier components of $Z(\varphi)$. This reduces the overall sensitivity of λ to the shape of the phase-response curve. On a more fundamental level, there are at least two reasons why λ is not a good predictor of reliability: (1) The stimulus typically fluctuates on a timescale τ of a few milliseconds, whereas the neuron requires a time $1/|\lambda|$ to converge back to the unperturbed trajectory, which is typically in the range of tens and hundreds of milliseconds, as we will see below. Thus, in practice, the stimulus does not allow the neuron to relax back to its unperturbed trajectory because before that happens the stimulus keeps forcing (perturbing) the neuron. In addition, recent theoretical work on stochastic phase oscillators (Teramae, Nakao, & Ermentrout 2009; Yoshimura & Arai 2008) demonstrates the importance of deviations along normal components of the limit cycle, which can be comparable in magnitude to deviations along the limit cycle (phase resetting). (2) The Lyapunov exponent (10.8) of the linearized dynamics (10.4) is an incomplete measure of the overall robustness of the dynamics, since it is calculated on the assumption that both phases are already close to each other. However, in order to display spike-time reliability, the convergence to the same trajectory from very different states is also relevant.

These limitations derive in part from expression (10.3), which considers an instantaneous perturbation (different initial conditions) but does not take into account the effect of continuous background noise. The latter case is more accurately modeled by

$$\begin{cases} \dot{\varphi}_1 = \omega + Z(\varphi_1)(\sigma I + \sigma_\eta \eta_1) \\ \dot{\varphi}_2 = \omega + Z(\varphi_2)(\sigma I + \sigma_\eta \eta_2) \end{cases}, \tag{10.9}$$

where $\eta_{1,2}$, represent ongoing background perturbations as uncorrelated white-noise processes. This system has recently been investigated by several authors in the limit of $\tau \to 0$, i.e., assuming that the input fluctuations $I(t)$ are very fast, so that they can be modeled as a white-noise process (Galán, Ermentrout, & Urban 2007a, 2007b;

Nakao, Arai, & Kawamura 2007). In this case, the problem reduces to investigating two identical oscillators driven by correlated white noise. To see this, define $\xi_1 = \sigma I + \sigma_\eta \eta_1$ and $\xi_2 = \sigma I + \sigma_\eta \eta_2$, where σ_η is the mean amplitude (standard deviation) of the background noise. The cross-correlation coefficient of ξ_1 and ξ_2 is then

$$c = \frac{\langle \xi_1 \xi_2 \rangle}{\sqrt{\langle \xi_1 \xi_1 \rangle \langle \xi_2 \xi_2 \rangle}} = \frac{\sigma^2}{\sigma^2 + \sigma_\eta^2}.$$

No correlation between the inputs ($c = 0$) corresponds to the case with $\sigma = 0$, and perfect correlation between the inputs ($c = 1$) corresponds to the case with $\sigma_\eta = 0$. The numerical solution of the Fokker–Planck equation for system (10.9) reveals that the shape of the phase-response curve, as well as the correlation of the inputs affect the probability of finding the system in any combination of the phases (Galán et al. 2007b). In particular, the probability distribution of the phase difference, $P(\Delta)$, becomes sharper and sharper around $\Delta = 0$ as the input correlation increases. When the inputs are identical (perfect correlation), $P(\Delta) = \delta(\Delta)$. Furthermore, for any value of the input correlation between zero and one, $P(\Delta)$ is sharper for a canonical type-II phase response, $Z(\varphi) = -\sin(\varphi)$, than for a canonical type-I phase response, $Z(\varphi) = 1 + \cos(\varphi + \pi)$. More generally, $P(\Delta)$ has been calculated analytically for the case of oscillators driven by correlated white noise (Nakao et al. 2007) as well as for the case of oscillators driven by correlated Poisson noise (Ermentrout et al. 2009; Marella & Ermentrout 2008). Interestingly enough, for small perturbations, $P(\Delta)$ has the same form in both scenarios, namely,

$$P(\Delta) = \frac{N}{U(0) - cU(\Delta)},$$

where N is the normalization constant, such that $\int_{-\pi}^{\pi} P(\Delta)\mathrm{d}\Delta = 1$, c is the cross-correlation coefficient of the stochastic inputs, and U is the autocorrelation of the phase-response curve,

$$U(y) = \int_0^{2\pi} Z(x)Z(x + y)\mathrm{d}x.$$

Note that $U(\Delta) = U(-\Delta)$ and $P(\Delta) = P(-\Delta)$. Then, the variance of Δ,

$$\mathrm{var}(\Delta) = \int_{-\pi}^{\pi} \Delta^2 P(\Delta)\mathrm{d}\Delta \qquad (10.10)$$

informs us about the reliability of the neuron: the larger the variance, the lower the reliability. Note that the variance is bounded between zero, when $P(\Delta) = \delta(\Delta)$

(perfect correlation between the inputs $\rightarrow$ highest synchrony or reliability), and $\pi^2/3 \approx 3.3$, when $P(\Delta) = 1/(2\pi)$ (no correlation between the inputs $\rightarrow$ lowest synchrony or reliability).

In summary, from the dynamical (10.1) of the single neuron dynamics, including the membrane conductances, one can calculate its phase-response curve and in turn, via (10.8) and (10.10) calculate the Lyapunov exponent and the variance of the relative phase lag. Analogously, for real neurons, the phase-response curve can be estimated experimentally and then, these two parameters can also be calculated. *For a negative Lyapunov exponent, the neuron's response to a fluctuating stimulus will be more reliable, the smaller this variance*. Based on this theoretical framework, one can investigate how membrane conductances contribute to spike-time reliability and synchronization with the phase-response curve, combining simulations and experiments, as described below.

3 Results from Simulations and Experiments

The strategy is as follows: first, a conductance-based model of the neuron is constructed. From the model we first calculate the phase-response curve and then, λ and var(Δ). In parallel, we run computer simulations with the model, and use the correlation coefficient of the membrane potential in successive repetitions of a stimulus in the presence of background noise as the actual measure of spike time reliability. Similar quantifications are possible, but they yield similar results (Galán, Fourcaud-Trocmé, Ermentrout, & Urban 2006; Hunter et al. 1998; Schreiber et al. 2004). We then compare spike-time reliability with λ and var(Δ) to validate these parameters as predictors of reliability. Then, we "block" a specific conductance in the model and recalculate all these parameters: λ and var(Δ) directly from the phase-response curve of the model and spike-time reliability from simulations. Finally, we proceed to do the experiments with real neurons along the lines of the computational simulations.

To exemplify this strategy, we start with a conductance-based model adapted from Connor and Stevens, which includes a potassium I_A current (Connor & Stevens 1971). This choice is motivated by two observations: (1) potassium currents which are not fully inactivated around the firing threshold make the slope of the membrane potential (dV/dt) shallower, thereby making the neuron less likely to fire in response to upstrokes of a fluctuating stimulus, and hence, making the spike times less reproducible across repetitions of the stimulus; (2) A-like currents have been recently shown to regulate spike-timing and modulate firing patterns in mitral cells (Balu & Strowbridge 2007). The results of the simulations and the comparison with our theoretical predictions are shown in Fig. 10.3. The stimulus is simulated as an input current that fluctuates in time quickly without a preferred frequency. Technically, this is obtained by convolving white noise with an alpha function (Galán et al. 2008; Galán et al. 2006), which mimics the superposition of a large number of input barrages and the low-pass filter effect exerted by the synaptic kinetics. The stimulus generated this way looks like the inputs drawn in Fig. 10.1, with an autocorrelation time, τ of three to five milliseconds. The same stimulus is then applied in two successive repetitions in the presence of 50% background noise (modeled as white noise uncorrelated across repetitions) to a neuron with the A-current, I_A (Fig. 10.3a). In this case, the traces of the membrane potential for the two repetitions overlap considerably, but not perfectly, indicating that spike-time reliability is good but not optimal. I_A is then blocked in the model by setting its conductance to zero. The bias current is then decreased to make the cell fire with the same frequency as before the blockade. Then, the same stimulus is applied with the same protocol as before (Fig. 10.3b). In this case, the traces of the membrane potential for different repetitions overlap almost perfectly, indicating that reliability is (close to) optimal. Figure 10.3d displays the phase-response curve calculated directly from the model, before and after blocking I_A. Note that the curve becomes more sinusoidal after the blockade.

The knowledge of the phase-response curve also allows us to calculate var(Δ) as explained above (Fig. 10.3d). The variance of the phase difference with I_A is clearly larger (13.4%) than without I_A. This suggests that spike time reliability in the former case must be substantially lower than in the latter case. Indeed, that is what one observes when spike-time reliability is quantified as the cross-correlation coefficient between the voltage traces in successive repetitions of a fluctuating stimulus. Figure 10.3e shows the distribution of reliability values for ten different stimuli with and without I_A. Note that after blocking I_A, the values of spike-time reliability increase significantly (Wilcoxon's test for equal medians: $p = 0.001 \ll 0.05$). The Lyapunov exponent for the Connor–Stevens model also predicts an increase in reliability. With I_A the value is $\lambda = -0.0007\,\mathrm{ms}^{-1}$and without I_A the value is $\lambda = -0.0015\,\mathrm{ms}^{-1}$ (114% increase in absolute value). The small absolute values of λ are due to the fact that for the calculation of the Lyapunov exponent, we were assuming that $\sigma/\omega \ll 1$, which implies that the trajectories of the oscillators in (10.3) deviate very little from a limit cycle. Recall that the Lyapunov exponent along a limit cycle is zero.

A similar analysis is performed with a different neuronal model, the Wang–Buzsaki model (Wang & Buzsaki 1996). This model does not possess an A-current, so in this case one has to add it instead of removing it. The results are displayed in Fig. 10.4. Note that the phase-response curves in this case look very different from the ones in the previous model. In this case, the phase-response curve is flattened and the center of gravity shifts to the right after blocking the A-current. Interestingly enough, the role of the A-current with respect to spike-time precision is the same as in the Connor–Stevens model: the A-current increases the variance of the phase difference and hence, deteriorates spike-time reliability. The Lyapunov exponent for the Wang–Buzsaki model, however, fails to predict the role of I_A in spike-time reliability. With I_A the value is $\lambda = -0.0005\,\mathrm{ms}^{-1}$ and without I_A the value is $\lambda = -0.0002\,\mathrm{ms}^{-1}$ (60% decrease in absolute value).

After validating the variance of the phase difference as an indicator of reliability changes in computer simulations, we proceeded to carry out electrophysiological experiments in the same fashion (Fig. 10.5). Whereas many cells in the vertebrate's

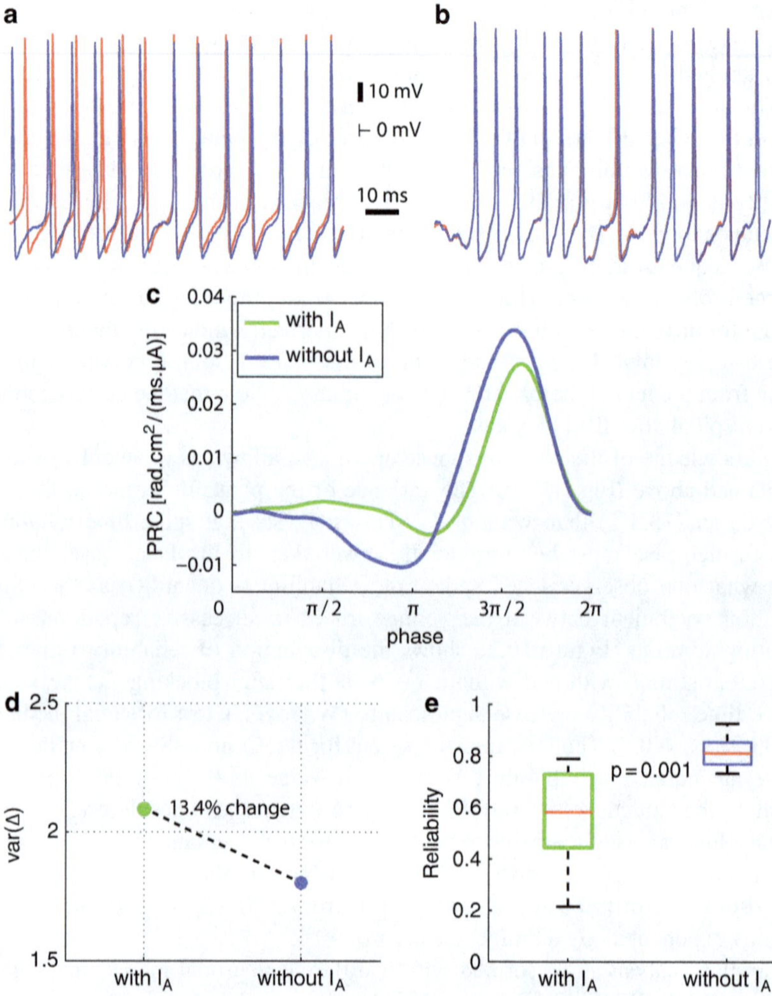

Fig. 10.3 Computer simulations of the Connor–Stevens model. (**a**) Responses to two repetitions (*red and blue*) of a rapidly fluctuating stimulus in a neuron with I_A: spike-time reliability is good but not perfect. (**b**) Responses to two repetitions of the same stimulus in the same neuron after blocking I_A: spike-time reliability is almost perfect. (**c**) Phase-response curves (PRC) of the neuron before and after blocking I_A. The peak and trough of the PRC becomes more pronounced in the latter case. (**d**) The variance of the phase difference predicted from the PRC decreases after blocking I_A, suggesting that the neuron's responses become more reliable. (**e**) Indeed, spike-time reliability calculated as the cross-correlation of the spike trains increases significantly. The small p-value reflects the reproducibility of this effect for ten different stimuli

brain, including the neocortex, hippocampus, and olfactory bulb, express A-like currents (Hille 2001), it has been in mitral cells of the olfactory bulb, where an A-like current has been demonstrated recently to regulate the timing of spike

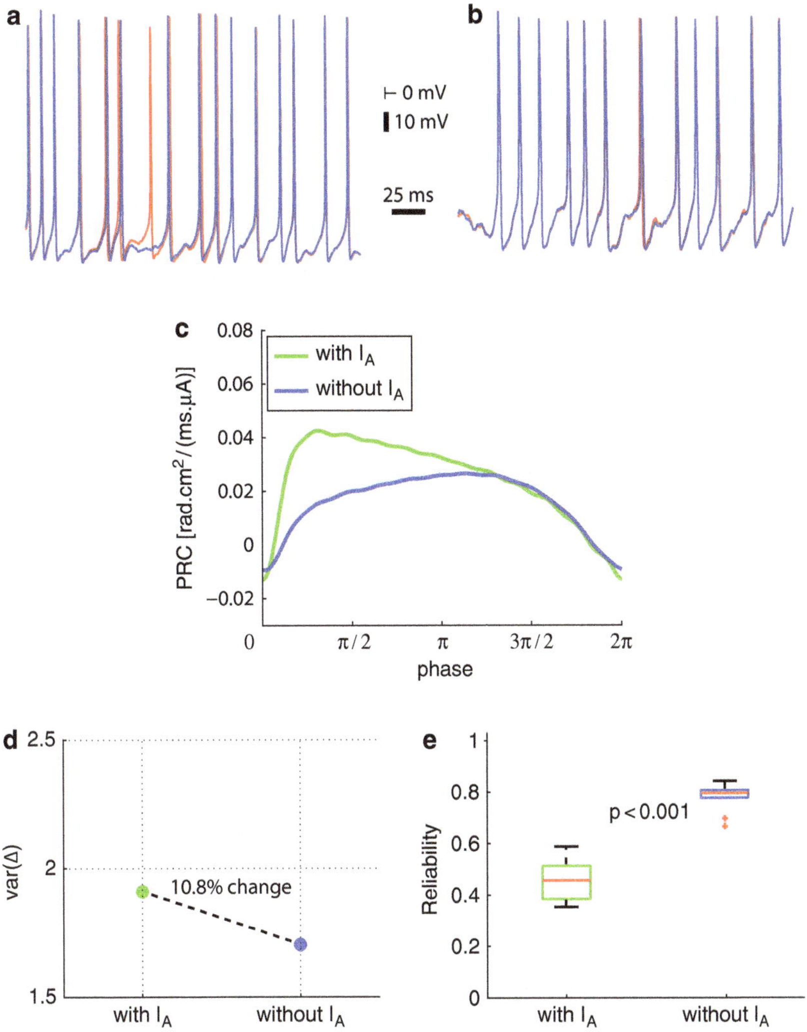

Fig. 10.4 Computer simulations of the Wang–Buzsaki model with A-current added. (**a**) Responses to two repetitions (*red and blue*) of a rapidly fluctuating stimulus in a neuron with I_A: spike-time reliability is good but not perfect. (**b**) Responses to two repetitions of the same stimulus in the same neuron after blocking I_A: spike-time reliability is almost perfect. (**c**) Phase-response curves (PRC) of the neuron before and after blocking I_A. Note the flattening of the PRC in the latter case. (**d**) The variance of the phase difference predicted from the PRC decreases after blocking I_A, suggesting that the neuron's responses become more reliable. (**e**) Indeed, spike-time reliability calculated as the cross-correlation of the spike trains increases significantly. The negligible p-value reflects the reproducibility of this effect for ten different stimuli

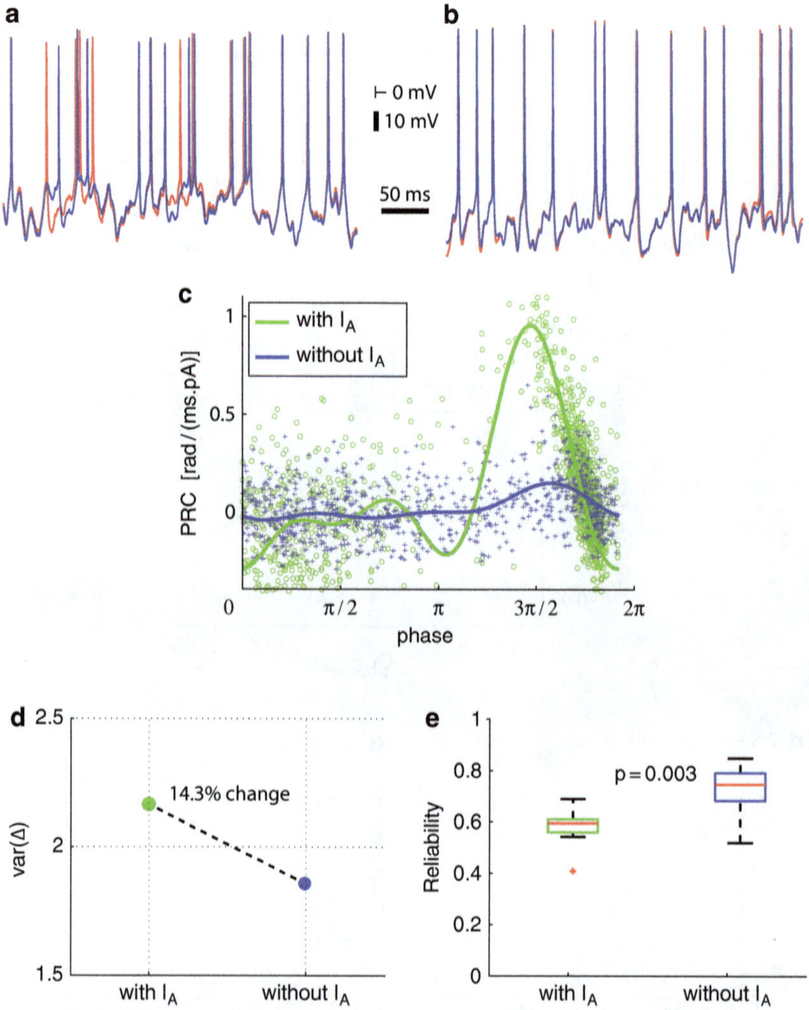

Fig. 10.5 Whole-cell patch-clamp experiments. (**a**) Responses to two repetitions (*red and blue*) of a rapidly fluctuating stimulus in a mitral cell with I_A: spike-time reliability is good but not perfect. (**b**) Responses to two repetitions of the same stimulus in the same mitral cell after blocking I_A: spike-time reliability is almost perfect. (**c**) Phase-response curves (PRC) of the neuron before and after blocking I_A. Note the flattening of the PRC in the latter case. Lines are best fit of data points (*green circles and blue crosses*) to a periodic function. (**d**) The variance of the phase difference predicted from the PRC decreases after blocking I_A, suggesting that the neuron's responses become more reliable. (**e**) Indeed, spike-time reliability calculated as the cross-correlation of the spike trains increases significantly. The small p-value reflects the reproducibility of this effect for ten different stimuli

discharges (Balu et al. 2004; Balu & Strowbridge 2007). We therefore performed our preliminary experiments in mitral cells of the olfactory bulb. The stimulus was generated off-line on the computer controlling the experimental equipment, in exactly the same fashion as for the simulations. The stimulus was then injected several times as an input current in current-clamp mode. Two successive responses are shown in Fig. 10.5a. As in the simulations, the overlap between the membrane potential traces is quite good, although not perfect, indicating that spike-time reliability is good but not optimal. Then, the A-current blocker 4AP (4-aminopyridine) was bath-applied at a concentration of 10 μM. The same stimulus was then injected into the cell using the same protocol as before. In this case and as in our simulations, the membrane potential traces for successive repetitions overlapped almost perfectly, suggesting that spike-time reliability is close to optimal (Fig. 10.5b).

In addition to the reliability experiments, we also measured the phase-response curves of the neuron experimentally, before and after blocking I_A (Fig. 10.5c). To this end, we used a protocol recently developed by the author and collaborators (Galán et al. 2005). Note that the change of the phase-response curve after blocking the A-current is quite different from the changes in both models. However, there are some reminiscent features: the phase-response curve preserves some sinusoidal shape as in the Connor–Stevens model; but it also preserves some sinusoidal shape while its center of gravity is shifted to the right, as in the Wang–Buzsaki model with an A-current. From the estimated phase-response curves we calculated var(Δ) (Fig. 10.5d). As in the simulations, var(Δ) clearly decreases (14.3%) after blocking I_A, which predicts an increase in spike-time reliability. Indeed, as in the simulations, the distribution of reliability values from the experimental data (Fig. 10.5e) is significantly shifted to higher values after blocking I_A (Wilcoxon's test for equal medians, $p = 0.003 \ll 0.05$). The Lyapunov exponent fails to predict this trend again. With I_A the value is $\lambda = -0.8261\,\text{ms}^{-1}$ and without I_A the value is $\lambda = -0.0226\,\text{ms}^{-1}$ (97% decrease in absolute value).

The overall agreement between the theoretical predictions based on var(Δ), which is estimated from the phase-response curve, suggests that this approach could be applied to other cell types and to other membrane currents to investigate their role on spike-time reliability and stochastic synchronization, along the lines of our analysis of A-like currents in mitral cells.

4 Discussion

We have shown in two computer models and in real neurons (mitral cells of the mouse olfactory bulb) that blockade of certain potassium currents, in particular, A and A-like currents, changes the phase-response curve of the neuron. The variance of the phase difference calculated from the phase-response curve decreases after blocking this current type, which is a good predictor of an increase in spike-time reliability and stochastic synchronization. The same approach could be applied to different membrane currents to systematically study their effect on spike timing. It is noteworthy that the blockade of A-like currents in our simulations and experiments consistently led to an increase in reliability, whereas the changes of the phase-response curve were quite different. This indicates that the entire shape of the phase-response curve matters and not just a single feature of it, like, e.g., the position of its peak or trough. In phase oscillator theory, the phase-response curve that minimizes (maximizes the absolute value of) the Lyapunov exponent has been recently calculated. This yields a sinusoidal phase-response curve when the L^2-norm of the solution (the integral of its square) is fixed (Abouzeid & Ermentrout 2009). A similar calculation may be performed for the variance of the phase difference. However, as can be inferred from Figs. 10.3c, 10.4c, and 10.5c, the blockade of membrane currents does not preserve the norm, since the shape and amplitude are clearly distorted. In the Wang–Buzsaki model with an A-current, as well as in the experiments, the amplitude of the phase-response curve is significantly decreased after blocking the A-current, whereas it increases in the Connor–Stevens model. Thus, the changes of reliability are due to the overall change of the phase-response curve, including its amplitude and specific shape.

In our experiments and simulations, the firing rate of the neurons after blocking A-like currents was adjusted to the same level as before. This was to ensure that the increase in spike-time reliability was not simply due to an increase in the firing frequency. Indeed, when the firing rate increases, neurons become less sensitive to input fluctuations but also to background noise. Since the background is in general smaller than the driving stimulus, a firing rate increase will generally increase the signal-to-noise ratio, and hence the reliability to the stimulus response.

The functional roles of spike-time reliability and neuronal synchronization are still debated (Ermentrout et al. 2008; Tiesinga et al. 2008). Perhaps the best argument supporting a critical role in brain function is that they are significantly altered in certain pathological conditions (Uhlhaas & Singer 2006). Indeed, spike-time reliability in hippocampal cells is dramatically reduced in epileptic rat brains *in vitro* (Foffani, Uzcategui, Gal, & Menendez de la Prida 2007). Interestingly enough, this effect can be alleviated by reducing potassium currents with TEA (tetraethylammonium), as Foffani and colleagues recently demonstrated (Foffani et al. 2007). Whereas those currents are different from the A-like currents investigated here, it is worth noting the role of potassium conductances in controlling the level of spike-timing precision in normal and pathological brain function.

There are at least three other biologically relevant scenarios our approach may shed light on. These scenarios do not necessarily involve the membrane currents studied here, but could also involve other currents, like the h-current or muscarinic currents. In the first scenario, a fast neuromodulator would down regulate the activation of the membrane current, in our case the A-current, thereby increasing the overall firing rate and the reliability of the neuron to fast fluctuating stimuli. In this case, the presence of the neuromodulator might be a biological correlate of a higher cognitive function, like attention. In the second scenario, a mechanism for homeostatic plasticity would down regulate the expression of membrane current channels, in our case A-current channels, over several hours at the same time that the spontaneous firing rate of the neuron is preserved. The net result would make the neuron more reliable and susceptible to entrain other neurons receiving common inputs. The third scenario would consist of similar neurons, for instance, pyramidal cells in the cortex, some of them expressing fewer channels of the membrane current, in our case of the A-current, than the rest. Those with a lower expression of the corresponding channels may be specialized in encoding and relaying information with millisecond precision, whereas the neurons with larger conductances of the A-current, would fire less and would be likely to behave more as integrators, like the Wang–Buzsaki model with A-current (Fig. 10.4c, green trace), or as resonators, like the Connor–Stevens model (Fig. 10.3c, green trace) with respect to their synaptic inputs.

In summary, phase oscillator theory has contributed considerably to our understanding of spike-time reliability and stochastic synchronization in the last few years. As demonstrated here, a combination of computer models and electrophysiological experiments using the concept of phase-response curve can also help us elucidate the underlying cellular mechanisms of these remarkable phenomena.

Acknowledgments The author thanks three anonymous reviewers for constructive suggestions. The data displayed in Fig. 10.5a–c were collected, recorded and analyzed in Nathan Urban's lab at Carnegie Mellon University in collaboration with Aushra Abouzeid, a graduate student with Bard Ermentrout at the University of Pittsburgh. The author's work is currently supported by a scholarship of The Mount Sinai Health Care Foundation and a fellowship of The Alfred P. Sloan Foundation.

References

Abouzeid, A., & Ermentrout, B. (2009). Type-II phase resetting curve is optimal for stochastic synchrony. *Phys Rev E: Stat Nonlinear Soft Matter Phys, 80*(1 Pt 1), 011911.

Balu, R., Larimer, P., & Strowbridge, B. W. (2004). Phasic stimuli evoke precisely timed spikes in intermittently discharging mitral cells. *J Neurophysiol, 92*(2), 743–753.

Balu, R., & Strowbridge, B. W. (2007). Opposing inward and outward conductances regulate rebound discharges in olfactory mitral cells. *J Neurophysiol, 97*(3), 1959–1968.

Bryant, H. L., & Segundo, J. P. (1976). Spike initiation by transmembrane current: a white-noise analysis. *J.Physiol, 260*(2), 279–314.

Connor, J. A., & Stevens, C. F. (1971). Prediction of repetitive firing behaviour from voltage clamp data on an isolated neurone soma. *J Physiol, 213*(1), 31–53.

Demir, A. (2002). Phase noise and timing jitter in oscillators with colored-noise sources. *IEEE Trans Circuits Syst I, 49*(12), 1782–1791.

Eckmann, J. P., Kamphorst, S. O., Ruelle, D., & Ciliberto, S. (1986). Liapunov Exponents from Time-Series. *Phys Rev A, 34*(6), 4971–4979.

Ermentrout, B. (1996). Type I membranes, phase resetting curves, and synchrony. *Neural Comput, 8*(5), 979–1001.

Ermentrout, B., Urban, N. N., & Galán, R. F. (2009). Stochastic synchrony in the olfactory bulb. In K. Josic, J. Rubin, M. Matías & R. Romo (Eds.), *Coherent Behavior in Neuronal Networks* (pp. 229–245). New York: Springer.

Ermentrout, G. B., Galán, R. F., & Urban, N. N. (2008). Reliability, synchrony and noise. *Trends Neurosci, 31*(8), 428–434.

Fellous, J. M., Tiesinga, P. H., Thomas, P. J., & Sejnowski, T. J. (2004). Discovering spike patterns in neuronal responses. *J Neurosci, 24*(12), 2989–3001.

Foffani, G., Uzcategui, Y. G., Gal, B., & Menendez de la Prida, L. (2007). Reduced spike-timing reliability correlates with the emergence of fast ripples in the rat epileptic hippocampus. *Neuron, 55*(6), 930–941.

Galán, R. F. (2009a). Analytical calculation of the frequency shift in phase oscillators driven by colored noise: implications for electrical engineering and neuroscience. *Phys Rev E: Stat Nonlinear Soft Matter Phys, 80*(3 Pt 2), 036113.

Galán, R. F. (2009b). The phase oscillator approximation in neuroscience: an analytical framework to study coherent activity in neural networks In R. Wennberg & J. L. Perez-Velazquez (Eds.), *Coordinated Activity in the Brain* (pp. 65–89). New York: Springer.

Galán, R. F., Ermentrout, G. B., & Urban, N. N. (2005). Efficient estimation of phase-resetting curves in real neurons and its significance for neural-network modeling. *Phys Rev Lett, 94*(15), 158101.

Galán, R. F., Ermentrout, G. B., & Urban, N. N. (2007a). Reliability and stochastic synchronization in type I vs. type II neural oscillators. *Neurocomp, 70*, 2102–2106.

Galán, R. F., Ermentrout, G. B., & Urban, N. N. (2007b). Stochastic dynamics of uncoupled neural oscillators: Fokker–Planck studies with the finite element method. *Phys Rev E: Stat Nonlinear Soft Matter Phys, 76*(5 Pt 2), 056110.

Galán, R. F., Ermentrout, G. B., & Urban, N. N. (2008). Optimal time scale for spike-time reliability: theory, simulations, and experiments. *J Neurophysiol, 99*(1), 277–283.

Galán, R. F., Fourcaud-Trocmé, N., Ermentrout, G. B., & Urban, N. N. (2006). Correlation-induced synchronization of oscillations in olfactory bulb neurons. *J Neurosci, 26*(14), 3646–3655.

Gerstner, W., van Hemmen, J. L., & Cowan, J. D. (1996). What matters in neuronal locking? *Neural Comput, 8*(8), 1653–1676.

Gleeson, J. P. (2006). Phase diffusion due to low-frequency colored noise. *IEEE Trans Circuits Syst II, 53*(3), 183–186.

Goldberg, J. M., Smith, C. E., & Fernandez, C. (1984). Relation between discharge regularity and responses to externally applied galvanic currents in vestibular nerve afferents of the squirrel monkey. *J Neurophysiol, 51*(6), 1236–1256.

Goldobin, D. S., & Pikovsky, A. (2006). Antireliability of noise-driven neurons. *Phys Rev E: Stat Nonlinear Soft Matter Phys, 73*(6 Pt 1), 061906.

Hille, B. (2001). *Ion channels of excitable membranes* (3rd ed.). Sunderland, Mass.: Sinauer.

Hodgkin, A. L., & Huxley, A. F. (1952). A quantitative description of membrane current and its application to conduction and excitation in nerve. *J Physiol, 117*(4), 500–544.

Hunter, J. D., & Milton, J. G. (2003). Amplitude and frequency dependence of spike timing: implications for dynamic regulation. *J Neurophysiol, 90*(1), 387–394.

Hunter, J. D., Milton, J. G., Thomas, P. J., & Cowan, J. D. (1998). Resonance effect for neural spike time reliability. *J Neurophysiol, 80*(3), 1427–1438.

Kuramoto, Y. (1984). *Chemical Oscillations, Waves, and Turbulence*: Springer Verlag.

Maffezzoni, P. (2007). Frequency-shift induced by colored noise in nonlinear oscillators. *IEEE Trans Circuits Systems II, 54*(10), 887–891.

Mainen, Z. F., & Sejnowski, T. J. (1995). Reliability of spike timing in neocortical neurons. *Science, 268*(5216), 1503–1506.

Mainen, Z. F., & Sejnowski, T. J. (1996). Influence of dendritic structure on firing pattern in model neocortical neurons. *Nature, 382*(6589), 363–366.

Marder, E., & Bucher, D. (2007). Understanding circuit dynamics using the stomatogastric nervous system of lobsters and crabs. *Annu Rev Physiol, 69*, 291–316.

Marder, E., & Goaillard, J. M. (2006). Variability, compensation and homeostasis in neuron and network function. *Nat Rev Neurosci, 7*(7), 563–574.

Marella, S., & Ermentrout, G. B. (2008). Class-II neurons display a higher degree of stochastic synchronization than class-I neurons. *Phys Rev E: Stat Nonlinear Soft Matter Phys, 77*(4 Pt 1), 041918.

Movshon, J. A. (2000). Reliability of neuronal responses. *Neuron, 27*(3), 412–414.

Nakao, H., Arai, K., & Kawamura, Y. (2007). Noise-induced synchronization and clustering in ensembles of uncoupled limit-cycle oscillators. *Phys Rev Lett., 98*(18), 184101.

O'Doherty, F., & Gleeson, J. P. (2007). Phase diffusion coefficient for oscillators perturbed by colored noise. *IEEE Transactions on Circuits and Systems II, 54*(5), 435–439.

Rieke, F., Warland, D., de Ryuter van Stevenick, R., & Bialek, W. (1997). *Spikes. Exploring the neural code*. Cambridge, Mass: MIT.

Rinzel, J., Ermentrout, B., Koch, C., & Segev, I. (1998). Analysis of Neural Excitability. In C. Koch & I. Segev (Eds.), *Methods in Neuronal Modeling* (Vol. 2nd, pp. 251–291). Cambridge, Mass: MIT.

Schreiber, S., Fellous, J. M., Tiesinga, P., & Sejnowski, T. J. (2004). Influence of ionic conductances on spike timing reliability of cortical neurons for suprathreshold rhythmic inputs. *J Neurophysiol, 91*(1), 194–205.

Schuster, H. G. (1988). *Deterministic chaos: an introduction* (2nd rev. ed. ed.). New York: VCH.

Stein, R. B. (1965). A theoretical analysis of neuronal variability. *Biophys J, 5*, 173–194.

Tateno, T., & Robinson, H. P. (2007). Phase resetting curves and oscillatory stability in interneurons of rat somatosensory cortex. *Biophys J, 92*(2), 683–695.

Teramae, J. N., Nakao, H., & Ermentrout, G. B. (2009). Stochastic phase reduction for a general class of noisy limit cycle oscillators. *Phys Rev Lett, 102*(19), 194102.

Tiesinga, P., Fellous, J. M., & Sejnowski, T. J. (2008). Regulation of spike timing in visual cortical circuits. *Nat Rev Neurosci, 9*(2), 97–107.

Tsubo, Y., Takada, M., Reyes, A. D., & Fukai, T. (2007). Layer and frequency dependencies of phase response properties of pyramidal neurons in rat motor cortex. *Eur J Neurosci, 25*(11), 3429–3441.

Uhlhaas, P. J., & Singer, W. (2006). Neural synchrony in brain disorders: relevance for cognitive dysfunctions and pathophysiology. *Neuron, 52*(1), 155–168.

Wang, X. J., & Buzsaki, G. (1996). Gamma oscillation by synaptic inhibition in a hippocampal interneuronal network model. *J Neurosci, 16*(20), 6402–6413.

Yoshimura, K., & Arai, K. (2008). Phase reduction of stochastic limit cycle oscillators. *Phys Rev Lett, 101*(15), 154101.

Chapter 11
Recovery of Stimuli Encoded with a Hodgkin–Huxley Neuron Using Conditional PRCs

Anmo J. Kim and Aurel A. Lazar

Abstract Understanding neural encoding/decoding mechanisms is one of the most fundamental problems in the field of sensory neuroscience. The Hodgkin–Huxley equations provide an explicit description of an encoding mechanism. However, the daunting complexity of the Hodgkin–Huxley equations makes the task of recovery of stimuli encoded with a Hodgkin–Huxley neuron particularly challenging. A highly effective strategy calls for reducing the Hodgkin–Huxley neuron to a project-integrate-and-fire (PIF) neuron. Using the reduced PIF model, we present three different recovery algorithms for stimuli encoded with a Hodgkin-Huxley neuron. All algorithms reconstruct the stimuli from the neuron's output spike train. The first algorithm is based on the assumption that the Hodgkin–Huxley neuron has a known PRC. The second algorithm assumes that the PRC is conditionally known on each interspike time interval. Finally, the third algorithm operates under the assumption that the conditional PRC is unknown and has to be estimated. We establish an estimate of the conditional PRC based upon the readily observable inter-spike time interval. We compare the performance of these algorithms for a wide range of input stimuli.

1 Introduction

Neural circuits in sensory systems encode continuous-time stimuli, such as the concentration of an odor, the intensity of an acoustic event or the brightness of a light source. Information is represented in these circuits as a sequence of action potentials, or a spike train. Yet information can be reliably transmitted and received between neurons. Thus, a spike train may faithfully encode information, and postsynaptic neurons readily decode it.

A.J. Kim • A.A. Lazar (✉)
Department of Electrical Engineering, Columbia University, New York, NY 10027, USA
e-mail: ak2263@columbia.edu; aurel@ee.columbia.edu

N.W. Schultheiss et al. (eds.), *Phase Response Curves in Neuroscience: Theory, Experiment, and Analysis*, Springer Series in Computational Neuroscience 6,
DOI 10.1007/978-1-4614-0739-3_11,

Generation of an action potential across a cell membrane fascinated the earliest neuroscientists and opened the dawn of electrophysiological studies in neuroscience (Schuetze 1983). Hodgkin and Huxley measured the change of ionic conductances for different membrane potential values by employing the voltage-clamp method, and their results led to a mathematical description of the action potential generation, the so called Hodgkin–Huxley neuron (Hodgkin & Huxley 1952). From an information theory perspective, the Hodgkin–Huxley neuron provides an explicit description of encoding a continuous-time stimulus into a discrete time sequence, or spike train.

The problem of reconstruction of continuous-time stimuli from a spike train has been investigated in the literature primarily using phenomenological rather than mechanistic models. For example, one of the most popular phenomenological encoding models is the linear-nonlinear-Poisson (LNP) model. In this model, the Poisson spike generation is preceded by linear or nonlinear processing blocks. For these models, a spike rate function is first recovered by constructing a peri-stimulus time histogram (PSTH) or by Gaussian filtering (Dayan & Abbott 2001; Perkel, Gerstein, & Moore 1967). The stimulus reconstruction is completed by taking the inverse of the processing blocks acting on the spike rate function. (Bialek & Rieke 1991; Arcas, Fairhall, & Bialek 2003).

However, these methods generally require many repeated trials to achieve an adequate temporal resolution and thereby average out the spike train variability across trials (Medina & Lisberger 2007). Moreover, the validity of the phenomenological models is limited and often strongly dependent on statistical parameters of stimuli (Naka, Chan, & Yasui 1979; Kim, Lazar, & Slutskiy 2010). Such phenomenological model limitations call for the derivation of stimulus reconstruction algorithms from mechanistic models of the stimulus encoding process. In this chapter, we derive stimulus recovery algorithms using the explicit Hodgkin–Huxley encoding model.

The dynamics of Hodgkin–Huxley neurons for simple injected current waveforms such as steps and ramps has been extensively investigated in the literature (Izhikevich 2007). A non-linear perturbation analysis showed that a Hodgkin–Huxley neuron with deterministic gating variables is I/O-equivalent with a project-integrate-and-fire (PIF) neuron with a variable threshold sequence (Lazar 2007, 2010). The PIF neuron integrates a projection of the stimulus onto the phase response curve that is, in turn, modulated by a phase shift process. The phase shift process is described by a differential equation that is stimulus driven. In the absence of the small perturbation term, the stimulus is tangentially coupled into the limit cycle of the Hodgkin–Huxley neuron and the PIF neuron reduces to an integrate-and-fire neuron.

Based on the I/O-equivalent PIF neuron a stimulus recovery algorithm was given in (Lazar 2007, 2010). The recovery works well provided that the stimulus restricts the PRC to a small parameter set. If, however, the Hodgkin–Huxley neuron sweeps across a broad set of PRCs, different recovery algorithms are needed for improved performance. The key limitation of the recovery algorithm in (Lazar 2007, 2010) is due to the static nature of the PRC. By introducing an adaptive sampling kernel, the PIF can provide a more faithful I/O description of the Hodgkin-Huxley neuron.

In order to achieve an accurate I/O description, the PIF is endowed with a sampling kernel conditioned on the stimulus on each interspike time interval. The kernel is, in effect, a conditional PRC that is parameterized by the interspike time interval. We shall also consider the case when the conditional sampling kernel is unknown and devise a simple kernel estimator. The proposed recovery algorithms are very flexible; they only require the inversion of a matrix. They can be used for the recovery of stimuli encoded by a large class of neurons in the limit cycle region. It remains an open problem of how the dendritic tree of a postsynaptic neuron is capable of recovering the information from its incoming spike sequence.

This paper is organized as follows. In Sect. 2, a Hodgkin–Huxley neuron is shown to be I/O-equivalent with the PIF neuron. The recovery of stimuli encoded with a Hodgkin–Huxley neuron is presented in Sect. 3. Three cases are considered. In Sect. 3.1, the stimulus is assumed to consist of a known base term and a weak bandlimited signal. In Sect. 3.2, the problem is extended to strong stimuli by exploiting the known base terms. In Sect. 3.3, the values of the base terms are unknown but, as shown, can be estimated. For all three cases, a recovery algorithm is presented. Examples are given in Sect. 4 and a brief discussion in Sect. 5.

2 Reduction of the Hodgkin–Huxley Neuron to the I/O-Equivalent Project-Integrate-and-Fire Neuron

Using a vector notation, the Hodgkin–Huxley neuron model (see also the Appendix) is described by the system of differential equations (Hodgkin & Huxley 1952)

$$\frac{\mathrm{d}\mathbf{x}}{\mathrm{d}t} = \mathbf{f}(\mathbf{x}) + \frac{1}{C}[I_{\mathrm{ext}}\ 0\ 0\ 0]^{\mathrm{T}},$$

where $\mathbf{x}$ and $\mathbf{f}$ are vectors of appropriate dimensions, and $\mathbf{x}(0) = \mathbf{x}_0$ is the initial condition. The state vectors can easily be identified from the set of Hodgkin–Huxley equations as $\mathbf{x} = (x_1, x_2, x_3, x_4) = (V, m, h, n)$. The expression for $\mathbf{f} = (f_1, f_2, f_3, f_4)$ can also be easily derived from the same set of equations (see the Appendix).

In what follows we shall assume that, if the (step) injected current I_{ext} is in the appropriate range, the essential dynamics of this set of equations are described by a limit cycle (Izhikevich 2007) (see Fig. 11.1a). In other words, when $I_{\mathrm{ext}} = b$ (μA/cm^2), these equations have a periodic solution $\mathbf{x}^o(t, b)$ (also called an orbital solution) with period $T = T(b)$ (see Fig. 11.1b).

The dynamics of the Hodgkin–Huxley neuron are rather complex. In order to devise reconstruction algorithms, it is advantageous to find an I/O-equivalent model of the Hodgkin–Huxley neuron that is mathematically tractable. For weak stimuli, the I/O-equivalent model is the PIF neuron. We shall first derive the phase equation of the Hodgkin–Huxley neuron. Since the state of the phase equation is not observable, we shall reduce it to the PIF neuron model whose spikes are observable.

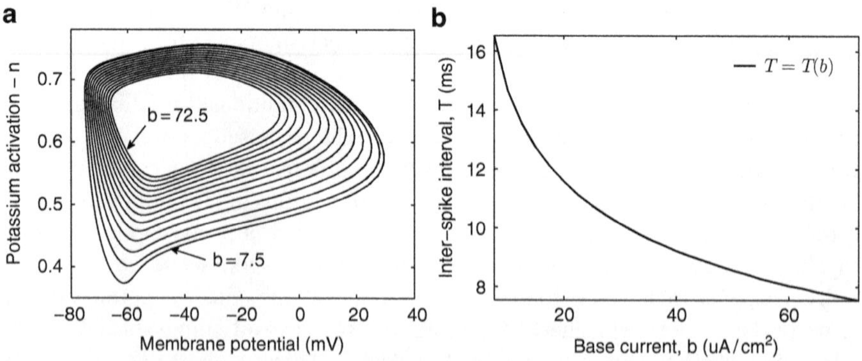

Fig. 11.1 The limit cycle and interspike time interval as a function of the base current. (**a**) The size of the trajectory of the limit cycle depends on the base current; it shrinks as the base current increases (step $= 5\,\mu\text{A/cm}^2$). (**b**) interspike time interval with respect to the base current

The Hodgkin–Huxley and PIF neuron models have the property that they are first order I/O equivalent. That means that, given the same stimulus, the spike times of these neurons are, to the first order, the same.

Because of the periodicity of the orbital solution (see Fig. 11.1a), the behavior of the Hodgkin–Huxley neuron can simply be described by its phase rather than the four state variables of the Hodgkin–Huxley equations. In this chapter, the phase of the system $\theta(t) \in [0, T(b)]$ is defined as

$$\theta(t) = t - t_k \bmod T(b),$$

where $T(b)$ is the period of the oscillation, and t_k is the last spike event ($t \geq t_k$). Note that the phase is in units of time and thus it retains the dependency on the base current b.

If a weak perturbation current $u = u(t), t \in \Re$, is injected in addition to the base current b, the system deviates from the orbit and results in either an advancement or a delay of its phase.

The phase shift in response to a current perturbation is determined by the vector space in the neighborhood of the limit cycle. The mapping of the delta-pulse current perturbation into the value of the phase shift of the orbital solution is known as the *infinitesimal Phase Response Curve (iPRC)*, or simply the PRC hereafter (Hastings & Sweeney 1958; Winfree 1967; Kuramoto 1984; Ermentrout 1996; Izhikevich 2007). Informally, the tangential component of the perturbation to the limit cycle contributes to a phase shift, i.e., the value of the phase response curve. In contrast, the normal component of the perturbation dies out over time and does not contribute to the value of the phase of the orbital solution. This simple analysis is valid provided that the stimulus is weak.

The PRC, denoted by $\psi(t, b)$ (see Fig. 11.2), is parameterized by b since the limit cycle $\mathbf{x}^o(t, b)$ depends on the base current. It can be obtained via a simulations

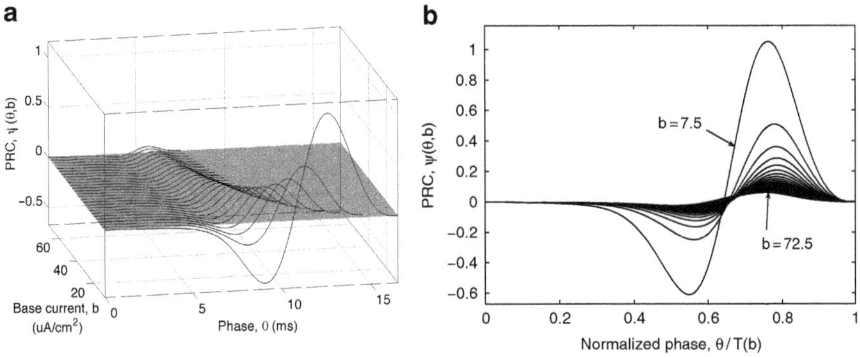

Fig. 11.2 Phase response curves of the Hodgkin–Huxley neuron for different base currents ($b \in [7.5, 72.5]\ \mu\text{A/cm}^2$).) The shape of the phase response curve depends on the base current; its amplitude decreases as the base current increases (step $= 2.5\ \mu\text{A/cm}^2$). (**a**) PRCs depicted in 3D. (**b**) Projection of the PRCs onto the b axis

method in which the change of the phase of the system is evaluated with a Dirac-delta pulse as its input. Based on Malkin's theorem (Hoppensteadt & Izhikevich 1997), one can also obtain the PRC by solving the adjoint equation

$$\dot{\psi}(t,b) = (J(\mathbf{f}(\mathbf{x}^o(t,b))))^{\mathrm{T}}\psi(t,b),$$

where $(J(\mathbf{f}(\mathbf{x}^o(t,b)))^{\mathrm{T}}$ is the transposed Jacobian of $\mathbf{f}$, with the boundary condition $\psi(0,b)(\mathbf{f}(\mathbf{x}^o(0,b)) + [b\ 0\ 0\ 0]^{\mathrm{T}}) = 1$. When using the simulation method, the resulting PRC is normalized by the amplitude and width of the Dirac-delta pulse. In what follows, all examples are based on PRCs evaluated by the direct simulation method.

When the dynamical system describing the Hodgkin–Huxley neuron runs along a limit cycle, its phase is, by definition, moving at a fixed velocity of 1, i.e.,

$$\dot{\theta}(t) = 1. \tag{11.1}$$

A weak perturbation $u(t)$ introduces an additive term to the right hand side of the equation above. This term consists of the perturbation weighted by the corresponding PRC amplitude (Winfree 1967; Izhikevich 2007; Kuramoto 1984; Yoshimura & Arai 2008), i.e.,

$$\dot{\theta} = 1 + \psi(\theta,b)u(t) + \varepsilon(t). \tag{11.2}$$

Here, $\varepsilon(t)$ represents the model reduction error due to the finite amplitude of $u(t)$ that renders the linear approximation of the local vector field inaccurate. Note, however, that $\varepsilon(t)$ becomes arbitrarily small if $u(t)$ is arbitrarily small.

Equation (11.2), also called a *phase equation*, is the result of the reduction of a multidimensional nonlinear dynamical system to a single-dimensional differential equation. When $\varepsilon(t)$ is small enough, the phase equation is approximated by

$$\dot{\theta} = 1 + \psi(\theta, b)u(t). \tag{11.3}$$

The state of the phase equation is not observable. However, spike times are observable. Integrating both sides of (11.3) on an interspike time interval $[t_k, t_{k+1}]$ yields

$$\theta(t_{k+1}) - \theta(t_k) = \int_{t_k}^{t_{k+1}} [1 + \psi(\theta(s), b)u(s)]\,\mathrm{d}s.$$

Since, to a first order, $\theta(t_{k+1}) - \theta(t_k) = T(b)$ (Lazar 2007, 2010),

$$\int_{t_k}^{t_{k+1}} [1 + \psi(\theta(s), b)u(s)]\,\mathrm{d}s = T(b). \tag{11.4}$$

This equation represents the t-transform of a *PIF neuron* (Lazar 2007, 2010). The t-transform provides a measurement or sampling of the input $u = u(t)$ on the time interval between two consecutive spikes. The input stimulus is multiplied with an encoding kernel $\psi(\theta, b)$, and the next spike is generated when the integration of these values reaches the threshold value $T(b) - (t_{k+1} - t_k)$. The PIF neuron is input/output equivalent to the Hodgkin–Huxley neuron under weak perturbation (Lazar 2007, 2010). Since the evaluation of $\theta(t)$ is mathematically intractable, the PIF neuron will be further simplified.

If $u(t)$ is small enough, the phase shift $\theta(t) - t$ becomes negligibly small within $[t_k, t_{k+1}]$, and thus $\psi(\theta(t), b)$ can be replaced with $\psi(t, b)$ (Lazar 2007, 2010; Ota, Omori, & Aonishi 2009). Equation (11.4) can then be rewritten as

$$\int_{t_k}^{t_{k+1}} \psi(s - t_k, b)u(s)\mathrm{d}s = T(b) - (t_{k+1} - t_k). \tag{11.5}$$

A neuron model whose t-transform is given by (11.5) is called *a reduced PIF neuron*.

Note that numerical solutions of (11.4) require the evaluation of $\theta(t)$. The latter is often unstable due to the accumulation of numerical errors. In contrast, the numerical computation of the reduced PIF model is a simple dot product of two vectors and it is not prone to the accumulation of errors.

In order to assess the approximations introduced by the I/O-equivalent neurons above, we ran numerical simulations of three neural encoders: the Hodgkin–Huxley neuron, the PIF neuron, and the reduced PIF neuron. Assuming the same initial condition, the occurrences of the first two spikes were recorded and the interspike time intervals compared when the latter were generated by the PIF neurons and the Hodgkin–Huxley neuron, respectively. The perturbation signal is bandlimited ($\Omega = 2\pi \cdot 50$ Hz) by construction

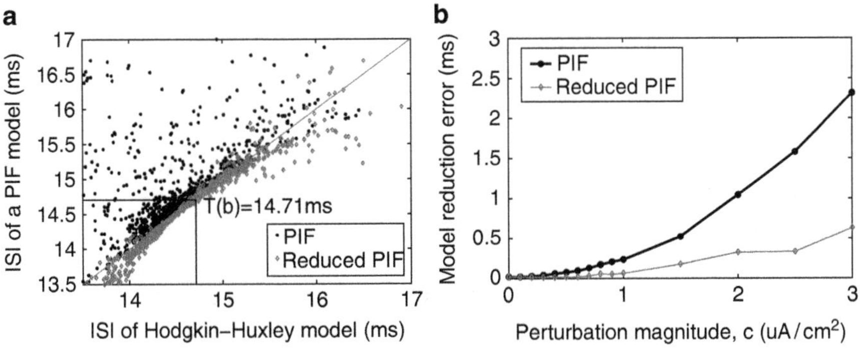

Fig. 11.3 Model reduction error of the project-integrate-and-fire neuron with respect to the Hodgkin–Huxley neuron (**a**) Interspike time intervals (ISI) of the PIF models with respect to the ISI of Hodgkin–Huxley neuron. (**b**) The model reduction error of the PIF models with respect to the perturbation magnitude. The model reduction error is defined as the difference of ISIs between the Hodgkin-Huxley neuron and the PIF models ($b = 10\,\mu\text{A/cm}^2$, $n = 50$, and $\Omega = 2\pi 50$)

$$u(t) = c \sum_{k \in \mathbf{Z}} u\left(\frac{k\pi}{\Omega}\right) \frac{\sin(\Omega t - k\pi)}{\Omega t - k\pi}, \tag{11.6}$$

where $(u(k\pi/\Omega))_{k\in\mathbf{Z}}$ is drawn randomly from a uniform distribution ranging between $[-1, 1]$, c represents the maximum magnitude of the perturbation.

Figure 11.3a depicts the interspike time intervals of the PIF models with respect to Hodgkin–Huxley model with $b = 10\,\mu\text{A/cm}^2$. The steady state interspike time interval $T(b)$ is 14.71 ms. The samples along the diagonal have zero model reduction error, and the vertically projected distance of each sample to the diagonal line is the amount of error for a particular perturbation signal. Figure 11.3b depicts the mean absolute errors with respect to different perturbation magnitudes. In general, increasing the magnitude of the perturbation increases the model reduction error; the reduced PIF model exhibits lower errors than the (standard) PIF neuron.

3 Recovery of Stimuli Encoded with Hodgkin–Huxley Neurons

In this section, we present algorithms for recovery of stimuli encoded with a Hodgkin–Huxley neuron. All algorithms recover the stimuli from the neuron's output spike train. As shown in the previous section and following (Lazar 2007, 2010), the multidimensional nonlinear Hodgkin–Huxley neuron can be reduced to an I/O-equivalent PIF neuron.

In what follows, we shall assume that the stimulus is a bandlimited signal of the form $u(t) = \sum_{k\in\mathbf{Z}} b_k 1_{[t_k, t_{k+1}]} + v(t)$, where b_k is the mean stimulus amplitude during the kth interspike time interval and $v = v(t)$ is a weak perturbation signal with zero average on each interspike time interval. We shall distinguish three cases: (1) $b_k = b$ is known to take a constant value (Sect. 3.1), (2) b_k is constant between two consecutive spike times $[t_k, t_{k+1}]$ and is known (Sect. 3.2) and (3) b_k is unknown (Sect. 3.3). In all cases, the range of values for b_k leads to limit cycle oscillations of the Hodgkin–Huxley neuron.

In Sect. 3.1, we assume weak stimuli with known base current b. Since the perturbation is weak, the encoding mechanism of the system can be described by a known PRC. In Sect. 3.2, we deal with stimuli with known base current b_k on $[t_k, t_{k+1}]$. Since the neuron oscillates on different limit cycles depending on the level of the base current, the conditional PRC characterizes the operation of the neuron on each inter-spike time interval. Finally in Sect. 3.3, the b_k's are unknown and have to be estimated on each interspike time interval.

3.1 Recovery of Stimuli Encoded with Known PRCs

Following the established decoding practice of stimuli encoded with the integrate-and-fire neuron (Lazar 2004; Lazar & Pnevmatikakis 2008; Lazar & Tóth 2004), the derivation of the recovery algorithm begins with the formulation of the t-transform that describes the mapping (or encoding) of the stimulus u into the spike time sequence (t_k).

As already mentioned, the t-transform of the reduced PIF is given by

$$\int_{t_k}^{t_{k+1}} \psi(s - t_k, b)u(s)\mathrm{d}s = T(b) - (t_{k+1} - t_k). \tag{11.7}$$

If the stimulus $u = u(t), t \in \Re$, lives in the space of bandlimited functions with frequency support $[-\Omega, \Omega]$ then (Lazar 2004),

$$\hat{u}(t) = \sum_{k \in \mathbf{Z}} c_k g(t - s_k), \tag{11.8}$$

where $g(t) = \frac{\sin \Omega t}{\pi t}$ and $s_k = \frac{t_k + t_{k+1}}{2}$. The above equality holds if the Nyquist-type rate condition (Lazar & Tóth 2004)

$$t_{k+1} - t_k \leq \frac{\pi}{\Omega}$$

is satisfied. Replacing $u(t)$ in (11.7) with the representation in (11.8), we obtain

$$\sum_{l \in \mathbf{Z}} c_l \int_{t_k}^{t_{k+1}} \psi(s - t_k, b)g(s - s_l)\mathrm{d}s = T(b) - (t_{k+1} - t_k). \tag{11.9}$$

Thus the coefficients $[\mathbf{c}]_l = c_l$ satisfy the matrix equation

$$\mathbf{G}\mathbf{c} = \mathbf{q},$$

where $[\mathbf{G}]_{kl} = \int_{t_k}^{t_{k+1}} \psi(s-t_k, b) g(s-s_l) \mathrm{d}s$, and $[\mathbf{q}]_k = T(b)-(t_{k+1}-t_k)$. Note that, entries of the matrix $\mathbf{G}$ can be obtained from the PRC $\psi(t,b)$ and the reconstruction kernel $g(t), t \in \mathfrak{R}$, and the vector $\mathbf{q}$ from the period $T(b)$ of the oscillation and the spike times $(t_k), k \in \mathbf{Z}$, of the neuron. Therefore, $\mathbf{c}$ can be computed by taking the Moore-Penrose pseudo-inverse of the $\mathbf{G}$ matrix (Penrose 1955), i.e.,

$$\mathbf{c} = \mathbf{G}^{+}\mathbf{q}.$$

Finally, the reconstruction of $u(t)$ is completed by evaluating (11.8) with the values of the vector $\mathbf{c}$.

We arrived, therefore, at the following result (Lazar 2007, 2010).

Theorem 1. *Let $(t_k), k \in \mathbf{Z}$, be the sequence of spike times generated by a Hodgkin–Huxley neuron with input $b + u(t)$. If b is known, the stimulus can be reconstructed using*

$$\hat{u}(t) = \sum_{k \in \mathbf{Z}} c_k g(t - s_k),$$

where $g(t) = \frac{\sin(\Omega t)}{\pi t}$, $s_k = \frac{t_k + t_{k+1}}{2}$, and the coefficients $[\mathbf{c}]_k = c_k$, $k \in \mathbf{Z}$, are given by

$$\boldsymbol{c} = \boldsymbol{G}^{+}\boldsymbol{q},$$

where $[\boldsymbol{G}]_{kl} = \int_{t_k}^{t_{k+1}} \psi(s - t_k, b) g(s - s_l) \mathrm{d}s$ and $[\boldsymbol{q}]_k = T(b) - (t_{k+1} - t_k)$.

An implementation of the encoding and decoding algorithms described in this section is shown in block diagram form in Fig. 11.4.

Remark 1. While the input to the Hodgkin-Huxley neuron is $b + u(t)$, we incorporated b into the PRC $\psi(t,b)$. Consequently, we only had to recover the weak perturbation signal $u(t), t \in \mathfrak{R}$.

3.2 Recovery of Stimuli Encoded with Known Conditional PRCs

In this section, we consider the recovery of a bandlimited stimulus of the form $u(t) = \sum_{k \in \mathbf{Z}} b_k 1_{[t_k, t_{k+1}]} + v(t)$, where b_k is the mean stimulus amplitude during the kth inter-spike time interval and $v(t)$ is the perturbation signal. Note that although $u(t)$ is a bandlimited stimulus, $v(t) = u(t) - \sum_{k \in \mathbf{Z}} b_k 1_{[t_k, t_{k+1}]}$ is, in general, not bandlimited, since $\sum_{k \in \mathbf{Z}} b_k 1_{[t_k, t_{k+1}]}$ typically has infinite bandwidth.

Since the mean stimulus value b_k may take a distinct value on each interspike time interval, the t-transform has to be established separately for every interval. This can be achieved by using a formulation akin to the one in (11.7) and requires a selection of the corresponding PRC for every interspike time interval. We formulate the t-transform as

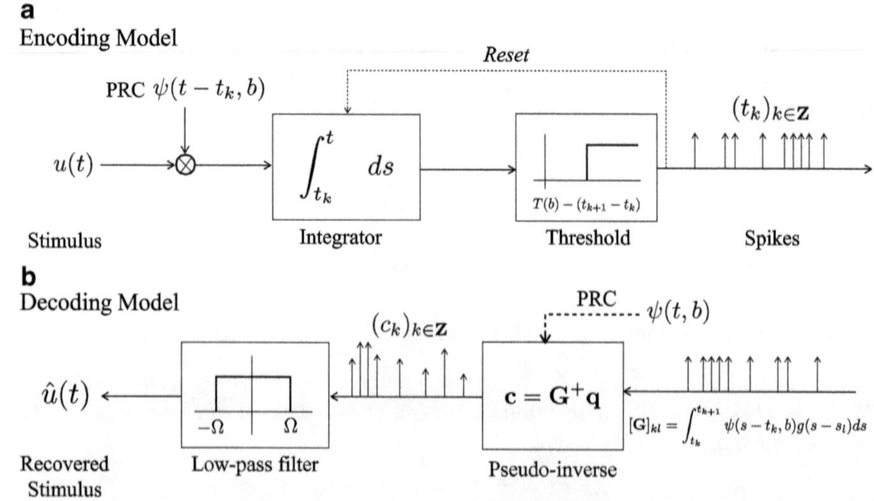

Fig. 11.4 Model of stimulus encoding/decoding using the project-integrate-and-fire neuron with known PRC. (**a**) The stimulus $u(t)$ is multiplied by the PRC determined by the known base current, and a spike is generated when the output of the integrator reaches the threshold value. (**b**) The known PRC is used to construct the **G** matrix, which generates an irregular spikes train, $(c_k), k \in \mathbf{Z}$, whose amplitudes are the solution of a system of linear equations. The stimulus is finally recovered by passing these spikes through a low-pass filter

$$\int_{t_k}^{t_{k+1}} \psi(s-t_k,b_k)v(s)\mathrm{d}s = T(b_k)-(t_{k+1}-t_k), \tag{11.10}$$

where b_k, is the mean stimulus amplitude for the kth inter-spike interval, $k \in \mathbf{Z}$. $\psi(t,b_k), t \in \Re$, for a given $k, k \in \mathbf{Z}$, is called the *conditional PRC* (see Fig. 11.2).

As we shall require the recovered signal to live in a bandlimited space, we will replace in equation (11.10) $v(t)$ with $u(t)-b_k$ on each interspike time interval $[t_k, t_{k+1}]$, and thereby obtain the equivalent t-transform

$$\int_{t_k}^{t_{k+1}} \psi(s-t_k,b_k)u(s)\mathrm{d}s = T(b_k)-(t_{k+1}-t_k)+b_k\int_{t_k}^{t_{k+1}} \psi(s-t_k,b_k)\mathrm{d}s. \tag{11.11}$$

As in the known PRC case, we have the following recovery algorithm.

Theorem 2. *Let* $(t_k), k \in \mathbf{Z}$, *be the sequence of spike times generated by a Hodgkin-Huxley neuron with bandlimited input* $u(t)=v(t)+\sum_{k\in\mathbf{Z}} b_k \mathbf{1}_{[t_k,t_{k+1}]}$. *If the mean stimulus amplitudes* $(b_k)_{k\in\mathbf{Z}}$ *are known, the stimulus can be reconstructed using*

$$\hat{u}(t)=\sum_{k\in\mathbf{Z}} c_k g(t-s_k),$$

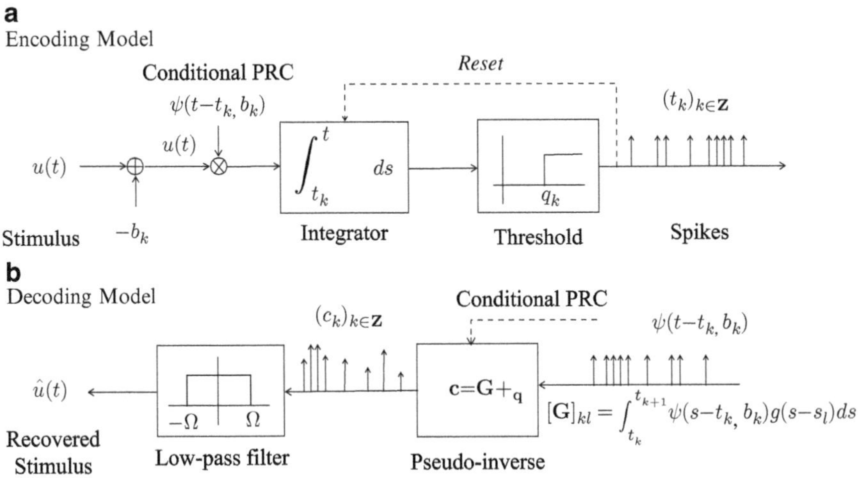

Fig. 11.5 Model of stimulus encoding/decoding using the project-integrate-and-fire neuron with known conditional PRC. (**a**) The stimulus $u(t)$ is weighted by the conditional PRC determined by the base current. (**b**) The sampling kernel (the known conditional PRC) is used to construct the **G** matrix. The stimulus is recovered as the response of a low-pass filter to the observed spike train

where $g(t) = \frac{\sin(\Omega t)}{\pi t}$ *and* $s_k = \frac{t_k + t_{k+1}}{2}$, $t \in \Re, k \in \mathbf{Z}$. *The coefficients* $[\mathbf{c}]_k = c_k$, $k \in \mathbf{Z}$, *are given by*

$$\boldsymbol{c} = \boldsymbol{G}^{+}\boldsymbol{q},$$

where $[\mathbf{G}]_{kl} = \int_{t_k}^{t_{k+1}} \psi(s - t_k, b_k) g(s - s_l) \mathrm{d}s$ *and* $[\boldsymbol{q}]_k = q_k = T(b_k) - (t_{k+1} - t_k) + b_k \int_{t_k}^{t_{k+1}} \psi(s - t_k, b_k) \mathrm{d}s$.

An implementation of the encoding and decoding algorithms described in this section is shown in block diagram form in Fig. 11.5.

Remark 2. Note that in this section the recovered signal is the sum of a perturbation $v(t)$ and a variable base level current, whereas, in the previous section, the recovered signal is only the perturbation signal.

3.3 Recovery of Stimuli Encoded with Unknown Conditional PRCs

As in the previous section, we consider the recovery of a bandlimited stimulus of the form $u(t) = \sum_{k\in\mathbf{Z}} b_k 1_{[t_k, t_{k+1}]} + v(t)$, where b_k is the mean stimulus amplitude during the kth interspike time interval. However, we shall assume here that the b_k's are unknown.

On each interspike time interval $[t_k, t_{k+1}]$

$$\int_{t_k}^{t_{k+1}} \psi(s-t_k, b_k)u(s)\mathrm{d}s = T(b_k) - (t_{k+1} - t_k) + b_k \int_{t_k}^{t_{k+1}} \psi(s-t_k, b_k)\mathrm{d}s. \tag{11.12}$$

Since for large values of b_k,

$$T(b_k) - (t_{k+1} - t_k) \ll b_k \int_{t_k}^{t_{k+1}} \psi(s-t_k, b_k)\mathrm{d}s,$$

equation (11.12) becomes

$$\int_{t_k}^{t_{k+1}} \psi(s-t_k, b_k)u(s)\mathrm{d}s = b_k \int_{t_k}^{t_{k+1}} \psi(s-t_k, b_k)\mathrm{d}s.$$

Normalizing by the right hand side of the above equation,

$$\int_{t_k}^{t_{k+1}} \chi(s-t_k, b_k, t_{k+1}-t_k)u(s)\mathrm{d}s = 1, \tag{11.13}$$

where $\chi(t-t_k, b_k, t_{k+1}-t_k) = \psi(t-t_k, b_k)/(b_k \int_{t_k}^{t_{k+1}} \psi(s-t_k, b_k)\mathrm{d}s)$. Since the value of b_k is unknown, we shall use on each interspike time interval $[t_k, t_{k+1}]$ the estimate $\hat{b}_k = T^{-1}(t_{k+1} - t_k)$ instead. Therefore, by abuse of notation,

$$\chi(t-t_k, t_{k+1}-t_k) = \psi(t-t_k, \hat{b}_k) \Bigg/ \left(\hat{b}_k \int_{t_k}^{t_{k+1}} \psi(s-t_k, \hat{b}_k)\mathrm{d}s\right)$$

with $\hat{b}_k = T^{-1}(t_{k+1} - t_k)$ for $t \in [t_k, t_{k+1}]$ and zero, otherwise. Equation (11.13) is the t-transform of the PIF neuron and $\chi(t-t_k, t_{k+1}-t_k)$ is called the *estimated conditional sampling kernel* (eCSK) (see Fig. 11.6). For the recovery algorithm, the same matrix formulation for the evaluation of the vector of coefficients $\mathbf{c}$ is used, i.e.,

$$\mathbf{c} = \mathbf{G}^{+}\mathbf{q},$$

where $[\mathbf{c}]_l = c_l$, $[\mathbf{G}]_{kl} = \int_{t_k}^{t_{k+1}} \chi(s-t_k, t_{k+1}-t_k)g(s-s_l)\mathrm{d}s$, and $[\mathbf{q}]_k = 1$. Evaluation of the right hand side of (11.8) with the resulting values for $\mathbf{c}$ completes the recovery of $u(t)$ (see also Fig. 11.7b).

Theorem 3. *Let* $(t_k), k \in \mathbf{Z}$, *be the sequence of spike times generated by a Hodgkin–Huxley neuron with bandlimited input* $u(t) = v(t) + \sum_{k\in\mathbf{Z}} b_k 1_{[t_k, t_{k+1}]}$. *If the mean stimulus amplitudes* $b_k, k \in \mathbf{Z}$, *are unknown, the stimulus can be reconstructed using*

$$\hat{u}(t) = \sum_{k\in\mathbf{Z}} c_k g(t-s_k),$$

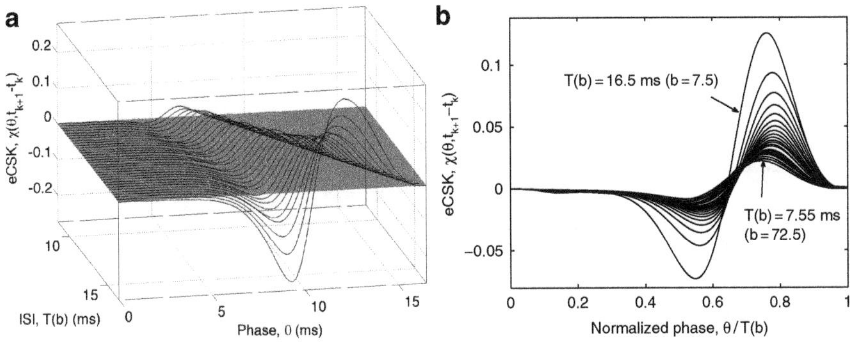

Fig. 11.6 The estimated conditional sampling kernels (eCSKs) for different interspike time intervals. The amplitude of eCSKs are monotonically decreasing with shorter interspike time intervals. (**a**) eCSKs in 3D space for varying interspike time intervals (ISIs). (**b**) 2D projection of (**a**) along the ISI axis

where $g(t) = \frac{\sin(\Omega t)}{\pi t}$ *and the coefficients* $[\boldsymbol{c}]_k = c_k$, $k \in \mathbf{Z}$, *are given by*

$$\boldsymbol{c} = \boldsymbol{G}^{+}\boldsymbol{q},$$

where $[\boldsymbol{G}]_{kl} = \int_{t_k}^{t_{k+1}} \chi(s - t_k, t_{k+1} - t_k) g(s - s_l)\mathrm{d}s$, *and* $[\boldsymbol{q}]_k = 1$. *Finally,*

$$\chi(t - t_k, t_{k+1} - t_k) = \psi(t - t_k, \hat{b}_k) / \left(\hat{b}_k \int_{t_k}^{t_{k+1}} \psi(s - t_k, \hat{b}_k)\mathrm{d}s \right)$$

with $\hat{b}_k = T^{-1}(t_{k+1} - t_k)$ *for* $t \in [t_k, t_{k+1}]$ *and zero, otherwise.*

eCSKs are plotted in Fig. 11.6 for different interspike time intervals. Each eCSK is a scaled version of a conditional PRC with a variable scale factor, and this makes the amplitude ratio of the lowest eCSK ($b = 72.5\,\mu\mathrm{A/cm}^2$ and $T(b) = 7.55\,\mathrm{ms}$) to the tallest eCSK ($b = 7.5\,\mu\mathrm{A/cm}^2$ and $T(b) = 16.5\,\mathrm{ms}$) smaller than that of conditional PRCs.

An implementation of the encoding and decoding algorithms described in this section is shown in block diagram form in Fig. 11.7.

4 Examples

The recovery algorithms for stimuli encoded with a Hodgkin–Huxley neuron were extensively tested. All three proposed algorithms were evaluated with the same set of randomly generated stimuli by visually comparing the sample paths of – as well as by measuring the root-mean-square (RMS) error between – the recovered signal and the original signal.

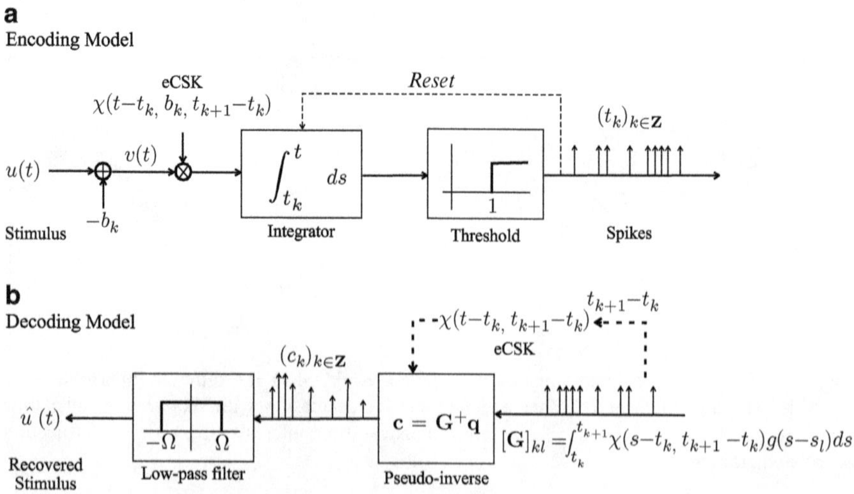

Fig. 11.7 Model of stimulus encoding/decoding using the project-integrate-and-fire neuron with unknown conditional PRC. (**a**) The stimulus $u(t)$ is weighted by the estimated conditional sampling kernel (eCSK) determined by an estimate of the base current. (**b**) The eCSK is used to construct the **G** matrix. The stimulus is recovered as the response of a low-pass filter to the weighted spike train

The input stimulus is a bandlimited signal generated using (11.6) with $\Omega = 2\pi \cdot 20$ Hz and $b = 25\,\mu\text{A/cm}^2$. In order to test the recovery algorithms with various waveform patterns, the samples $(u(k\pi/\Omega)), k \in \mathbf{Z}$, were generated from a uniform distribution between $[-1, 1]$. Thus, the parameter c in equation (11.6) determines the magnitude of the perturbation.

Our simulation results are depicted in Fig. 11.8. In each plot in Fig. 11.8, both the original stimulus and the recovered stimulus as a function of time are shown. The three rows show the sample paths of the recovery results of the three algorithms, i.e., for stimulus encoding with a known PRC, a known conditional PRC, and an unknown conditional PRC. In the left column a weak stimulus with $c = 0.5\,\mu\text{A/cm}^2$ was employed. In the right column, a strong stimulus was used with $c = 15\,\mu\text{A/cm}^2$.

The three algorithms show different levels of error recovery for the weak and the strong stimuli, respectively. Algorithm 1, based on Theorem 1, can recover a weak perturbation with a low recovery error (Fig. 11.8a). For strong perturbations, however, the error is markedly larger (Fig. 11.8b). Algorithm 3, based on Theorem 3, exhibits a low recovery error for strong stimuli (Fig. 11.8f) but fails to recover weak perturbations (Fig. 11.8e). Finally Algorithm 2, based on Theorem 2, displays median performance for both stimuli.

The differences in performance of the three recovery algorithms can be traced back to the assumptions about the three sampling kernels: known PRC, known conditional PRC, and unknown conditional PRC, respectively. The three sampling

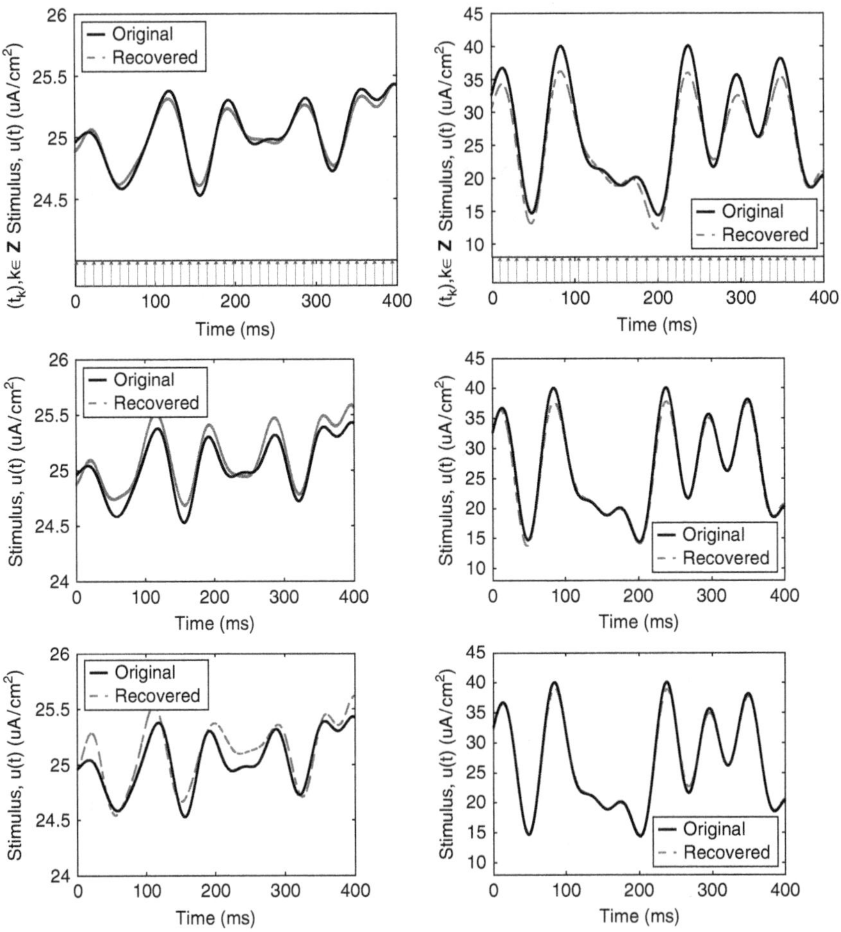

Fig. 11.8 Recovery of weak (*left column*) and strong (*right column*) stimuli. (**a**, **b**) Recovery of a weak stimulus ($c = 0.5\,\mu\text{A/cm}^2$) and of a strong stimulus ($c = 15\,\mu\text{A/cm}^2$), respectively, encoded with a known PRC. (**c**, **d**) Recovery of the weak and the strong stimulus, respectively, encoded with known conditional PRC. (**e**, **f**) Recovery of the weak and the strong stimulus, respectively, encoded with an unknown conditional PRC

kernels encoding the strong stimuli in Fig. 11.8b,d,f are shown in Fig. 11.9. The dashed vertical lines in Fig. 11.9 designate the spike times of the Hodgkin–Huxley neuron. Note that the sampling kernels act independently during each interspike time interval. In Fig. 11.9 they were stitched together at spike times and, consequently, appear as continuous-time signals. The base current $b = 25\,\mu\text{A/cm}^2$ was chosen throughout. While the overall pattern is quite similar, minute differences are observed in Fig. 11.9b,c (marked by arrows). The estimation of the base current level for a weak and a strong stimulus is shown in Fig. 11.10 and 11.11, respectively.

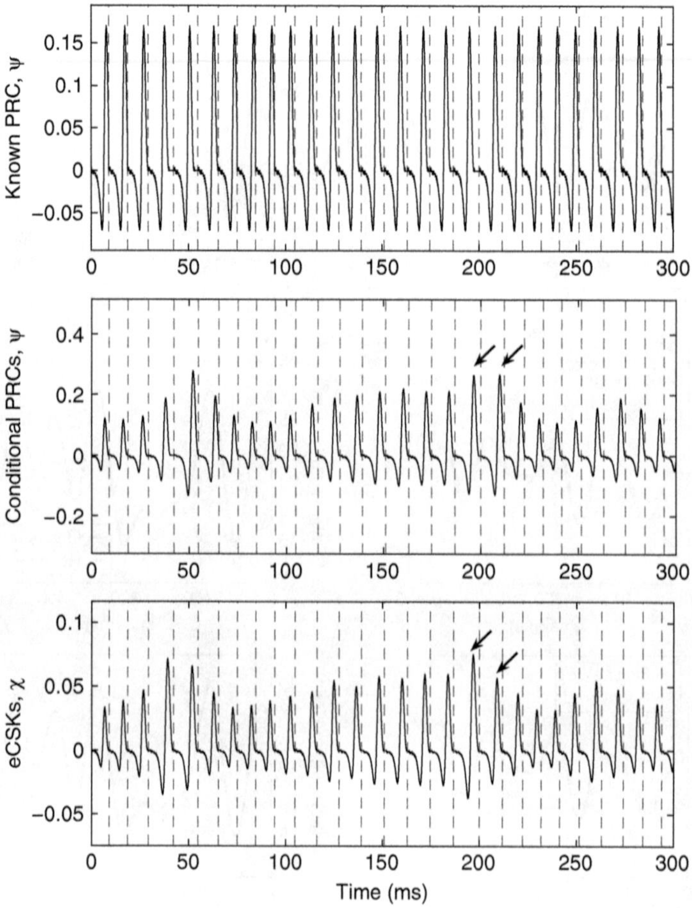

Fig. 11.9 Train of sampling kernels for the Hodgkin–Huxley neuron. (**a**) Train of sampling kernels for known PRC. (**b**) Train of sampling kernels for known conditional PRC. (**c**) Train of estimated conditional sampling kernels for unknown conditional PRC. The envelope of the sampling kernels in (**b**) and (**c**) are similar, but small differences are observed as well (*arrows*)

An extensive simulation of the three recovery algorithms was carried out in response to randomly generated signals of various magnitudes. Each stimulus was generated as in Fig. 11.8, and c was varied from $0.1\,\mu\text{A/cm}$ to $25\,\mu\text{A/cm}^2$ with $b = 25\,\mu\text{A/cm}^2$ and $\Omega = 2\pi \cdot 20$ Hz. 30 trials were run for each value of c. The RMS error defined as

$$\sqrt{\frac{1}{S}\int_0^S [u(s) - \hat{u}(s)]^2\,\mathrm{d}s},$$

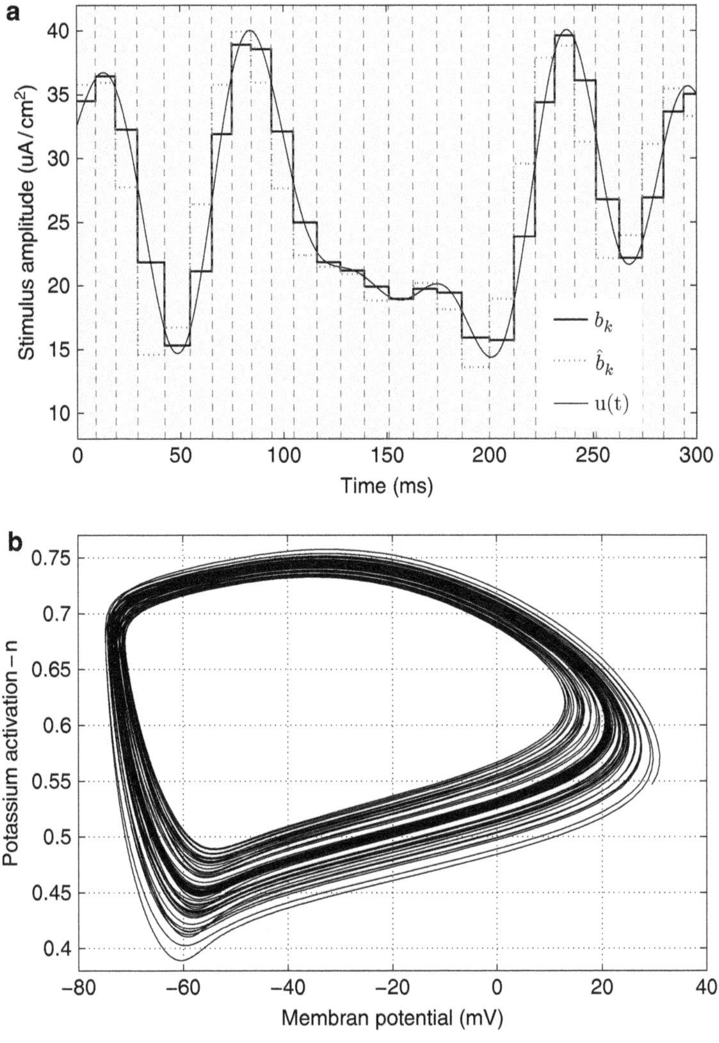

Fig. 11.10 The bandlimited stimulus and the base current for a strong stimulus. (**a**) The base current, the estimated base current and the stimulus $u(t)$ are in a close range. (**b**) Characterization of the stimulus in the phase plane. A large number of limit cycles corresponds to a strong stimulus

where $\hat{u}(t)$ is the recovered stimulus and S is the time interval of the simulation (400 ms), was used to characterize the distance between the stimulus and its estimate. In Fig. 11.12, the recovery errors of three different algorithms are compared with respect to the perturbation magnitude c. When the perturbation magnitude is weak (Fig. 11.12b), recovery Algorithm 1 and Algorithm 2 exhibit lower RMS than Algorithm 3. In the derivation of the Algorithm 3 in Sect. 3.3, we

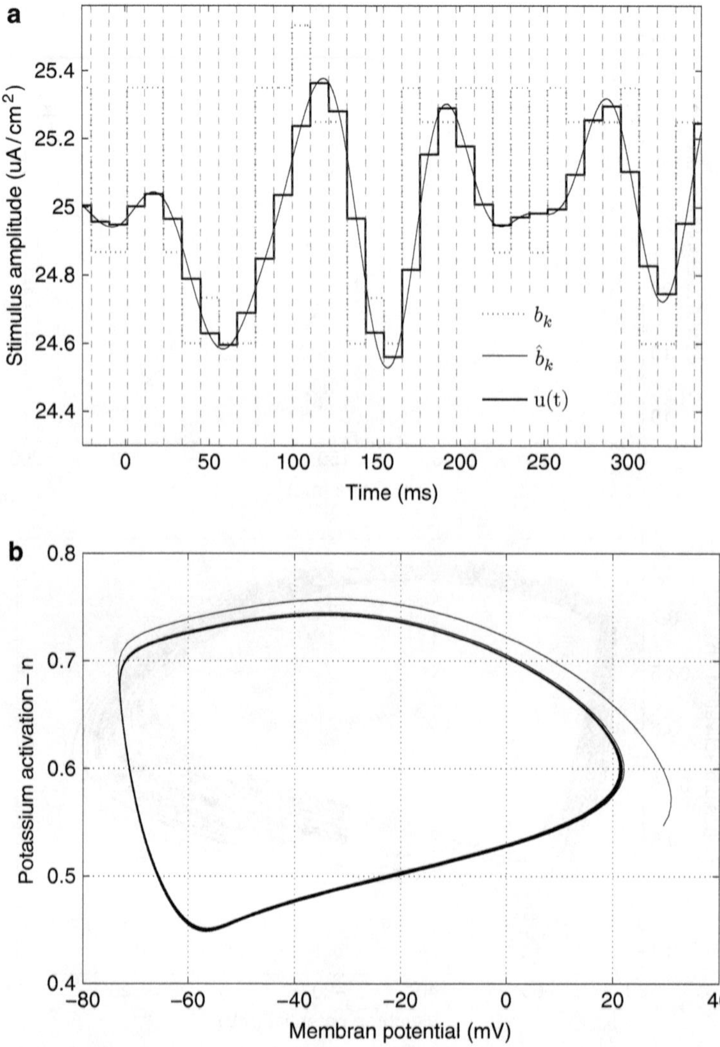

Fig. 11.11 The bandlimited stimulus and the base current for a weak stimulus. (**a**) The base current and the stimulus $u(t)$ are in a close range. The estimated base current does not faithfully estimate the base current. (**b**) Characterization of the stimulus in the phase plane. A small number of bundled limit cycles corresponds to a weak stimulus

essentially assumed that $T(b_k) - t_{k+1} - t_k = 0$. This assumption does not hold for weak perturbations (say, below $2\,\mu\text{A}/\text{cm}^2$ in Fig. 11.12b).

The RMS of Algorithm 1 increases rapidly with the perturbation magnitude and the associated changes of the phase portrait of the Hodgkin–Huxley neuron. The performance of Algorithm 2 matches the one of Algorithm 1 for weak stimuli and the one of Algorithm 3 for strong stimuli.

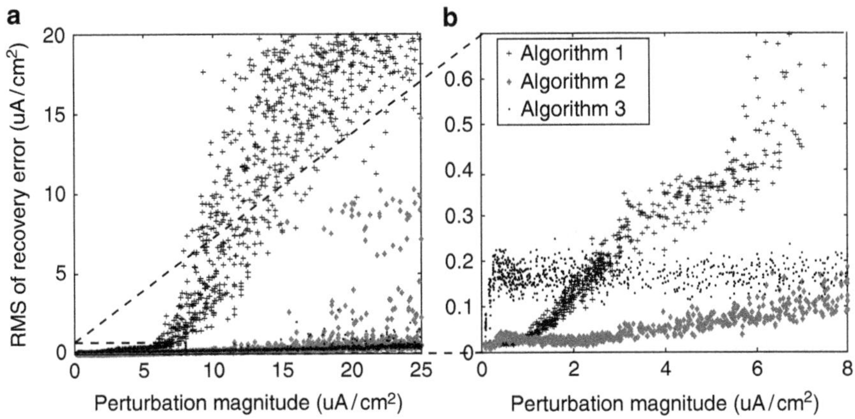

Fig. 11.12 Recovery errors of the three different algorithms. (**b**) zooms into the dashed box in (**a**). Algorithm 1 rapidly increases its recovery error as the perturbation magnitude increases; it shows the lowest level of error when the perturbation magnitude is below 1 $\mu A/cm^2$. Algorithm 2 shows the lowest level of errors on average across the weak/strong perturbation range. The recovery results with Algorithm 3 show the highest error level for weak perturbation levels; beyond 8 μA/cm^2 the least level of error is recorded

5 Discussion

We investigated the recovery of stimuli encoded with a Hodgkin–Huxley neuron. The daunting complexity of the Hodgkin–Huxley equations makes this task particularly challenging. A highly effective strategy pioneered in (Lazar 2007, 2010) calls for reducing the Hodgkin–Huxley neuron to a one-dimensional PIF neuron. The latter projects the input stimulus onto the phase response curve and compares the result with a threshold value. The PIF model is I/O-equivalent with the Hodgkin–Huxley neuron; given the same input and the same initial condition, the two neurons generate, to the first order, the same spike trains.

Using the reduced PIF model, we presented three different recovery algorithms for stimuli encoded with a Hodgkin–Huxley neuron. All algorithms reconstruct the stimuli from the neuron's output spike train.

The first algorithm is based on the assumption that the Hodgkin–Huxley neuron has a known PRC. It performs well for weak stimuli. However, if the base level of the stimulus changes, the recovery algorithm rapidly increases its error. The second algorithm assumes that the PRC is known on each interspike time interval. Since the base level is known, the sampling kernels of the t-transform that best describe the Hodgkin–Huxley neuron are conditionally known on each interspike time interval. The algorithm recovers both the strong drift of the base level and the weak perturbation of the stimulus. Finally, the third algorithm operates under the assumption that the conditional PRC is unknown and has to be estimated. We established an estimate of the conditional PRC based upon the readily observable inter-spike time interval. The third recovery algorithm shows excellent performance for strong base level signals.

The proposed algorithms are easy to implement and are very flexible; they only require the inversion of a matrix. These algorithms can also be used for the recovery of stimuli encoded by a large class of spiking neuron models in the limit cycle region. How the dendritic tree of a postsynaptic neuron recovers the information contained in the incoming spike sequence remains an open problem.

Appendix

The Hodgkin–Huxley Equations

The Hodgkin–Huxley equations are as follows. They differ from the original formulation in that the membrane potential is shifted by −65 mV.

$$
\begin{aligned}
C\frac{\mathrm{d}V}{\mathrm{d}st} &= -\left[\bar{g}_{Na}m^3h(V-E_{Na}) + \bar{g}_K n^4(V-E_K) + \bar{g}_L(V-E_L)\right] + I_{\text{ext}} \\
\frac{\mathrm{d}sn}{\mathrm{d}st} &= \alpha_n(V)(1-n) - \beta_n(V)n \\
\frac{\mathrm{d}sh}{\mathrm{d}st} &= \alpha_h(V)(1-h) - \beta_h(V)h \\
\frac{\mathrm{d}sm}{\mathrm{d}st} &= \alpha_m(V)(1-m) - \beta_m(V)m,
\end{aligned}
$$

where

$$
\begin{aligned}
\alpha_n(V) &= -0.01(V+55)/\left(\mathrm{e}^{-(V+55)/10}-1\right) \\
\beta_n(V) &= 0.125\mathrm{e}^{-(V+65)/80} \\
\alpha_h(V) &= 0.07\mathrm{e}^{-(V+65)/20} \\
\beta_h(V) &= 1/\left(\mathrm{e}^{-(V+35)/10)}+1\right) \\
\alpha_m(V) &= -0.1(V+40)/\left(\mathrm{e}^{-(V+40)/10}-1\right) \\
\beta_m(V) &= 4\mathrm{e}^{-(V+65)/18}
\end{aligned}
$$

and

$$
\begin{array}{lll}
\bar{g}_{Na} = 120\,\mathrm{mS/cm^2} & \bar{g}_K = 36\,\mathrm{mS/cm^2} & \bar{g}_L = 0.3\,\mathrm{mS/cm^2} \\
E_{Na} = 50\,\mathrm{mV} & E_K = -77\,\mathrm{mV} & E_L = -54.387\,\mathrm{mV} \\
C = 1\,\mathrm{uF/cm^2}. & &
\end{array}
$$

References

Arcas, B. A. y, Fairhall, A. L., & Bialek,W. (2003). Computation in a single neuron: Hodgkin and Huxley revisited. *Neural Computation, 15(8)*, 1715–49.

Bialek, W., & Rieke, F. (1991). Reading a neural code. *Science, 252*(5014), 1854–1857.

Dayan, P., & Abbott, L. (2001). *Theoretical Neuroscience: Computational and Mathematical Modeling of Neural Systems*. MIT Press.

Ermentrout, B. (1996). Type I membranes, phase resetting curves, and synchrony. *Neural Computation, 8*(5), 979–1001.

Hastings, J., & Sweeney, B. (1958). A Persistent Diurnal Rhythm of Luminescence in Gonyaulax Polyedra. *The Biological Bulletin, 115*(3), 440–458.

Hodgkin, A. L., & Huxley, A. F. (1952). A Quantitative Description of Membrane Current and its Application to Conduction and Excitation. *J. Physiol, 117*, 500–557.

Hoppensteadt, F., & Izhikevich, E. (1997). *Weakly Connected Neural Networks*. Springer.

Izhikevich, E. M. (2007). *Dynamical Systems in Neuroscience: The Geometry of Excitability and Bursting*. The MIT Press.

Kim, A. J., Lazar, A. A., & Slutskiy, Y. B. (2010, Aug). System identification of drosophila olfactory sensory neurons. *Journal of Computational Neuroscience*. (published online doi:10.1007/s10827-010- 0265-0)

Kuramoto, R. (1984). *Chemical Oscillations, Waves, and Turbulence*. Springer- Verlag.

Lazar, A. A. (2004). Time Encoding with an Integrate-and-Fire Neuron with a Refractory Period. *Neurocomputing, 58–60*, 53–58.

Lazar, A. A. (2007). Recovery of Stimuli Encoded with Hodgkin-Huxley Neurons. In *Cosyne'07* (p. III-94).

Lazar, A. A. (2010). Population encoding with hodgkin-huxley neurons. *IEEE Transactions on Information Theory*, 56(2), 821–837. Special Issue on Molecular Biology and Neuroscience.

Lazar, A. A., & Pnevmatikakis, E. A. (2008). Faithful Representation of Stimuli with a Population of Integrate-and-Fire Neurons. *Neural Computation, 20*(11), 2715–2744.

Lazar, A. A., & Tóth, L. T. (2004). Perfect Recovery and Sensitivity Analysis of Time Encoded Bandlimited Signals. *Circuits and Systems I: Regular Paper, IEEE Transactions on, 51*(10), 2060–2073.

Medina, J. F., & Lisberger, S. G. (2007). Variation, Signal, and Noise in Cerebellar Sensory-Motor Processing for Smooth-Pursuit Eye Movements. *Journal of Neuroscience, 27*(25), 6832.

Naka, K. I., Chan, R. Y., & Yasui, S. (1979). Adaptation in catfish retina. *Journal of Neurophysiology, 42*(2), 441–54.

Ota, K., Omori, T., & Aonishi, T. (2009). MAP Estimation Algorithm for Phase Response Curves based on Analysis of the Observation Process. *Journal of Computational Neuroscience, 26*(2), 185–202.

Penrose, R. (1955). A Generalized Inverse for Matrices. In *Proceedings of the Cambridge Philosophical Society* (Vol. 51, pp. 406–413).

Perkel, D. H., Gerstein, G. L., & Moore, G. P. (1967). Neuronal spike trains and stochastic point processes. I. The single spike train. *Biophysical Journal, 7*(4), 391–418.

Schuetze, S. M. (1983). The discovery of the action potential. *Trends in Neurosciences, 6*, 164–168.

Winfree, A. (1967). Biological Rhythms and the Behavior of Populations of Coupled Oscillators. *Journal of Theoretical Biology, 16*(1), 15–42.

Yoshimura, K., & Arai, K. (2008, Jan). Phase Reduction of Stochastic Limit Cycle Oscillators. *Physical Review Letters, 101*, 15401-1–15401-4.

Chapter 12
Cholinergic Neuromodulation Controls PRC Type in Cortical Pyramidal Neurons

Klaus M. Stiefel and Boris S. Gutkin

Abstract The phase response curve (PRC) reflects the dynamics of the interplay between diverse intrinsic conductances that lead to spike generation. PRCs measure the spike time shift caused by perturbations of the membrane potential as a function of the phase of the spike cycle of a neuron. A purely positive PRC is a signature of type I (saddle-node) dynamics while type II (subcritical Hopf dynamics) yield a biphasic PRC with both negative and positive lobes. Previous computational work hypothesized that cholinergic modulation of M-type potassium current can switch a neuron with type II dynamics to type I dynamics. We recorded from layer 2/3 pyramidal neurons in cortical slices, and found that cholinergic action, consistent with downregulation of slow voltage-dependent potassium currents such as the M-current, indeed changed the PRC from type II to type I. We then explored the potential specific K-current-dependent mechanisms for this switch using a series of computational models. In all of these models, we show that a decrease in spike-frequency adaptation due to downregulation of the M-current is associated with the switch in PRC type. Interestingly spike-dependent I-AHP is downregulated at lower Ach concentrations than the M-current. Our simulations showed that type II nature of the PRC is amplified by low Ach level, while the PRC became type I at high Ach concentrations. We further explored the spatial aspects of Ach modulation in a compartmental model. This work suggests that cholinergic modulation of slow potassium currents may shape neuronal responding between "resonator" to "integrator."

K.M. Stiefel (✉)
Theoretical and Experimental Neurobiology Unit, Okinawa Institute of Science and Technology, Okinawa, Japan
e-mail: stiefel@oist.jp

B.S. Gutkin
Group for Neural Theory, Laboratoire de Neuroscience Cognitive, INSERM U960, Paris, France

Departement des Etudes Cognitives, ENS, 75005 Paris, France

CNRS, Campus Gérard-Mégie, 3 rue Michel-Ange - F-75794 Paris cedex 16, France
e-mail: boris.gutkin@ens.fr

N.W. Schultheiss et al. (eds.), *Phase Response Curves in Neuroscience: Theory, Experiment, and Analysis*, Springer Series in Computational Neuroscience 6,
DOI 10.1007/978-1-4614-0739-3_12,

1 Introduction

The ability of neurons to synchronize their spiking activity depends on their mutual synaptic connectivity and their intrinsic properties. The intrinsic properties can be revealed by the phase response curve (PRC, (Ermentrout 1996)), defined as the spike time shift caused by a small perturbation of the membrane potential as a function of the time of the perturbation during the spike cycle. The PRC allows neurons to be grouped into two fundamentally different classes with regard to their spiking behavior (Ermentrout 1996) that has implications for the stability of synchronous oscillations in networks.

In neurons with a purely positive, or *type I*, PRC, perturbations at every phase of the oscillatory cycle cause an acceleration of the next spike, or spike advance. In such neurons, the transition from rest to tonic spiking comes about by a *saddle-node bifurcation* (Ermentrout & Rinzel 1998; Izhikevich 2007). The resulting dynamics imply a general absence of subthreshold oscillations and the onset frequency of firing is arbitrarily low. The firing frequency – injected current relationship (the FI-curve) is nearly linear at higher frequencies; the action potentials are all-or-none and well separated from subthreshold responses. No bistability is apparent for this type of membrane excitability. The majority of biophysical models of cortical pyramidal neurons has type I excitability (Ermentrout 1996) (Ermentrout et al. 2001) (Hansel & Mato 1995).

In neurons with a biphasic, or *type II*, PRC, perturbations at early phase positions in the firing cycle cause a delay, whereas late perturbations cause an acceleration, of the next spike. These neurons show subthreshold oscillations and a finite onset frequency of firing. The transition from rest to tonic spiking occurs at a *subcritical Hopf bifurcation* (Rinzel and Ermentrout 1998; Izhikevich 2007). The firing FI-curve has a discontinuity, thus being strongly nonlinear, and the action potential amplitude is graded. For certain parameter regimes bistability between a quiescent and a sustained firing behavior can appear. The Hodgkin–Huxley equations with the original parameters describing spiking in the squid giant axon (Hodgkin and Huxley 1952) are a classical example of type II.

The shape of the PRC depends on the shape and amplitude of the perturbation as well as on the intrinsic properties of the neuron (or the oscillator more generally as the PRC can be defined for any oscillator not just neuronal). When the amplitude of the perturbation becomes infinitesimally small, the PRC converges to the infinitesimal PRC (iPRC), which can be derived numerically for all, and analytically for some, types of oscillators (Ermentrout & Kopell 1998; Izhikevich 2007). The PRC of a neuron is an important quantity from a theoretical point of view as it provides insight into the neurons synchronization behavior when interacting with other neurons. For example, excitatory coupling tends to desynchronize networks of neurons with a positive PRC (Ermentrout 1996), and synchronize neurons that have significant negative regions or strong right skew in the PRC. (Ermentrout et al. 2001; Gutkin et al. 2005). Hence, the PRC links the biophysics of spike generation in neurons and their behavior in large networks.

Over the last several years, neuronal PRCs have been measured experimentally in a number of cortical in vitro preparations. Probably the earliest such measurement was by Reyes and Fetz (1993a,b) who measured PRCs in layer V cortical pyramidal neurons. The authors found that these neurons had a characteristically skewed shape and appeared to be mostly type I (no negative portions in the PRC). It is interesting to note that Reyes and Fetz (1993a,b) did their measurements without knowing the implications of the PRC for theoretical work. More recently, Netoff et al. (2005) identified PRCs in stellate cells in the entorhinal cortex using the dynamic clamp technique; their results show PRCs qualitatively similar to those found by Reyes and Fetz (1993a,b) but with a small negative component at the beginning of the firing cycle. Galan et al. (2005) measured PRCs from mitral cells in the olfactory bulb and found PRCs with clear negative portion at the beginning of the cells firing cycle. Further recent data show that certain CA3 pyramidal neurons appear to be type II (Lengyel et al. 2005). In vitro recordings in the somatosensory cortex indicate that fast-spiking interneurons are of type II (Tateno et al. 2004). Thus, motor-cortical neurons appear to be of type I, entorhinal neurons are "weakly" type II, olfactory bulb neurons, fast-spiking interneuron and CA3 pyramids are solidly type II. A recent in vitro study showed that pyramidal neurons in layer V are predominantly type I, while layer 2/3 pyramids are predominantly type II (Tsubo et al. 2007) depending on the basal firing rate of the neuron as predicted in Gutkin et al. (2005). These earlier studies, with the notable exception of the last, have not asked the question of what intrinsic cellular and functional mechanisms might control the change between the two PRC classes in a given cell or in mathematical terms what might cause a transition from one bifurcation class to another.

As we noted above, the PRC of a neuron is determined by its intrinsic excitability properties. When such properties are modified, by changes in the various biophysical parameters of the neuronal membrane (e.g., see Prescott et al. 2008) the PRC shape can change. A prominent example of such is the increase in levels of slow potassium currents that underlie spike-frequency adaptation, that can convert neurons from type I to type II. One such current is the muscarine sensitive I_M that activates at voltages below the threshold. Classically, I_M is considered to play a role in spike-frequency adaptation as well as control of the resting potential of the neuron (Cox et al. 1994). Theoretical studies have also shown that I_M may affect the dynamics of spike generation, the shape of the firing rate transfer function of the neuron, and the dynamics of repetitive oscillatory firing (Ermentrout 1998; Ermentrout et al. 2001). Furthermore, modeling studies (Crook et al. 1998a,b; Ermentrout et al. 2001) suggest that cholinergic neuromodulation, acting through muscarinic Ach-receptors to downregulate I_M, can affect the mechanisms leading to synchronous and asynchronous oscillations. It was shown that reduction of I_M changes the shape of the neuronal PRC in such a way as to decrease the possibility of synchronization with excitation but favor synchrony with inhibition (Ermentrout et al. 2001). In fact, a parametric decrease of I_M strength in a reduced model (the theta neuron) was shown to (1) decrease the skew of the PRC and (2) convert type II PRCs to type I (Gutkin et al. 2005). Taken together, previouswork suggests that modulating I_M has a significant effect on the intrinsic spike-generating properties of cortical neurons with possible important consequences on both cortical network dynamics and function.

Focusing on modulation of I_M and its effects on the PRC is particularly functionally relevant since slow potassium channels are under tight control by intrinsic neuromodulators notably acetylcholine (Ach). In general acetylcholine, a central nervous system neuromodulator, plays an important role in several aspects of cognitive function. For example, pharmacological manipulations of the cholinergic system have a significant effect on cognitive abilities (e.g., Bartus et al. 1982). Cholinergic modulation of the cortex, due to inputs emanating from several subcortical nuclei, have strong effects on attention and vigilance (Everitt and Robbins 1997), memory and learning as well as arousal levels (McCormick 1990, Steriade & Buzsáki 1990, Hasselmo 2006). The level of acetylcholine is elevated during awake, vigilant states, and it is associated with a globally desynchronized EEG, increased synaptic plasticity (Brocher et al. 1992, Stiefel et al. 2005) and increased power in higher frequency bands (Steriade 2004). At the same time, cholinergic modulation appears to control the frequency content and the coherence of cortical oscillations (Steriade 2004). For example, gamma-band synchronized oscillations are observed under increased cholinergic levels (Metherate et al. 1992) whereas in vitro, cholinergic antagonists appear to lead to synchronized high-frequency oscillatory activity. More specifically, such oscillations and synchrony have been tied to actions of Ach through muscarinic receptors coupled to various cation channels (McCormick 1993). Such currents, and in particular of the muscarine sensitive K-current (I_M) control cellular excitability in cortical pyramidal neurons. Ach acts through muscarinic metabotropic receptors to decrease conductance of the I_M (as well as a number of other K-currents).

In our experiments, we corroborated theoretical predictions and explicitly demonstrated a change in PRC type due to cholinergic modulation. While it is not practical to use Ach directly in experiments, a muscarinic Ach agonist carbachol is readily used in in vitro preparations to mimic the Ach action. Hence, we decided to test the predictions about the I_M modulation effects on spike-generating dynamics in vitro by measuring muscarinic effects on the shape of the phase response curves of cortical pyramidal neurons using carbachol. We then used computational models to corroborate our experimental observations with multiple effects of carbachol and to show their compatibility with the I_M reduction hypothesis.

2 Electrophysiological Experiments to Determine the PRC and its Modulation by Carbachol

We recorded from layer II/III pyramidal neurons in the slices of mouse visual cortex. We followed standard in vitro slice preparation procedures and these are described in full detail in Stiefel, Gutkin and Sejnowski (2008).

2.1 Determining the PRCs from the Recordings

We determined the spike times by setting a putative spike threshold of −20 mV and searching for upward crossings of the voltage trace. From these spike times (t_n), the interspike intervals (ISIs) were determined as the difference between t_n and t_{n+1}. In order to determine the PRC, we followed the following procedure that is illustrated in Fig. 12.1. First, we identified the ISI that included the perturbing pulse. Then, we calculated the average duration of the 10 ISIs that preceded the tagged ISI. These 10 ISIs did not contain any perturbations; hence, we used this average ($<\mathrm{ISI}_{10}>$) as the control ISI duration. We then determined the change of the ISI length (Δt ISI) for the ISIs containing a perturbing pulse by subtracting its length from this average. We determined the PRC by plotting Δt ISI/$<\mathrm{ISI}_{10}>$ of all ISIs containing a perturbing pulse as a function of the relative time of the pulse, with ISI duration normalized to 0 to 2π (Fig. 12.1c). This gave the PRC values in terms of a phase. We then binned the values for Δt ISI/$<\mathrm{ISI}_{10}>$, according to the relative pulse time (10 bins/2π).

We should note that the ISI tended to vary within a neuron (Fig. 12.1a,b), thus yielding a significantly noisy PRCs. We were a bit concerned if such noisy PRCs indeed were decent estimates of the PRC that would be defined in a neuron without noise. Our simulations (Stiefel et al. 2009) as well as theoretical work by others (e.g., Galan et al. 2005, Ermentrout and Saunders 2006) showed that noisy PRCs can give a good estimate of the underlying noiseless PRC.

2.2 Main Experimental Results

We recorded from 12 cortical layer II/III pyramidal neurons and determined their PRCs. On average, 2020 ISIs without and 570 ISIs with perturbing current pulses were used for the determination of the PRCs. Of these neurons, 5 initially had type I (purely positive) and 7 type II (biphasic) PRC. An example of individual cell results is given in Fig. 12.2.

Out of 7 type II neurons, 4 changed to type I after bath application of the cholinergic agonist carbachol (see Fig. 12.2a). In these 4 neurons, carbachol increased the PRC at early phases (bins $1/5\pi$ to $3/5\pi$) and decreased it at late phases ($>9/5\pi$) (Figs. 12.2a and 12.3a for averaged data). In the other 3 neurons, carbachol on average decreased the PRC at most phase angles ($1/5\pi$ to $9/5\pi$) and never increased the PRC (Fig. 12.3b and see Fig. 12.2b for a representative cell example). Thus, two groups of initially type II neurons expressed changes of different polarity in response to carbachol.

The PRC of a neuron depends on a large number of factors, including the stimulus amplitude; possibly the conductance loads (Prescott et al. 2008) and the neurons firing rate (Gutkin et al. 2005). When the firing frequency is increased, the PRC of a neuron can shift from type II to type I. This occurs when the time

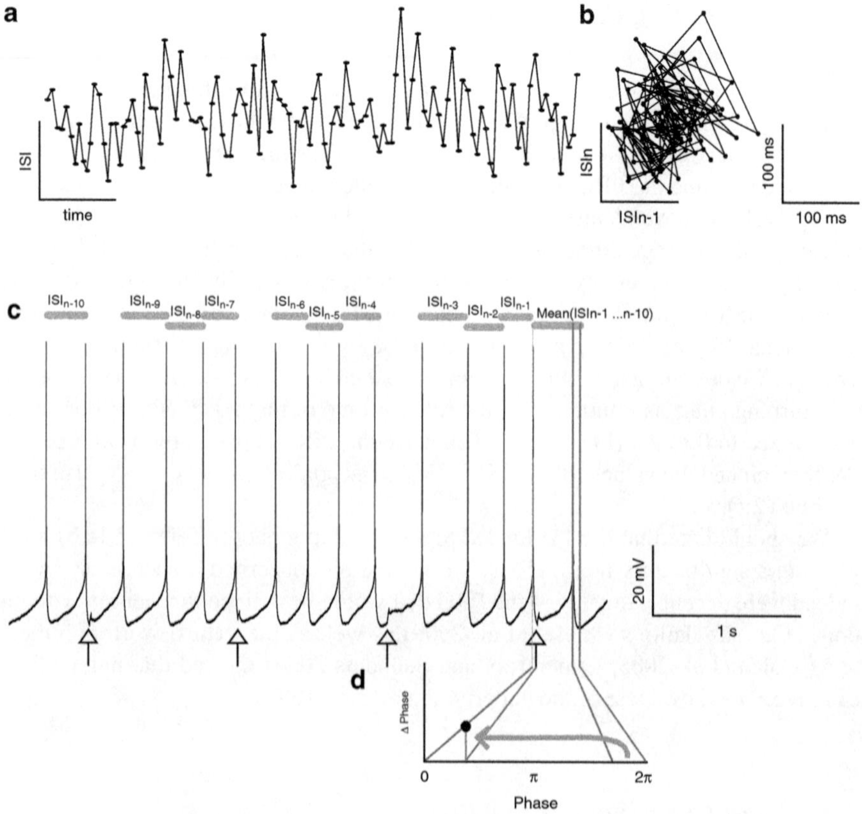

Fig. 12.1 Determination of PRCs from the perturbation of tonically spiking neurons. (**a**) ISIs as a function of time and (**b**) return map (x-axis: ISI_n y-axis: ISI_{n+1}). The lack of a geometrical structure points to the absence of a higher-order periodicity giving rise to the ISI fluctuations. (**c**) Representative spike train with perturbations (marked by *arrows*). The ISI_n with the perturbing pulse (*arrow*) was subtracted from the mean of the previous 10 ISIs ($\mathrm{ISI}_{n-1}\ldots\mathrm{ISI}_{n-10}$), excluding other ISIs containing pulses. (**d**) This value divided by the mean ISI was plotted against the relative time of the pulse after the spike. We note that the time was normalized to give the phase, i.e., duration of the control ISI as 2π. Figure modified from Stiefel et al. 2008

constants of the adaptation current responsible for type II behavior became too slow in relation to the spiking dynamics (Gutkin et al. 2005). We thus had to assure that the observed switch in PRC type was not due to a higher firing frequency during the PRC determination under carbachol (2.6 vs. 4.9 Hz of the ISIs with a perturbation). In order to do so, we needed to restrict our PRC computation only to data where the firing rate was not significantly increased by carbachol. Hence, we repeated the analysis including only ISIs ranging from 2.5 (400 ms) to 5 Hz (200 ms), or in other words largely constant regardless of carbachol conditions. The average firing frequencies of the ISIs included in this analysis was 3.8/4.1 Hz, and the qualitative switch of PRC type was still observed (see Stiefel et al. 2008).

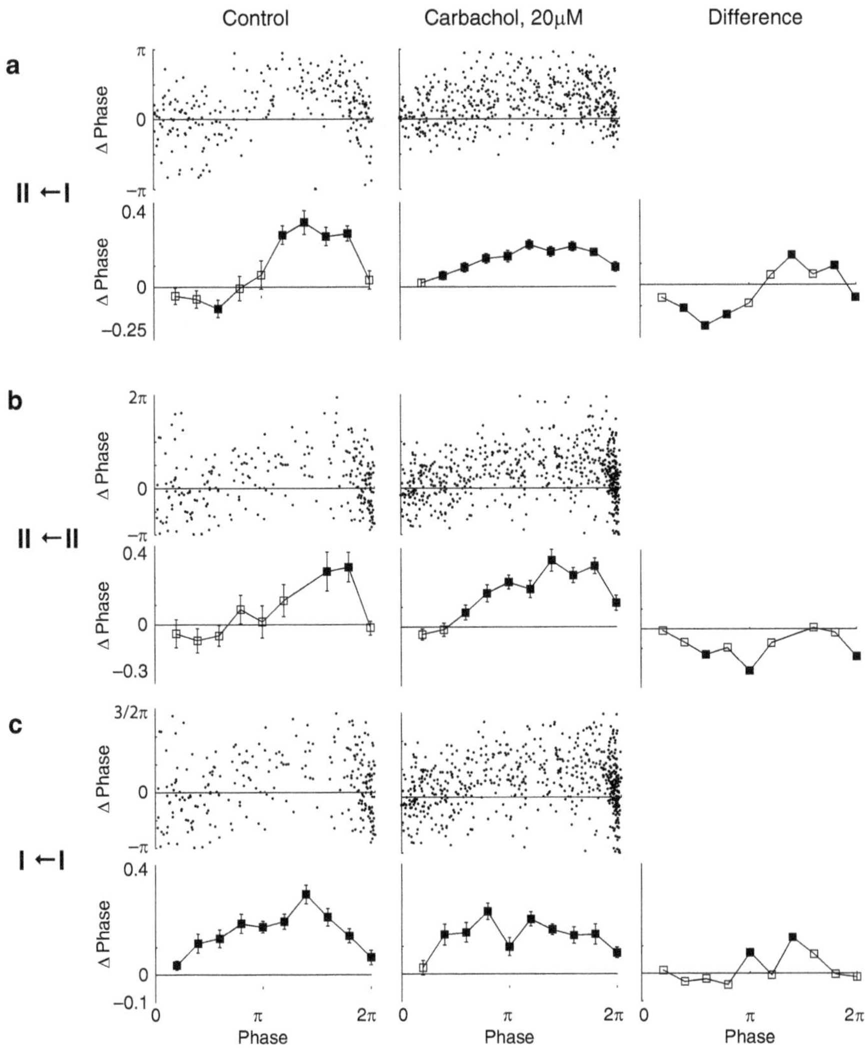

Fig. 12.2 Change of PRCs in individual neurons due to cholinergic neuromodulation. *Left column*: control conditions. Example cells showing a transition from a type II to a type I PRC (**a**) or remaining with a type II (**b**) or type I PRC (**c**). *Center column*: bath application of carbachol. *Right column*: difference. *Top plots*: raw data. *Bottom plots*: mean ± s.e.m. of binned data. *Filled symbols*: significant changes in comparison to the ISIs without a perturbation. Figure modified from Stiefel et al. 2008

Carbachol did not change the PRC type in any of the 5 type I neurons (see Fig. 12.2c for a representative example and Fig. 12.3c for averages). The increases ($>9/5\pi$) and decreases ($1/5\pi$ to $2/5\pi$, $4/5\pi$ to π, $6/5\pi$ to $7/5\pi$) of the PRC as a result of carbachol application did not show a phase-dependent trend. We do

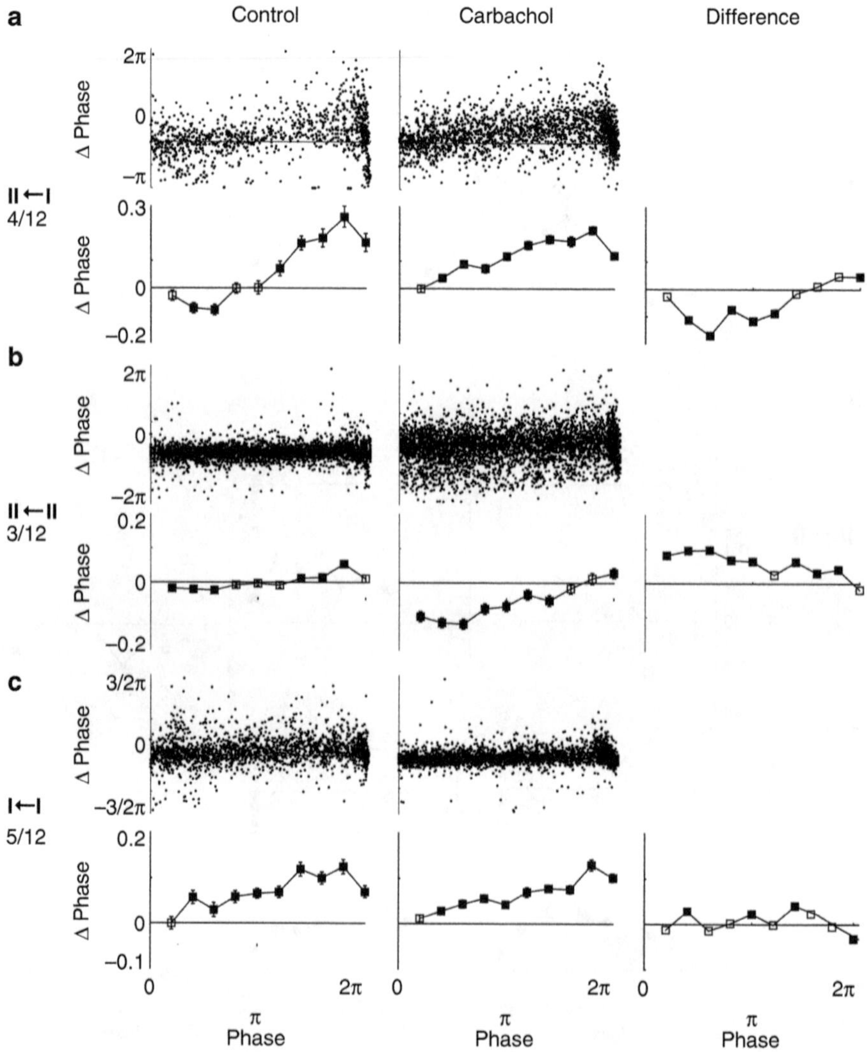

Fig. 12.3 Averages of all cells showing a transition from a type II to a type I PRC (**a**) or remaining with a type II (**b**) or type I PRC (**c**). Figure conventions as in Fig. 12.2. Figure modified from Stiefel et al. 2008

note though that visually the PRCs in such neurons did tend to have less skew with carbachol (change not statistically significant). This kind of shift is consistent with the theoretical predictions in Gutkin et al. (2005).

The application of carbachol changed the input resistance (R_{in}), resting potential (V_r), and adaptation (defined as ISI_1/ISI_3 in a train of at least 4 spikes in response to the first 500 ms current pulse). In the type II neurons that changed to type I in response to carbachol, R_{in} on average increased from 195 to 296 MΩ, V_r from

−76.5 to −63.5 mV, and the adaptation decreased from 3.05 to 1.04. In the type II neurons that remained type II in response to carbachol, R_{in} increased from 244 to 425 MΩ, V_r from −68.4 mV to above firing threshold, and the adaptation remained virtually constant at 0.99. Finally, in neurons with a type I PRC before and after the application of carbachol, R_{in} increased from 249 to 269 MΩ and V_r and the adaptation decreased from −69.9 to −74 mV and 1.72 to 0.99, respectively. When pooling all data, R_{in} increased from 229 to 318 MΩ, V_r from −71.8 to −67 mV, and the adaptation decreased from 1.99 to 1.09.

In summary, the experimental results show that the cholinergic switch from type II to type I PRC is accompanied by an increase in R_{in}, increase in V_r and decrease in levels of Spike-frequency adaptation (see Stiefel et al. 2008 for further details). These changes are compatible with the theoretically predicted mechanisms based on cholinergic down regulation of the I-M (Gutkin et al. 2005, Ermentrout et al. 2001). We should note that our results also indicate that the neurons that maintained their PRC type in face of cholinergic modulation were not insensitive to carbachol, but rather that the changes in cellular properties caused by carbachol were insufficient switch their PRC type.

3 Computational Analysis of Cholinergic Modulation of PRCs

The goal of the computational analysis was, first to show a generic basis of the experimental effects by analysis of a canonical model of spike generation in cortical neurons. We further used detailed computational models to understand how changes induced by carbachol on specific cross-membrane currents would affect the type of the PRC.

Previous analysis suggested that the shape of the PRC depends critically on the slow negative feedback on the neural membrane excitability produced by potassium currents underlying spike-frequency adaptation. It was suggested the I_{M}, a muscarine-sensitive potassium current that has graded voltage dependence for the activation, is sufficient to induce negative portions in the PRC. Hence, blocking such a current should be sufficient to remove the negative portion of the PRC and convert it from type II to type I. In order to corroborate that such a change in the experiments could indeed be due to the carbachol-dependent block of the $I\!-\!M$, we used several complimentary computational models: a reduced model combining a spike nonlinear spike-generating mechanism with a simple model for voltage-dependent adaptation (the so-called theta neuron with adaptation), and single- and multicompartmental conductance-based models of a pyramidal neuron.

3.1 Computing the PRCs in the Models

We determined the PRCs in all models using the same method as in the experiments above by first simulating spiking until the spike discharge frequency reached stationarity ($\mathrm{ISI}_n/\mathrm{ISI}_{n-1} < 1.01$). Then, we perturbed the voltage by injecting a small (<1 mV) and short (<20 ms) exponentially decaying depolarizing synaptic event. The Δt of the ISI as a result of the synaptic input was determined and normalized to the control (unperturbed) ISI duration. This was repeated at 64 phase positions spanning the entire period of the oscillation. Changing conductances to mimic cholinergic effects can alter (increase) the firing frequency of the neuron. When we intended to keep the firing frequency of the model neuron constant during a parameter sweep, we determined a firing frequency/current curve (FI curve) for each parameter value, and used it as a lookup table to determine the appropriate current for the desired frequency.

3.2 Modulation of Low-threshold Adaptation Accounts Generically for PRC-Type Switch: Analysis of the Adapting Theta neuron

The results of the experiments are mirrored in a low-dimensional canonical neuronal model: the adaptive theta neuron, showing their generality. Without the spike-frequency adaptation, this is a canonical model for type I saddle-node dynamics, the theta-neuron, or equivalently the quadratic integrate-and-fire (QIF). We then augment it with a slow negative feedback implementing I_{M}-like spike-frequency adaptation (see Gutkin et al. 2005 for further details). The adaptive theta neuron (Fig. 12.4a) is:

$$\frac{\mathrm{d}\Theta}{\mathrm{d}t} = 1 - \cos\Theta + (1 + \cos\Theta)(\beta + s\overline{g}_{\mathrm{M}} + S) \tag{12.1}$$

$$\tau_{\mathrm{M}}\frac{\mathrm{d}s}{\mathrm{d}t} = (k_{\mathrm{M}}\Theta)(1 - s) - s \tag{12.2}$$

$$k_{\mathrm{M}} = a\mathrm{e}^{-b-\cos\Theta-\Theta\mathrm{th}} \tag{12.3}$$

where θ is the phase around the cells firing cycle, with the top of the spike at $\theta = \pi$; s governs the voltage-dependent adaptation; $K_{\mathrm{M}}(\theta)$ is the "activation" function for s, with θth being the activation threshold. Here, we set θth quite low ($\theta_{\mathrm{th}} = 2$) to ensure that the adaptation is partially active subthreshold and heuristically reflects such slow K-currents as I_{M}. Note that the other alternative is the spike-dependent adaptation that could be modeled by setting a high activation threshold, as, for example, due to calcium-activated I_{AHP}. Other parameters were set to: a $=$ 2, $b = 8$; $\tau_M = 400$. β is a free parameter that governs the amplitude of the constant bias. Hence, the term β controls the firing rate of the cell, provided that it is above the rheobase (here: above zero). Strength of I_M, g_M, was changed as indicated below. Parameters of the model were adjusted to yield basic behavior qualitatively similar to the experiment (XPPAUT scripts are available upon request). Runge–Kutta method was used for numerical integration.

The nonadapting model is the normal form for the saddle-node bifurcation whose PRC can be computed analytically (see Ermentrout 1996), and is strictly positive. Turning on spike-frequency adaptation can induce the PRC to become biphasic. We modeled the effect of cholinergic modulation by decreasing the adaptation current (g_M from 1 to 0.7) and by increasing the drive (β from 0.5 to 0.52). These manipulations mimicked the experimentally observed decrease in I_M, the depolarization, and the increase in the input resistance. In response to these manipulations, the steady-state firing frequency increased from 8.6 Hz to 69.1 Hz and the PRC of the theta neuron switched from type II to type I (Fig. 12.4b).

Parameter sweeps over g_M (1 to 0.6, Fig. 12.4c) and β (0.49 to 1, Fig. 12.4d) showed that this switch was observed both when lowering g_M and when increasing β. During these parameter sweeps, the firing frequency increased from 10.2 Hz to 93 Hz and from 5.8 Hz to 196 Hz. High firing frequencies can mask type II dynamics when the frequency is so high that the dynamics of the adaptation current become much slower than the spiking dynamics (for further discussion, see Gutkin, Ermentrout and Gutkin 2005). We wanted to disassociate the changes in the PRC due to the firing rate instability from effects of I_M modulation. We thus performed a parameter sweep over g_m (1 to 0.6) with β adjusted to keep the firing frequency at 10 ± 2 Hz (Fig. 12.4c, lower panel). With decreasing g_M, the negative part of the PRC decreased and its maximum, minimum and zero crossing shifted to the left (earlier parts of the phase). The zero crossing, when present, also shifted to the left. This shows that a change in g_M is sufficient to alter the PRC of the theta neuron.

The simulations of the theta neuron show that the cholinergic effect on the PRC is a generic property of the interplay between the spike-generating mechanisms and the adaptation currents. Interestingly, somatic adaptation currents that have a steep voltage dependence in the activation, e.g., that require a spike in order to be activation such as the Ca-dependent I_{AHP}, cannot induce a negative region in the PRC (see Ermentrout et al. (2001) for further discussion of this point). Hence, the reduced model results suggested that the blockade of the I_M by carbachol is most likely to be responsible for the change of the PRC type.

This hypothesis is supported by formal mathematical arguments. There is a relationship between the I_M-dependent appearance of the negative region in the PRC and the bifurcation structure leading to onset of spiking (see Ermentrout et al. 2001). As the strength of the adaptation increases, a subcritical Hopf bifurcation appeared instead of the saddle-node bifurcation, thus changing in the PRC shape. Here, it was critically important that the adaptation current was dependent on the voltage and was partially active at subthreshold voltage levels. These requirements make I_M a plausible mechanism for such change.

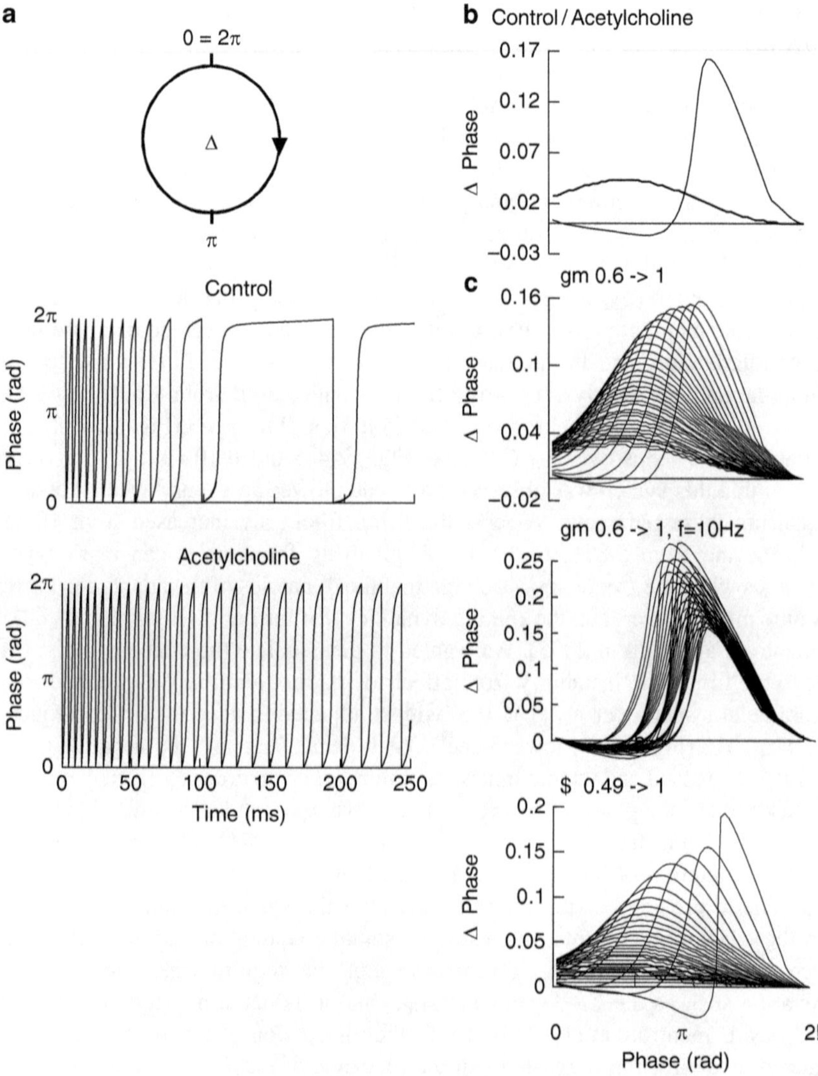

Fig. 12.4 Theta neuron. (**a**) Schematic depiction of the model (*top*). Voltage traces with standard parameters (*middle panel*, $g_{\mathrm{m}} = 1$, $\beta = 0.5$) and with parameter modifications simulating cholinergic effects (*bottom panel*, $g_{\mathrm{m}} = 0.7$, $\beta = 0.52$). (**b**) Phase response curves (PRCs) with standard (*thin line*) and cholinergic (*thick line*) parameters. (**c**) (*Top*): g_{m} varied from 0.6 to 1.0; (*middle*): g_{m} varied from 0.6 to 1.0 with β adjusted to keep the firing rate at 10 ± 2 Hz. (**d**) β varied from 0.49 to 1.0. Figure modified from Stiefel et al. 2009

3.3 I_M Mechanism for PRC Switch: Support from Conductance-Based Models

It is well known that at the cellular level cholinergic modulation affects a variety of both ionotropic (nicotinic) and metabotropic (muscarinic) receptor targets (Schoebeberger et al. 2000). Acetylcholine modulation through the muscarinic receptors down regulates several slow potassium currents (I_M, I_{AHP} (Madison et al. 1987, McCormick 1993)). In addition muscarinic cholinergic modulation increases the input resistance of the neurons by down regulating the leak conductance. Taken together, these currents control cellular excitability and repetitive firing in cortical pyramidal neurons.

Given the diverse acetylcholine actions at the single cell level, an interesting issue is which of the effects mediated by the muscarinic acetylcholine receptor modulation is key for the emergence of collective behaviors, such as synchrony and oscillations. Previous theoretical work has hinted that suppression of slow cation channels can bias pyramidal neurons toward synchrony with excitatory synaptic coupling by modulating the dynamics of spike generation (Crook et al. 1998a, 1998b, Ermentrout et al. 2002). As we argued above the key here is the change in the type of the PRC. We thus set out to systematically examine the various mechanisms of cholinergic action and to identify those that are more likely to affect the dynamics of spike generation in cortical neurons. Above we showed that I_M modulation is most likely to be responsible. In order to study if there are other cholinergic effects that could be complementary to these mechanisms, we simulated more complex models. In particular, we studied how the carbachol induced changes in I_{AHP}, I_{NAP} interplay with I_M in shaping the PRC.

In order to further corroborate the proposed mechanism for the experimentally observed PRC type switch, we used a single compartment model for a representative cortical pyramidal neuron (expanded version of model in Golomb & Amitai, 1997; see Fig. 12.5a,b). This model included the M-type K^+-current I_M, as well as I_{Nap}, I_{KAHP}, I_{KA}, and I_H to give a more detailed representation of the currents found in layer II/III pyramidal neurons. Because the I_{AHP} is Ca^{2+} dependent, an L-type Ca^{2+}-current and an exponential Ca^{2+} decay representing a Ca^{2+}-buffer were also included. Further details on the model equations are in Appendix B and parameters are given in Stiefel et al. (2008). The Neuron code for the model is available upon request.

We analyzed the PRC for the conductance-based model described above. Under standard parameter regime, the model showed a significant level of spike-frequency adaptation. The calculated PRC was biphasic with a negative portion at the beginning of the ISI. The peak of the PRC showed a shift to the right and occurred just before 2π phase (see Fig. 12.5c, thin black line).

The effect of cholinergic neuromodulation was modeled by altering the conductance (g_{leak}, from 2×10^{-5} to 1.8×10^{-5} S cm^{-2}) and reversal potential (E_{leak}, from -60 to 58 mV) of the leak current and the conductance of I_M and I_{AHP}. Madison et al. (1987) showed that in hippocampal pyramidal neurons, these two currents are

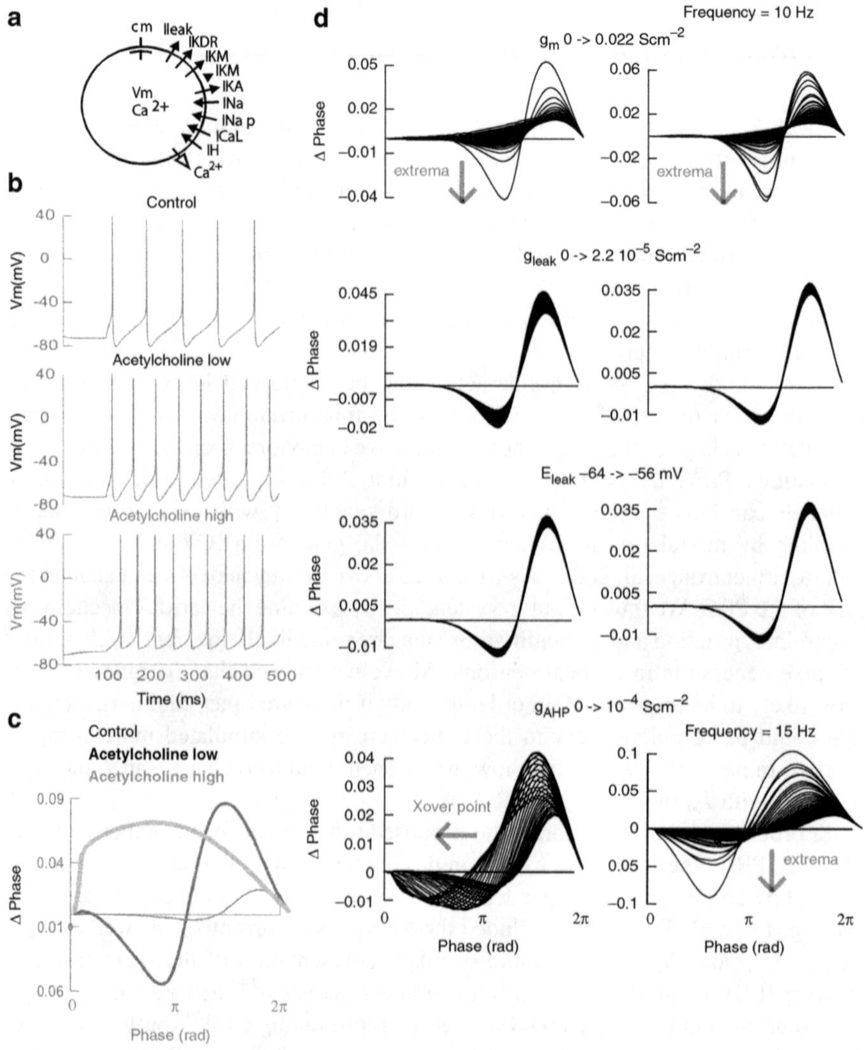

Fig. 12.5 Complex single-compartment neuron. (**a**) Schematic depiction of the model. (**b**) Voltage traces with standard parameters ($E_{\text{leak}} = -60\,\text{mV}$, $g_{\text{leak}} = 2 \times 10^{-5}\,\text{S cm}^{-2}$, $g_{\text{KM}} = 5 \times 10^{-4}\,\text{S cm}^{-2}$, $g_{\text{AHP}} = 1 \times 10^{-4}\,\text{S cm}^{-2}$), with parameter modifications simulating low-concentration cholinergic effects ($E_{\text{leak}} = -58\,\text{mV}$, $g_{\text{leak}} = 1.8 \times 10^{-5}\,\text{S cm}^{-2}$, $g_{\text{KM}} = 4.5 \times 10^{-4}\,\text{S cm}^{-2}$, $g_{\text{AHP}} = 2.5 \times 10^{-5}\,\text{S cm}^{-2}$) and parameters simulating low-concentration cholinergic effects ($E_{\text{leak}} = -58\,\text{mV}$, $g_{\text{leak}} = 1.8 \times 10^{-5}\,\text{S cm}^{-2}$, $g_{\text{KM}} = 2.5 \times 10^{-5}\,\text{S cm}^{-2}$, $g_{\text{AHP}} = 0\,\text{S cm}^{-2}$). (**c**) PRCs with standard (*thin line*), low-concentration cholinergic (*thick black line*) and high-concentration cholinergic (*thick gray line*) parameters. (**d**) PRCs with g_{KM} varied from 0 to $0.0006\,\text{S cm}^{-2}$ (*top*), g_{leak} varied from 1.6×10^{-5} to $2.2 \times 10^{-5}\,\text{S cm}^{-2}$ (*middle*) and E_{leak} varied from -64 to $-58\,\text{mV}$ (*bottom*), g_{AHP} varied from 0 to $1 \times 10^{-4}\,\text{S cm}^{-2}$. *Left column*: all other parameters kept constant. *Right column*: PRCs with I_{inject} adjusted to keep the firing rate at 10 ± 1 Hz (15 ± 1 Hz in the last row, as the minimum firing frequency with $g_{\text{AHP}} = 0$ is >13 Hz). The *gray arrows* depict the direction of the trends along the parameter sweeps. Figure modified from Stiefel et al. 2009

both cholinergically modulated, albeit with different sensitivities. I_{AHP} is inhibited by cholinergic modulation at about ten times lower concentrations than I_M (IC50 = 0.3 and 5 ? M for carbachol). We assumed that superficial neocortical neurons would react similarly. Therefore we were able to simulate two different conditions: low-concentration cholinergic modulation (I_M down regulated by 10% to 4.5×10^{-4} and I_{AHP} down regulated by 75% to $2.5 \times 10^{-5}\,\mathrm{S\,cm^{-2}}$) and high-concentration cholinergic modulation (I_M down regulated by 95% to 5×10^{-5} and I_{AHP} down regulated by 100% to $0\,\mathrm{S\,cm^{-2}}$). Interestingly, a simulation of low concentration cholinergic modulation lead to a PRC that was even more pronouncedly biphasic than under control conditions (Fig. 12.5c, light gray line). Only the simulation of high-concentration cholinergic modulation lead to the transition from a type II to a type I PRC (Fig. 12.5c thick black line); the firing frequencies in these simulations were kept at 12 ± 1 Hz in order to ensure comparability).

Parameter sweeps g_{leak}, E_{leak}, g_M, and g_{AHP} (g_{leak}: from 1.6×10^{-5} to $2.2 \times 10^{-5}\,\mathrm{S\,cm^{-2}}$, E_{leak}: from -64 to 56 mV, g_M: from 0 to $6 \times 10^{-4}\,\mathrm{S\,cm^{-2}}$, g_{AHP} 0 to $10^{-4}\,\mathrm{S\,cm^{-2}}$) revealed that the qualitative change in the PRC was solely a consequence of changes in g_M (see Fig 12.5d). This occurred regardless of whether the firing frequency was kept constant by adjusting current injection. No switch from a biphasic to a purely positive PRC occurred when g_{leak}, I_{leak} or g_{AHP} were varied. The sweeps over the parameters of the leak current had little effect on the PRCs. Increasing g_{AHP} moved the crossover point between the negative and the positive part of the PRC to the left (toward earlier phases). When the firing frequency during the determination of the PRCs was kept constant, this effect was greatly reduced and the extrema of the PRCs decreased with an increasing amount of g_{AHP}.

In conclusion, simulations of the conductance-based single compartmental model showed that the effect of cholinergic neuromodulation on the PRC is biphasic. An increase of the negative part of the PRC mediated by low modulator concentration is followed by a decrease and disappearance of this negative part mediated by a high acetylcholine concentration. The switch from a type II to a type I PRC is due to the decrease of I_M. However, under control conditions, the I_{AHP} masks a part of the negative portion of the PRC.

3.4 Global Effects of Local I_M Cholinergic Modulation on PRC Type: Analysis of Multicompartment Model

We wanted to know if the global and/or spatially localized modulation of I_M would lead to the PRC switch. To do so we analyzed a multicompartment model of a cortical layer II/III pyramidal neuron ((Mainen & Sejnowski 1995). Note that this cell type matches our in vitro experiments (Stiefel et al. 2003). The details of the model are given in Appendix C as well as Stiefel et al. (2008) and the Neuron code (Hines and Carnevale 1997, 2000) is available upon request.

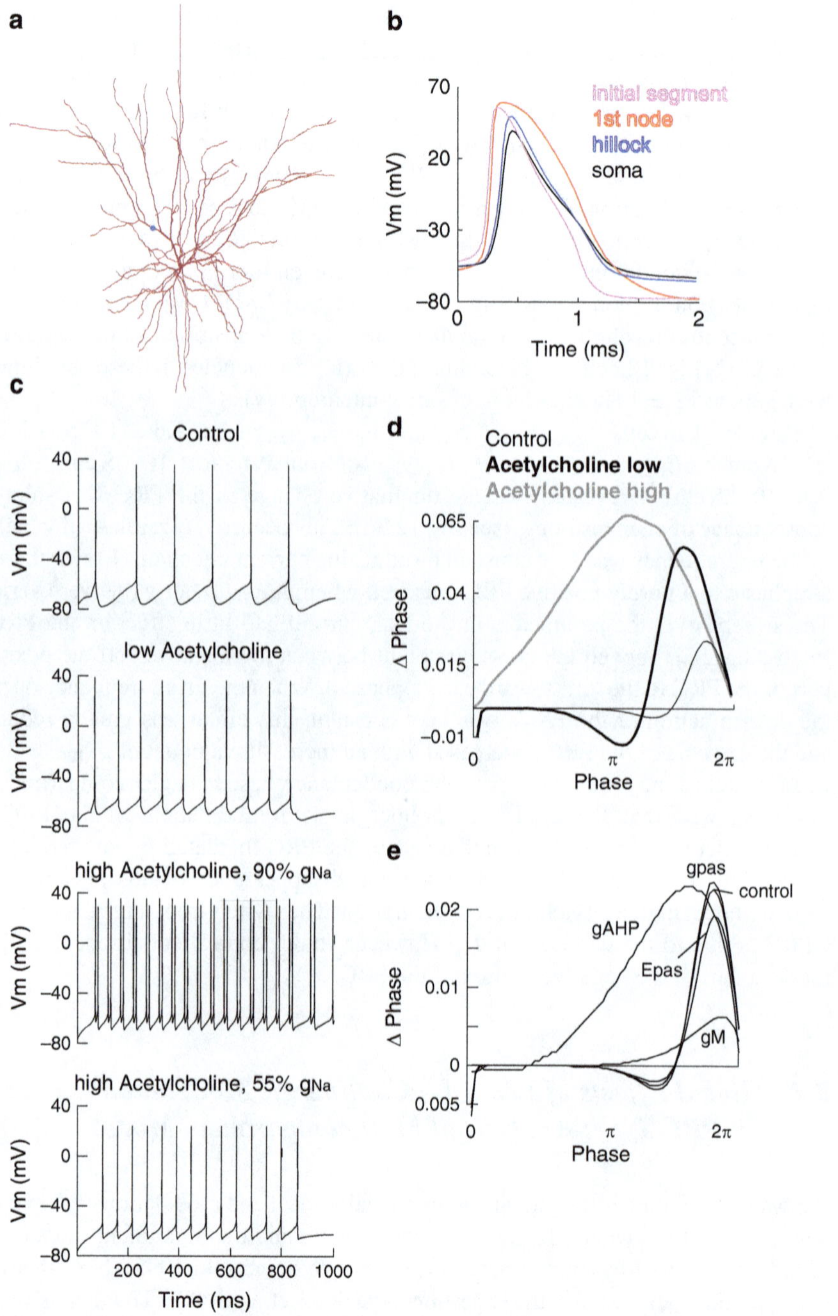

Fig. 12.6 Complex multicompartment neuron. (**a**) Model neuron morphology and synapse location, indicated by *circles*. (**b**) Timecourse of action potential initiation at different cellular

The multicompartmental model had not only a spatial extension (explicit simulation of soma, dendrites and axon, see Fig. 12.6a) but also a number of ionic conductances distributed heterogeneously over the different parts of the neuron. In addition, in accordance with the experimental literature, the spike is initiated in the axon (Stuart et al. 1997, Fig. 12.6b). Using this model, we simulated the effect of low and high concentrations of cholinergic neuromodulation by implementing the following parameter changes in the dendrites: to mimic low concentrations, we changed $E_{\mathrm{leak}} = -32$ to $-30\,\mathrm{mV}$, $g_{\mathrm{leak}} = 2 \times 10^{-5}$ to $1.810^{-5}\,\mathrm{S\,cm}^{-2}$, $g_{\mathrm{M}} = 2 \times 10^{-4}$ to $1.8 \times 10^{-4}\,\mathrm{S\,cm}^{-2}$ and g_{AHP} 3 10–4 to 7.5 10–5 $\mathrm{S\,cm}^{-2}$. To mimic high concentrations, we changed $E_{\mathrm{leak}} = -32$ to $-30\,\mathrm{mV}$, $g_{\mathrm{leak}} = 2\,10^{-5}$ to $1.8\,10^{-5}\,\mathrm{S\,cm}^{-2}$, $g_{\mathrm{M}} = 2\,10^{-4}$ to $0.1\,10^{-4}\,\mathrm{S\,cm}^{-2}$. Cantrell et al. (1996) found a significant muscarinic acetylcholine-receptor mediated down regulation of sodium current conductance. Interestingly, they found two populations of neurons, in which this current was reduced to 90% and 55%, respectively. Hence, we globally (dendrites, soma, and axon) decreased g_{Na} to 55% of its original value. Reducing g_{Na} to only 90% of its initial value caused the model neurons to spike in doublets (Fig. 12.6c). We chose the larger reduction since we focused on PRCs of individual spikes and not of doublets.

Initially, we determined the PRC by transient activation of a synapse located on an intermediate part of the apical dendritic tuft (Fig. 12.6d). The PRC under control conditions was biphasic (type II), and became even more biphasic when low concentrations of cholinergic neuromodulation (weakly reducing I_{M} and strongly reducing I_{AHP}) were simulated. When simulating high concentrations of cholinergic neuromodulation (strongly reducing I_{M} and abolishing I_{AHP}), the PRC became monophasic (type I). This switch was observed when the value of g_{M} was changed in isolation, but not when the values of g_{leak}, E_{leak}, or g_{AHP} were changed (Fig. 12.6e). Whereas the effects of changing the parameters of the passive conductance were minimal, reducing $\mathrm{I_{AHP}}$ reduced the negative part of the PRC and shifted it to an earlier part of the phase. These observations further strengthen the importance of I_{M} in determining the type of PRC displayed by the neuron.

In order to explore the role of the spatial location of the synapse, we determined the PRC by synapses on three different locations, ranging from the proximal primary to a tertiary dendrite (Fig. 12.7a, synapse 2 is identical to the previously investigated synapse). As the spike initiation takes place in the axon, the investigated synapses

Fig. 12.6 (continued) locations. Soma: *black*, 1st node of Ranvier: *red*, axon hillock: *blue*, axon initial segment: *pink*. (**c**) Voltage traces with standard parameters and with parameter modifications simulating low and high concentration cholinergic effects, with a strong and moderate reduction in g_{Na} ($I_{\mathrm{injected}} = 0.12, 0.12, 0.03, 0.3\,\mathrm{nA}$). (**d**) PRCs with control (*thin line*), low concentration cholinergic (*thick line*) and high concentration cholinergic parameters (*gray line*). Firing frequencies were kept at $7 \pm 1\,\mathrm{Hz}$, $I_{\mathrm{injected}} = 0.15, 0.11, 0.02\,\mathrm{nA}$). (**e**) PRCs with individual parameters affected by cholinergic neuromodulation modified. Parameter modifications for the respective curves are indicated in the figure ($I_{\mathrm{injected}} = 0.15\,\mathrm{nA}$, Hz, firing frequencies 6.7, 8.1, 23.6, 6.6, 7.0 Hz). Figure modified from Stiefel et al. 2009

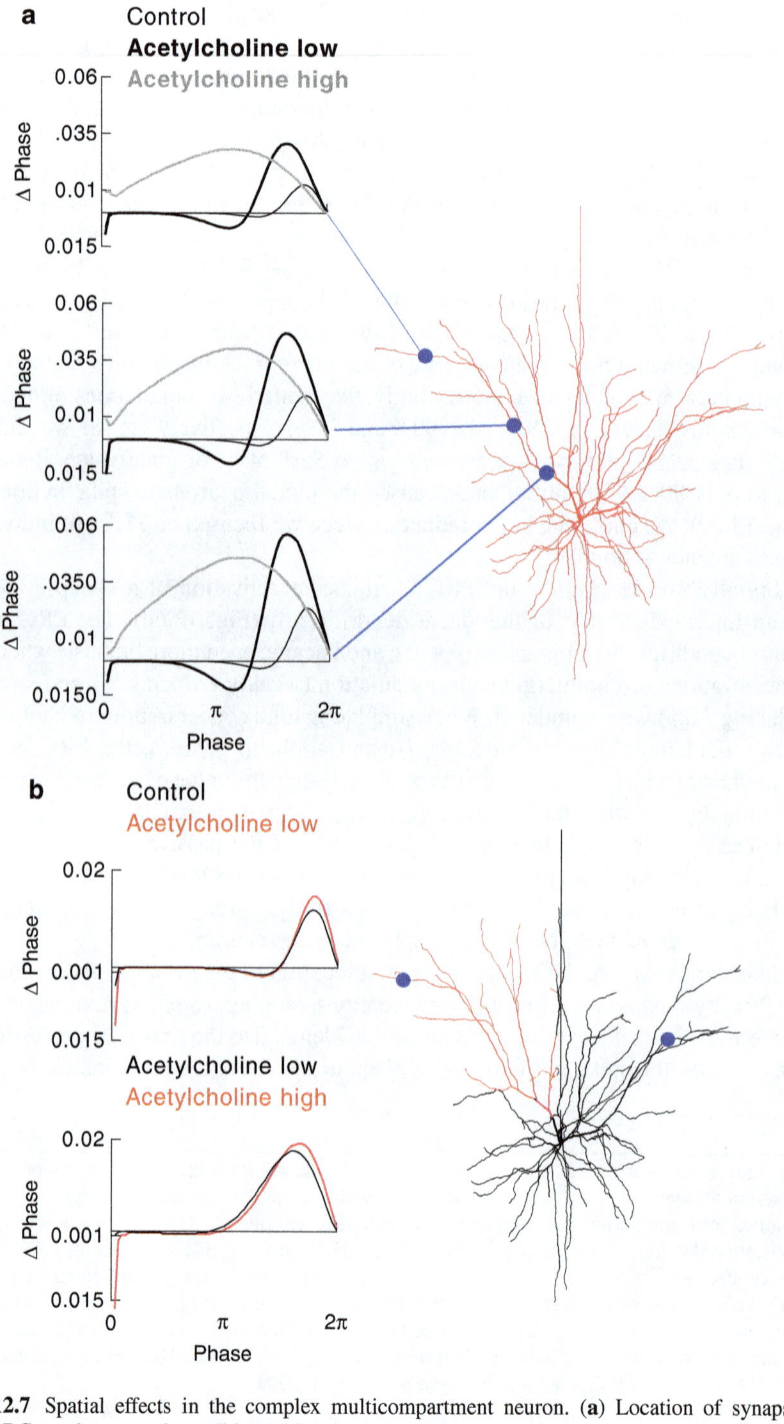

Fig. 12.7 Spatial effects in the complex multicompartment neuron. (**a**) Location of synapses and PRCs under control conditions (*thin line*), low concentration cholinergic (*thick line*), and

are thus located progressively further from the initiation site. The currents from the dendritic synapses have to pass through active and passive cable filtering by the dendrites and the soma (a large sink), before reaching this spike initiation site. All these factors can potentially influence the PRCs associated with the investigated synapses (Remme et al. 2009, Goldberg et al. 2007). We observed that the PRCs determined for all three synapses are qualitatively similar. The PRCs under control conditions were biphasic (type II), even more pronouncedly biphasic when simulating low concentrations, and monophasic (type I) when simulating high concentrations of cholinergic neuromodulation (Fig. 12.7a). The amplitudes of the PRCs decreased the more distant the synapse was; this is in accord with the reduced amplitude of the EPSPs at the site of the spike initiation caused by progressively stronger dendritic cable filtering.

We next asked how a spatially nonuniform parameter change simulating cholinergic neuromodulation could change the influence of a synaptic input on the phase of neuronal spiking. The ion channel proteins carrying I_M (KCNQ) are known to be distributed relatively uniformly over the dendritic surface in related types of neurons (hippocampal CA1 neurons in culture (Shah et al. 2002)). The muscarinic acetylcholine receptors mediating the decrease in I_M conductance are also found on the dendritic surface, and the G-protein mediates signal transduction between muscarinic receptors. On the other hand, KCNQ is highly local (spread of the signal $\sim 1\,\mu m$). Also, localized cholinergic projections arising from the nucleus basalis or from subtypes of cholinergic interneurons could activate muscarinic receptors in a spatially limited dendritic region. Taken together, the spatial proximity of the cholinergic receptors, ion channels, the signaling cascades mediating the signaling between them, and the nature of the cholinergic projections make subcellularly localized cholinergic effects a possibility. Similar reasoning holds for the modulation of the ion channels responsible for the leak current (GIRK).

We thus simulated spatially heterogeneous cholinergic modulation of dendritic conductances (Fig. 12.7b). At first, we simulated a low concentration cholinergic modulation in a single dendritic tree (18% of the neuron's surface area) and no cholinergic effects in the remaining neuron. Under these conditions, we determined the PRC for a synapse located within and outside of the region experiencing changes. The PRCs were intermediate between the PRCs determined for homogeneous parameter changes. The PRCs were biphasic, with the crossover point shifted

Fig. 12.7 (continued) high concentration cholinergic parameters (*gray line*). (**b**) PRCs with control (*thin line*), low concentration cholinergic (*thick line*), and high concentration cholinergic parameters (*gray line*) for synapses at three different locations. Firing frequencies were kept at $7 \pm 1\,\text{Hz}$, $I_{\text{injected}} = 0.15, 0.11, 0.02\,\text{nA}$). (**c**) Nonuniform changes of parameters simulating cholinergic effects. PRCs were determined for two synapses at the positions indicated by *circles*. The parameters of the dendritic conductances were changed in the *black/red* parts of the dendrite to simulate control conditions/low acetylcholine (*top*, $I_{\text{injected}} = 0.03\,\text{nA}$, firing frequency 7.1 Hz) and low/high acetylcholine (*bottom*, $I_{\text{injected}} = 0.15\,\text{nA}$, firing frequency 10.4 Hz). Figure modified from Stiefel et al. 2009 to earlier phases as compared to a neuron under homogeneous control conditions. This was expected, as the dendritic tree with cholinergic parameters contained a significant fraction of the neuron's membrane surface. However, surprisingly the PRCs determined for the synapse within and outside the region with the cholinergically modified parameters were quite similar. The PRC of the synapse inside the dendritic region modulated by acetylcholine had a slightly higher positive peak.

Next, we modified the parameters in the same dendritic tree to simulate a high, and in the remaining neuron a low concentration of cholinergic modulation. Again, we determined the PRCs of a synapse within and outside of this dendritic tree. As before, the PRCs were intermediate between the PRCs obtained with each parameter regime homogenously implemented over the complete dendritic tree. The negative part of the PRCs was strongly reduced. Equally as in the previous case, the PRCs were very similar for the synapses inside and outside of the dendritic tree with different parameter modifications. The positive peak of the former PRC was slightly higher in amplitude and shifted to later phases.

The modulation of PRC thus seems to be global to a neuron. Changes of membrane properties in small dendritic regions are insufficient to switch the type of PRC of a synapse, whereas changes in large regions cause enough difference in the total currents to affect synapses outside of the region of change.

4 Discussion

Acetylcholine is a powerful neuromodulator that has effects at the behavioral, circuit, and cell levels. In particular, some of the more prominent effects of acetylcholine at the cell level are to reduce the spike-frequency adaptation by blocking currents responsible for it, as well as to increase the resting potential of the cell, possibly by a partial block of leak currents. Such modulation of cell excitability has been hypothesized to lead to changes in the structure of collective network activity in the cortex (Ermentrout 2001, Crook et al. 1998b). Namely, theoretical work suggested that blocking adaptation can change the synchronization properties of various frequency bands (e.g., gamma and beta, etc.) by promoting different synaptic mechanisms for synchronization (Ermentrout 2001). It is interesting to note that EEG studies in attentive vs. nonattending conditions have given an indirect support to this idea (Fan et al. 2006), while in vitro studies pointed out how ach modulation can lead to onset of coherent oscillations in local cortical circuits (Buhl et al. 1998, Fisahn et al. 1998).

We examined how acetylcholine affects the shape of the neurons PRC. Our experimental results shown that cortical pyramidal neurons can switch their PRC type, and thus most likely the type of bifurcation leading to spiking in response to cholinergic neuromodulation. In 4/7 neurons that initially had a type II PRC it changed to type I after the application of carbachol. It remained unchanged at type

II in the remaining three of these neurons, and it did not change in the five neurons with initially a type I PRC. No case of a switch from type I to type II in response to carbachol was observed.

We observed heterogeneity in the properties of cortical pyramidal neurons. Not all the recorded neurons were type II and some of them switched to type I under cholinergic neuromodulation. A similar heterogeneity was also observed in the values of the neurophysiological parameters R_{in}, V_{r}, and adaptation and their changes induced by acetylcholine.

The fact that only the type II neurons that showed a strong decrease in spike-frequency adaptation switched their PRC type is consistent with the theoretical finding that decreasing an adaptation current can switch the type of bifurcation leading from rest to tonic spiking. We suggest that the cholinergic decrease in a potassium current, possibly I_{M}, was responsible for the transition from type II to type I and that in the type II neurons that remained unchanged, acetylcholine did not decrease I_{M} sufficiently to switch the PRC type, also seen by the lack of decrease of adaptation. Finally, in the type I neurons a decrease of adaptation was observed, but they were already of the PRC type associated with a small amount of adaptation current and their PRC type also did not change. Should more than half of the pyramidal neurons in a cortical column (58%) switch their PRC type due to cholinergic neuromodulation, this would be likely to influence the synchronization properties of the network. Acetylcholine is known to cause oscillations at a higher frequency and in a more localized spatial extent. The switch in the spike-generating dynamics, observed experimentally as a change in the PRC type, may be an important factor in this change.

The modeling results support the experimental findings in two critical ways. First, the analysis of the theta neuron model supported our hypothesis by showing the generic nature of mechanism proposed. Modulating a low-threshold spike-frequency adaptation in this model led to a change in the bifurcation structure and appearance of type II PRC as seen in the experiments. Second, the simulations of the conductance-based models specifically point to the I_{M} block as the mechanism responsible for the change in the PRC. Using single- and multicompartmental simulations, we demonstrated that other cholinergic effects, such as changes in input resistance and resting potential were not responsible for the from type II to type I (unpublished observations). Furthermore, these results rest on a previous analysis of spikegeneration in similar conductance-based models showing how I_{M} strength can determine the spike-generating dynamics: strong I_{M} leads to type II dynamics with a subcritical Hopf bifurcation yielding repetitive firing, while weak I_{M} leads to Type I saddle-node dynamics (for further discussion see Ermentrout et al. 2001). Our simulations of the multicompartmental model showed the spatial dependence of the PRC modulation within the cell, pointing out that it is a global cell-wide phenomenon. In addition we show that nontrivial interactions between the various K-currents modulated can actually strengthen the biphasic nature of the PRC at low concentrations of acetylcholine, and convert the PRC to type I at high Ach concentration.

Based on our experiments and simulations, a picture thus emerges that cortical networks may synchronize through excitation at low and moderate Ach levels. Hence, pyramidal neurons may be synchronized directly by the mutual interactions. When Ach levels are high, synchronization may be due more to the inhibitory coupling with pyramidal neurons entrained to the gamma rhythm by the interneuronal network. Given that excitatory connections are not only local but also long range, the low Ach synchronization may have a large scale, possibly interhemispheral nature, while the high Ach gamma would be more like to be local. This hypothetical picture is consistent with the patchy and local nature of gamma oscillations seen during wakeful states and a more global nature of low frequency-band (delta, theta) activity seen in states of rest and sleep.

In conclusion, our results raise the possibility of a link between cholinergic neuromodulation of the spike-generating process of individual neurons and changes in the large-scale network behavior and associated cognitive states in the mammalian cortex.

Appendix A: Experimental Preparation for PRC experiments

We recorded from layer II/III pyramidal neurons in the slices of the mouse visual cortex. All animal experiments were done in accordance with ethical guidelines of the Salk Institute. We recorded using whole-cell patch clamp at 32 degrees with electrodes ranging from 5 to 8 MΩ. In our preparation, glutamatergic ionotropic synaptic transmission was blocked with DNQX (20 μM) and APV (50 μM) in all experiments. GABAergic ionotropic transmission was blocked with biccuculine (10 μM) in the majority of experiments.

Synaptic transmission in the experiments was blocked and the neurons were put into oscillatory regime with a predetermined firing frequency by injecting DC current via the patch-electrode, from −10 pA in the case of spontaneous spiking caused by the application of carbachol, to 150 pA. Continuous stretches spanning 32 seconds were recorded. In addition, brief (10 to 20 ms) depolarizing (10 to 30 pA) current pulses were injected every 100 to 1,000 (in most cases) ms to perturb the regular spiking process (see below as well for explanation of PRC measurement). Experiments were continued at least 5 minutes after bath application carbachol (10, 20, or 50 μM). DC current injection was adjusted to achieve similar firing rates as before drug application.

To ensure that the neurons from which we recorded did not change their physiological condition and recording quality, we monitored the input resistance, resting potential, and spike amplitude throughout the experiment. Data acquired after significant changes in these parameters (independently of pharmacological manipulations) were discarded. Also, when entrainment of the spikes by the current pulses was observed, all the data from that neuron was discarded.

Appendix B: Single Compartment Conductance-Based Model

We adapted a cortical pyramidal neuron model (Golomb & Amitai 1997). This model had identical equations for the current kinetics as the original (Golomb & Amitai 1997) including the fast Na^+, leak, delayed rectifier K^+ currents, and I_M and I_{AHP} K^+ currents are included to generate an after-hyperpolarization (AHP) and responsible for spike-frequency adaptation. Because the I_{AHP} is Ca^{2+}-dependent, an I_L Ca^{2+}-current and an exponential Ca^{2+} decay representing a Ca^{2+}-buffer were also included. The mixed cation hyperpolarization activated current I_H, a persistent sodium current I_{NAP} and an inactivating potassium current I_A were also included. All these currents are known to occur in cortical pyramidal neurons (Johnston & Wu 1995), and we wanted to see how their interactions may impact the effect of cholinergic modulation.

The voltage equation is:

$$\begin{aligned} C_m \frac{dV}{dt} &= -\bar{g}_{Na} m^3 h (V - E_{Na}) - \bar{g}_{KDR} n (V - E_K) - \bar{g}_M s (V - E_K) \\ &\quad - g_{leak}(V - E_{leak}) - \bar{g}_{AHP} u (V - E_K) - \bar{g}_H (w_1 0.8 + w_2 0.2)(V - E_H) \\ &\quad (V - E_H) - g_L x_L^2 (V - E_{Ca}) - \bar{g}_A y^3 z (V - E_K) - \bar{g}_{NaP} m_{NaP}^3 (V - E_{Na}) \end{aligned} \tag{12.4}$$

With $\bar{g}$ the maximum conductances, *m, n, h, w, x, y,* and *u* the state variables of the currents, and E the ionic reversal potentials. The equations governing the state variables are of the form:

$$\tau_h \frac{dh}{dt} = h_\infty - h$$

$$h_\infty = 1/(1 + e^{(V+c)/d})$$

The persistent Na^+ current is identical to the fast Na^+ current, but for the absence of the inactivation state variable h. The values for $\bar{g}$, a, b, and τ for the different state variables are given in Golomb & Amitai, 1997, and in the simulation files available on ModelDB (http://senselab.med.yale.edu/modeldb).

Note that this model simulates neuronal dynamics of both the membrane potential and the Ca^{2+} concentration. The formalism for the Ca^{2+} dynamics follows:

$$\frac{dCa_i^{2+}}{dt} = \frac{2I_{Ca^{2+}}}{\delta} + \frac{(Ca_0^{2+} - Ca_i^{2+})}{\tau_{Ca^{2+}}} \tag{12.5}$$

with F the Faraday constant (96,500 Coulomb) and Δ the fraction of the intracellular space affecting the Ca^{2+} dependent current (a 1 μm shell). This term couples the Ca^{2+} current to the Ca^{2+} concentration, which in the absence of any influx relaxes back to ${Ca^{2+}}_0$ with the time constant $\tau_{Ca}{}^{2+} = 35$ ms.

The state variables for the Ca^{2+}-dependent I_{KAHP} current are governed by the equations:

$$\tau_u \frac{du}{dt} = u_\infty - u$$

$$u_\infty = \left(\frac{Ca_i^{2+}}{Ca_o^{2+}}\right)^2 \Bigg/ \left(1 + \left(\frac{Ca_i^{2+}}{Ca_o^{2+}}\right)^2\right)$$

$$\tau_u = 1 \Bigg/ \beta \left(1 + \left(\frac{Ca_i^{2+}}{Ca_o^{2+}}\right)^2\right)$$

with $\beta = 0.03$.

Appendix C: Multicompartmental Model

This model of a cortical layer II/III pyramidal neuron is based on the model of Mainen & Sejnowski, 1995. The neuronal morphology was originally discretized into 257 compartments. When investigating the effects of local dendritic changes, the number of compartments was increased to 1393, to assure that each compartment each sufficiently small (<8 μm) to be approximately isopotential.

The voltage equation for each compartment in addition to the active voltage-dependent currents (as described above for the unicompartmental model) contains an additional term for current flow from the neighboring compartments (for the dendritic compartment n):

$$C_m \frac{dV_n}{dt} = \sum I - I_{inj} - g_{ax}(V_n - V_{n-1}) - g_{ax}(V_{n+1} - V_n) \quad (12.6)$$

where g_{ax} is the axial conductance and ΣI the sum of all intrinsic ionic currents. The model consisted of axonal, somatic, and dendritic regions with distinct properties. These membrane properties were adjusted from their original values in order to more realistically reproduce the spike and AHP-shape and pattern seen in our recordings. Specifically, we increased the somatodendritic conductances $g_M = 2 \times 10^{-4}\,S\,cm^{-2}$ and $g_{leak} = 2 \times 10^{-5}\,S\,cm^{-2}$ and decreased the somatic and somatodendritic conductances to $g_{AHP(Ca)} = 1 \times 10^{-2}\,S\,cm^{-2}$ and $g_{CaL} = 1 \times 10^{-5}\,S\,cm^{-2}$. In addition, the time constant of the Ca^{2+}-decay was decreased to 35 ms to reproduce the shape of the recorded AHP more closely and to match the complex single-compartment model. g_{KDR} in the soma was decreased from 200 to $100\,S\,cm^{-2}$. The axonal conductances remained unchanged.

References

Bartus RT, Dean RL, Beer B, Lippa AS (1982) The cholinergic hypothesis of geriatric memory dysfunction. *Science* 217:408–414.

Brocher S, Artola A, Singer W (1992) Agonists of cholinergic and noradrenergic receptors facilitate synergistically the induction of long-term potentiation in slices of rat visual cortex. *Brain Res* 573:27–36.

Buhl EH, Tamas G, Fisahn A (1998) Cholinergic activation and tonic excitation induce persistent gamma oscillations in mouse somatosensory cortex in vitro. *J Physiol* 513:117–126.

Cantrell AR, Ma JY, Scheuer T, Catterall WA (1996) Muscarinic modulation of sodium current by activation of protein kinase C in rat hippocampal neurons *Neuron* 16:1019–1026.

Cox CL, Metherate R, Ashe JH (1994) Modulation of cellular excitability in neocortex: muscarinic receptor and second messenger-mediated actions of acetylcholine. *Synapse* 16:123–136.

Crook SM, Ermentrout GB, Bower JM (1998a) Dendritic and synaptic effects in systems of coupled cortical oscillators. *J Comput Neurosci* 5(3):315–329.

Crook SM, Ermentrout GB, Bower JM. (1998b) Spike frequency adaptation affects the synchronization properties of networks of cortical oscillations. *Neural Comput* 10(4):837–854.

Ermentrout B (1998) Linearization of F–I curves by adaptation. *Neural Comput* 10:1721–1729.

Ermentrout B, Pascal M, Gutkin B (2001) The effects of spike frequency adaptation and negative feedback on the synchronization of neural oscillators. *Neural Comput* 13:1285–1310.

Ermentrout GB Kopell N (1998) Fine structure of neural spiking and synchronization in the presence of conduction delays. *Proc Natl Acad Sci USA* 95:1259–1264.

Ermentrout B (1996) Type I membranes, phase resetting curves, and synchrony.*Neural Comput* 8(5):979–1001.

Ermentrout GB Rinzel J (1998) Analysis of neural excitability and oscillations. Methods in Neuronal Modeling. In C. Koch and I. Segev (eds.). Cambridge, MIT.

Ermentrout B, Saunders D (2006) Phase resetting and coupling of noisy neural oscillators. *J Comput Neurosci* 20(2):179–190.

Everitt BJ, Robbins TW (1997) Central cholinergic systems and cognition. *Annu Rev Psychol* 48:649–684.

Fan J, Byrne J, Worden MS, Guise KG, McCandliss BD, Fossella J, Posner MI (2006) The relation of brain oscillations to attentional networks. *J Neurosci* 27(23):6197–6206.

Fisahn A, Pike FG, Buhl EH, Paulsen O (1998) Cholinergic induction of network oscillations at 40 Hz in the hippocampus in vitro. *Nature* 394:186–189.

Galan R, Ermentrout GB, Urban NN (2005) efficient estimation of phase-resetting curves in real neurons and its significance for neural modeling. *Phys Rev Lett* 94:158101.

Goldberg JA, Deister CA, Wilson CJ. (2007) Response properties and synchronization of rhythmically firing dendritic neurons *J Neurophysiol* 97:208–219.

Golomb D Amitai Y (1997) Propagating neuronal discharges in neocortical slices: computational and experimental study. *J Neurophysiol*, 78:1199–1211.

Gutkin BS Ermentrout GB (1998) Dynamics of membrane excitability determine interspike interval variability: a link between spike generation mechanisms and cortical spike train statistics *Neural Comput* 10(5):1047–1065.

Gutkin BS, Ermentrout GB, Reyes A (2005). Phase response curves determine the responses of neurons to transient inputs. *J Neurophysiol* 94(2):1623–1635.

Hansel D Mato G (1995) Synchrony in excitatory neural networks. *Neural Comput* 7:307–337.

Hasselmo ME (1999) Neuromodulation: acetylcholine and memory consolidation. *Trends Cog Sci* 3:351–359.

Hasselmo ME (2006) The role of acetylcholine in learning and memory. *Curr Opin Neurobiol* 16(6):710–715.

Hines ML Carnevale NT (1997) The NEURON simulation environment. *Neural Comput* 9:1179–1209.

Hines ML Carnevale NT (2000) Expanding NEURON's repertoire of mechanisms with NMODL. *Neural Comput*, 12:995–1007.

Hodgkin AL Huxley AF (1952) A quantitative description of membrane current and its application to conduction and excitation in nerve. *J Physiol* 117:500–544.

Izhikevich EM. (2007) Dynamical systems in neuroscience: the geometry of excitability and bursting. MIT.

Keck T, Netoff TI, White JA (2003) Comparing dynamic responses of neurons to dendritic and somatic inputs. *Society for Neuroscience Abstracts*.

Lengyel M, Kwang J, Paulsen O, Dayan P. (2005) Matching storage and recall: hippocampal spike timing-dependent plasticity and phase response curves. *Nat Neurosci* 8:1677–1683.

Mainen ZF Sejnowski TJ (1995) Reliability of spike timing in neocortical neurons. *Science*, 268:1503–1506.

Madison DV, Lancaster B Nicoll RA (1987) Voltage clamp analysis of cholinergic action in the hippocampus *J Neurosci*, 7(3):733–741.

Magee JC (1999) Dendritic Ih normalizes temporal summation in hippocampal CA1 neurons. *Nat Neurosci* 2(9):848.

McCormick DA (1990) Cellular mechanisms of cholinergic control of neocortical and thalamic neuronal excitability. In: *Brain cholinergic system.*, edited by Steriade M, Biesold, D. Oxford: Oxford University Press pp. 236–264.

McCormick DA (1993) Actions of acetylcholine in the cerebral cortex and thalamus and implications for function. *Prog Brain Res* 98:303–308.

Metherate R, Cox CL, Ashe JH (1992) Cellular bases of neocortical activation: modulation of neural oscillations by the nucleus basalis and endogenous acetylcholine. *J Neurosci* 12:4701–4711.

Netoff TI, Acker CD, Bettencourt JC, White JA (2005) Beyond two-cell networks: experimental measurement of neuronal responses to multiple synaptic inputs. *J Comput Neurosci* 18.

Prescott SA, De Koninck Y, Sejnowski TJ (2008) Biophysical basis for three distinct dynamical mechanisms of action potential initiation. *PLoS CB*, 4(10):e1000198.

Remme MWH, Lengyel M, Gutkin BS (2009) The role of ongoing dendritic oscillations in single-neuron dynamics. *PLoS CB*, 5(9): e1000493.

Reyes AD Fetz EE. (1993a) Effects of transient depolarizing potentials on the firing rate of cat neocortical neurons. *J Neurophysiol* 69:1673–1683.

Reyes AD Fetz EE. (1993b) Two modes of interspike interval shortening by brief transient depolarizations in cat neocortical neurons. *J Neurophysiol* 69: 1661–1672.

Schobesberger H, Wheeler DW, Horn JP (2000) A model for pleiotropic muscarinic potentiation of fast synaptic transmission. *J Neurophysiol* 83(4):1912–1923.

Shah MM, Mistry M, Marsh SJ, Brown DA Delmas P. (2002) Molecular correlates of the M-current in cultured rat hippocampal neurons. *J Physiol* 544:29–37.

Steriade M, Buzsáki G (1990) Parallel activation of thalamic and cortical neurons by brainstem and basal forebrain cholinergic systems. In: *Brain cholinergic system*, edited by Steriade M, Biesold, D. Oxford: Oxford University Press, pp. 236–264.

Steriade M (2004) Acetylcholine systems and rhythmic activities during the waking–sleep cycle. Prog Brain Res 145:179–196.

Stiefel KM, Gutkin BS, Sejnowski TJ (2008) Cholinergic neuromodulation changes phase response curve shape and type in cortical pyramidal neurons. *PLoS One*, 3(12):e3947.

Stiefel KM, Gutkin BS, Sejnowski TJ (2009) The effects of cholinergic neuromodulation on neuronal phase-response curves of modeled cortical neurons. *J Comput Neurosci* 26(2):289–301.

Stiefel KM, Tennigkeit F, Singer W (2005) Synaptic plasticity in the absence of backpropagating spikes of layer II inputs to layer V pyramidal cells in rat visual cortex. *Eur J Neurosci* 21(9):2605–2610.

Stiefel K, Gutkin BS, Sejnowski TJ (2003) Cholinergic modulation of spike generating dynamics in cortical pyramidal neurons. *Soc Neurosci Abstr*, 29.

Tateno T, Harsch A, Robinson HP (2004) Threshold firing frequency-current relationships of neurons in rat somatosensory cortex: type 1 and type 2 dynamics. *J Neurophysiol* 92: 2283–2294.

Tsubo Y, Takada M, Reyes AD Fukai T (2007) Layer and frequency dependencies of phase response properties of pyramidal neurons in rat motor cortex, Euro. *J Neurosci* 25:3429–3441.

Chapter 13
Continuum of Type I Somatic to Type II Dendritic PRCs; Phase Response Properties of a Morphologically Reconstructed Globus Pallidus Neuron Model

Nathan W. Schultheiss

Abstract A neuron's phase response curve (PRC) describes how inputs at different times during the spike cycle affect the timing of subsequent spikes, and PRC analysis is a powerful technique for predicting and interpreting the emergence of synchronous modes in coupled networks and neuronal populations. However, neuron models whose PRCs are used to study network dynamics typically consist of a minimal set of dynamic variables and often lack a realistic dendritic structure. In this chapter I describe the phase response properties of a fully reconstructed morphological model of a globus pallidus (GP) neuron during intrinsic pacemaking and during faster spiking driven by somatic current injection. This approach allows investigation of how intrinsic conductances contribute to neuronal responses in a model that preserves the complex staptiotemporal interactions between active properties of the model. I demonstrate that a single neuron can possess both type I and type II PRCs for inputs delivered to different places within the neuronal morphology. Whereas the somatic PRC for the GP model is type I, the distal dendritic PRC is type II as a consequence of the high voltage activated calcium current (Ca_{HVA}) and the small conductance calcium-activated potassium current (SK) in distal dendritic segments. The precise shapes of these somatic and dendritic PRCs are highly sensitive to the spike frequency of the model such that during faster spiking the skewness of the primary somatic PRC is reduced, the second-order somatic PRC contains a significant negative lobe, and the negative lobe in the primary dendritic PRC is shifted to the second-order PRC. These results illustrate the complex spatiotemporal interactions of synaptic inputs with intrinsic neuronal mechanisms and highlight the need to consider network connectivity on a scale that distinguishes inputs to different regions of the neuronal morphology.

N.W. Schultheiss (✉)
Center for Memory and Brain, Psychology Department, Boston University, Boston, MA, USA
e-mail: nschult@bu.edu

N.W. Schultheiss et al. (eds.), *Phase Response Curves in Neuroscience: Theory, Experiment, and Analysis*, Springer Series in Computational Neuroscience 6,
DOI 10.1007/978-1-4614-0739-3_13, © Springer Science+Business Media, LLC 2012

1 A Morphologically Reconstructed Globus Pallidus (GP) Neuron Model

As described throughout this text, phase response curves (PRCs) are a compact device relating the timing of synaptic inputs within a neuron's spike cycle to the consequent shifts in the timing of output spike initiation. Experimental and theoretical investigations of neuronal phase response properties, many of which are also described in this text, have demonstrated the utility of PRCs for prediction of synchronization phenomena in connected networks of neurons and populations of neurons receiving correlated inputs. Generally, these approaches represent a reductionist perspective of neurons and networks where only the essential dynamics are preserved. In network models, for example, the complexity of individual neurons is typically reduced to a minimal set of dynamic variables and heterogeneity between neurons is sometimes neglected. Experimental estimations of neuronal PRCs are typically limited to somatic recordings and stimulation protocols and are typically carried out in slice preparations which may not fully represent neuronal dynamics under in vivo conditions. Furthermore, connectivity between neurons is often assumed to be very weak such that the effects of individual inputs on spike timing sum linearly. In many cases, such reductions of neuronal or network complexity are necessary to allow neuronal PRCs to be related analytically to network dynamics. However, excluding these complexities eliminates higher order interactions between neuronal morphology, spatially distributed active conductances, and complex network connectivity patterns that may be important features of neural computation. In this chapter, I will describe a nearly opposite approach to PRC analysis wherein, instead of minimizing neuronal complexity, a great degree of physiological realism is incorporated into a single neuron model. The model is then treated as if it were a real neuron for which perfect stimulus control and complete access to all sources of data are available.

In point neuron models, all dynamic variables interact creating an n-dimensional dynamical system in which the membrane currents feed back onto a single voltage term. This is essentially like assuming that all the computational machinery of a neuron resides in the soma, and that the entire neuron engages as a unit in one computation at a time. A central advantage of modeling the spatially extended processes of a neuron is that it allows for multiple voltage terms reflecting differences in membrane voltage at different places within the neuronal morphology at any given time. Because it contains separate voltage terms for each compartment, a multicompartmental model allows for the coexistence of multiple dynamical subsystems within the neuronal morphology such that perisomatic inputs may have very different effects on neuronal output than distal dendritic inputs, and different regions of dendrite can potentially be involved in distinct but concurrent integrative processes. Furthermore, different contingents or distributions of active membrane conductances in the dendrite and soma may contribute substantially to differences in local synaptic integration taking place in real neurons that cannot be captured in a point neuron model.

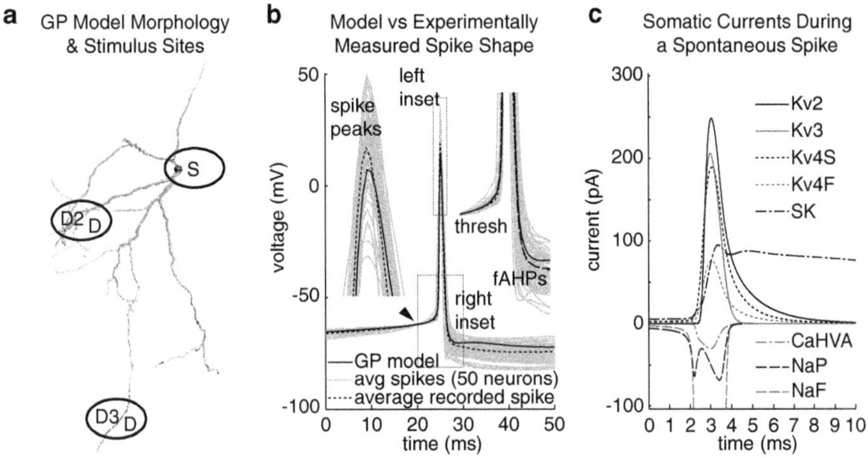

Fig. 13.1 GP model morphology and spontaneous spiking (modified from Schultheiss et al. 2010). (**a**) Morphology of the GP model, and stimulation sites: soma (S), the distal tip of the second dendrite ($D2_D$), and the distal tip of the third dendrite ($D3_D$). (**b**) Shape of a spontaneous spike generated by the model (*solid black trace*) compared to average spikes from 50 GP neurons recorded in slice (*grey traces*) and the average across recorded neurons (*dashed black trace*). The voltage traces are anchored 5 ms prior to the spike peaks (*arrowhead*) to allow comparison of fast after-hyperpolarizations (fAHPs) and medium AHPs. (**c**) Somatic currents during a spontaneous spike of the model

All of the simulations that I will describe were carried out using a 585 compartment model of a globus pallidus (GP) neuron built using the GENESIS simulation package (Fig. 13.1a) (Hanson et al. 2004; Gunay et al. 2008; Edgerton et al. 2010; Schultheiss et al. 2010). The GP model contains an axon with a highly excitable initial segment, a soma, and 511 dendritic compartments that reproduce the morphology of a GP neuron for which a battery of electrophysiological measures were made in vitro. One calcium-activated conductance, the small-conductance calcium-activated potassium current (SK), modeled using the Hill equation, and eight voltage-gated membrane conductances, modeled using Hodgkin–Huxley formalism, were implemented in the model based on experimental evidence of their presence in GP neurons. The voltage-gated conductances include: fast-transient and persistent sodium currents, NaF & NaP; fast and slow delayed-rectifier (K_{dr}) potassium currents, K_V3 & K_V2; an A-type potassium current, K_V4; an M-type potassium current, KCNQ; the hyperpolarization-activated mixed cation current, or h-current; and a high-voltage-activated calcium current (Ca_{HVA}) representing a mixture of L-, N-, and P/Q-type currents. Based on data from an immunocytochemical study of Ca_{HVA} channel proteins in GP neurons, the Ca_{HVA} conductance was distributed in the dendrite of the model with higher densities in thinner dendritic segments (Hanson and Smith 2002). The conductance densities of the membrane currents in the model were tuned such that the model reproduces the mean electrophysiological behaviors of GP neurons recorded in slice (Gunay et al. 2008; Schultheiss et al. 2010), including: spike shape (Fig. 13.1b), the injected current to spike frequency relationship (FI curve), spike height attenuation, h-current dependent sag, and latency to the first spike following the offset of hyperpolarizing current injections. Figure 13.1c shows the somatic currents activated during a spontaneous action potential and illustrates the complex dynamics of such high dimensional systems as neurons.

2 Phase Response Analysis of GP

The globus pallidus (GP) is a critical component of the indirect pathway of basal ganglia circuitry. GP neurons are intrinsic oscillators spiking rhythmically at <10 Hz in slice preparations, but in vivo GP neurons are driven by ongoing synaptic activity to spike with a mean frequency of ~ 30 Hz in the rodent (Ruskin et al. 1999; Urbain et al. 2000; Mallet et al. 2008). The involvement of GP neurons in different modes of network activity in normal animals and disease models makes them a particularly interesting target for phase response analyses. In normal animals, GP neurons are markedly desynchronized (Bar-Gad et al. 2003) perhaps in part as a consequence of perisomatically terminating inhibitory collaterals among GP neurons. However, in models of Parkinson's disease (PD), GP neurons transition to a synchronous, rhythmic spike pattern (Nini et al. 1995; Raz et al. 2000). This transition potentially reflects increased susceptibility to entrainment to rhythmic excitatory inputs from the subthalamic nucleus (STN) which is the primary source of glutamatergic synaptic innervation of GP (Magill et al. 2001; Mallet et al. 2008), and/or feedback activity within the reciprocally connected GP-STN microcircuit (Plenz and Kitai 1999; Bevan et al. 2002; Terman et al. 2002). The underlying mechanisms for the emergence of rhythmic, synchronized activity within GP are not well understood, but are likely to involve a combination of cellular and network factors that are sensitive to the degeneration of dopaminergic modulation accompanying PD.

In this chapter, I will focus on depolarizing inputs to the GP model, to address the consequences of excitatory STN inputs to GP for synaptic integration and network synchronization and entrainment. I will discuss two approaches to phase response analysis of the GP model. First, I will describe the phase response properties of the GP model when stimuli are delivered to different regions of the neuronal morphology during intrinsic pacemaking, and second, I will describe how driving the model to higher spike frequencies using somatic applied current affects the phase response properties of the model. I will also outline the cellular mechanisms underlying the pattern of PRCs observed across different morphological regions of the model and how these mechanisms influence spike timing during somatically driven faster spiking.

3 Useful Variants of the PRC: Single-cycle PRCs and Permanent PRCs

Throughout this chapter, PRCs are plotted as the magnitude of stimulus-evoked shifts in spike timing against the phase of the inputs within the ongoing spike cycle such that positive values reflect advancements of the spiking rhythm. To highlight the features of the phase response properties of the GP model in the different sets of simulations, I will at times use variants of this definition of the phase-response curve. For very weak stimuli, it is common that stimuli only affect the ongoing spike cycle; however, physiologically realistic stimuli can influence the timing of the next few spikes. In simulations where the model is allowed to spike rhythmically at its intrinsic spike frequency or is driven to spike faster by tonic applied current, in addition to the standard PRC, I will also use single-cycle PRCs and another type of PRC, the *permanent* PRC, to illustrate how shifts in spike timing evolve across successive spike cycles subsequent to stimulation.

Single-cycle PRCs represent the independent effects of stimulation on the individual spike cycles affected by the stimulus. Thus, a stimulus that shortens the ongoing spike cycle but lengthens the subsequent spike cycle will yield a positive value in the first single-cycle PRC (F1) (referred to as the primary or first-order PRC) and a negative value in the second-order PRC (F2). The single-cycle PRCs for spike cycles whose durations are neither shortened nor lengthened by the stimulus, i.e., interspike intervals several cycles subsequent to stimulation, contain zero values, because although the phase of the spiking rhythm may remain shifted, there is no effect of the stimuli on the length of those intervals. Using the convention of Preyer and Butera, higher-order single-cycle PRCs are plotted at negative phase values, i.e., to the left of lower-order PRCs (F3 before F2 before F1, etc.) (Preyer and Butera 2005). This convention yields a single continuous curve from the individual single-cycle PRCs, containing zero values at the most negative phases for the higher-order PRCs that do not show stimulus effects. Thus, the curve composed of the set of single-cycle PRCs attenuates from right to left reflecting the diminishing effect of inputs occurring during earlier spike cycles with respect to a reference spike (the shifted spike) which is assigned a phase of one.

The permanent PRC is the sum of single-cycle PRCs and illustrates the net effect of stimuli over all affected spike cycles (Prinz et al. 2003). (For example, note that the permanent PRCs in Fig. 13.5 are the sums of single-cycle PRCs in Fig. 13.4.) Therefore, the permanent PRC is a compact way to summarize the total effects of stimuli on shifts in spike timing, representing the lasting reset of spiking phase that will persist indefinitely for an isolated oscillator.

4 Phase Response Curves During Intrinsic Pacemaking

Like GP neurons in vitro, the GP model is an intrinsic pacemaker spontaneously firing action potentials at ~8 Hz. In order to assess whether the location of synaptic inputs within the neuronal morphology plays a significant role in the phase response properties of the model, PRCs were derived by delivering either weak, brief current injections or dual exponential synaptic conductance inputs at 72 time points during the intrinsic spike cycle. Stimuli were delivered either to the soma or one of several sites throughout the dendritic tree. (I will focus primarily on the two distal dendritic sites indicated in Fig. 13.1a.)

Based on seminal investigations of the phase response properties of the Connor (Connor et al. 1977) and Hodgkin–Huxley (HH) models (Hodgkin and Huxley 1952a, b, c, d), PRCs containing only positive values are termed type I, whereas PRCs containing a significant negative region are termed type II (Hansel et al. 1995). This classification of PRCs differentiates models for which excitatory inputs can only advance subsequent spiking (type I), such as the Connor model, from those for which excitatory inputs can either advance or delay subsequent spiking depending on input phase (type II), i.e., the HH model. This distinction in turn reflects the models' classes of membrane excitability (Hodgkin 1948; Rinzel and Ermentrout 1998) and the types of bifurcations leading from quiescence to rhythmic spiking (Ermentrout 1996; Rinzel and Ermentrout 1998; Izhikevich 2006), as well as the ability of networks to synchronize when coupled with excitation (Hansel et al. 1993, 1995; Kopell and Somers 1995; Izhikevich 2006).

Somatic PRCs. All PRC stimuli delivered to the soma of the GP model resulted in advancements of the spiking rhythm. Thus, the somatic PRC for the model was type I, containing only positive values for stimuli delivered at all times within the spontaneous spike cycle. For very weak depolarizing current injections, the somatic PRC scaled linearly in amplitude with the strength of the inputs. However, this relationship held only for 3 ms current injections of less than 10 pA (data not shown, but see (Schultheiss et al. 2010)). The peak conductance of a single AMPAergic synaptic event is on the order of 250 pS to 1 nS producing a peak synaptic current of approximately 15–60 pA (Hernandez et al. 2006). Thus, in the GP model synaptic currents due to the activation of even a single synapse lie outside the domain of weak coupling (introduced in Chap. 1 and discussed widely throughout this text as a basis for prediction of network states).

To address the question of how the input phase of physiologically realistic synaptic input affects spike timing, stronger current injections or dual exponential AMPA-type conductance inputs were delivered to the model. Like the weaker inputs discussed above, somatic current pulses of up to 300 pA or synaptic conductance inputs of up to 4 nS yielded type I PRCs (Fig. 13.2a). The peak of the somatic PRC occurred earlier in the spike cycle for stronger depolarizing inputs, reflecting that maximal sensitivity to somatic inputs occurs just before a spike, and also that a depolarizing input cannot advance the subsequent spike so much that it occurs before the input itself. This latter phenomenon is sometimes referred to as the

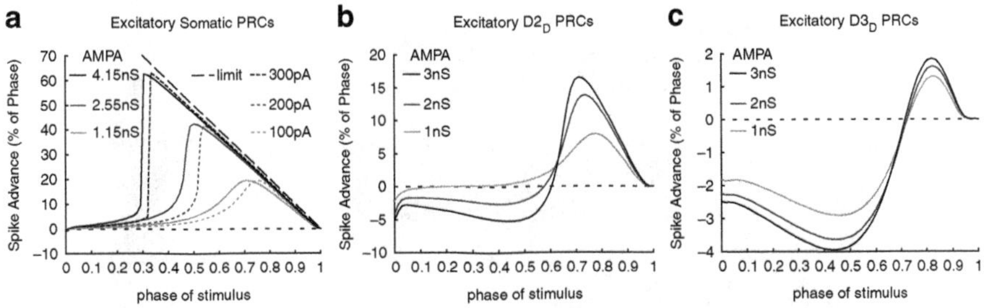

Fig. 13.2 Phase response curves for inputs to different regions of the GP model morphology (modified from Schultheiss et al. 2010). (**a**) Type I somatic PRCs for current and synaptic conductance inputs of physiological amplitude. (**b**) PRCs transition from type I to type II for synaptic inputs of increasing strengths to $D2_D$. (**c**) Type II PRCs for synaptic inputs to ($D3_D$)

causal limit of spike advancement (Fig. 13.2a, long-dashed line). The causal limit is reached when a spike is initiated immediately with delivery of the stimulus, and as shown in Fig. 13.2a, stronger somatic inputs caused the PRC to approach the limit of spike advancement over a larger range of phase.

Dendritic PRCs. To test how the strength and timing of physiologically realistic inputs to different sites of the dendritic tree affect output spike timing, PRCs were constructed for seven dendritic sites using synaptic inputs of a range of peak conductance amplitudes. In each case, PRCs for proximal dendritic inputs were qualitatively similar to somatic PRCs. Surprisingly, larger AMPAergic inputs to distal dendritic regions yielded PRCs with increasing type II character. PRCs for the stimuli delivered to the distal tips of the second and third dendritic branches ($D2_D$& $D3_D$) are shown in Fig. 13.2b,c, respectively. Whereas $D2_D$ PRCs for excitatory inputs of 1 nS or less were type I, inputs stronger than 1 nS significantly delayed the next spike when delivered during the first half of the spike cycle. For stronger synaptic inputs to $D2_D$, the negative regions of the corresponding type II PRCs were deeper reflecting greater delays of the spiking rhythm. Thus, paradoxically, strengthening "excitatory" synaptic input to $D2_D$ changed the net effect of the input from phase-advancing to phase-delaying for a significant portion of the spike cycle. The positive peaks of PRCs for $D2_D$ stimuli were also larger for stronger excitatory inputs (Fig. 13.2b) resulting in biphasic, type II PRCs of greater amplitude. Excitatory synaptic inputs to $D3_D$ yielded type II PRCs for all input strengths tested (Fig. 13.2c), but as was the case for $D2_D$, stronger AMPA inputs to $D3_D$ increased the amplitude of the corresponding PRCs.

It is worth noting that the PRCs for the GP model are not easily dichotomized as type I for somatic inputs and type II for dendritic inputs. Rather, as Fig. 13.3 illustrates, the precise shapes of PRCs for different regions along the extent of a given dendrite compose a continuum from type I somatic to type II dendritic PRCs. The precise topography of such a phase response "landscape" as is illustrated in Fig. 13.3 will differ for different dendritic branches even within a given neuron (and

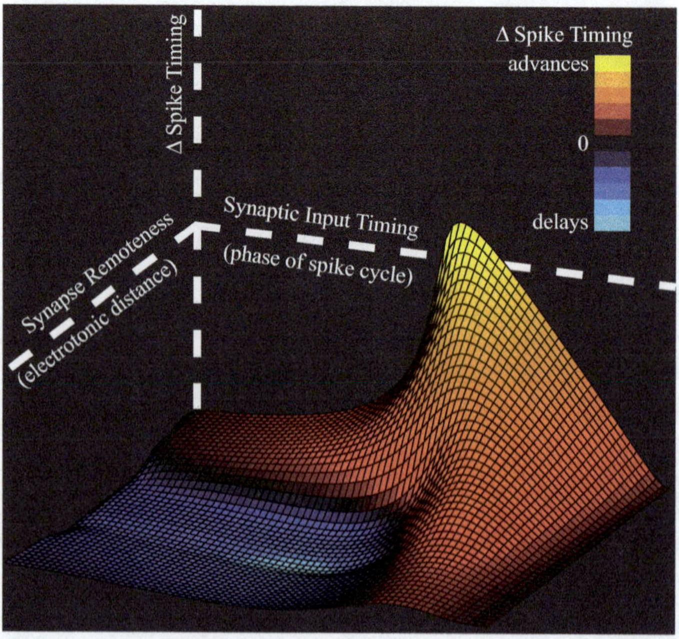

Fig. 13.3 Continuum from type I somatic to type II dendritic PRCs. PRCs for stimulus locations along the extent of the third dendritic branch of the GP model (D3)

does for the other dendrites in the GP model), as a consequence of differences in branching structure and dendritic segment dimensions, as well as nonuniformities in the spatial distribution of dendritic membrane conductances such as those discussed below.

These data illustrate that identical synaptic inputs to different regions of the neuronal morphology can have very different effects on output spike timing. Such a pattern of effects could not be experimentally revealed using purely somatic stimulation and recording protocols and cannot be captured by a single compartment model. Furthermore, the qualitative difference between somatic and dendritic PRCs for the GP model suggests that the targeting of synaptic inputs to different regions within the neuronal morphology might play a significant role in shaping network activity, and this level of organization should be considered in network models particularly in systems where projections are known to target particular regions of the postsynaptic neuronal morphology, e.g., perforant pathway projections arising from the entorhinal cortex and terminating on the distal dendrites of neurons in the CA1 region of the hippocampus.

Active conductances responsible for type II dendritic PRCs. Analysis of membrane currents in the model indicated that local dendritic activation of the small conductance calcium-activated potassium current, SK, an outward current activated by calcium influx, was responsible for the delays evoked by "excitatory" synaptic input to the distal dendrite (Schultheiss et al. 2010). These delays were a consequence of the fact that when inputs were delivered sufficiently early within the spontaneous spike cycle, a net current sink was generated local to the site of stimulation, i.e., the sum of the inward synaptic current and the net outward intrinsic membrane currents (primarily SK) was positive (outward). This distal current sink, coupled to the soma by axial current along the dendritic length, dragged on the depolarization of the soma during the spontaneous spike cycle delaying the next spike. For stimuli delivered later in the spike cycle, however, there was insufficient time for the SK current to counterbalance the inward synaptic current, and subsequent spiking was advanced because the net current across the membrane local to the stimulus was inward. Therefore, the type II character of dendritic PRCs was a consequence of the balance between the synaptic current and evoked SK current which differed for stimuli delivered at different times within the spike cycle.

The SK current is calcium activated, and so the high-voltage-activated calcium channel (Ca_{HVA}, the calcium source for SK activation in the model) in the dendrite is also a major part of the mechanism responsible for type II dendritic PRCs. As mentioned earlier, based on immunocytochemical evidence (Hanson and Smith 2002), Ca_{HVA} is distributed in the GP model such that thinner dendritic segments possess higher densities of the calcium source for SK activation. Distal dendritic compartments are thinner than more proximal dendritic segments and the distal tips of each dendritic branch are closed-ended, resulting in high local input resistance. Thus, synaptic inputs to the distal dendrite evoke larger voltage transients than at more proximal regions, driving the local voltage more strongly into the activation range for Ca_{HVA}, and activate a larger pool of calcium channels leading to greater local activation of the SK conductance. Note, however, that Fig. 13.3 illustrates that the greatest delays of spiking were not evoked by inputs to the most distal segments of the third dendritic branch, i.e., the deepest negative lobes in D3 PRCs were obtained for stimulation sites somewhat interior to the distal extremity of that dendritic branch, even though local activation of SK was highest in that location. This is a consequence of the electrotonic remoteness of the tip of the third dendritic branch leading to lesser impact on spike timing of stimuli delivered to that location. The phase response topography for stimulus locations along the *shorter* second dendritic branch (D2) (data not shown), differed from the D3 topography such that the deepest negative lobes were at the distal extremity where SK activation was maximal.

Manipulations of distal dendritic densities of SK conductance and Ca_{HVA} conductance allowed direct demonstration of the role of the SK conductance in shaping dendritic PRCs. Importantly, the intrinsic spiking rhythm of the model was not disturbed, because these manipulations were limited to the relatively few (25) distal dendritic compartments composing the sites of stimulation, $D2_D$ or $D3_D$. Local up- or downregulation of Ca_{HVA} and SK in tandem directly scaled the magnitude of SK current evoked at the distal dendrite (Fig. 13.4a,c) and resulted in type II distal dendritic PRCs with deeper or shallower negative regions, respectively (Fig. 13.4b,d). Removal of *either* Ca_{HVA} *or* SK from the distal dendrite almost completely eliminated the type II character of the corresponding PRCs, resulting in visually identical PRCs with predominantly type I character (Fig. 13.4b,d).

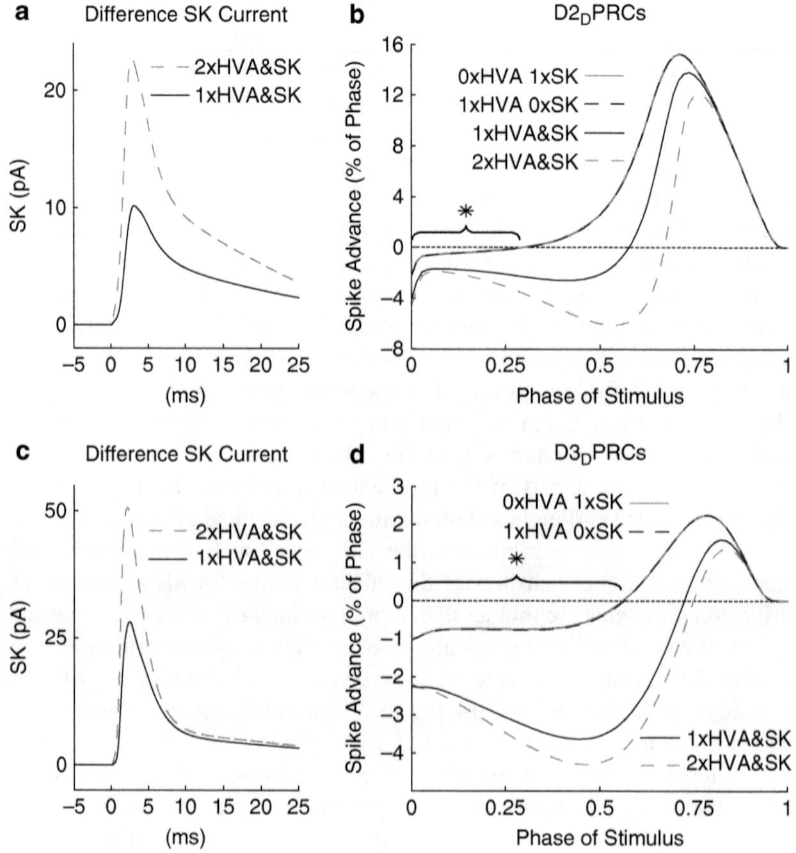

Fig. 13.4 SK current evoked locally determines the depth of negative regions in distal dendritic PRCs (modified from Schultheiss et al. 2010). (**a**) SK current evoked locally by a $D2_D$ stimulus in models with the base conductance density or twice the base conductance density of SK and Ca_{HVA}. (The evoked or "difference" SK current was calculated by subtracting SK current during the control simulation from that during a simulation containing a stimulus.) (**b**) $D2_D$ PRCs for models where the local SK and/or Ca_{HVA} conductance densities were manipulated. (**c**) SK current evoked locally by a ($D3_D$) stimulus in models with the base or twice the base conductance densities of SK and Ca_{HVA}. (**d**) ($D3_D$) PRCs for models where the local SK and/or Ca_{HVA} conductance densities were manipulated

However, despite the local removal of the Ca_{HVA} or SK conductance, these PRCs still contained shallow negative regions (Fig. 13.4b,d, asterisks) as a consequence of SK activation in compartments adjacent to the stimulated site. The scaling of the negative region of dendritic PRCs with varying amounts of SK conductance can be considered an elaboration of results from Gutkin and colleagues (2005), who showed that the addition of increasing amounts of a spike-frequency adaptation (SFA) potassium current to a point neuron model (θ-neuron model) resulted in increasing skew of the corresponding PRCs (Gutkin et al. 2005).

Interestingly, increasing the dendritic density of fast sodium conductance increased the amplitude of type II dendritic PRCs (Schultheiss et al. 2010): larger advancements of the spiking rhythm were elicited by stimuli delivered in the second half of the spike cycle for models with higher dendritic NaF conductance. However, for stimuli delivered early in the spike cycle, the additional inward sodium current elicited by PRC stimuli was further counterbalanced by activation of SK current mediated by additional calcium influx through Ca_{HVA}, resulting in greater delays of the subsequent spike (Schultheiss et al. 2010). The amplitude of dendritic type II PRCs was maximal when dendritic NaF conductance was sufficient to support dendritic spikes that propagated to the soma but were not capable of triggering a somatic spike. These dendritic action potentials maximally activated the SK conductance throughout the membrane of the stimulated dendritic branch causing maximal delays of subsequent spiking for early phase "excitatory" inputs. Sodium channel density has been shown to be greater near excitatory synapses on GP neuron dendrites (Hanson et al. 2004), suggesting that perisynaptic microdomains of active dendritic properties may contribute strongly to the integration of excitatory synaptic inputs. The relationship between amplification of synaptic depolarizations by perisynaptic sodium current and the resultant augmentation of calcium influx and SK activation represents an unexpected potential role for dendritic sodium channels. This complex spatially-localized interaction between dendritic conductances illustrates a mechanism that might not be predicted given the common assumption that dendritic sodium channels primarily function to amplify dendritic inputs offsetting passive attenuation along the dendritic extent.

5 Phase Response Curves During Somatically Driven Faster Spiking

Experimental estimation of PRCs is commonly done in slice preparations where the measured neuron is allowed to spike spontaneously as in the simulations described so far. One potential shortcoming of this approach is that neurons in slice may fire spontaneously at lower spike frequencies than in intact animals in vivo, and output spike frequency has been shown to affect the shape of neuronal PRCs. For example, Gutkin and colleagues (2005) showed that the PRC for the θ-neuron model (plus a spike frequency adaptation current) is type I, but its precise shape shows a strong dependence on the frequency of spiking: For low spike frequencies, the PRC for the θ-neuron model with SFA current was strongly skewed suggesting that the model acts as a coincidence detector when spiking slowly, whereas at higher spike frequencies the PRC became nearly symmetrical such that the model behaved as an integrator of synaptic inputs arriving at different times throughout an ongoing spike cycle (Gutkin et al. 2005). Applying tonic somatic current to the morphologically reconstructed GP model (Schultheiss et al. 2010), allowed the further investigation of how the cellular mechanisms underlying the *dendritic* phase response properties of the model function when the model is engaged in faster spiking. Furthermore, using the convention of Butera and colleagues described earlier, plotting the effects of phasic PRC stimuli on multiple spike cycles following stimulation (Preyer and Butera 2005) provides a more complete picture of the relationship between synaptic input timing and output spiking.

During 7.9 Hz spontaneous spiking of the GP model, only the ongoing ISI was affected by 2 nS AMPA input to the soma, indicating the absence of higher-order effects of synaptic excitation. Therefore, at this spike frequency, only the first-order PRC contained nonzero values, and the somatic PRC for the second interval (F2) was flat (Fig. 13.5a). Driving the GP model to higher spike frequencies (15 Hz and 30 Hz) revealed a similar pattern of spike frequency dependence for somatic PRCs as shown by Gutkin et al., such that the skewness of the first-order somatic PRC (Fig. 13.5a; F1) was reduced during faster spiking.

During 30 Hz spiking, the soma was depolarized by >5 mV compared to the average voltage during spontaneous spiking, and the activation profile of somatic currents was substantially changed, i.e., depolarization gated membrane conductances maintained higher levels of activation. While spike initiation immediately after AMPA stimulation of the soma corresponded to a large phase advance (Fig. 13.5; F1), a part of this advance was lost in the second spike cycle following stimulation (F2), because this second ISI was longer than the baseline ISIs. This was a consequence of somatic accumulation of SK activation taking place at higher spike frequencies (Schultheiss et al. 2010). Therefore, at in vivo spike frequencies, an important effect of somatic SK conductance may be to resist lasting advancements of the spiking rhythm. This effect of somatic SK conductance appears in the single-cycle somatic PRCs as a negative peak in the F2 that opposes nearly symmetrically the positive peak in the F1 (Fig. 13.5a).

Distal dendritic voltage and current transients elicited by AMPA stimuli were generally unaffected by faster somatic spike rates, e.g., the SK current elicited by input to $D2_D$ was nearly identical during slow spontaneous spiking and during 30 Hz firing (Schultheiss et al. 2010). Nevertheless, the PRC for $D2_D$ stimulation showed a dramatic shift at faster somatic spike rates. Like the reduction of skewness of the F1 somatic PRC at higher spike frequencies, the F1 PRC for $D2_D$ inputs showed a left shift of the positive peak, which resulted in increasing type I character of the F1 PRC for spike frequencies of or above 30 Hz (Fig. 13.5b; F1). Furthermore, because the time course of SK current elicited by distal dendritic stimuli was invariant across spike frequencies, the SK current showed a different phase relationship with respect to the somatic spike cycle. The waveform of AMPA-evoked SK current lasted more than 50 ms, and therefore, at spike frequencies above 15 Hz, the effect of SK current evoked by late-phase AMPA inputs to the distal dendrite impinged primarily on the second spike cycle (F2) following the stimulus. Consequently, the second ISI was prolonged, and the F2 PRC contained a significant negative peak for late-phase inputs (Fig. 13.5b).

Stimulation of $D2_D$, which is substantially closer to the soma than $D3_D$, showed a combination of the effects seen at the soma and $D2_D$. The voltage at these locations was somewhat more depolarized during faster spiking (as at the soma), but local

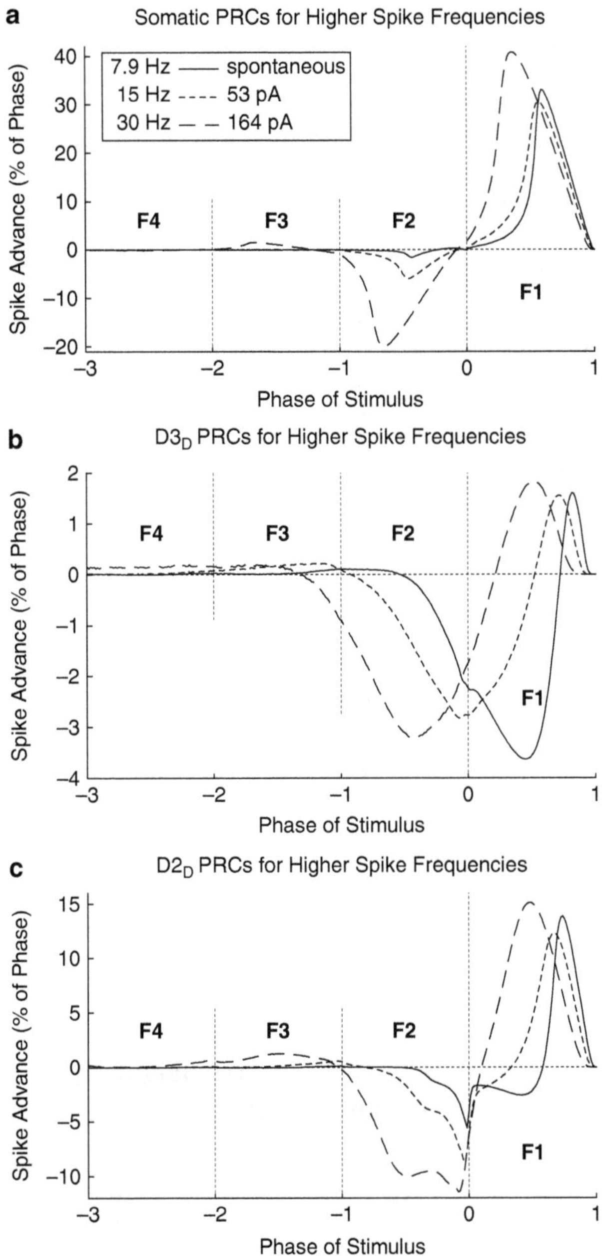

Fig. 13.5 Single-cycle PRCs for higher spike frequencies driven by somatic applied current (modified from Schultheiss et al. 2010). (**a**) Somatic PRCs for spontaneous spiking, 15 Hz, and 30 Hz. Applied current is indicated in legend. (**b**) $D3_D$ PRCs during faster spiking. (**c**) $D2_D$ PRCs during faster spiking

SK transients evoked by AMPA inputs were still relatively invariant across spike frequencies (as at the more-distal dendrite). Like the $D3_D$ case, AMPA stimuli applied late in the spike cycle led to a negative region that shifted from the F1 to the F2 PRC for higher spike rates (Fig. 13.5c). Thus, the F2 PRC for $D2_D$ contained two negative peaks (Fig. 13.5c; 30 Hz trace): one occurring at earlier phases of the spike cycle, opposing advancements elicited in the first spike cycle, and resulting from elevated activation of *somatic* SK conductance; and a second, occurring for stimuli delivered late in the spike cycle and reflecting the impact of *distal dendritic* SK current on the timing of the second spike subsequent to stimulation.

Permanent PRCs. As described earlier, the permanent PRC (Prinz et al. 2003) reflects the lasting shift of the spike train compared to the control oscillation, i.e., the total effect across spike cycles of phasic PRC stimuli, and is calculated as the sum of all single-cycle PRCs that show stimulus effects. For somatic AMPA stimulation, the permanent PRC for the GP model remained type I for all spike frequencies from 10–60 Hz (Schultheiss et al. 2010) (15 & 30 Hz shown in Fig. 13.6a), indicating that the delay of spiking exerted by *somatic* SK current on the second spike cycle (F2) never exceeded the original spike advance in the stimulated spike cycle. In the case of $D3_D$ stimulation, the delayed spikes in F2 and F3 intervals more than compensated any spike advance occurring in the stimulated spike cycle. Thus, the corresponding permanent PRC became entirely negative at spike frequencies of 15 Hz or greater, i.e., a strong AMPA stimulus at any phase of the spike cycle ultimately resulted in a delay of the spike oscillation. These simulations predict that at in vivo spike rates, strong AMPA input to the distal dendrites of GP neurons will advance the first subsequent spike but ultimately reduce spike frequency. At 30 Hz, the lasting effects of AMPA inputs on spike timing were similar for $D2_D$ and $D3_D$, such that the positive peak of the permanent PRC shifted leftward and was reduced, and the deepest negative regions of the permanent PRCs occurred for stimuli near the beginning and end of the spike cycle (Fig. 13.6b,c). However, for $D2_D$ AMPA stimuli, the PRC retained a positive region for all spike frequencies, and thus remained type II.

6 Phase Response Analyses of a Morphologically Reconstructed GP Neuron Model

Although rigorous analysis of the somatic electrophysiology of GP neurons (Gunay et al. 2008) does not directly support previous classifications of GP neuron subtypes (Nambu and Llinas 1994, 1997; Cooper and Stanford 2000), staining for calcium binding proteins clearly differentiates a population ($\sim 2/3$ of GP neurons) of larger, parvalbumin-positive neurons from smaller, PV-negative neurons (Kita et al. 1990; Kita 1994). It remains a possibility that differences in the *dendritic* properties of GP neurons may be present and important for synaptic integration that are not reliably accessible to somatic patch clamp studies. Interestingly, activation

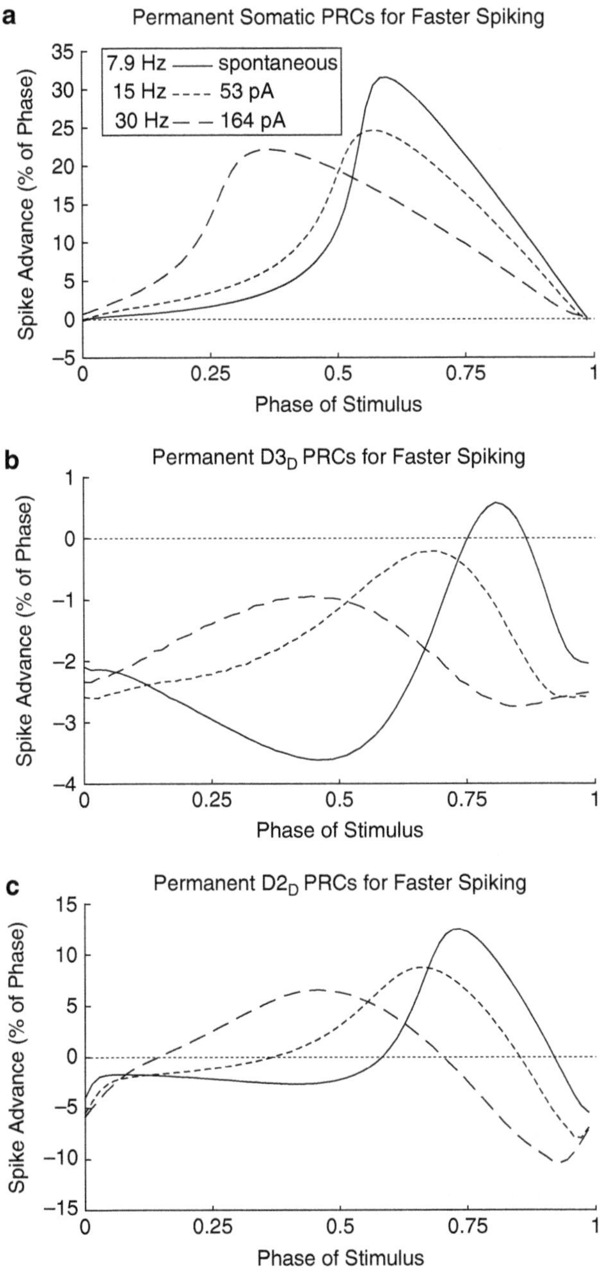

Fig. 13.6 Permanent PRCs for higher spike frequencies driven by somatic applied current (modified from Schultheiss et al. 2010). (**a**) Permanent somatic PRCs. (**b**) Permanent D3$_D$ PRCs. (**c**) Permanent D2$_D$ PRCs

of $GABA_B$ receptors has been shown to strongly modulate the Ca_{HVA} calcium current in small, but not large, GP neurons (Stefani et al. 1999) in agreement with the observation that the Ca_{HVA} conductance density is greater in thinner dendritic segments (Hanson and Smith 2002). If the SK conductance mechanism underlying the dendritic phase response properties of GP neurons contributes to the entrainment of pallidal networks, it is possible that differences in dendritic calcium channel conductance densities or modulatory state could support different modes of network interactions between GP and STN.

The abilities to directly measure membrane currents and to manipulate conductance densities in particular sets of dendritic compartments in these simulations allowed the exploration of how individual currents and interactions between currents can contribute to neuronal responses amidst the highly complex array of active properties responsible for neuronal dynamics. It is important to note that the bulk of synaptic input is on the dendrites of GP neurons as well as most other types of neurons. The active properties of dendrites are therefore of primary importance for synaptic integration and can potentially underlie surprising and significant relationships between synaptic inputs, spike timing, and network activity. Morphological reconstruction and tuning of a large multicompartmental model require a significant investment of time and energy, and rigorous parameter space exploration is necessary to characterize the robustness of any finding with such a model. However, the results described in this chapter using the GP neuron model are potentially representative of phenomena in other neuron types, and the morphological modeling approach represents an important contributor to understanding what the "essential dynamics" of neurons really are, particularly when the critical experiments in vitro or in vivo are not realistically possible. The simulations described in this chapter incorporate a considerable degree of physiological realism providing an important bridge of the gap between theoretical investigations of simplified systems which may not fully capture the dynamics of real neurons, and experimental physiology which is limited by technical constraints. This approach allows verification of predictions made from reduced models and generalization of such findings to provide insight into in vivo neuronal dynamics.

I have emphasized that inputs to different regions of the neuronal morphology can yield qualitatively different phase response properties. In the GP model, perisomatic PRCs are type I, whereas dendritic PRCs are type II. This pattern of phase response properties in the GP model is a consequence of spatially distributed membrane conductances, i.e., the high-voltage-activated calcium current, Ca_{HVA}, and the small-conductance calcium-activated potassium current, SK. This finding provides a striking example of how membrane conductances that are not directly involved in a neuron's spiking mechanism, i.e., the neuron's class of excitability, can contribute to the integration of inputs, shape the neuron's phase response curve, and potentially influence network dynamics. Furthermore, the differential effects of inputs targeted to different regions of the neuronal morphology highlight that for network models, it can be critical to consider the connectivity between neurons at a finer spatial scale than is usually incorporated in networks composed of highly-simplified neuronal units. I have also described that at in vivo spike frequencies, which are commonly higher than those observed in vitro, the phase relationships between the time courses of the interspike interval and the activation of intrinsic membrane conductances by physiologically realistic synaptic inputs are shifted such that the effects of inputs can span multiple spike cycles. In the GP model, this shifts the rhythm-delaying effect of dendritic SK current into the second spike cycle following excitatory dendritic input yielding advancements of the first spike subsequent to stimulation and delays of the second spike. Taken together these findings provide a foundation for further investigations of neuronal phase response properties during in vivo high conductance states (Destexhe et al. 2003) generated by considerable ongoing synaptic activity (Schultheiss et al. submitted 2011) and of the complex network interactions between neurons whose dendrites serve as fundamental computational intermediates between synaptic input and spiking output.

References

Bar-Gad I, Heimer G, Ritov Y, Bergman H (2003) Functional correlations between neighboring neurons in the primate globus pallidus are weak or nonexistent. Journal of Neuroscience 23:4012–4016.

Bevan MD, Magill PJ, Terman D, Bolam JP, Wilson CJ (2002) Move to the rhythm: oscillations in the subthalamic nucleus-external globus pallidus network. Trends in Neurosciences 25: 525–531.

Connor JA, Walter D, McKown R (1977) Neural repetitive firing - modifications of Hodgkin–Huxley axon suggested by experimental results from crustacean axons. Biophysical Journal 18:81–102.

Cooper AJ, Stanford IM (2000) Electrophysiological and morphological characteristics of three subtypes of rat globus pallidus neurone in vitro. Journal of Physiology 527:291–304.

Destexhe A, Rudolph M, Pare D (2003) The high-conductance state of neocortical neurons in vivo. Nature Reviews Neuroscience 4:739–751.

Edgerton JR, Hanson JE, Gunay C, Jaeger D (2010) Dendritic sodium channels regulate network integration in globus pallidus neurons: a modeling study. Journal of Neuroscience 30: 15146–15159.

Ermentrout B (1996) Type I membranes, phase resetting curves, and synchrony. Neural Comput 8:979–1001.

Gunay C, Edgerton JR, Jaeger D (2008) Channel density distributions explain spiking variability in the globus pallidus: a combined physiology and computer simulation database approach. Journal of Neuroscience 28:7476–7491.

Gutkin BS, Ermentrout GB, Reyes AD (2005) Phase-response curves give the responses of neurons to transient inputs. Journal of Neurophysiology 94:1623–1635.

Hansel D, Mato G, Meunier C (1993) Phase dynamics for weakly coupled Hodgkin–Huxley neurons. Europhys Lett 23:367–372.

Hansel D, Mato G, Meunier C (1995) Synchrony in excitatory neural networks. Neural Computation 7:307–337.

Hanson JE, Smith Y (2002) Subcellular distribution of high-voltage-activated calcium channel subtypes in rat globus pallidus neurons. Journal of Comparative Neurology 442:89–98.

Hanson JE, Smith Y, Jaeger D (2004) Sodium channels and dendritic spike initiation at excitatory synapses in globus pallidus neurons. Journal of Neuroscience 24:329–340.

Hernandez A, Ibanez-Sandoval O, Sierra A, Valdiosera R, Tapia D, Anaya V, Galarraga E, Bargas J, Aceves J (2006) Control of the subthalamic innervation of the rat globus pallidus by D-2/3 and D-4 dopamine receptors. Journal of Neurophysiology 96:2877–2888.
Hodgkin AL (1948) The local electric changes associated with repetitive action in a non-medullated axon. Journal of Physiology-London 107:165–181.
Hodgkin AL, Huxley AF (1952a) The components of membrane conductance in the giant axon of Loligo. The Journal of physiology 116:473.
Hodgkin AL, Huxley AF (1952b) The dual effect of membrane potential on sodium conductance in the giant axon of Loligo. The Journal of physiology 116:497.
Hodgkin AL, Huxley AF (1952c) A quantitative description of membrane current and its application to conduction in the giant axon of Loligo. J Physiol 117:500–544.
Hodgkin AL, Huxley AF (1952d) Currents carried by sodium and potassium ions through the membrane of the giant axon of Loligo. The Journal of physiology 116:449.
Izhikevich EM (2006) Dynamical systems in neuroscience: The geometry of excitability and bursting: The MIT press.
Kita H (1994) Parvalbumin immunopositive neurons in rat globus pallidus - a light and electron microscopic study. Brain Research 657:31–41.
Kita H, Kosaka T, Heizmann CW (1990) Parvalbumin-immunoreactive neurons in the rat neostriatum - a light and electron microscopic study. Brain Research 536:1–15.
Kopell N, Somers D (1995) Antiphase solutions in relaxation oscillators coupled through excitatory interactions. J Math Biol 33:261–280.
Magill PJ, Bolam JP, Bevan MD (2001) Dopamine regulates the impact of the cerebral cortex on the subthalamic nucleus-globus pallidus network. Neuroscience 106:313–330.
Mallet N, Pogosyan A, Marton LF, Bolam JP, Brown P, Magill PJ (2008) Parkinsonian beta oscillations in the external globus pallidus and their relationship with subthalamic nucleus activity. Journal of Neuroscience 28:14245–14258.
Nambu A, Llinas R (1994) Electrophysiology of globus pallidus neurons in vitro. Journal of Neurophysiology 72:1127–1139.
Nambu A, Llinas R (1997) Morphology of globus pallidus neurons: Its correlation with electrophysiology in guinea pig brain slices. Journal of Comparative Neurology 377:85–94.
Nini A, Feingold A, Slovin H, Bergman H (1995) Neurons in the globus pallidus do not show correlated activity in the normal monkey, but phase-locked oscillations appear in the MPTP model of parkinsonism. Journal of Neurophysiology 74:1800–1805.
Plenz D, Kitai ST (1999) A basal ganglia pacemaker formed by the subthalamic nucleus and external globus pallidus. Nature 400:677–682.
Preyer AJ, Butera RJ (2005) Neuronal oscillators in Aplysia californica that demonstrate weak coupling in vitro. Physical Review Letters | Physical Review Letters 95:138103/138101–138104.
Prinz AA, Thirumalai V, Marder E (2003) The functional consequences of changes in the strength and duration of synaptic inputs to oscillatory neurons. Journal of Neuroscience 23:943–954.
Raz A, Vaadia E, Bergman H (2000) Firing patterns and correlations of spontaneous discharge of pallidal neurons in the normal and the tremulous 1-methyl-4-phenyl-1,2,3,6-tetrahydropyridine vervet model of parkinsonism. Journal of Neuroscience 20:8559–8571.
Rinzel JM, Ermentrout GB (1998) Analysis of neural excitability and oscillations. In: Methods in Neuronal Modeling, 2nd Edition (Koch C, Segev I, eds), pp 251–291: MIT Press.
Ruskin DN, Bergstrom DA, Kaneoke Y, Patel BN, Twery MJ, Walters JR (1999) Multisecond oscillations in firing rate in the basal ganglia: Robust modulation by dopamine receptor activation and anesthesia. Journal of Neurophysiology 81:2046–2055.
Schultheiss NW, Edgerton JR, Jaeger D (2010) Phase response curve analysis of a full morphological globus pallidus neuron model reveals distinct perisomatic and dendritic modes of synaptic integration. Journal of Neuroscience 30:2767–2782.

Schultheiss NW, Edgerton JR, Jaeger D (submitted 2011) Robustness, variability, phase-dependence, and longevity of individual synaptic inputs' effects on spike timing during in vivo high conductance states; A simulation study of globus pallidus neuron phase response properties.

Stefani A, Spadoni F, Giacomini P, Lavaroni F, Bernardi G (1999) The modulation of calcium current by GABA metabotropic receptors in a sub-population of pallidal neurons. European Journal of Neuroscience 11:3995–4005.

Terman D, Rubin JE, Yew AC, Wilson CJ (2002) Activity patterns in a model for the subthalamopallidal network of the basal ganglia. Journal of Neuroscience 22:2963–2976.

Urbain N, Gervasoni D, Souliere F, Lobo L, Rentero N, Windels F, Astier B, Savasta M, Fort P, Renaud B, Luppi PH, Chouvet G (2000) Unrelated course of subthalamic nucleus and globus pallidus neuronal activities across vigilance states in the rat. European Journal of Neuroscience 12:3361–3374.

Part IV
Prediction of Network Activity with Phase Response Curves

Introduction

The first three parts of this text have introduced the theoretical foundations of phase response analysis; outlined strategies for estimation of neuronal PRCs; detailed the cellular mechanisms underlying phase response properties of a number of neuronal systems of varying complexity and under a variety of conditions representative of diverse neural functions; and in many cases introduced relationships between PRCs and patterning of network behavior. The content of the last part of the book builds upon these earlier materials to extend PRC analysis to a number of current foci in the field. The chapters in Part IV describe uses of PRCs for prediction or interpretation of network activity and tie the network level of neural processing to its cellular substrates.

Chapter 14 by Lewis and Skinner considers electrical coupling between pairs of integrate-and-fire models, conductance-based models, and multicompartmental models to address how phase response curves (PRCs) and weakly coupled oscillator theory can be used to understand patterned activity in networks of spiking cortical neurons. In Chap. 15, Hansel et al. extend this analysis to larger networks and consider coupling through both chemical synapses and electrical synapses. They show that synchronization mediated by electrical synapses is more versatile than previously considered and depends on the intrinsic currents and morphology of neurons as well as on the combination of electrical and inhibitory synapses. Inhibitory synapses are ubiquitous in networks of oscillatory neurons, including central pattern-generating circuits. In Chap. 16, Nadim et al. use the concept of a synaptic PRC to examine a possible role of feedback inhibition as a promoter of oscillation stability in the face of extrinsic perturbations.

Chapter 17 by Oprisan details how the existence and stability of phase-locked modes in ring networks can be predicted using phase-resetting curves and spike time resetting curves of the individual network components. PRC theory is used by Achuthan et al. in Chap. 18 to analyze phase-locked patterns in pulse-coupled all-to-all networks of neurons that receive multiple inputs per cycle. They show that the existence and stability of synchronous clusters of neurons in the network can to a large extent be explained by factors that influence the slope of the PRC.

In Chap. 19, Fink et al. investigate the differential effects of type I and type II PRCs on network synchrony by simulating large-scale networks of Morris–Lecar neurons with excitatory coupling. They demonstrate that as the intrinsic frequency of component neurons is increased, type I phase response properties promote increased network synchrony, while type II properties lead to reduced network synchrony.

Finally, Chap. 20 by Zeberg et al. uses a phase-resetting model based on the synaptic phase-resetting function to show how the intrinsic properties of cortical fast-spiking interneurons and their synaptic connections allow synchronization of firing over a wide gamma frequency band and to provide insights into the onset and stability of patterns of gamma oscillation in the cortex.

Taken together, the chapters in Part IV are pushing the boundaries of phase response analysis by challenging it with multiple different neuron types, combinations of different types of synaptic connections, large-scale networks of different connectivity topology, and multiple inputs per cycle of a neuronal oscillator. We have likely only seen the beginning of efforts at linking PRC analysis with intrinsic and synaptic biophysical mechanisms. PRC methods are likely to be increasingly utilized to fathom the underpinnings of functional and pathophysiological oscillations in neural systems.

Chapter 14
Understanding Activity in Electrically Coupled Networks Using PRCs and the Theory of Weakly Coupled Oscillators

Timothy J. Lewis and Frances K. Skinner

Abstract In this chapter, we describe in detail how phase-locking in electrically coupled networks of spiking neurons can be understood using the framework of phase response curves (PRCs) and weak coupling theory. We provide the necessary mathematical background and biological context to allow the reader to acquire an understanding of the network dynamics. We present the work using neuronal representations that include general integrate-and-fire and conductance-based models as well as spatially distributed, compartmental models.

1 Introduction

Neurons communicate with each other by electrical and chemical synapses. Chemical communication involves the release of neurotransmitters from presynaptic neurons into a synaptic cleft that separates presynaptic and postsynaptic neurons. Electrical communication on the other hand occurs by direct intercellular channels between presynaptic and postsynaptic neurons. These intercellular channels consist of two hemichannels called connexons, and each hemichannel is composed of six specialized protein molecules called connexins. A gap junction is a cluster of these channels (see Meier & Dermietzel 2006 for more detail). Several types of connexin gap junction proteins are expressed in neurons, but connexin36 is expressed exclusively in neurons. In particular, connexin36 mediates the extensive

T.J. Lewis (✉)
Department of Mathematics, University of California, Davis, CA, USA
e-mail: tjlewis@ucdavis.edu

F.K. Skinner
Toronto Western Research Institute, University Health Network
and University of Toronto, Ontario, Canada
e-mail: frances.skinner@gmail.com

N.W. Schultheiss et al. (eds.), *Phase Response Curves in Neuroscience: Theory, Experiment, and Analysis*, Springer Series in Computational Neuroscience 6,
DOI 10.1007/978-1-4614-0739-3_14, © Springer Science+Business Media, LLC 2012

electrical coupling between many cortical inhibitory neurons (Connors & Long 2004; Galarreta & Hestrin 2001; Söhl et al. 2005). Although chemical synapses are more common in neural systems and have been studied more extensively, electrical synapses are more advanced from an evolutionary perspective (Bennett 2000) and provide connection characteristics that are different from those of chemical synapses. Specifically, they provide fast, bidirectional coupling and include coupling of subthreshold voltages. Functionally, they are thought to play a role in synchronization of neuronal activity, but the details are far from clear.

The goal for this chapter is to discuss how synchronization patterns depend on a neuron's intrinsic properties and the electrical coupling. Specifically, we will show how phase response curves (PRCs) in conjunction with the theory of weakly coupled oscillators (see Chap. 1) can provide important insight into this issue.

1.1 Basic Model and Actions of Gap Junction Coupling

Gap junctions can sometimes exhibit rectification and voltage gating (Hormuzdi et al. 2004), and models of electrical coupling can involve considerable complexity (e.g., Vogel & Weingart 2002). However, gap junctions largely act like electrical resistors (Bennett 1977). Therefore, we will only consider mathematical models of gap junctions as simple ohmic resistors. Specifically, the basic current balance equation that we will use for a pair of electrically coupled neurons is:

$$C\frac{\mathrm{d}V_j}{\mathrm{d}t} = -I_{\mathrm{ionic},j} + I_{\mathrm{app},j} + g_{\mathrm{gap}}\left(V_k - V_j\right), \tag{14.1}$$

where t is time, C is the membrane capacitance, V_j is the membrane potential of the cell j, $I_{\mathrm{app},j}$ is a constant applied current (injected DC current) to cell j, and $I_{\mathrm{ionic},j}$ is the ionic current of cell j. The electrical coupling current flowing from cell k into j is modeled as $I_{\mathrm{gap},i,k} = g_{\mathrm{gap}}(V_k - V_j)$, where g_{gap} is the gap junction coupling conductance.

It can be seen from (14.1) that the electrical coupling current between cells flows down the electrical gradient and tends to bring the cells' membrane potentials closer together. As such, we intuitively expect electrical coupling to synchronize activity. This naturally synchronizing action of gap junctions (due to their rapid, bidirectional natures) has been discussed in several reviews (Bennett & Zukin 2004; Connors & Long 2004; Galarreta & Hestrin 2001). However, modeling work has shown that electrical coupling can support both synchronous and asynchronous activities (Bem et al. 2005; Chow & Kopell 2000; Cymbalyuk et al. 1994; Lewis & Rinzel 2003; Sherman & Rinzel 1992). Furthermore, stable antiphase activity has specifically been found in many conductance-based models of electrically coupled pairs of cortical inhibitory interneurons (Di Garbo et al. 2005; Lewis & Rinzel 2004; Mancilla et al. 2007; Nomura et al. 2003; Pfeuty et al. 2003; Skinner et al. 1999). Gap junctions in conjunction with the cell membrane act as low pass filters, and, therefore, a "presynaptic" spike gives rise to an attenuated and delayed voltage response in the "postsynaptic" cell. Measured postsynaptic responses (spikelets) vary in size and can include both depolarizing and hyperpolarizing components. This depends on the details of the spike shape, such as spike width and afterhyperpolarization, as well as where the gap junctions are located and where postsynaptic responses are measured. Moreover, spike frequency would be expected to affect the resulting network dynamics. These aspects make it difficult to intuit the output from electrically coupled networks,

How can we more fully understand the actions of gap junction coupling? Only a limited amount of information and mechanistic understandings can be obtained using direct numerical simulations. We, therefore, turn to PRCs and the theory of weakly coupled oscillators (Kuramoto 1984).

1.2 PRCs and the Theory of Weakly Coupled Oscillators

As described in the introductory Chap. 1, the PRC, or the phase resetting curve describes whether a perturbation advances or delays the phase of a periodically firing neuron.[1] PRCs have been used to predict phase-locking behaviors for periodic, pulsatile stimuli (Glass & Mackey 1988; Rinzel & Ermentrout 1998) and for cell pairs connected by fast, pulsatile chemical synapses (see Chap. 4). PRCs can also be used to predict phase-locking in networks of oscillators with general coupling, including electrical coupling, provided that the coupling is sufficiently weak. This latter case employs the theory of weak coupling (Kuramoto 1984). The PRC used in this theory measures the response to brief, small current pulses[2] and is often referred to as the infinitesimal PRC (iPRC).

PRCs are often divided into two classifications: Type I PRCs are nonnegative, meaning that spikes can only be advanced with perturbations, and type II PRCs are biphasic, usually with negative aspects at early phases and positive aspects at later phases (i.e., spikes can be advanced or delayed depending on the timing of the perturbation relative to the spike). Note that phase zero is typically defined as the peak of the spike at the soma, and we follow this definition in this chapter. Whether a neuron exhibits type I or type II PRCs can be important because it has implications for synchrony in networks coupled with chemical synapses (e.g., see Crook et al. 1998b; Hansel et al. 1995), and this has been taken advantage of in examining how cholinergic neuromodulation might affect whether type I or type II PRC characteristics are present (Stiefel et al. 2009).

[1]These phase shifts are usually assessed by spike times.

[2]Theoretically, a delta-function current stimulus with small amplitude (in terms of total charge injected) is used to compute the iPRC. The iPRC can also be computed by linearizing the system about the limit cycle and solving the adjoint problem (see Chap. 1).

The basic mathematical equations for weakly coupled oscillator theory have been described in Chap. 1, and the same notation is followed in this chapter. When coupling between oscillators is sufficiently weak, the state of each oscillator can be captured by its relative phase, and the dynamics of a pair of identical cells connected by electrical coupling is governed by

$$\frac{d\psi_j}{dt} = \frac{1}{T}\int_0^T Z(t) g_{\text{gap}}(V_{LC}(t-(\psi_j-\psi_k)) - V_{LC}(t))d\tilde{t} = g_{\text{gap}} H(-(\psi_j-\psi_k)), \tag{14.2}$$

where ψ_j is the relative phase of the jth cell, $V_{LC}(t)$ is the T-periodic membrane potential of the unperturbed (uncoupled) neurons, and $Z(t)$ is the iPRC of the neurons. The evolution of the phase difference between the cells, $\varphi = \psi_1 - \psi_2$, is described by the scalar differential equation

$$\frac{d\varphi}{dt} = g_{\text{gap}}(H(-\varphi) - H(\varphi)) = g_{\text{gap}} G(\varphi), \tag{14.3}$$

Phase differences φ^* with $G(\varphi^*) = 0$ are phase-locked states. A phase-locked state is stable if $G'(\varphi^*) < 0$ and unstable if $G'(\varphi^*) > 0$. Note that the coupling strength g_{gap} simply scales the G-function. It does not affect the existence or stability of the phase-locked states; it only affects how fast the system approaches or diverges from the phase-locked state and how robust the phase-locked states are to noise and heterogeneities (see Chap. 1 for details). Note also that when the electrically coupled cells are identical, both the synchronous state $\varphi^* = 0, T$ and the antiphase state $\varphi^* = \frac{T}{2}$ are phase-locked states, as can be seen by (14.3). Furthermore, the stability of the antiphase state is given by the sign of $G'(\frac{T}{2}) = -2H'(\frac{T}{2})$. These observations will be useful in our analysis.

The form of the G-function relates the existence, stability, and robustness of phase-locked states to the iPRC, Z, and the membrane potential, V_{LC}. Z, and V_{LC} are strictly properties of the individual neurons, and they are relatively easy to obtain for model cells and for real cells. Thus, the theory of weak coupling provides a two step method to obtain insight into both the dynamical and biophysical mechanisms underlying phase-locking:

Step 1: Determine how the shapes of Z and V_{LC} affect phase-locking, i.e., study how the shapes of the functions Z and V_{LC} combine (according to the "convolution" integral in the phase equations (14.2) to influence the phase-locked states.

Step 2: Explore how the shape of Z and V_{LC} are altered by changes in frequency or individual membrane conductances or by the presence of neuromodulators (or effects of anesthetics, pharmacological agents, temperature, pH, and so on), i.e., if we understand how the shapes of Z and V_{LC} alter phase-locking, then the problem of "understanding how membrane conductances affect phase-locking" is reduced to "understanding how they affect Z and V_{LC}." While this latter problem is still nontrivial, it is a much more manageable problem than the former one.

1.3 Chapter Outline and Limitations

In this chapter, we describe how PRCs and weakly coupled oscillator theory can be used to understand activities produced by electrically coupled cells. For space reasons, we take a narrow focus, only considering *electrical coupling* between *pairs* of *spiking* neurons in cortical systems and related theoretical works.[3] As such, important work on electrical coupling involving bursting cells, nonneuronal cells (such as glia), nonspiking cells and invertebrate systems are not included in this chapter. We examine mathematical models that include the basic features of spiking neurons and coupling dynamics. With simpler neuronal caricatures, more extensive analyses can be performed. In general, analyses provide a theoretical framework with which to understand the behavior of more complicated models and experimental preparations, as well as the in situ physiological system.

In what follows, we describe the work that has been done using mathematical models of electrically coupled neuronal networks and the insights into the mechanisms underlying synchronization patterns in these networks. We start by describing simple two-cell networks that do not include a biophysical description for the individual neuron and then continue with two-cell networks that do include a biophysical description and that consider gap junctions that are not located close to the cell body. Finally, we provide a short summary and discussion.

2 Single-Compartment Integrate-and-Fire Models for Electrically Coupled Cell Pairs

In this section, we consider phase-locking of pairs of one-variable single-compartment integrate-and-fire neurons that are connected by electrical coupling. The dynamics of the cell pairs are given by

$$C\frac{\mathrm{d}V_j}{\mathrm{d}t} = -I_{\mathrm{ionic}}(V_j) + I_{\mathrm{app}} + g_{\mathrm{gap}}(V_k - V_j). \tag{14.4}$$

Note that the ionic current $I_{ionic}(V_j)$ depends only on the membrane potential of the cell. These dynamics are augmented by a fire-and-reset rule: When V_j reaches a threshold potential V_{th}, cell j fires a "spike" and V_j is reset to the potential V_{reset}.[4] The suprathreshold portion of a spike is modeled as a delta function. Each time a cell fires, $\beta\delta(t)$ is added to its membrane potential (Lewis & Rinzel 2003; Pfeuty et al. 2003).

---

[3]Chapter 15 extends the analysis presented in this chapter to larger networks and considers coupling through both chemical synapses and electrical synapses.

[4]These models could be augmented by including an explicit refractory period and a time-dependent threshold.

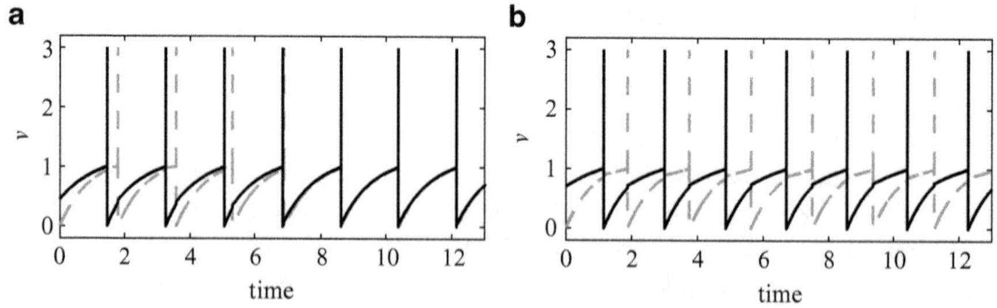

Fig. 14.1 Bistable activity in a pair of electrically coupled LIF cells. $I = 1.2$, $\beta = 0.2$, $g_{\text{gap}} = 0.2$. *Black solid curves* and *grey dashed curves* represent the membrane potentials of the two cells. Different initial conditions lead to different phase-locked states: (**a**) synchrony and (**b**) antiphase

If $I_{\text{app}} \geq I_{\text{ionic}}(V_{\text{reset}})$ and $I_{\text{app}} \leq I_{\text{ionic}}(V_j)$ for some $V_{\text{reset}} < V_j \leq V_{\text{th}}$, then the isolated cells eventually approach a stable resting potential and do not fire. If $I_{\text{app}} > I_{\text{ionic}}(V_j)$ for all $V_{\text{reset}} \leq V_j \leq V_{\text{th}}$, then V_j increases monotonically from V_{reset} until it reaches the threshold potential V_{th} at which point the cell "fires." V_j is then reset to V_{reset}, and the process is repeated. Thus, for sufficiently large applied current, integrate-and-fire cells display periodic oscillations. The frequency of the oscillations increases with I_{app}.

Electrically coupled cell pairs that are modeled using standard leaky-integrate-and-fire (LIF) dynamics (i.e., without the delta-function spike) exhibit only anti-phase activity for all suprathreshold applied currents and coupling strengths.[5] This highlights the fact that electrotonic effects of the spikes play an essential role in determining phase-locking patterns (Chow & Kopell 2000; Lewis & Rinzel 2003; Pfeuty et al. 2003), and, therefore, they must be included in any adequate model of electrically coupled cells. As mentioned above, we model the suprathreshold portion of a spike as a delta-function $\beta\delta(t)$.[6] Thus, a spike in one cell will instantaneously increase the membrane potential of the other cell toward threshold by a fixed amount $g_{\text{gap}}\beta$. With this spike effect included, the LIF model can exhibit both synchrony and antiphase (Fig. 14.1).

Below, we consider leaky integrate-and-fire (LIF) and quadratic integrate-and-fire (QIF) dynamics as specific examples, but we also provide some results for general integrate-and-fire (GIF) dynamics, i.e., integrate-and-fire with any $I_{\text{ionic}}(V)$. Many of these results are presented in Lewis & Rinzel (2003) and Pfeuty et al. (2003). The results highlight the importance of the shape of the iPRC and the components of the membrane potential waveform.

---

[5]In fact, it can be shown that, as $g_{\text{gap}} \to \infty$, any single-variable IF model will support only antiphase activity, whereas any single-compartment conductance-based model will support only synchronous activity (Lewis, unpublished results).

[6]For certain IF models such as QIF, the spike could also be modeled by tuning V_{th}, so that the membrane potential rises rapidly to a high level before being reset (as in Latham et al., 2000).

2.1 LIF, QIF, and GIF Dynamics

LIF dynamics: For the LIF model, the ionic current is described only by the leakage current

$$I_{\text{ionic}}(V) = g_m(V - V_r),$$

where V_r is the reversal potential of the leakage current, and g_m is the membrane leakage conductance of the cell. In nondimensional form (see Appendix 1), the LIF model is

$$\frac{\mathrm{d}v_j}{\mathrm{d}t} = -v_j + I$$

with a threshold potential of 1 and a reset potential 0. For $I > 1$, LIF cells fire periodically with a membrane potential waveform of

$$v_{LC}(t) = I(1 - \mathrm{e}^{-t}) + \beta\delta(t - T), \quad t \in [0, T]$$

and a period of oscillation

$$T = \ln\left(\frac{I - v_{\text{reset}}}{I - v_{\text{th}}}\right) = \ln\left(\frac{I}{I - 1}\right).$$

The infinitesimal phase-response curve $Z(t)$ for the LIF model (Hansel et al. 1995; Neltner et al. 2000) is

$$Z(t) = \begin{cases} \frac{1}{I}\exp(t), & 0 < t < T \\ 0, & t = 0, T \end{cases}.$$

Several ways to derive $Z(t)$ for LIF cells are covered in the appendix of Lewis & Rinzel 2003.

QIF dynamics: For the QIF model, the ionic current is

$$I_{\text{ionic}}(V) = g_m \frac{(V_T - V)(V - V_R)}{(V_T - V_R)},$$

where V_R and V_T are the resting potential and "resting threshold" potential[7] of the neuron, respectively, for $I_{\text{app}} = 0$, and g_m is related to the resting membrane conductance. $I_{\text{ionic}}(V)$ in the QIF model can be thought of as a combination of currents that capture the effects of the leakage current and some of the threshold features of the fast sodium current.

In nondimensional form (see Appendix 1), the QIF model is

$$\frac{\mathrm{d}v_j}{\mathrm{d}t} = v_j^2 + I,$$

[7]Note that the "resting threshold" potential V_T is different than the spike-reset threshold v_{th}.

with a threshold potential ν_{th} and a reset potential ν_{reset}. For $I > 0$, the T-periodic solution of the QIF model is

$$\nu_{LC}(t) = \sqrt{I}\tan\left(\sqrt{I}(t + \gamma(\nu_{\text{reset}}))\right) + \beta\delta(t - T), \quad t \in [0, T]$$

where

$$\gamma(\nu) = \frac{1}{\sqrt{I}}\arctan\left(\frac{\nu}{\sqrt{I}}\right),$$

with a period of

$$T = \gamma(\nu_{\text{th}}) - \gamma(\nu_{\text{reset}}).$$

The infinitesimal phase-response curve $Z(t)$ for the QIF model is

$$Z(t) = \begin{cases} \frac{1}{I}\cos^2\left(\sqrt{I}(t + \gamma(\nu_{\text{reset}}))\right), & 0 < t < T \\ 0 & t = 0, T \end{cases}.$$

GIF dynamics: The periodic solution for GIF dynamics cannot be obtained explicitly, but it can be shown that iPRC $Z(t)$ for any single-variable integrate-and-fire model is equal to the reciprocal of the derivative of the membrane potential during the oscillations

$$Z(t) = \begin{cases} \left(\nu'_{LC}(t)\right)^{-1}, & 0 < t < T \\ 0, & t = 0, T \end{cases}.$$

During the spike-and-reset phase at $t = 0, T$, we set $Z(t)$ to 0 in all cases. This assumes that during the spike, the cell is not affected by external perturbations (e.g., due to coupling). This is a reasonable assumption because the input conductance is extremely high during spikes, and this renders the cell insensitive to external perturbations.

Note that, outside of the spike, the iPRC $Z(t)$ for any single-variable integrate-and-fire model is strictly positive, and, therefore, positive currents will always phase advance the cell. This is a characteristic of many neuronal oscillators, especially at low firing rates, and is referred to as type I phase resetting (Ermentrout 1996; Hansel et al. 1995).

2.2 Phase Models for Integrate-and-Fire Cells

Plugging the expressions for $Z(t)$ and $\nu_{LC}(t)$ into the equations for the phase model ((14.2) and (14.3)) yields the G-function $G(\varphi)$ for the IF models. Note that because of the threshold discontinuity in the "periodic" solution, the integral must be defined in a piecewise manner

$$G(\varphi) = H(-\varphi) - H(-(T - \varphi)) \tag{14.5}$$

where

$$H(-\varphi) = \frac{1}{T}\int_0^{\varphi^-} Z(t)(v_{LC}(t+T-\varphi) - v_{LC}(t))\mathrm{d}t$$
$$+\frac{1}{T}\int_{\varphi^+}^{T} Z(t)(v_{LC}(t-\varphi) - v_{LC}(t))\mathrm{d}t + \frac{1}{T}\beta Z(\varphi), \qquad (14.6)$$

Note that $H(-\varphi)$, and, therefore, $G(\varphi)$, is a linear combination of terms describing the subthreshold effects and the suprathreshold spike effects: (1) The terms with the integrals accounts for all subthreshold activity including the reset, and (2) the term that is scaled by β accounts for the effect of the suprathreshold portion of the spike, i.e.,

$$H_{\text{spike}}(-\varphi) = \frac{1}{T}\int_{\varphi^-}^{\varphi^+} Z(t)(\beta\delta(t-\varphi))\mathrm{d}t = \frac{1}{T}\beta Z(\varphi)$$

As stated in Sect. 1.2, the stability of the phase-locked states φ^* is given by the sign of $G'(\varphi^*)$. However, for $\beta \neq 0$, $G(\varphi)$ is discontinuous at $\varphi = 0, T$. In this case, $G(0) < 0$ (and $G(T) > 0$) implies stability of the synchronous state.

2.3 *Phase-Locking in a Pair of Electrically Coupled LIF Cells*

For the LIF model, the G-function is

$$G(\varphi) = \begin{cases} \frac{2}{T}(\varphi\sinh(T-\varphi) - (T-\varphi)\sinh(\varphi)) + \frac{\beta}{TI}(\mathrm{e}^{\varphi} - \mathrm{e}^{T-\varphi}), & 0 < \varphi < T \\ 0, & \varphi = 0, T \end{cases} \qquad (14.7)$$

Figure 14.2 plots voltage profiles (v_{LC}), iPRCs (Z), and the corresponding G-functions for $I = 1.15$ and $I = 1.5$ with $\beta = 0.1$. The G-function reveals that that both the synchronous state $\varphi = 0, T$ and the antiphase state $\varphi = \frac{T}{2}$ are stable at $I = 1.15$, whereas only the synchronous state $\varphi = 0, T$ is stable at $I = 1.5$. These results can be understood by considering (14.7).

The phase-locked states and their stability are determined by a competition between the subthreshold portion and suprathreshold portion of the G-function (the solid and dashed grey curves in Fig. 14.2, respectively). The subthreshold portion of the G-function (obtained by setting $\beta = 0$ in (14.7)) is always positive in $0 < \varphi < \frac{T}{2}$ and negative in $0 < \varphi < \frac{T}{2}$. This implies that the subthreshold dynamics always have a desynchronizing effect, forcing the coupled LIF cells toward the antiphase state. Conversely, the suprathreshold (spike) portion of the G-function is always negative in $0 < \varphi < \frac{T}{2}$ and positive in $0 < \varphi < \frac{T}{2}$, and, therefore, the spike always acts to stabilize the synchronous state $\varphi = 0, T$ and destabilize the antiphase state $\varphi = \frac{T}{2}$. The delta-function spike causes the synchronous state to be always stable. However, at low frequencies (I just above 1), the subthreshold term,

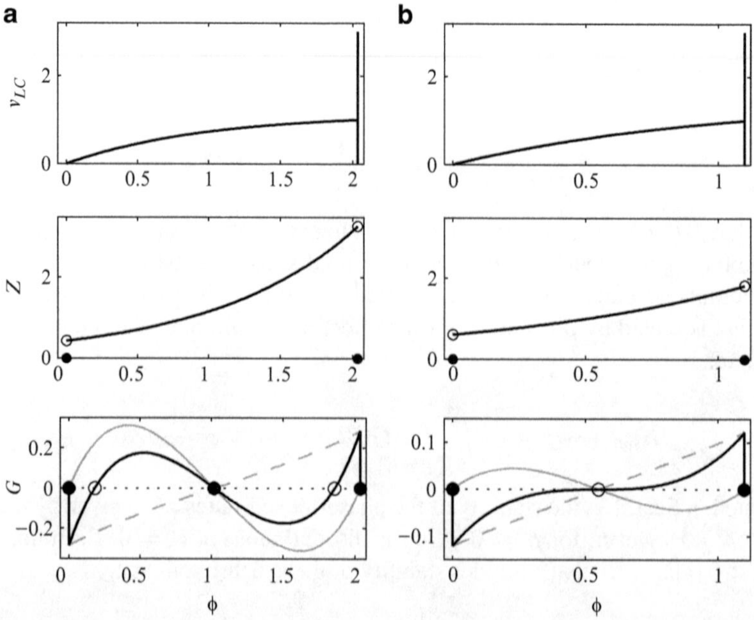

Fig. 14.2 Membrane potential (v_{LC}), iPRC (Z), and G-functions for LIF cell pair with weak electrical coupling. (**a**) $I = 1.15$, $\beta = 0.1$, (**b**) $I = 1.5$, $\beta = 0.1$. In G-function panels, *black solid lines* are the full G-functions, and the *filled circles* (*open circles*) represent stable (unstable) phase-locked states. The *dashed grey lines* are the portion of the G-function accounting for the effects of the spike, which always tend to synchronize activity of LIF cells. The *solid grey lines* are the portion of the G-function accounting for subthreshold activity (obtained by setting $\beta = 0$), which always tend to desynchronize the activity of LIF cells

and therefore the antiphase state, dominates, i.e., almost all initial conditions evolve to the antiphase state. As the frequency increases (with increased I), the relative magnitude of the spike effect increases and eventually destabilizes the antiphase state.

The phase-locked states φ^* over an interval of I for $\beta = 0.1$ are plotted in Fig. 14.3. As pointed out in Fig. 14.2, both the synchronous and the antiphase states are stable in the case of $I = 1.15$, but if I is increased to $I = 1.5$ only the synchronous state is stable. In general, at high values of I, the only stable state is the synchronous state, and at sufficiently low I, stable synchronous and antiphase states coexist. The antiphase state loses its stability via a subcritical pitchfork bifurcation at the critical value I^*, which depends upon the value of β.

By considering $G'(\frac{T}{2})$, one can obtain the relationship between the critical value I^* and β analytically

$$\beta = \left(I^* - \frac{1}{2}\right) \ln\left(\frac{I^*}{I^* - 1}\right) - 1 = \frac{\frac{T}{2}}{\tanh\left(\frac{T}{2}\right)} - 1 \tag{14.8}$$

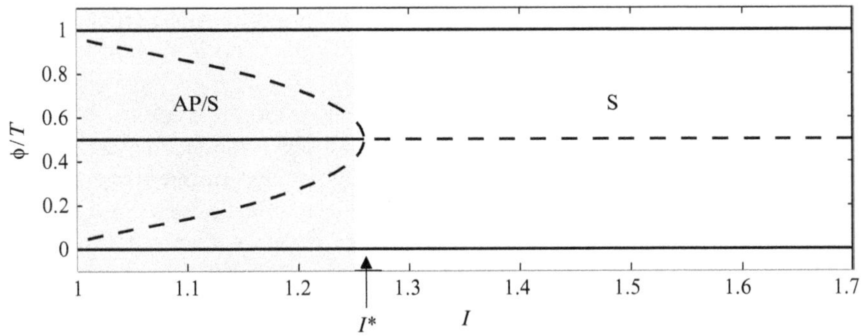

Fig. 14.3 Bifurcation Diagram for LIF cell pair with weak electrical coupling, $\beta = 0.1$. *Solid and dashed lines* indicate stable and unstable phase-locked states, respectively. I^* is the critical value of I at which antiphase activity ($\frac{\varphi}{T} = 0.5$) changes stability. For $I > I^*$, only the synchronous activity (S) is stable; for $1 < I < I^*$, both the synchronous (S) and antiphase (AP) states are stable. (I^* decreases as β increases)

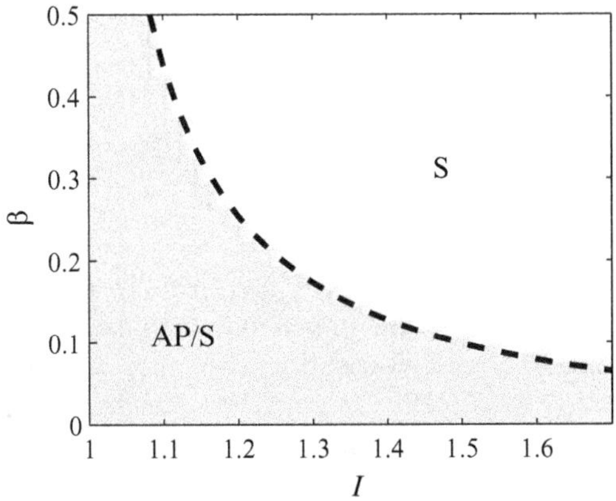

Fig. 14.4 Two-parameter (β vs. I) bifurcation diagram for the LIF cell pair connected by weak electrical coupling. The dashed curve plots the critical value I^* in relation to β (see (14.8)). Above the curve, the only stable phase-locked behavior is synchrony (S). Below the curve, the cells can exhibit either stable synchrony (S) or stable antiphase activity (AP). Note that increased spike strength β and/or increased applied current I promotes synchronous activity

I^* increases as β decreases, as is seen in Fig. 14.4. This implies that, when the electrotonic effect of the delta-function spikes is weak, antiphase persists for a larger range of intrinsic frequencies. As the spike effect vanishes ($\beta \to 0$), the critical current at which the antiphase state loses stability I^* goes to infinity and the non-antiphase unstable steady states approach the synchronous state. This effectively leaves the antiphase state as the only stable solution for all I (as was the case for $\beta = 0$).

The asynchrony at low frequencies can be understood in terms of the effect of the fast repolarization (reset) of the LIF cells and the shape of their iPRC $Z(t)$. Consider two coupled cells with a small phase difference. During the slow depolarization toward threshold, the membrane potential in the lagging cell is slightly less than that in the leading cell. Consequently, there is a small positive (depolarizing) current in the lagging cell that flows from the leading cell due to the electrical coupling. This current speeds up the lagging cell and slows down the leading cell, which acts to synchronize the cells. However, because the voltage differences are small, the electrotonic current is small and the synchronizing effect of this portion of the subthreshold activity is small. Once the leading cell reaches threshold, it fires and is immediately reset. If the spike does not instantaneously synchronize the cells, the voltage of the leading cell is now below that of the lagging cell, and the electrical coupling current switches direction. Now, coupling acts to impede the increase to threshold of the lagging cell and to speed up the leading cell. Because the potential difference between the cells is large and the sensitivity of the cells is highest around threshold, the lagging cell is substantially delayed before it fires. This desynchronizing effect is so strong at low firing rates (for which the cells are very sensitive at threshold) that it overcomes the previous synchronizing effect and, therefore, causes the net effect of coupling over an entire period to be desynchronizing.

2.4 *Phase-Locking of a Pair of Electrically Coupled GIF/QIF Cells*

The LIF model highlights how the iPRC and membrane potential (spike, reset, and subthreshold portions) combine to generate the phase-locking characteristics of electrically coupled cells. However, LIF dynamics are somewhat limited. In particular, the iPRCs of LIF cells are always increasing and concave up with a maximum at $\varphi \to T$, i.e., as the membrane potential increases toward threshold, the cell is increasingly sensitive to perturbations and the rate of this increasing sensitivity is itself increasing. Therefore, it is not possible to look for general rules for how the shapes of the iPRC and the membrane potential affect synchronization patterns. We will now consider the GIF models, which allow for a richer set of iPRC shapes. In particular, we will examine how the "skewness" of the PRC (as measured by the location of peak) affects phase-locking. The QIF model will be used to provide specific examples.

Figure 14.5 plots the voltage profiles (v_{LC}), the iPRCs (Z) and the corresponding G-functions for the QIF model with three different parameter sets. In all cases, $I = 0.1, \beta = 0.13$, and $v_{\text{th}} - v_{\text{reset}} = 3.0$, but v_{reset} is -2.85, -1.50, and -0.15 in panels a, b, and c, respectively. Note that the iPRC changes from having a rightward skew of the peak in (a) to being symmetric in (b) to having a leftward-skewed peak in (c). Associated with these changes in the iPRCs are changes in the stability of the phase-locked states. The G-function associated with the rightward-skewed iPRC in panel (a) shows that both the synchronous state $\varphi = 0, T$ and the antiphase state $\varphi = \frac{T}{2}$ are stable, although the antiphase state is dominant. The G-function

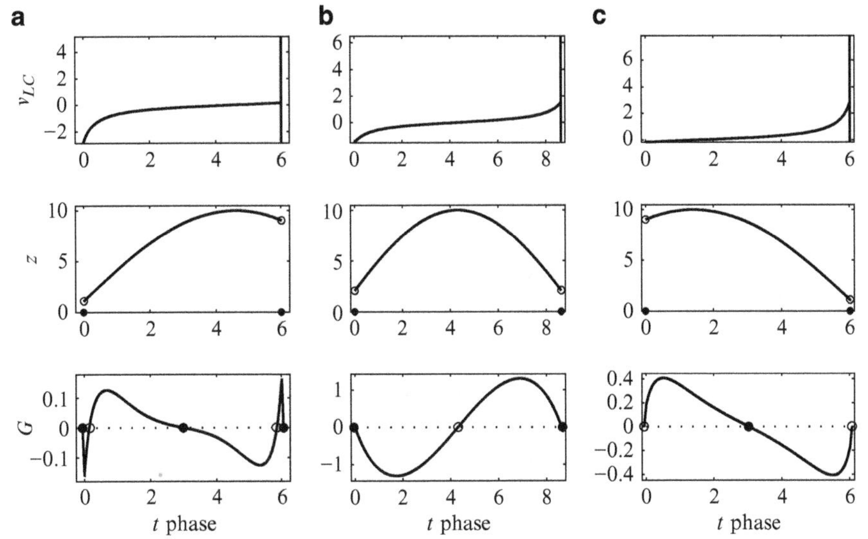

Fig. 14.5 Membrane potential (v_{LC}), iPRC (Z), and G-functions for QIF cell pairs with weak electrical coupling. $I = 0.1$, $\beta = 0.13$. (**a**) $v_{\text{reset}} = -2.85$, $v_{\text{th}} = 0.15$. Both the synchronous and antiphase states are stable. (**b**) $v_{\text{reset}} = -1.50$, $v_{\text{th}} = 1.50$. Only the synchronous state is stable. (**c**) $v_{\text{reset}} = -0.15$, $v_{\text{th}} = 2.85$. Only the antiphase state is stable

associated with the symmetric iPRC in panel (b) shows that only the synchronous state $\varphi = 0, T$ is stable. The G-function associated with the leftward-skewed iPRC in panel (c) shows that only the antiphase state $\varphi = \frac{T}{2}$ is stable.

Pfeuty et al. (2003) noted that the skewness of the iPRC has a substantial influence on the stability of the phase-locked states, and they extensively examined how the skewness affected the stability of the antiphase state of the QIF model. Recall that the stability of the antiphase state is given by the sign of $G'(\frac{T}{2}) = -2H'(\frac{T}{2})$. Therefore, to see how the above changes in the stability of the antiphase state arise, we can examine how the shape of the iPRC affects $G'(\frac{T}{2})$. By differentiating equation (14.5) with respect to φ and evaluating at $\varphi = \frac{T}{2}$, we see that

$$G'\left(\frac{T}{2}\right) = G'_{\text{sub}}\left(\frac{T}{2}\right) + G'_{\text{reset}}\left(\frac{T}{2}\right) + G'_{\text{spikes}}\left(\frac{T}{2}\right) \tag{14.9}$$

where

$$G'_{\text{sub}}\left(\frac{T}{2}\right) = 2\left(-\frac{1}{T}\int_0^{\frac{T}{2}} Z(t)v'_{LC}\left(t+\frac{T}{2}\right)\mathrm{d}t - \frac{1}{T}\int_{\frac{T}{2}}^{T} Z(t)v'_{LC}\left(t-\frac{T}{2}\right)\mathrm{d}t\right),$$

$$G'_{\text{reset}}\left(\frac{T}{2}\right) = 2\left(\frac{1}{T}Z\left(\frac{T}{2}\right)(v_{\text{th}} - v_{\text{reset}})\right),$$

$$G'_{\text{spike}}\left(\frac{T}{2}\right) = 2\left(\frac{T}{2}\beta Z'\left(\frac{T}{2}\right)\right).$$

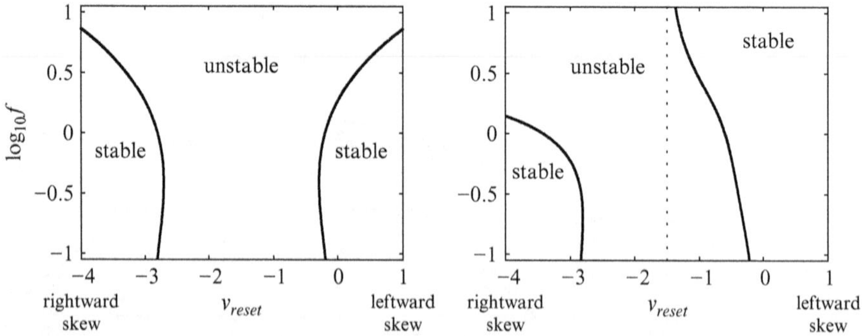

Fig. 14.6 Stability of the antiphase state: Two-parameter (ν_{reset} vs. $\log_{10}$(frequency)) bifurcation diagrams for a QIF cell pair connected by weak electrical coupling. The solid curves plot the critical values at which the antiphase state changes stability as determined by (14.9). (**a**) $\beta = 0$, $\nu_{\text{th}} - \nu_{\text{reset}} = 3$, (**b**) $\beta = 0.25$, $\nu_{\text{th}} - \nu_{\text{reset}} = 3$

G'_{sub} contains the influence of the intervals of subthreshold depolarization, G'_{reset} contains the influence of the reset from ν_{th} to ν_{reset}, and G'_{spike} contains the influence of the delta-function spike of strength β. Because $Z \geq 0$ and $\nu'_{LC} \geq 0$ for all IF models, G'_{sub} is always negative, indicating that the subthreshold depolarization phase in the oscillations always acts to stabilize antiphase activity. On the other hand, the reset term G'_{reset} is always positive, and, therefore, the instantaneous reset always acts to destabilize antiphase activity. Note that this was the case with the LIF model. The spike term G'_{spike} shows that the delta-function spikes act to stabilize antiphase activity if and only if the iPRC (Z) has a negative slope at $\varphi^* = \frac{T}{2}$. For LIF, $Z'(\frac{T}{2}) > 0$ and spikes always act to synchronize activity. For the QIF, $Z'(\frac{T}{2}) > 0$ whenever the iPRC has a rightward skew (i.e., when $|\nu_{\text{reset}}| > |\nu_{\text{th}}|$), and $Z'(\frac{T}{2}) < 0$ whenever the iPRC has a leftward skew (i.e., when $|\nu_{\text{reset}}| > |\nu_{\text{th}}|$). Thus, spikes act to destabilize antiphase activity whenever the iPRC has a rightward skew and stabilize antiphase whenever the iPRC has a leftward skew.

Figure 14.6 depicts the stability of antiphase activity for an electrically coupled QIF cell pair in ν_{reset} vs frequency (f) parameter space with $\nu_{\text{th}} - \nu_{\text{reset}} = 3.0$ Fig. 14.6a shows the case for $\beta = 0$, which is $G'_{\text{sub}} + G'_{\text{reset}}$. It shows that the net subthreshold activity acts to destabilize the antiphase state at all frequencies when the iPRC is symmetric, i.e. when $\nu_{\text{reset}} = -\nu_{\text{th}}$. However, as the magnitude of $\nu_{\text{th}} + \nu_{\text{reset}}$ increases, or equivalently the skewness of the iPRC increases in either direction, the influence of the net subthreshold activity on the antiphase state becomes stabilizing. Figure 14.6b shows the results of adding the spike term G'_{spike} to the subthreshold terms. Recall that spikes act to stabilize the antiphase state when iPRCs skewed to the left ($\nu_{\text{reset}} > -1.5$, i.e., to the right of the dotted line) and destabilize the antiphase state when iPRCs skewed to the right ($\nu_{\text{reset}} < -1.5$, i.e., to the left of the dotted line). Furthermore, the spike has a greater effect at higher frequencies. Thus, for iPRCs skewed to the left, antiphase is unstable at low frequencies, but it becomes stable as the frequency increases, or for iPRCs skewed highly to the left, antiphase is stable for all frequencies. For iPRCs skewed to the right, antiphase is stable at low frequencies, but it becomes unstable as the frequency increases. Note that the LIF model has iPRCs similar to those in this latter parameter regime and exhibits equivalent behavior.

A similar analysis can be implemented to determine stability of the synchronous state (Lewis, unpublished results). When the spike effect is included ($\beta > 0$), the synchronous state is stable if $G(\varphi) < 0$, which is the case whenever $Z(0) < Z(T)$. This always holds for the LIF model. It also holds for the QIF model when the iPRC is skewed to the right, but the spike effect acts to destabilize the synchronous state when the iPRC is skewed to the left.

2.4.1 Summary

Analytical tractability of the IF models has allowed us to establish a clear link between the phase-locking of coupled cells and the shape of the iPRC and to identify the relative contributions of the various different phases in the oscillations. The LIF model showed that only synchrony was stable at higher frequencies, but at low frequencies, the sharp reset (repolarization) and the high sensitivity to input in the late phases had strong desynchronizing effects that led to a stable antiphase state. The QIF model generalized this result. When the iPRC is skewed to the right (with a peak late in the oscillation cycle), antiphase was typically stable at low frequencies but destabilized at higher frequencies. Furthermore, increased magnitude of the spike hindered antiphase activity in this case. On the other hand, the QIF model showed that when the iPRC is skewed to the left (with a peak early in the oscillation cycle), antiphase was stable at higher frequencies but destabilized at low frequencies. In this case, increased magnitude of the spike promoted antiphase activity. In terms of the effects of the spike, the sign of $Z'(\frac{T}{2})$ was the key quantity.

IF models are highly idealized and allow fairly extensive mathematical analyses. However, they also have considerable limitations in terms of linking phase-locking back to biophysical mechanisms. For example, unlike real cells or conductance-based models, the PRC and the membrane potential always have the relationship $Z(t) = \frac{1'}{\nu_{LC}}(t)$, and the iPRCs of IF models are strictly positive (i.e., they are always type I and never type II). Thus, it is necessary to consider conductance-based models to acquire insight into the biophysical mechanisms of synchronization.

3 Single-Compartment Conductance-Based Models for Electrically Coupled Cell Pairs

A major goal is to determine the role that specific intrinsic membrane conductances play in shaping the synchronization patterns of electrically coupled cells. To do this, we must consider phase-locking in conductance-based models of neurons. However, key insights provided by IF models can be used to guide the examination of these more complex, biophysically based models.

The conductance-based models that we will consider in this section are single-compartment models given by the Hodgkin–Huxley formalism

$$C\frac{\mathrm{d}V_j}{\mathrm{d}t} = -I_{\mathrm{ionic},j}(V_j, m_j, h_j, n_j, \ldots) + I_{\mathrm{app},j} + g_{\mathrm{gap}}(V_k - V_j),$$
$$\frac{\mathrm{d}y_j}{\mathrm{d}t} = \frac{1}{\tau(V_j)}(y_\infty(V_j) - y_j), \quad y_j = m_j, h_j, n_j, \ldots \tag{14.10}$$

where $I_{\mathrm{ionic},j}$ is the sum of the various different ionic conductances in the model neuron and $y_j = m_j, h_j, n_j, \ldots$ are the gating variables for the conductances.

Pfeuty et al. (2003) is perhaps the most extensive modeling study on how intrinsic conductances affect phase-locking of electrically coupled cells. In a set of numerical simulations of a conductance-based model of a spiking neuron, Pfeuty et al. found that the persistent sodium conductance promoted antiphase activity (or more generally asynchronous activity), whereas a fast potassium conductance and a slow potassium conductance both promoted synchrony. Using the insight gained from the QIF modeling results, Pfeuty et al. noted that these trends could be understood in terms of their effects on the skew of the iPRC. They showed that increasing the potassium conductances led to a rightward shift in the peak of the iPRC (see also Crook et al. 1998b; Ermentrout et al. 2001), whereas the persistent sodium conductance shifted the peak to the left. Pfeuty et al. pointed out that these effects on the iPRCs can be understood in the following way: the potassium conductances increase refractoriness of a neuron after the spike, and, therefore, they reduce the responsiveness of the neuron to external perturbations during the first portion of the period, shifting the maximum of the iPRC toward latter phases. On the other hand, the persistent sodium conductance is activated near rest and depolarizing perturbations boost its activation, and thus it increases the responsiveness of the neuron after a spike and shifts the maximum of the iPRC toward first half of the period. The upshot of Pfeuty et al. (2003) was that the refractory effects of the potassium conductances and the boosting effects of the persistent sodium conductance are responsible for how these conductances alter the phase-locking dynamics of electrically coupled neurons.

Curiously, Mancilla et al. (2007) found that both potassium conductances in the Erisir et al. (1999) model for neocortical fast-spiking (FS) cells (i.e., a Kv1.3-conductance and a Kv3.2/3-conductance) actually promoted antiphase behavior. This apparent disagreement between the Mancilla et al. (2007) and Pfeuty et al. (2003) can be resolved by reconsidering the results from the QIF model. Figure 14.6b indicates that a decrease in v_{reset} (a rightward skew of the iPRC) will destabilize antiphase at higher frequencies, but it will stabilize antiphase at lower frequencies. Thus, assuming that it will cause a rightward skew of the iPRC, an increase in the potassium conductances should have different effects on antiphase in different frequency regimes. Indeed, the antiphase that Pfeuty et al. considered was at higher firing rates, whereas the antiphase that Mancilla et al. considered was at lower frequencies. This implies a modified rule of thumb for the effects of potassium conductances: potassium conductances promote synchrony at higher frequencies, but they can promote antiphase at lower frequencies.

The above explanation, which was motivated by the IF modeling, provides some understanding of how certain conductances affect phase-locking; however, it appears not to be the entire story. The actions of these and other conductances can sometimes be more subtle. For example, potassium conductances can have various influences on phase-locking of electrically coupled neurons. Nomura et al. (2003) studied a model of FS interneurons with a strong Kv3.1 conductance and found that it exhibited only stable synchrony over the entire frequency range studied ($\sim$5 Hz to 250 Hz). They also examined the Hodgkin–Huxley model, which has a large Kv1.1 delayed-rectifier conductance, and found that stable antiphase and synchronous states co-existed for frequencies up to $\sim$100 Hz. On the other hand, Di Garbo et al. (2005) studied another FS interneuron model (modified from Durstewitz et al. 2000) that also had a fast and strong Kv3-like conductance, but they found bistability of the antiphase state and the synchronous state up to 80 Hz. Ermentrout and Wechselberger (2009) found that the iPRC of the Erisir et al. model for cortical FS interneurons was highly oscillatory at very low frequencies, which gave rise to multistability in cell pairs and clustering in larger networks of cells. They showed that the slow potassium current (Kv1.3) was at the root of this behavior. Finally, in Mancilla et al. (2007), while both potassium conductances in the FS cell model promoted antiphase behavior at low frequencies, these potassium conductances had different effects on the iPRC. At 25 Hz, the Kv1 conductance shifted the iPRC peak to the right (and had a visually unperceivable change on the membrane potential), whereas Kv3 conductance shifted the peak in the iPRC to the left (and substantially decreased the spike width and increased the AHP).

In essence, while much insight can be obtained from the explanation based on the skewness of the iPRC, predicting electrically coupled network dynamics from iPRCs alone can be tricky, especially because predictions from the weakly coupled oscillator theory involve both iPRCs and the voltage trajectories via the coupling function. Mancilla et al. (2007) showed that the voltage trajectory can sometimes have a predominant effect over the iPRC in predicting the stability of the antiphase state in electrically coupled networks of conductance-based models. In particular, by explicitly manipulating the voltage trajectory, they found that an increased spike width and a decreased after-hyperpolarization (AHP) destabilized the antiphase state. Thus, spike details as described by conductance-based models also need to be taken into consideration in predicting the output from electrically coupled networks.

One way to start addressing the role of the membrane potential is to exploit the linearity of the integral in the G-function by decomposing v_{LC} into two or more components. We can then consider the corresponding components of $G'(\frac{T}{2})$ separately. For example, we can consider the influence of the AHP by adding an AHP component to the baseline membrane potential, $v_{LC}(t) + v_{\mathrm{AHP}}(t)$. The integral for $G'(\frac{T}{2})$ will separate into two components, and by integrating by parts, we see that

$$G'\left(\frac{T}{2}\right) = -\frac{2}{T}\int_0^T Z(t)\left(\nu'_{LC}\left(t+\frac{T}{2}\right) + \nu'_{\text{AHP}}\left(t+\frac{T}{2}\right)\right)dt$$

$$= -\frac{2}{T}\int_0^T Z(t)\nu'_{\text{baseline}}\left(t+\frac{T}{2}\right)dt + \frac{2}{T}\int_0^T Z'(t)\nu_{\text{AHP}}\left(t+\frac{T}{2}\right)dt.$$

The second integral term contains the influence of the added AHP on the stability of antiphase state. It reveals that the AHP will have a stabilizing effect if $Z'(t) > 0$ during the bulk of the AHP (note $\nu_{\text{AHP}}(t) < 0$).

The above discussion clearly demonstrates that, in order to identify the conductances responsible for promoting synchronization, we need to understand how intrinsic membrane conductances shape both the PRCs and membrane potential trajectory of individual cells (Acker et al. 2003; Crook et al. 1998b; Ermentrout et al. 2001; Mancilla et al. 2007; Pfeuty et al. 2003; Stiefel et al. 2009). More work needs to be done to sort this out. Specific questions that need to be addressed are: What do negative lobes do? How do details of activation curves and time constants determine a conductance's effect on phase-locking behavior? How important is the interactions with other conductances?

Another issue to consider is that neurons are not "single somatic compartments," and some gap junctions are known to be located on dendrites, often quite far from the cell body (Fukuda 2007). In the next section, we describe work with multi-compartment models that takes into consideration the effect on network dynamics of different gap junction coupling locations.

4 Multi-Compartment Models: Effects of Coupling Location and Dendritic Conductances

In this section, we first summarize what has been done and predicted with multi-compartment models in which the dendrites are either passive or active. We then describe the equations and analyses using a three-compartment model for illustrative purposes.

Crook et al. (1998a) were the first to use iPRCs and weakly coupled oscillator theory to understand how the presence of dendrites on individual neurons could affect the activity of neuronal networks. The low-pass filtering effects of dendrites on both membrane potential and iPRCs were described in Lewis & Rinzel (2004), and Goldberg et al. (2007) derived a closed form expression for iPRCs generated by dendritic perturbation (dPRC). Specifically, the passive properties of the dendrites cause a leftward shift and attenuation of the somatic iPRC. What this means is that a negative lobe present in a somatic iPRC (due to a type II model) will no longer exist in the dPRC at a far enough dendritic location, so that one gets a type I response in the dendrite. The leftward shift and attenuation of PRCs can be seen in

Pfeuty et al. (2005) for their two-compartment conductance-based model, as well as in Lewis & Rinzel (2004) where a type II response in the soma and a type I response in the dendrite is also apparent in the two-compartment model that they used.

4.1 Passive Dendrites

Using a ball-and-stick neuron model, Crook et al. (1998a) extended a somatic neuronal oscillator model to include a thin dendrite. Using the weakly coupled oscillator theory with *chemical* coupling, they showed that stable, phase-locked, synchronous solutions depend on the length of the dendrite and its passive biophysical properties (membrane time constant and length constant). They also used an 11-compartment model of a pyramidal cell (somatic compartment, basal dendrite compartment and nine apical dendrite compartments) and applied the theory to predict phase-locked states. With dendritic coupling, asynchronous, but non-antiphase, stable phase-locking is predicted and shown in their network simulations.

For electrical coupling, Lewis & Rinzel (2004) used a two-compartment model of a FS cortical neuron (active soma, passive dendrite) to demonstrate how dendritic filtering and the location of gap junctions could alter phase-locking. They showed that stable antiphase patterns can occur at higher frequencies when coupling is in dendritic regions, but when the electrical coupling is somatic, antiphase patterns do not occur at the higher frequencies. As in the chemical coupling case, asynchronous, but non-antiphase, phase-locking is predicted to occur with dendritic coupling. In a series of simulations involving more realistic neuronal morphologies, Saraga & Skinner (2004) built a 372-compartment model of a hippocampal basket cell and examined two-cell networks that were electrically coupled at proximal, middle or distal dendritic locations, in which the dendrites were passive. Asynchronous phase-locked states occurred when the cells were coupled at middle or distal locations, but not when coupled at proximal locations. Moreover, synchrony never occurred in these model neurons at any frequency when the weak electrical coupling occurred at distal locations (Saraga & Skinner 2004; Saraga et al. 2006).

4.2 Active Dendrites

It is clear that with the addition of passive dendrites, electrically coupled network output changes. However, the dendrites of neurons contain active voltage-gated channels (Johnston & Narayanan 2008) and this would also be expected to affect the network output. Goldberg et al. (2007) showed in their analysis that dendritic conductances could be classified as either regenerative (boosting dPRCs and giving more of a leftward shift) or restorative (high pass filtering dPRCs possibly creating a negative portion in the dPRC so that a type II response can occur). Regenerative conductances would include persistent sodium currents, and restorative conductances would include delayed rectifier potassium currents. Note that these classifications correspond to the results of Pfeuty et al. (2003) that were discussed in the previous section. In two-cell network simulations performed with 372-compartment models in which the dendrites included voltage-gated sodium and delayed rectifier potassium currents, the dPRCs exhibited negative lobes – type II PRCs (Saraga et al. 2006). However, as pointed out by Goldberg et al. (2007), care needs to be taken in making such statements because adding active conductances in the dendrite can affect whether the somatic response is type I or II, and the density of the active conductances would also have an effect. In other words, it becomes more difficult to predict the changes in PRCs as more realistic neurons with active dendrites are considered. Indeed, in examining the effects of cholinergic neuromodulation on PRCs (type II to type I) of modeled cortical neurons, which included a 257-compartment model of a layer II pyramidal neuron, Stiefel et al. (2009) found that the modulation was independent of the perturbation location. That is, the location was not a determining factor in changing the PRC characteristics.

If one takes the next step of applying weakly coupled oscillator theory using the dPRCs, then predictions can be made. This was done by Crook et al. (1998a) in showing that inward and outward currents affected the size of stable phase-lags in two-cell networks coupled with excitatory synapses. Saraga et al. (2006) examined how including different densities of voltage-gated sodium and potassium channels in the dendrites affected phase-locked states in electrically coupled networks. They applied weak coupling theory using a three-compartment model, which was a reduced version of their 372-compartment model of a hippocampal basket cell. The density of voltage-gated channels in the dendrites was shown to have a dramatic effect on the phase-locked state, and this could be somewhat correlated with a negative lobe in the dPRC. Specifically, negativity in the dPRC was predicted to give phase-locked states that were asynchronous when electrically coupled at these distal sites.

In moving toward multi-compartment models of neurons that represent both the appropriate morphology as well as biophysical characteristics of the particular cell, Zahid & Skinner (2009) built multi-compartment models of hippocampal interneurons that matched the spike attenuation characteristics measured in these cells, by adjusting the dendritic conductances. These neurons are of particular interest because it is known that gap junctions are present at dendritic locations (Fukuda & Kosaka 2000). They generated dPRCs and used the theory to predict phase-locked states. By directly comparing theoretical predictions and simulations, they suggested that quantifying the amount of skewness in measured PRCs was a good indicator of whether electrically coupled networks would produce synchronous or asynchronous output. However, as noted earlier (see previous section), one needs to be careful in predicting network dynamics from PRCs alone as details of the voltage trajectory also play a role. Nevertheless, it is helpful to try to make predictions based on PRCs alone.

We now describe the procedure that one can follow in using weakly coupled oscillator theory with multi-compartment models to predict the output from two-cell networks. We first note that with a larger number of compartments in multi-compartment models, PRCs can be obtained by directly perturbing the model. If there are only a few compartments then the iPRC, $Z(t)$, can be obtained by computing the solution to the linearized model system's adjoint equations (see Chap. 1), e.g., using XPPAUT (Ermentrout 2002). However, in working with a reduced model with a minimal number of compartments, it is important to note that the full model's electrotonic characteristics cannot be completely captured (Saraga et al. 2006; Crook et al. 1998a), and this could be critically important in determining the network output.

To illustrate the application of the weak coupling oscillator theory process for a multi-compartment model, let us assume that we have a three-compartment model consisting of a soma (s), a proximal dendrite (pd) and a distal dendrite (dd). The voltage components of the three-compartment model system are given by the following equations for cell 1 (subscript for cell number and superscript for the compartment).

$$C\frac{\mathrm{d}V_1^{\mathrm{s}}}{\mathrm{d}t} = \gamma(V_1^{\mathrm{pd}} - V_1^{\mathrm{s}}) - I_{\mathrm{ionic},1}^{\mathrm{s}}$$

$$C\frac{\mathrm{d}V_1^{\mathrm{pd}}}{\mathrm{d}t} = \gamma(V_1^{\mathrm{s}} - V_1^{\mathrm{pd}}) + \gamma(V_1^{\mathrm{dd}} - V_1^{\mathrm{pd}}) - I_{\mathrm{ionic},1}^{\mathrm{pd}}$$

$$C\frac{\mathrm{d}V_1^{\mathrm{dd}}}{\mathrm{d}t} = \gamma(V_1^{\mathrm{pd}} - V_1^{\mathrm{dd}}) - I_{\mathrm{ionic},1}^{\mathrm{dd}}$$

where γ is the coupling conductance between connected compartments (assumed to be uniform) in the multi-compartment model cell. Other parameters are defined as in the basic equation (14.1) given in the Introduction. Cell 2 equations would be the same with adjusted cell number subscript. An electrically coupled two-cell network is formed when gap junctions are present between compartments of two different cells (cell 1 and cell 2). In that case, an additional current would be added to the right-hand side of the equation for the appropriate compartment of each cell. If gap junctions are present between the kth compartments of cell 1 and cell 2, then

$$I_{\mathrm{gap},1:2}^{k} = g_{\mathrm{gap}}(V_1^k - V_2^k)$$

$$I_{\mathrm{gap},2:1}^{k} = g_{\mathrm{gap}}(V_2^k - V_1^k)$$

would be added to the cell 1 and cell 2 model equations for the kth compartment, respectively.

To apply the weakly coupled oscillator theory to predict what phase-locked states would be present in an electrically coupled two-cell network of multi-compartment model cells, one needs to use the PRC and voltage for the compartment where the coupling occurs. (The process has been described for single compartment models in Chap. 1.)

For example, consider the situation where the gap junction coupling is between the dd's of cell 1 and cell 2. Let $Z^{\mathrm{dd}}(t)$ be the iPRC at that location. Then the dynamics of a pair of electrically coupled cells is given by:

$$\frac{\mathrm{d}\psi_q}{\mathrm{d}t} = \frac{1}{T}\int_0^T Z^{\mathrm{dd}}(\tilde{t}+\psi_q)(g_{\mathrm{gap}}[V^{\mathrm{dd}}(\tilde{t}+\psi_p) - V^{\mathrm{dd}}(\tilde{t}+\psi_q)])\mathrm{d}\tilde{t}$$
$$= g_{\mathrm{gap}} H\left(-\left(\psi_q - \psi_p\right)\right), \tag{14.11}$$

where ψ_q is the phase of the qth cell and

$$H\left(-\left(\psi_q - \psi_p\right)\right) = \frac{1}{T}\int_0^T Z^{\mathrm{dd}}(s)[V^{\mathrm{dd}}(s-(\psi_q-\psi_p)) - V^{\mathrm{dd}}(s)])\mathrm{d}s.$$

Thus, the G-function $G(\phi) = g_{\mathrm{gap}}[H(-\phi) - H(\phi)]$, which describes the evolution of the phase difference between the two cells ($\phi = \psi_q - \psi_p$), includes properties of the uncoupled cells, the voltage at the dd location, the PRC at the dd location, and the expression of the gap junction coupling. If coupling is at the soma (s) or proximal dendrite (pd) location instead, then Z^{s} and V^{s} or Z^{pd} and V^{pd}, respectively, would be used instead in the above equations.

In Fig. 14.7, we plot the voltage at all three locations and show the resulting Z function (iPRCs) at each of these locations. The iPRC at time zero corresponds to the spike peak in the soma, which does not correspond to peaks in the dendritic locations. This is made clear by the plot in the far right panel of the first row in which voltages at all three locations are plotted. Comparison of the iPRCs at the three locations is shown in the far right panel of the second row. The predicted leftward shift and attenuation can be seen. Also apparent is the development of a negative portion in the PRC with non-somatic perturbations. The dramatic effect on the resulting G-function as the coupling location changes can be seen in the third row of the figure. Parameter values and additional equations are given in Appendix 2.

Example voltage output from a two-cell network in which weak electrical coupling is between dd's of three-compartment neurons is given in the bottom row of Fig. 14.7. Voltage output from each of the three compartments (soma, pd, dd) is shown. For the parameters used in this example in which voltage-gated channels are present in the dendrites, asynchronous output occurs with about a 20% phase lag between the two cells (i.e., about a 10 ms lag for a 47 ms period). This phase lag is predicted by the zero crossing of the G-function which has a negative slope, as given in the third row, third panel of Fig. 14.7. If coupling occurred at a pd or soma location instead, then the G-function would give a prediction of either about 5 ms lag or synchrony, respectively – note the zero crossings with negative slope of the G-function in the first and second panels of the third row of Fig. 14.7. Thus, the location of gap junctions can have a dramatic effect on phase-locked states, and this effect can be predicted using weakly coupled oscillator theory.

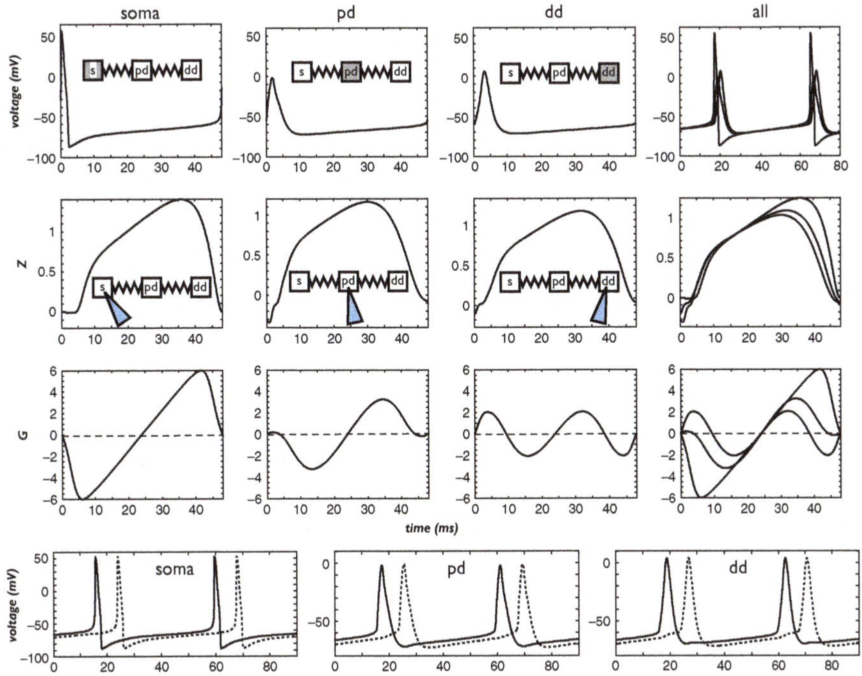

Fig. 14.7 Three-compartment model voltages, iPRCs and G-functions.*Top row*: one cycle of the voltage at the soma, proximal dendrite (pd) and distal dendrite (dd) locations. Rightmost plot shows two cycles at all three locations (thickest line is the soma voltage and *thinnest line* is the dd voltage). *Second row:* iPRCs obtained for perturbations at the soma, pd and dd, with the rightmost plot showing all three (thickest line is soma and *thinnest* is dd). Note the leftward shift of the iPRC with more distal locations. Zero time refers to the spike peak at the soma. *Third row: G*-function for electrical coupling at the soma, pd and dd, with the rightmost plot showing all three (*thickest line* is for soma coupling and thinnest line is for coupling at dd). Somatic coupling predicts stable synchrony (negative slope at zero), and pd and dd coupling predict stable nonzero phase lags (note zero crossing value for negative slopes). *Bottom row*: two-cell simulation for electrical coupling at the dd location. Plots show voltages at soma, pd and dd locations. Note that the phase lag is the same at all three locations and corresponds to about 10 ms, as predicted by the stable phase lag (zero crossing in third plot of above row of G-function). Note that the scale of the vertical axes for all three plots are not the same

5 Summary and Discussion

In this chapter, we have shown how the theory of weakly coupled oscillators can provide insight into the phase-locking behavior of electrically coupled spiking neurons, and we described how synchronization patterns depend on details of both the PRC and voltage trajectory, both being part of the G-function derived in the weak coupling theory. For some idealized models (e.g., integrate-and-fire), detailed analyses can be done to make precise statements about phase-locking behaviors. However, even when the models become more complex with biophysical conductances, multi-compartment structures and active dendrites, phase-locking behaviors can still be predicted from the G-functions. We note that the STRC method, which is another method that uses PRCs to predict network behavior (introduced in Chaps. 4 and 5), cannot be applied to networks with electrical coupling. This is because the coupling current that flows between the cells is not pulsatile and does not have a characteristic shape. Instead, the current flowing between the cells is present throughout the oscillation cycle and is different for any given phase-difference between the cells.

In this chapter, we have focused on networks of two identical noiseless cells. However, as discussed in introductory Chap. 1, the theory of weak coupling can be extended to include heterogeneity, noise and large numbers of cells. The maxima of G-functions determine the robustness of the phase-locked states to heterogeneity (e.g., Mancilla et al. 2007), and the effects of additive white noise on the distribution of phase-differences is given by a formula that involves integrals of the G-functions (Pfeuty et al. 2005; see also Chap. 1). Therefore, by examining how the shapes of Z and V_{LC} affect these quantities, one can explore the effects of noise and heterogeneity in electrically coupled networks. Perhaps not so surprisingly, the basic conclusions from the identical noiseless cell pairs largely carry over to these more complicated situations. For example, Pfeuty et al. (2003) observed that large heterogeneous networks of QIF neurons displayed the same phase-locking regimes that they found for QIF cell pairs, except for regimes that occurred for particularly small regions of parameter space. We should note, however, that fundamentally new behavior can be observed in large networks with spatially localized connectivity. For instance, Kazanci & Ermentrout (2007) showed transitions from global synchrony to travelling wave activity in a network with localized weak electrical coupling.

Strictly speaking, weakly coupled oscillator theory only applies when the coupling is sufficiently weak. However, qualitative results and general insights obtained using this theory often hold for moderate coupling strengths. For example, weakly coupled oscillator theory was used to determine three different regions of phase-locking behaviors for varying levels of active dendrites in multi-compartment models. When larger electrical coupling strengths were explored with simulations, these different regions were still apparent (Saraga et al. 2006). Furthermore, we note that techniques to explore strong electrical coupling are being developed using piecewise linear models (Coombes 2008).

5.1 Other Considerations, Biological Details, and Future Work

Ultimately, we would like to obtain a functional understanding of the roles played by gap junctional coupling in cortical systems. The goal of this chapter was to characterize synchronization patterns brought about by electrically coupled networks using the theory of weakly coupled oscillators. The use of theory is to expand our understanding and insight over that obtained from direct numerical simulations alone, to help obtain a functional understanding. However, it is sometimes the case that biologically detailed and realistic simulations can provide functional insights that are strengthened by subsequent theoretical work or can be explained by previous theoretical work. For example, work by Traub et al. (1999) using large, detailed network simulations predicted the need for axo-axonic gap junctions for the generation of high frequency population oscillations. Subsequent theoretical work by Lewis & Rinzel (2000) using cellular automata was able to explain how the resulting network frequencies occurred. In a large, detailed network of striatal FS cells, Hjorth et al. (2009) suggested that gap junctions played more of a shunting rather than a synchronizing role. This shunting aspect of gap junction coupling in large interconnected networks was examined by Amitai et al. (2002) in neocortical interneurons. These detailed electrically coupled network models included other details involving chemical synaptic input characteristics relevant for the particular situation.

The presence of both electrical and chemical synapses has motivated several modeling and theoretical works (e.g., Gao & Holmes 2007; Kazanci & Ermentrout 2007; Kopell & Ermentrout 2004; Lewis & Rinzel 2003; Pfeuty et al. 2005; Skinner et al. 1999). These works show the increased richness that emerges in these networks. What remains to be explored is how realistic spatial architecture of electrically coupled networks (Amitai et al. 2002) affects resulting synchronization patterns.

Appendix 1

Leaky Integrate-and-Fire (LIF)

For the standard leaky integrate-and-fire (LIF), the ionic current is described only by the leakage current

$$I_{\text{ionic}}(V) = g_m(V - V_r),$$

where V_r is the reversal potential of the leakage current, g_m is the membrane leakage conductance of the cell. For $I_{\text{app}} > g_m(V_{\text{th}} - V_r)$, the T-periodic solution is

$$\begin{aligned} V_{LC}(t) &= \left(\frac{I_{\text{app}}}{g_m} + V_r\right) - \left(\frac{I_{\text{app}}}{g_m} + V_r - V_{\text{reset}}\right) \mathrm{e}^{-\frac{g_m}{C}t}. \\ &= V_{\text{ss}} - (V_{\text{ss}} - V_{\text{reset}})\mathrm{e}^{-\frac{t}{\tau}} \end{aligned}$$

where

$$V_{\text{ss}} = \left(\frac{I_{\text{app}}}{g_m} + V_r\right), \qquad \tau = \frac{C_m}{g_m},$$

and the period is

$$T = \tau \ln\left(\frac{V_{ss} - V_{reset}}{V_{ss} - V_{th}}\right).$$

The infinitesimal phase-response curve for the LIF model (Hansel et al. 1995; Neltner et al. 2000) is

$$Z(t) = \begin{cases} \frac{\tau}{V_{ss} - V_{reset}} \exp\left(\frac{t}{\tau}\right), & 0 < t < T \\ 0, & t = 0, T \end{cases}.$$

Nondimensionalization: By taking

$$v_j = \frac{V_j - V_r}{V_{th} - V_r}, \quad \tilde{t} = \frac{t}{C_m/g_m},$$

and

$$I = \frac{I_{app}}{g_m(V_{th} - V_r)}, \quad \tilde{\beta} = \frac{\beta}{g_m(V_{th} - V_r)}, \quad \tilde{g}_{gap} = \frac{g_{gap}}{g_m}, \quad v_{reset} = \frac{V_{reset} - V_r}{V_{th} - V_r},$$

we obtain the following nondimensional LIF model for electrically coupled cells

$$\frac{dv_j}{d\tilde{t}} = -v_j + I + \tilde{g}_{gap}(v_k - v_j).$$

When v_j reaches a threshold potential 1, cell j fires a δ-function spike of magnitude $\tilde{\beta}$ after which v_j is reset to the potential 0. In the main text, the tildes on $\tilde{t}$, $\tilde{\beta}$, and $\tilde{g}_{gap}$ are omitted for notational convenience.

Quadratic Integrate-and-Fire (QIF)

For the QIF, the ionic current is

$$I_{ionic}(V) = g_m \frac{(V_T - V)(V - V_R)}{(V_T - V_R)} = g_m \frac{(V - \bar{V})^2}{V_-} - I_b,$$

where V_R and V_T are the resting potential and "threshold" potential of the neuron, respectively, for $I_{app} = 0$, and g_m is related to the resting membrane conductance. $\bar{V} = \frac{V_T + V_R}{2}$ and $V_- = V_T - V_R$. $I_b = \frac{g_m \bar{V}^2}{V_-}$ acts as an intrinsic bias current. $I_{ionic}(V)$ can be thought of as a combination of currents that captures the effects of the leakage current and some of the threshold aspects related to the fast sodium current. For $I_{app} > I_b$, the T-periodic solution is

$$V_{LC}(t) = \bar{V} + V_{-}\sqrt{I}\tan\left(\sqrt{I}\frac{t}{\tau} + \gamma(V_{\text{reset}})\right),$$

where

$$I = \frac{I_{\text{app}} - I_b}{g_m V_{-}}, \quad \gamma(V) = \frac{1}{\sqrt{I}}\arctan\left(\frac{\frac{V-\bar{V}}{V_{-}}}{\sqrt{I}}\right),$$

with a period of

$$T = \tau(\gamma(V_{\text{th}}) - \gamma(V_{\text{reset}})).$$

The infinitesimal phase-response curve $Z(t)$ for the QIF model is

$$Z(t) = \begin{cases} \dfrac{\tau}{V_{-}}\dfrac{1}{\sqrt{I}}\cos^2\left(\sqrt{I}\dfrac{t}{\tau} + \gamma(V_{\text{reset}})\right), & 0 < t < T \\ 0, & t = 0, T \end{cases}.$$

Nondimensionalization: By taking

$$v_j = \frac{V_j - \bar{V}}{V_{-}}, \quad \tilde{t} = \frac{t}{C_m/g_m},$$

and

$$\tilde{\beta} = \frac{\beta}{g_m V_{-}}, \quad \tilde{g}_{\text{gap}} = \frac{g_{\text{gap}}}{g_m}, \quad v_{\text{reset}} = \frac{V_{\text{reset}} - \bar{V}}{V_{-}}, \quad v_{\text{th}} = \frac{V_{\text{th}} - \bar{V}}{V_{-}},$$

we obtain the following nondimensional QIF model:

$$\frac{dv_j}{d\tilde{t}} = v_j^2 + I + \tilde{g}_{\text{gap}}(v_k - vj),$$

When v_j reaches a threshold potential v_{th}, cell j fires a δ-function spike of magnitude β after which v_j is reset to the potential v_{reset}. In the main text, the tildes on $\tilde{t}$, $\tilde{\beta}$, and $\tilde{g}_{\text{gap}}$ are omitted for notational convenience.

Appendix 2

Additional Equations and Parameter Values Used for Three-Compartment Model

The following equations are the same for each compartment in each of cell 1 and 2, with parameter values as given in the above table.

Parameter	Value	Units
γ	0.5	mS/cm^2
C	0.8	$\mu F/cm^2$
g^{s}_{Na}	184	mS/cm^2
g^{dd}_{Na}, g^{pd}_{Na}	2.76	mS/cm^2
g^{s}_{K}	140	mS/cm^2
g^{dd}_{K}, g^{pd}_{K}	2.1	mS/cm^2
g_L	0.0245	mS/cm^2
V_{Na}	55	mV
V_K	−90	mV
V_L	−60	mV
g_{gap}	0.02	mS/cm^2

$$I_{\text{ionic}} = g_{\text{Na}} m^3 h (V - V_{\text{Na}}) + g_K n^4 (V - V_K) + g_L (V - V_L)$$

$$\frac{\mathrm{d}m}{\mathrm{d}t} = \alpha_m (1 - m) - \beta_m m$$

$$\frac{\mathrm{d}h}{\mathrm{d}t} = \alpha_h (1 - h) - \beta_h h$$

$$\frac{\mathrm{d}n}{\mathrm{d}t} = \alpha_n (1 - n) - \beta_n n$$

where

$$\alpha_m = -0.1(V + 35)/(\exp[-0.1(V + 35)] - 1)$$

$$\beta_m = 4\exp[-(V + 60)/18]$$

$$\alpha_h = 0.07\exp[(V + 58)/20]$$

$$\beta_h = 1/(\exp[-0.1(V + 28)] + 1)$$

$$\alpha_n = -0.01(V + 34)/(\exp[-0.1(V + 34)] - 1)$$

$$\beta_n = 0.125\exp[-(V + 44)/80]$$

Acknowledgments TJL was supported by the National Science Foundation under grants DMS-09211039 and DMS-0518022. FKS thanks the Natural Sciences and Engineering Research Council of Canada for support.

References

Acker, C.N., Kopell, N., & White, J.A. (2003). Synchronization of strongly coupled excitatory neurons: Relating network behavior to biophysics. *Journal of Computational Neuroscience*, 15(1), 71–90.

Amitai, Y., Gibson, J.R., Beierlein, M., Patrick, S.L., Ho, A.M., Connors, B.W., & Golomb, D. (2002). The spatial dimensions of electrically coupled networks of interneurons in the neocortex. *Journal of Neuroscience*, 22(10), 4142–4152.

Bem, T., Le Feuvre, Y., Rinzel, J., & Meyrand, P. (2005) Electrical coupling induces bistability of rhythms in networks of inhibitory spiking neurons. *European Journal of Neuroscience*, 22, 2661–2668.

Bennett, M.V.L. (1977). Electrical transmission: a functional analysis and comparison to chemical transmission. In: *Handbook of Physiology. The Nervous System. Cellular Biology of Neurons.* (Sect. 1,Vol. I, pt. 1, pp. 357–416). Bethesda, MD: American Physiological Society.

Bennett, M.V.L. (2000). Seeing is relieving: electrical synapses between visualized neurons. *Nature*, 3(1), 7–9.

Bennett, M.V.L., & Zukin, R.S. (2004). Electrical coupling and neuronal synchronization in the mammalian brain. *Neuron*, 41, 495–511.

Chow, C.C., & Kopell, N. (2000). Dynamics of spiking neurons with electrical coupling. *Neural Computation*, 12, 1643–1678.

Coombes S. (2008). Neuronal networks with gap junctions: A study of piece-wise linear planar neuron models. *SIAM Journal of Applied Dynamical Systems*, 7, 1101–1129.

Connors, B.W. & Long, M.A. (2004). Electrical synapses in the mammalian brain. *Annual Reviews of Neuroscience*, 27, 393–418.

Crook, S.M., Ermentrout, G.B., & Bower, J.M. (1998a) Dendritic and synaptic effects in systems of coupled cortical oscillators. *Journal of Computational Neuroscience*, 5(3), 315–329.

Crook, S.M., Ermentrout, G.B., & Bower, J.M. (1998b) Spike frequency adaptation affects the synchronization properties of networks of cortical oscillators. *Neural Computation*, 10, 837–854.

Cymbalyuk, G.S., Nikolaev, E.V., & Borisyuk, R.M. (1994). In-phase and antiphase self-oscillations in a model of two electrically coupled pacemakers. *Biological Cybernetics*. 71, 153–160.

Di Garbo, A., Panarese, A., & Chillemi, S. (2005). Gap junctions promote synchronous activities in a network of inhibitory interneurons. *Biosystems* 79, 91–99.

Durstewitz, D., Seamans, J.K., & Sejnowski, T.J. (2000). Dopamine-mediated stabilization of delay-period activity in a network model of prefrontal cortex. *Journal of Neurophysiology*, 83,1733–1750.

Erisir, A., Lau, D., Rudy, B., & Leonard, S. (1999). Function of specific K+ channels in sustained high-frequency firing of fast-spiking neocortical cells. *Journal of Neurophysiology*, 82, 2476–2489.

Ermentrout, G.B. (1996). Type I membranes, phase resetting curves, and synchrony. *Neural Computation*, 8(5), 979–1001.

Ermentrout, G.B., Pascal, M., & Gutkin, B. (2001). The effects of spike frequency adaptation and negative feedback on the synchronization of neural oscillators. *Neural Computation*, 8(6), 1285–1310.

Ermentrout, G.B. (2002). *Simulating, Analyzing, and Animating Dynamical Systems: A Guide to XPPAUT for Researchers and Students*. Philadelphia: SIAM.

Ermentrout, G.B., & Wechselberger, M. (2009). Canards, clusters, and synchronization in a weakly coupled interneuron model. *SIAM Journal of Applied Dynamical Systems*, 8(1), 253–278.

Fukuda, T. (2007). Structural organization of the gap junction network in the cerbral cortex. *Neuroscientist*, 13, 199–207.

Fukuda, T., & Kosaka, T. (2000). Gap junctions linking the dendritic network of GABAergic interneurons in the hippocampus. *Journal of Neuroscience*, 20, 1519–1528.

Galarreta, M., & Hestrin, S. (2001). Electrical synapses between GABA-releasing interneurons. *Nature Reviews Neuroscience*, 2, 426–433.

Gao, J., & Holmes, P. (2007). On the dynamics of electrically-coupled neurons with inhibitory synapses. *Journal of Computational Neuroscience*, 22, 39–61.

Glass, L., & Mackey, M.C. (1988). *From Clocks to Chaos: The Rhythms of Life*. Princeton University Press.

Goldberg, J.A., Deister, C.A., & Wilson, C.J. (2007). Response properties and synchronization of rhythmically firing dendritic neurons. *Journal of Neurophysiology*, 97, 208–219.

Hansel, D., Mato, G., & Meunier, C. (1995). Synchrony in excitatory neural networks. *Neural Computation*, 7, 307–337.

Hjorth, J., Backwell, K.T., & Kotaleski, J.H. (2009). Gap junctions between striatal fast-spiking interneurons regulate spiking activity and synchronization as a function of cortical activity. *Journal of Neuroscience*, 29(16), 5276–5286.

Hormuzdi, S.G., Filippov, M.A., Mitropoulou, G., Monyer, H., & Bruzzone, R. (2004). Electrical synapses: a dynamic signaling system that shapes the activity of neuronal networks. *Biochimica et Biophysica Acta*, 1662, 113–137.

Johnston, D., & Narayanan, R. (2008). Active dendrites: colorful wings of the mysterious butterflies. *Trends in Neuroscience*, 31(6), 309–16.

Kazanci, F.G., & Ermentrout, G.B. (2007). Pattern Formation in an Array of Oscillators with Electrical and Chemical Coupling. *SIAM Journal of Applied Mathematics*, 67, 512–529.

Kopell, N., & Ermentrout, G.B. (2004). Chemical and electrical synapses perform complementary roles in the synchronization of interneuronal networks. *Proceedings of the National Academy of Science USA*, 101(43), 15482–15487.

Kuramoto, Y. (1984). *Chemical Oscillations, Waves, and Turbulence*. New York: Springer-Verlag.

Latham, P.E., Richmond, B.J., Nelson, P.G., & Nirenberg, S. (2000). Intrinsic dynamics in neuronal networks. I. Theory. *Journal of Neurophysiology*, 83(2), 808–827.

Lewis, T.J., & Rinzel, J. (2000). Self-organized synchronous oscillations in a network of excitable cells coupled by gap junctions. *Network: Computation in Neural Systems*. 11(4), 299–320,.

Lewis, T.J., & Rinzel, J. (2003). Dynamics of spiking neurons connected by both inhibitory and electrical coupling. *Journal of Computational Neuroscience*, 14, 283–309.

Lewis, T.J., & Rinzel, J. (2004). Dendritic effects in networks of electrically coupled fast-spiking interneurons. *Neurocomputing*, 58–60, 145–150.

Mancilla, J.G., Lewis, T.J., Pinto, D.J., Rinzel, J., & Connors, B.W. (2007). Synchronization of electrically coupled pairs of inhibitory interneurons in neocortex. *Journal of Neuroscience*, 27(8), 2058–2073.

Meier, C., & Dermietzel, R. (2006). Electrical synapses – gap junctions in the brain. *Results and Problems in Cell Differentiation*, 43, 99–128.

Neltner, L., Hansel, D., Mato, G., & Meunier, C. (2000). Synchrony in heterogeneous networks of spiking neurons. *Neural Computation*, 12, 1607–1641.

Nomura, M., Fukai, T., & Aoyagi, T. (2003). Synchrony of fast-spiking interneurons interconnected by GABAergic and electrical synapses. *Neural Computation*, 15, 2179–2198.

Pfeuty, B., Golomb, D., Mato, G., & Hansel, D. (2003). Electrical synapses and synchrony: The role of intrinsic currents. *Journal of Neuroscience*, 23(15), 6280–6294.

Pfeuty, B., Golomb, D., Mato, G., & Hansel, D. (2005). The combined effects of inhibitory and electrical synapses in synchrony. *Neural Computation*, 17, 633–670.

Rinzel, J., & Ermentrout, G.B. (1998). Analysis of neural excitability and oscillations (chp.7, pp251–191). In: *Methods in Neuronal Modeling: From Ions to Networks* (2nd ed., C. Koch, I. Segev eds). Cambridge MA: MIT Press.

Saraga, F., & Skinner, F.K. (2004). Location, location, location (and density) of gap junctions in multi-compartment models. *Neurocomputing*, 58–60, 713–719.

Saraga, F., Ng, L., & Skinner, F.K. (2006). Distal gap junctions and active dendrites can tune network dynamics. *Journal of Neurophysiol*, 95, 1669–1682.

Sherman, A., & Rinzel, J. (1992). Rhythmogenic effects of weak electrotonic coupling in neuronal models. *Proceedings of the National Academy of Science USA*, 89, 2471–2474.

Skinner, F.K., Zhang, L., Perez Velazquez, J., & Carlen, P. (1999). Bursting in inhibitory interneuronal networks: A role for gap-junctional coupling. *Journal of Neurophysiology*, 81, 1274–1283.

Söhl, G., Maxeiner, S., & Willecke, L. (2005). Expression and functions of neuronal gap junctions. *Nature Reviews Neuroscience*, 6, 191–200.

Stiefel, K.M., Gutkin, B.S., & Sejnowski, T.J. (2009). The effects of cholinergic neuromodulation on neuronal phase-response curves of modeled cortical neurons. *Journal of Computational Neuroscience*, 26(2), 289–301.

Traub, R.D., Schmitz, D., Jefferys, J.G.R., & Draguhn, A. (1999). High-frequency population oscillations are predicted to occur in hippocampal pyramidal neuronal networks interconnected by axoaxonal gap junctions. *Neuroscience*, 92, 407–426.

Vogel, R., & Weingart, R. (2002). The electrophysiology of gap junctions and gap junction channels and their mathematical modelling. *Biology of the Cell*, 94, 501–510.

Zahid, T., & Skinner, F.K. (2009). Predicting synchronous and asynchronous network groupings of hippocampal interneurons coupled with dendritic gap junctions. *Brain Research*, 1262, 115–129.

Chapter 15
The Role of Intrinsic Cell Properties in Synchrony of Neurons Interacting via Electrical Synapses

David Hansel, Germán Mato, and Benjamin Pfeuty

Abstract The study of the synchronization of neural activity has to deal with the existence of a great diversity of neuronal and synaptic properties, which both may influence the collective behavior of neurons. In this chapter, we present an approach that allows to disentangle the respective role of neuronal and synaptic properties in synchronization. It relies on the theory of weakly coupled oscillators and on the concept of infinitesimal Phase Response Curve (iPRC) which is relevant for systems where neurons fire periodically. Applying this theory to networks of neurons coupled with electrical synapses provides a unifying framework explaining how their synchronization behavior depends on the shape of the iPRC. It reveals that synchronization mediated by electrical synapses is more versatile than previously considered. It depends on the intrinsic currents and morphology of neurons as well as on their combination with inhibitory synapses.

D. Hansel (✉)
Neurophysique et Physiologie, CNRS UMR8119, Université Paris Descartes,
45 rue des Saint-Pères 75270 Paris, France
e-mail: david.hansel@univ-paris5.fr

G. Mato
Centro Atómico Bariloche and Instituto Balseiro (CNEA and CONICET),
8400 San Carlos de Bariloche, Argentina
e-mail: matog@cab.cnea.gov.ar

B. Pfeuty
PhLAM, CNRS UMR8523, Université Lille I, F-59655 Villeneuve d'Ascq, France
e-mail: pfeuty@phlam.univ-lille1.fr

N.W. Schultheiss et al. (eds.), *Phase Response Curves in Neuroscience: Theory, Experiment, and Analysis*, Springer Series in Computational Neuroscience 6,
DOI 10.1007/978-1-4614-0739-3_15,

1 Introduction

Synchronous electrical activity in the nervous system is known since the discovery of electroencephalography in the 1930s (Berger 1929) and the first studies on the α rhythm in mammals. Following these pioneering works, many other examples of synchronous activity have been discovered in various species and areas of the nervous system. They differ greatly in the power spectrum and the spatiotemporal patterns of the activity as well as in the conditions of occurrence, which can be pathological or physiological. The tremendous progress in experimental techniques which has occurred in the last 20 years, has opened new horizons to establish the relationship between synchronous electrical activities and cellular properties.

The observed synchronous patterns of activity in the nervous system raise several theoretical issues regarding the relationship between the structure of activity patterns and the cellular properties of the neurons, the nature of their interactions, the local and global architectures of the networks, and their interconnections with other brain areas. Furthermore, understanding how noise, heterogeneities, and spatial fluctuations of connectivity patterns affect synchrony is a fundamental question for neuroscientists, and more specifically for theoreticians. A suitable framework to address these issues is the physics of extended dynamical systems. One possible approach is to study the cooperative dynamical states in simplified network models, which within some limits can allow in-depth analytical investigations based on statistical mechanics of pure and disordered systems, the theory of stochastic systems and the physics of non-linear dynamical systems. Subsequently, one extends the analytical investigations by numerical simulations beyond the limits which can be treated analytically. This approach has lead to substantial progress in our understanding of general mechanisms of neuronal synchrony.

Electrical synapses have long been known to exist in invertebrates (Watanabe 1958; Furshpan and Potter 1959) but it is only recently that evidence of their ubiquity has been unequivocally found in the mammalian brain. Remarkably, electrical synapses in the central nervous system are frequently found to connect GABAergic interneurons. This is, for instance, the case in the striatum (Kita et al. 1990), the hippocampus (Venence et al. 2000), the cerebellum (Mann-Metzer and Yarom 1999), the reticular thalamic nucleus (Landisman et al. 2002), or in the neocortex (Gibson et al. 1999; Galarreta and Hestrin 1999; Fukuda and Kosaka 2000). In the neocortex, electrical as well as inhibitory synapses exist between Fast–Spiking (FS) interneurons (Gibson et al. 1999; Galarreta and Hestrin 2001) or between Multipolar Bursting neurons (Blatow et al. 2003). In contrast, low-threshold-spiking (LTS) interneurons interact mostly via electrical synapses, the inhibitory synapses between them being sparse (Gibson et al. 1999; Amitai et al. 2002). In the FS population, cells also interact via inhibition. FS and LTS neurons interact conversely via inhibition but not via electrical synapses.

Recent experiments have shown that electrical synapses may be involved in synchronizing neural activity. Blocking of inhibition and excitation does not reduce the synchronization of interneurons in the molecular layer of the cerebellum between which electrical synapses have been identified (Mann-Metzer and Yarom 1999). In slices of neocortex, in which LTS cells are activated by ACPD, the synchronous activity of LTS remains after blocking inhibition (Beierlein et al 2000). Reciprocally, blockade of gap-junctions through pharmacological agent (Perez-Velazquez et al. 1994; Traub et al. 2001; Friedman and Strowbridge 2003) or genetic manipulations (Deans et al. 2001; Hormuzdi et al. 2001; Blatow et al. 2003) reduces the synchronized activity in cortex or in hippocampus. Other studies suggest that electrical synapses alone are not sufficient for getting synchronous activity but that they are necessary. This has been shown in cortical slice preparations where electrical synapses or inhibitory ones are not able to synchronize neuronal activity (Tamas et al. 2000; Blatow et al. 2003). In contrast with these results, it has been reported that in inspiratory motoneurons synchronous activity depends on inhibition and that it is strongly enhanced in presence of carbenoxolone, a blocker of electrical synapses (Bou-Flores and Berger 2001). Therefore, electrical synapses may also *desynchronize* neural activity.

These experimental facts raise several issues. How do intrinsic cellular properties affect synchrony of neurons coupled via electrical synapses? How does synchrony depends on the localization of the synapses on the dendrite or the soma (or between axons) (Traub et al. 2001)? What is the significance of the combination of electrical and inhibitory couplings for neuronal dynamics? How does synchrony depends on the firing rates of the neurons? What is nature of the interplay between chemical and electrical synapses in promoting synchrony, in particular, how do inhibition and electrical coupling combine their effects in the dynamics of small and large neural systems?

In this chapter, we address these issues in the framework of the theory of weakly coupled oscillators. The intrinsic dynamics of the neuron can be described in terms of the infinitesimal Phase Response Curve also called infinitesimal Phase Response Function (iPRC). The iPRC can be defined for any dynamical system that displays a stable limit cycle (Kuramoto 1984). It is evaluated by applying a small perturbation at a given phase of the cycle and recording the change of the phase induced by the perturbation. Combining this concept with a weak coupling approximation one can develop very powerful tools to analyze complex dynamical systems with a large number of degrees of freedom. Not only the dimensionality of the problem is reduced, but the approximation gives rise to a theoretical framework in which synchronization properties can be analyzed in a very natural way.

The outline of the chapter is as follows: in the next section, we present the theory of weakly coupled oscillators following an approach developed by Kuramoto (1984). We then apply this theory in the third section to study the dynamics of neuronal networks in which neurons interact via electrical synapses. The results obtained in that framework emphasize how the synchronizing effect of electrical synapses depends on the intrinsic properties of the neurons. The various regimes of the interplay between electrical and inhibitory synapses are discussed in the fourth section. We conclude the chapter by discussing the relevance of the results derived in the framework of weakly coupled oscillators to explain the synchronization behavior observed in more realistic numerical simulations as well as in physiological experiments.

2 Phase Models for Networks of Weakly Coupled Neurons

2.1 Phase Response Function and iPRC of Intrinsic Oscillators

Let us consider a nonlinear oscillator whose state is described by the vector $\mathbf{X}$. For instance for the Hodgkin–Huxley (HH) (Hodgkin and Huxley 1952) model these components are V, m, n, h, where V is the membrane potential and m, n, h are the gating variables of the potassium and sodium currents. For a multicompartmental neuronal model the dimension of vector $\mathbf{X}$ is larger, because it incorporates variables related to each compartment. This vector evolves according to:

$$\frac{\mathrm{d}\mathbf{X}}{\mathrm{d}t} = \mathbf{F}(\mathbf{X}). \tag{15.1}$$

We assume that this system has a limit cycle solution, with period T ($\mathbf{X}(t+T) = \mathbf{X}(t)$) which is asymptotically stable, i.e., that there is a neighborhood $\mathcal{G}$ of the periodic solution such that the dynamical system with any initial condition in $\mathcal{G}$ eventually converges to this solution. For any point on the limit cycle we define a phase function $\phi(\mathbf{X})$. We first choose an arbitrary point $\mathbf{X}_0$ and we assign to it a phase variable $\phi(\mathbf{X}_0) = 0$. The phase of the other points on the limit cycle is defined via the differential equation:

$$\frac{\mathrm{d}\phi(\mathbf{X})}{\mathrm{d}t} = 1. \tag{15.2}$$

This definition can be extended to points sufficiently close to the limit cycle as follows. Let us consider a point $\mathbf{Y}_0$ in $\mathcal{G}$. This point evolves according to (15.1), giving rise to the trajectory $\mathbf{Y}(t)$. By definition of $\mathcal{G}$, this trajectory eventually converges to the limit cycle. Therefore, there is a unique point $\mathbf{X}_Y$ on the limit cycle, whose trajectory $\mathbf{X}_\mathbf{Y}(t)$ is such that: $\lim_{t\to\infty} |\mathbf{Y}(t) - \mathbf{X}_Y(t)| \to 0$. Hence, one can associate a phase variable to $\mathbf{Y}$ as: $\phi(\mathbf{Y}) \equiv \phi(\mathbf{X}_Y)$ and this phase evolves with the same velocity for points on the limit cycle or in its neighborhood, i.e.,

$$\frac{\mathrm{d}\phi(\mathbf{Y})}{\mathrm{d}t} = 1. \tag{15.3}$$

Let us now consider the perturbed system

$$\frac{\mathrm{d}\mathbf{X}}{\mathrm{d}t} = \mathbf{F}(\mathbf{X}) + \varepsilon\mathbf{p}(\mathbf{X}). \tag{15.4}$$

If the perturbation is weak enough, $\varepsilon \ll 1$, the trajectory stays in the neighborhood of the unperturbed limit cycle where the phase is still defined. This phase evolves according to:

$$\frac{\mathrm{d}\phi(\mathbf{X})}{\mathrm{d}t} = \frac{\partial\phi}{\partial\mathbf{X}}\cdot\frac{\mathrm{d}\mathbf{X}}{\mathrm{d}t} = 1 + \varepsilon\frac{\partial\phi}{\partial\mathbf{X}}\cdot\mathbf{p}(\mathbf{X}). \tag{15.5}$$

This equation controls the evolution of the phase in the basin of attraction of the limit cycle. It is exact but useless for determining the time evolution of the phase, because to evaluate the right hand side of the equation we need to know the point $\mathbf{X}$ at which to evaluate the functions $\frac{\partial \phi}{\partial \mathbf{X}}$ and $\mathbf{p}(\mathbf{X})$. In general, it is not possible to recover that information from the phase ϕ alone. However, since $\varepsilon \ll 1$, we can assume that the trajectory remains at a distance of order ε from the limit cycle. The right hand side can then be estimated *on the limit cycle*, i.e.,

$$\frac{\mathrm{d}\phi}{\mathrm{d}t} = 1 + \varepsilon \mathbf{Z}(\phi) \cdot \mathbf{p}(\phi) + O(\varepsilon^2) \tag{15.6}$$

with:

$$\mathbf{Z}(\phi) = \frac{\partial \phi(\mathbf{X})}{\partial \mathbf{X}} \tag{15.7}$$

and $\mathbf{p}(\phi) = \mathbf{p}(\mathbf{X})$ for a point $\mathbf{X}$ whose phase is ϕ.

The vector $\mathbf{Z}(\phi)$ is the infinitesimal phase response function of the oscillator. It characterizes the change of phase induced by a weak perturbation. Typically, the synaptic input from other neurons $\mathbf{p}$ affects only the voltage component of the state space. Therefore, we will usually consider only the voltage component of the phase response function, which defines an *iPRC Curve*. There are only few examples of nonlinear oscillators for which the limit cycle and the iPRC can be computed analytically (see some examples below). In general, numerical methods have to be used. One approach is to integrate numerically the dynamics to compute the limit cycle over one period and subsequently the derivative of the phase is evaluated by estimating the dephasing induced at large time by small initial perturbations at different points covering the limit cycles. This approach which is directly related to the definition of the iPRC is intuitive but not rigorous. In particular, it is necessary to check that the amplitude of the perturbation which is used is sufficiently small. This method was applied for instance in Hansel et al. (1993). Another method, more rigorous mathematically but less intuitive, is introduced in Ermentrout and Kopell (1992); Kopell (1998). It is implemented in the software XPP (Ermentrout 2003).

Knowing this, iPRC allows one to compute how the phase of the oscillator evolves under small perturbations. For any time-dependent function x, we define the time average over one period

$$< x > (t) = \frac{1}{T} \int_{t-T}^{t} x(t')\mathrm{d}t'. \tag{15.8}$$

Let us note that for a weakly perturbed oscillator, the time average differs from the function itself by an amount of order ε (Kuramoto 1984). The main reason for introducing the time average is that, as we will see below, in terms of this new variable the dynamics of a system of coupled oscillators depends only on the phase differences. Integrating (15.6) between $[t - T, t]$, we obtain the evolution of the time averaged phase:

$$\frac{\mathrm{d} < \phi > (t)}{\mathrm{d}t} = 1 + \frac{\varepsilon}{T} \int_{t-T}^{t} \mathbf{Z}(\phi(t')) \cdot \mathbf{p}(\phi(t'))\mathrm{d}t' + O(\varepsilon^2). \tag{15.9}$$

As ϕ and $< \phi >$ differ at most by an amount of order ε and we can replace ϕ by $< \phi >$ in the right hand side:

$$\frac{\mathrm{d} < \phi > (t)}{\mathrm{d}t} = 1 + \frac{\varepsilon}{T} \int_{t-T}^{t} \mathbf{Z}(< \phi(t') >) \cdot \mathbf{p}(< \phi(t') >) \mathrm{d}t' + O(\varepsilon^2). \quad (15.10)$$

If we change units such that the phase ϕ goes from 0 to 2π, we have

$$\frac{\mathrm{d}\phi}{\mathrm{d}t} = \omega + \frac{\varepsilon}{2\pi} \int_{0}^{2\pi} \mathbf{Z}(\phi).\mathbf{p}(\phi)\mathrm{d}\phi + O(\varepsilon^2), \quad (15.11)$$

where $\omega = 2\pi/T = 2\pi f$ and $\mathbf{Z}(\phi)$ and $\mathbf{p}(\phi)$ are 2π-periodic functions of ϕ. We have also dropped the brackets for simplicity. In the following we will be always writing the dynamics in terms of time averaged quantities.

2.2 *Examples of iPRC of Neurons*

The iPRC, which characterizes how neurons respond to small perturbations that affect its voltage component, is a key concept to understand the relationship between intrinsic properties of neurons and their collective dynamics (Hansel et al. 1993; van Vreeswijk et al. 1994; Kopell and Ermentrout 2002). The iPRC depends on the excitability properties of the neurons and therefore on the intrinsic currents involved in their dynamics.

2.2.1 iPRC of Conductance-Based Neurons

Figure 15.1 plots the iPRCs for the HH model and the Wang-Buzsáki (WB) model (two models which are frequently used in neural modeling). Both models involve a potassium delayed rectifier current and a sodium current. However, the parameters of these currents are not the same in the two models leading to excitability properties which differ qualitatively. In particular, the onset of firing is a *subcritical Hopf* bifurcation in the HH model whereas in the WB model it is a *saddle-node* bifurcation. Accordingly, the HH neuron has a minimal firing frequency in contrast to the WB neuron that is capable of firing at arbitrary low frequency in response of an external steady state current sufficiently close to the rheobase. The two models display also qualitative differences in their iPRCs. The iPRC of the WB neurons is positive everywhere on the limit cycle except in a very small region which coincides with the action potential. This means that the WB neuron responds to small depolarizations by advancing its phase. This is in contrast to what is found in the HH neuron where on a substantial range of the limit cycle the iPRC is negative. Another difference is that in a region of the limit cycle the iPRC of the HH neuron is close to zero indicating that the neuron is insensitive to perturbations in that region.

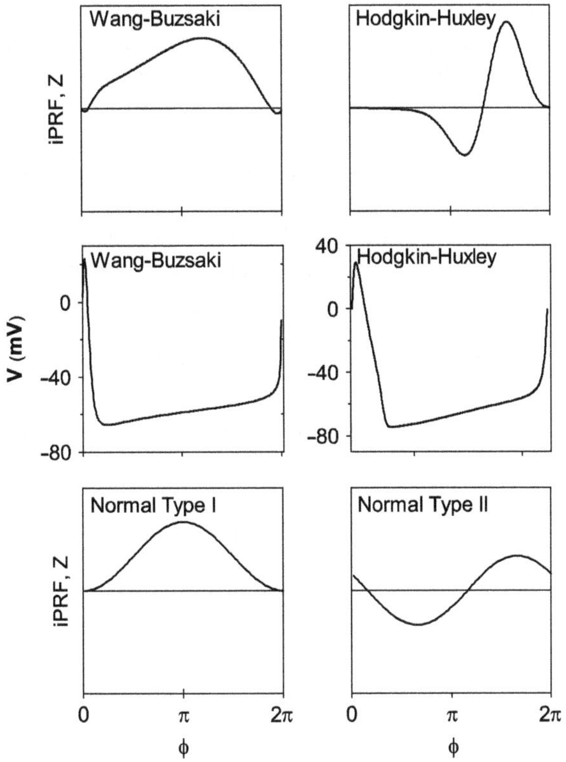

Fig. 15.1 iPRCs (*top panels*) and voltage traces (*middle panels*) during one period of oscillation for Wang–Buzsáki neuron (*left*) and Hodgkin–Huxley neuron (*right*). *Bottom panels* show iPRCs for the normal form of Type I neuron (*left*) and Type II neuron (*right*). $\phi = 0$ corresponds to the phase at which the membrane potential crosses $V = 0$ upwards

This last feature is related to the fact that the action potential is substantially broader for the HH model than for the WB model. In Fig. 15.4, we display other examples of iPRC obtained in conductance-based neurons showing how intrinsic currents shape the response to small perturbation.

The role of the intrinsic currents in shaping the iPRC can be studied analytically when the external current induces periodic spiking near the firing onset. This is because near this onset the neural dynamics can be assimilated to their *normal forms*, which depend only on the bifurcation structure. In Brown et al. (2004), the iPRCs have been evaluated for the four codimension one bifurcations. The more typical cases found in neural modeling are the *saddle node*, or Type I, and the *subcritical Hopf*, or Type II, bifurcations. One can show that for Type I models, the iPRC is given by:

$$Z(\phi) \propto 1 - \cos(\phi), \tag{15.12}$$

while for Type II models it is

$$Z(\phi) \propto \sin(\phi - \phi_H), \tag{15.13}$$

where that phase variable ϕ goes from 0 to 2π and the constant ϕ_H can be derived from the original equations of the model. It is important to remark that in the second case the iPRC of the full model will approach to the result of (15.13) only if the radius of the oscillation near the bifurcation is small (Brown et al. 2004). This is not the case in the HH model, therefore the relation between the iPRC of the full model and the iPRC of the normal form is only qualitative. In contrast, for the Type I model, the two iPRCs agree more at lower firing rates. However, the iPRCs of the HH model and the normal form of the Type II models share an important feature: the iPRC has a significant negative region before the positive region (see Fig. 15.1).

2.2.2 iPRC of Integrate-and-Fire Neurons

In integrate-and-fire model of neurons, the subthreshold membrane potential follows the dynamics equation:

$$\tau_0 \frac{\mathrm{d}v}{\mathrm{d}t} = f(v) + I_{\mathrm{ext}}, \tag{15.14}$$

where τ_0 is the time constant of the dynamics. These dynamics are supplemented by the condition that whenever the membrane potential reaches a threshold V_{T}, an action potential is fired and v is reset to a potential V_{r}. The membrane potential of the neurons can be written as the sum of its subthreshold component $v(t)$ and a δ-function which represents the suprathreshold component (i.e., the action potential):

$$V(t) = v(t) + \theta \sum_{\mathrm{spikes}} \delta(t - t_{\mathrm{spike}}), \tag{15.15}$$

where θ measures the integral over time of the supra-threshold part and t_{spike} satisfies $v(t_{\mathrm{spike}}) = V_{\mathrm{T}}$.

For one-dimensional models, the trajectory unequivocally determines the iPRC. It can be written as (Kuramoto 1984; Hansel et al. 1995):

$$Z(\phi) = \left(\frac{\mathrm{d}v(\phi)}{\mathrm{d}\phi}\right)^{-1}. \tag{15.16}$$

In the standard leaky integrate-and-fire (LIF) model:

$$f(v) = -(v - V_{\mathrm{L}}), \tag{15.17}$$

where V_{L} is the reversal potential of the leak current. The iPRC can be easily calculated. One finds:

$$Z(\phi) = \frac{2\pi\tau_0}{T(I_{\mathrm{ext}} + V_{\mathrm{L}} - V_{\mathrm{r}})} \exp\left(\frac{\phi T}{2\pi\tau_0}\right), \tag{15.18}$$

where the firing period is defined as 2π. The iPRC of LIF neurons is a monotonically increasing exponential function, which does not capture nonmonotonic behavior of iPRC observed in type-I conductance-based neuronal model.

Nonlinear integrate-and-fire models may provide a more accurate description of the membrane subthreshold dynamics that take into account some basic non-linearities induced by voltage-dependent currents. A special case is the quadratic integrate-and-fire (QIF) model whose membrane time course and iPRC can be computed analytically. Its membrane voltage evolves in time:

$$\tau_0 \frac{dv}{dt} = (v - V_s)^2 + I_{ext}, \tag{15.19}$$

where V_s is the value of potential membrane at the equilibrium point ($I_{ext} = 0$).

Integrating (15.19), one finds that the general solution for the subthreshold potential is for $I_{ext} > 0$:

$$v(t) = V_s + \sqrt{I_{ext}} \tan\left(\frac{t\sqrt{I_{ext}}}{\tau_0} + \alpha\right), \tag{15.20}$$

where α is a constant of integration. The condition that at $t = 0$ the membrane potential of the neuron is at its reset value, V_r, determines α:

$$\alpha = \tan^{-1}\left(\frac{V_r - V_s}{\sqrt{I_{ext}}}\right). \tag{15.21}$$

The function $v(t)$ increases monotonically to infinity. Therefore, starting from the resetting potential V_r, $v(t)$ reaches the threshold after some time T. The condition $v(T) = V_T$ determines the firing period. One finds:

$$T = \tau_0 \frac{\tan^{-1}\left(\frac{V_T - V_s}{\sqrt{I_{ext}}}\right) - \tan^{-1}\left(\frac{V_r - V_s}{\sqrt{I_{ext}}}\right)}{\sqrt{I_{ext}}}. \tag{15.22}$$

The f-I curve of the QIF neuron is given by $f = 1/T(I_{ext})$. Let us note that a continuous f-I curve for small $I - I_{ext}$ requires that the parameters $V_T - V_s$ and $V_r - V_s$ have opposite sign. As V_T is always larger than V_s we must take $V_T > V_s$ and $V_r < V_s$.

In the limit $V_T \to \infty$, $V_r \to -\infty$ or in the limit of small firing rate, the QIF model reduces to the theta model (Ermentrout and Kopell 1986; Ermentrout 1996). This model describes the normal form of the oscillatory dynamics appearing after a saddle-node bifurcation (Brown et al. 2004; Ermentrout 1996) that characterizes Type I neurons (see Hansel and Mato 2003 for the details of the derivation).

The iPRC of QIF neurons can be computed analytically by combining (15.16) and (15.19) with $\phi = 2\pi t/T$:

$$Z(\phi) = \frac{\tau_0}{\left[v\left(\frac{\phi T}{2\pi}\right) - V_s\right]^2 + I_{ext}}. \tag{15.23}$$

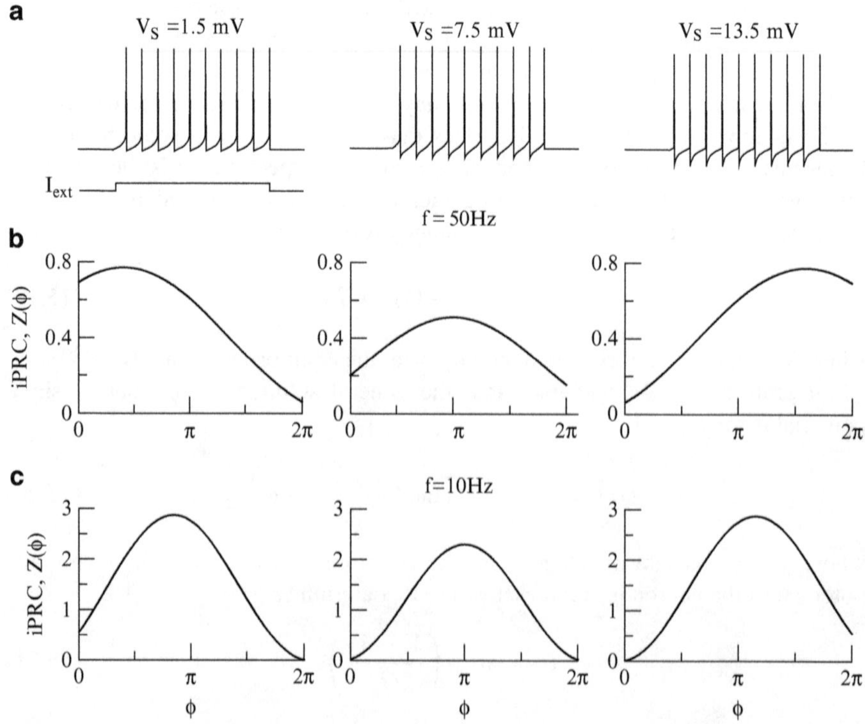

Fig. 15.2 Voltage traces and iPRCs of the QIF neuron. The voltage traces and the iPRCs were computed using (15.15) and (15.24), respectively (with $\tau_0 = 10\,\mathrm{ms}$). In (**a**) and (**b**), the external current is such that the firing rate is 50 Hz. (**a**), The voltage trace of the neuron in response to a constant injected current for three values of the ratio V_s . The reset depth is the same in all three cases: $V_T - V_r = 15\,\mathrm{mV}$. Note the changes in the concavity of the voltage trace between the spikes when V_s. (**b**), The iPRC, Z, for the same cases as in (**a**). (**c**), Same as in (**b**) but for a firing rate of 10 Hz. The phase response changes less than in (**b**) when V_s varies. Similar cases are also shown in Chap. 14

Replacing the membrane potential from (15.20), we finally find

$$Z(\phi) = \frac{\tau_0}{I_{\mathrm{ext}}} \cos^2\left(\frac{\phi T\sqrt{I_{\mathrm{ext}}}}{2\pi\tau_0}\right). \tag{15.24}$$

In the following sections, we always take $V_r = 0$ and $V_T = 15\mathrm{mV}$. The parameter V_s controls the shape of the subthreshold trajectory and the iPRC, as shown in Fig. 15.2. Figure 15.2a plots the membrane potential of a QIF neuron firing at 50 Hz for different values of V_s. For $V_s = (V_T + V_r)/2$, the concavity changes from upward to downward at the middle of the period. Taking more (resp. less) positive values of V_s this inflexion point shifts to the left (resp. right).

Figure 15.2b plots the iPRC for the same values of V_s and I_{ext}. As it is obvious from (15.24), $Z(\phi) > 0$ for all ϕ and has only one maximum. The location of the maximum of Z depends on V_s and I_{ext}. For $V_s < (V_T + V_r)/2$, it is located in the first half of the firing period, whereas for $V_s > (V_T + V_r)/2$ it is located in the second half. The larger the firing rate, the stronger the dependence of Z on V_s. For $V_s = (V_T + V_r)/2$, $Z(\phi)$ is symmetric and its maximum is always at $\phi = \pi$.

Expressions for the iPRC of LIF and QIF neurons are also given in Chap. 14, although in that case there is no rescaling of the period from T to 2π.

2.3 *Phase Dynamics of a Networks of Weakly Coupled Neurons*

We consider now a network of N identical neurons interacting in a pairwise manner. The dynamics of neuron l has the form:

$$\frac{\mathrm{d}\mathbf{X}_l}{\mathrm{d}t} = \mathbf{F}(\mathbf{X}_l) + g\sum_{j=1}^{N} J_{lj}\mathbf{G}(\mathbf{X}_l, \mathbf{X}_j) + \mathbf{s}_l(t) \tag{15.25}$$

where J_{lj} is the connectivity matrix: $J_{lj} = 1$ if the presynaptic neuron j interacts with the neuron l and $J_{lj} = 0$, otherwise. $\mathbf{G}$ is the synaptic interaction which depends on the state of presynaptic and postsynaptic neurons. The strength of the synapses is denoted by g. We also assume that the neurons receive a Gaussian white noise input that affects only the voltage component ($(\mathbf{s}_l)_j = 0$) for $j \neq 1$ with: $< (\mathbf{s}_l)_1 >= 0$, $< (\mathbf{s}_l(t))_1(\mathbf{s}_m(t'))_1 > = 2\sigma^2\delta_{l,m}\delta(t-t')$, where $\delta_{l,m} = 1$ for $l = m$ and $\delta_{l,m} = 0$ for $l \neq m$. This means that the noise is uncorrelated in time and across neurons. This equation has the form of (15.4). When the coupling strength is sufficiently weak ($g \ll 1$ and $\sigma^2 \approx g$), one can use the weak coupling theory developed above to reduce the dynamics of (15.25) to the dynamics of a network of phase oscillators. One finds:

$$\frac{\mathrm{d}\phi_l}{\mathrm{d}t} = \omega + \sum_{j=1}^{N} J_{lj}\Gamma(\phi_l - \phi_j) + \eta_l(t), \tag{15.26}$$

where

$$\Gamma(\phi) = \frac{g}{2\pi}\int_0^{2\pi} \mathbf{Z}(\phi' + \phi).\mathbf{p}(\phi' + \phi, \phi')\mathrm{d}\phi', \tag{15.27}$$

and $\eta_l(t)$ is the effective noise in the phase dynamics.

The function $\Gamma(\phi)$ is the phase interaction function. It is a convolution between the phase response function $\mathbf{Z}(\phi)$ and the perturbation induced by the interactions, $\mathbf{p}(\phi_1, \phi_2) = \mathbf{G}(\mathbf{X}(\phi_1), \mathbf{X}(\phi_2))$ evaluated on the limit cycle trajectory of presynaptic and postsynaptic neurons. It is a periodic function with period 2π because both $\mathbf{Z}(\phi)$ and $\mathbf{p}(\phi, \phi')$ are periodic with that period.

Generalizing the approach developed by Kuramoto (1991) in the context of the LIF model, one can show that $\eta(t)$ is white Gaussian noise with zero mean and variance D:

$$D = \frac{\sigma^2}{2\pi} \int_0^{2\pi} [(\mathbf{Z}(\phi))_1]^2 \mathrm{d}\phi. \tag{15.28}$$

Therefore, the effective noise in the phase dynamics of a neuron is proportional to the average of the square of its iPRC over one period.

In the following, we consider two types of neuronal systems, (1) a pair of coupled neurons and (2) a network of N all-to-all coupled neurons. Let us note that the first case is also covered in Chap. 14, but not the second one. They also do not analyze the effect of the noise.

2.3.1 Cross-Correlation of a Pair of Neurons in Presence of Weak Noise

The cross-correlation (CC) of the firing activity of pairs of interacting neurons measures the probability that the two neurons fire with some given time delay. This is a standard tool for evaluating the synchronization between pairs of neurons. Here, we derive an analytical expression for the CC of a pair of spiking neurons under the assumptions of weak interactions and weak noise. It provides valuable information about the way how noise and intrinsic and synaptic properties affect the shape of the CC. In the presence of weak noise, the dynamics of two coupled identical oscillators is given by

$$\frac{\mathrm{d}\phi_1}{\mathrm{d}t} = \omega + \Gamma(\phi_1 - \phi_2) + \eta_1(t), \tag{15.29}$$

$$\frac{\mathrm{d}\phi_2}{\mathrm{d}t} = \omega + \Gamma(\phi_2 - \phi_1) + \eta_2(t). \tag{15.30}$$

The time elapsed at time t since the last spike fired by neuron i is simply the firing period, T, multiplied by the phase, $\phi(t)$, computed modulo 2π ($\mathrm{mod}(\phi_i(t), 2\pi)$) and divided by 2π. Similarly, the time elapsed between the last spikes fired by neurons i and j is: $\tau = T \ \mathrm{mod}(\phi_i(t) - \phi_j(t), 2\pi)/(2\pi)$. In the weak coupling limit, the CC can be evaluated in terms of the probability distribution function of phase shifts. We define the joint probability $P_{1,2}(t_1, t_2)$ as the probability that neuron 1 fires between t_1 and $t_1 + \delta$ and neuron 2 fires between t_2 and $t_2 + \delta$. The spike train of neuron 1 is $S_1(t)$, i.e., $S_1(t) = 1$ if there is a spike in the bin $[t, t + \delta]$ and $S_1(t) = 0$ if there is no spike. The time average (over the whole spike train) is $< S_1(t) > \equiv P_1(t)$. $P_1(t)$ does not depend on time and it is equal to $f = \frac{\omega}{2\pi}$. Using Bayes theorem, we find

$$< S_1(t_1) S_2(t_2) > \ = P_{1,2}(t_1, t_2) = P_{2|1}(t_2|t_1) \, P_1(t_1), \tag{15.31}$$

where $P_{2|1}$ is the conditional probability that neuron 2 fires between t_2 and $t_2 + \delta$ given that neuron 1 fired between t_1 and $t_1 + \delta$. Assuming stationarity, this function depends only on the difference $\tau = t_2 - t_1$: $P_{2|1}(t_2|t_1) = \tilde{P}_{2|1}(\tau)$, and hence

$$\frac{< S_1(t_1)S_2(t_1+\tau) >}{< S_1(t) >< S_2(t) >} = \frac{\tilde{P}_{2|1}(\tau)}{\omega}. \tag{15.32}$$

Switching from a function of time to a function of phase, we obtain

$$C(\tau) \equiv \frac{< S_1(t_1)S_2(t_1+\tau) >}{< S_1(t) >< S_2(t) >} = 2\pi P_0(\phi), \tag{15.33}$$

where P_0 is the probability distribution of the phase shift between the two neurons.

Using (15.29) and (15.30) one finds that $\Delta = \phi_1 - \phi_2$ satisfies the Langevin equation:

$$\frac{\mathrm{d}\Delta}{\mathrm{d}t} = \Gamma_-(\Delta) + \eta(t), \tag{15.34}$$

where we have defined $\eta(t) \equiv \eta_1(t) - \eta_2(t)$ and $\Gamma_-(\Delta) \equiv (\Gamma(\Delta) - \Gamma(-\Delta))$.

The noise $\eta(t)$ is the difference of two Gaussian white noises, with zero average and variance D. Therefore, it is a Gaussian white noise with zero average and variance $2D$. Hence, the phase-shift distribution, $P(\Delta, t)$, satisfies a Fokker–Planck equation (van Kampen 1981):

$$\frac{\partial P(\Delta,t)}{\partial t} = D\frac{\partial^2}{\partial \Delta^2}P(\Delta,t) - \frac{\partial}{\partial \Delta}[\Gamma_-(\Delta)\, P(\Delta,t)]. \tag{15.35}$$

The stationary distribution of the phase shifts, P_0, is the solution of this equation which satisfies the additional constraint

$$\frac{\partial P_0(\Delta,t)}{\partial t} = 0 \tag{15.36}$$

that is:

$$P_0(\Delta) = \mathrm{e}^{G(\Delta)}\left[A\int_0^{\Delta} \mathrm{e}^{-G(\Delta')}\,\mathrm{d}\Delta' + B\right], \tag{15.37}$$

where:

$$G(\Delta) = \int_0^{\Delta} \frac{\Gamma_-(\Delta')}{D}\,\mathrm{d}\Delta' \tag{15.38}$$

and A and B are two integration constants which are determined by the periodicity of P_0, i.e., $P_0(0) = P_0(2\pi)$, and its normalization, i.e., $\int_0^{2\pi} P_0(\Delta)\mathrm{d}\Delta = 1$. One finds:

$$P_0(\Delta) = \frac{\mathrm{e}^{G(\Delta)}}{\int_0^{2\pi} \mathrm{e}^{G(\Delta')}\,\mathrm{d}\Delta'}. \tag{15.39}$$

One can show that P_0 has an extremum for phase shifts Δ that are solutions of:

$$\Gamma_-(\Delta) = 0. \tag{15.40}$$

These solutions for which

$$\frac{\mathrm{d}\Gamma_-}{\mathrm{d}\phi}(\Delta) = \Gamma'(\Delta) < 0 \ \ (\text{resp.} > 0) \tag{15.41}$$

are maxima (resp. minima) of P_0. Equations (15.40) and (15.41) can be easily interpreted. Indeed, in the absence of noise, the two neurons are phase locked at large time and the possible phase shifts are the stable fixed point of (15.34) with $\eta = 0$. It is straightforward to see that these phase shifts are the solutions of (15.40) and that (15.41) is the condition for their stability. In presence of noise, the phase shifts between the two neurons have an increase (resp. decrease) in probability density to be around the stable (resp. unstable) fixed points of the noiseless dynamics. In the limit $D \to 0$, $P_0(\Delta)$ is nonnegligible only for values of Δ in the vicinity of the maxima of $G(\Delta)$. In general, $G(\Delta)$ is continuously differentiable at least up to the second order. Expanding G around its maxima, one finds that P_0 is a sum of Gaussians centered around these points, with a width proportional to $1/\sqrt{-\Gamma'(\Delta)}$. Therefore, the greater the stability is of a fixed point in absence of noise, the sharper the probability distribution is when noise is present.

2.3.2 Stability of the Asynchronous State in Large Neuronal Network in Presence of Noise

We now consider the phase dynamics (15.26), with $N \gg 1$ and all-to-all coupling. The stability analysis of the asynchronous state can be investigated for a network of phase oscillators in presence of white noise. The dynamics is given by:

$$\frac{\mathrm{d}\phi_l}{\mathrm{d}t} = \omega + \sum_{j=1}^{N} \Gamma(\phi_l - \phi_j) + \eta_l(t), \tag{15.42}$$

where η_i are independently distributed white noise with Gaussian distribution (zero mean and variance D). Note that in order to have a well-defined limit, $N \to \infty$, the interaction strength has to scale as $1/N$.

Let us define $n(\phi, t)$ as the fraction of oscillators that have a given phase at a given time. This quantity evolves according to a *Fokker–Planck* equation (van Kampen 1981):

$$\frac{\partial n(\phi,t)}{\partial t} + \frac{\partial (v(\phi,t) n(\phi,t))}{\partial \phi} = D \frac{\partial^2 n(\phi,t)}{\partial^2 \phi} \tag{15.43}$$

with

$$v(\phi,t) = \omega + \int_0^{2\pi} \Gamma(\phi - \phi') n(\phi', t) \mathrm{d}\phi'. \tag{15.44}$$

$v(\phi, t)$ represents the velocity of the oscillators that have a phase ϕ at time t. This equation always has a solution $n(\phi, t) = N/(2\pi)$. This solution corresponds to the *asynchronous state* of the network in which the probability of firing of the neurons in the network is constant in time. In this state, the temporal fluctuation of any population average observable vanishes in the large N limit.

To analyze the stability of this state, we expand $n(\phi, t)$ in the vicinity of the asynchronous state solution:

$$n(\phi, t) = \frac{N}{2\pi} + \varepsilon(\phi, t). \tag{15.45}$$

Inserting this expression into the Fokker–Planck equation we obtain (up to order ε^2):

$$\frac{\partial \varepsilon(\phi, t)}{\partial t} + \frac{\partial((\omega + N\Gamma_0)\varepsilon(\phi, t) + u)}{\partial \phi} = D\frac{\partial^2 \varepsilon(\phi, t)}{\partial^2 \phi} \tag{15.46}$$

with $u = \frac{N}{2\pi}\int_0^{2\pi} \Gamma(\phi - \phi')\varepsilon(\phi', t)\mathrm{d}\phi'$ and $\Gamma_0 = \frac{1}{2\pi}\int_0^{2\pi} \Gamma(\phi)\mathrm{d}\phi$. This is a linear partial differential equation that can be transformed in a set of uncoupled linear ordinary differential equations by introducing a Fourier representation. Defining

$$\varepsilon_l(t) = \frac{1}{2\pi}\int_0^{2\pi} \varepsilon(\phi, t)\exp(il\phi)\,\mathrm{d}\phi, \tag{15.47}$$

$$\Gamma_l = \frac{1}{2\pi}\int_0^{2\pi} \Gamma(\phi)\exp(il\phi)\,\mathrm{d}\phi, \tag{15.48}$$

we find

$$\frac{\mathrm{d}\varepsilon_l(t)}{\mathrm{d}t} - il\left((\omega + N\Gamma_0)\varepsilon_l(t) + N\varepsilon_l(t)\Gamma_l\right) = -l^2 D\varepsilon_l(t). \tag{15.49}$$

Looking for solution of the form $\varepsilon_l(t) \propto \exp(\lambda_l t)$, one finds that the eigenvalues λ_l are given by:

$$\lambda_l = il\left((\omega + N\Gamma_0) + N\Gamma_l\right) - l^2 D. \tag{15.50}$$

The asynchronous state is stable if any small perturbation of it eventually decays to zero. The condition for this stability is therefore:

$$Re(\lambda_l) = -lN Im(\Gamma_l) - l^2 D \leq 0 \tag{15.51}$$

for all l.

3 Dynamics of Neurons Coupled with Electrical Synapses

An electrical synapse induces a synaptic current proportional to the difference of the membrane potentials of the two coupled neurons. Therefore, one may expect that electrical interactions always tend to equalize the membrane potentials favoring synchrony. However, the actual effect of electrical synapses is more subtle. For instance, it has been found that for pairs of identical neurons coupled with electrical interactions both in-phase and antiphase states can be stable (Han et al. 1995; Chow and Kopell 2000; Lewis and Rinzel 2003). One reason for this complexity is that the current generated by the interaction is filtered in the postsynaptic neuron. Depending on this effect it is possible that the positive current induced by a sharp spike could have less effect than the negative current induced by the after-spike hyperpolarization. Therefore, the temporal course of the membrane potential and the location of the synapse (that affects the filtering properties) can have a strong influence on the net effect of the interaction. In the following, we apply the phase reduction approach to clarify how electrical synapses affect synchrony of neurons.

3.1 Synchrony of a Pair of Neurons Electrically Coupled

As shown in Sect. 2, the phase interaction function can be evaluated as a convolution between the iPRC and the synaptic current:

$$\Gamma(\phi_l - \phi_j) = \frac{1}{2\pi}\int_0^{2\pi} Z(u+\phi_l) I_{\mathrm{syn}}(u+\phi_l, u+\phi_j)\mathrm{d}u, \tag{15.52}$$

where Z is the iPRC and $I_{\mathrm{syn}}(\phi_l, \phi_j)$ is the synaptic current coming from the presynaptic neuron j to the postsynaptic neuron l. For electrical synapses, this current is:

$$I_{\mathrm{syn}}(\phi_l, \phi_j) = -g(V(\phi_l) - V(\phi_j)). \tag{15.53}$$

According to (15.52), the effective phase interaction between two neurons coupled electrically is therefore:

$$\Gamma(\phi) = \frac{g}{2\pi}\int_0^{2\pi} Z(u)(V(u-\phi) - V(u))\mathrm{d}u. \tag{15.54}$$

Using this function, we can evaluate the quantities needed to determine the stability of in-phase and antiphase locking of a pair of neurons as well as the asynchronous state of a large network.

The stability of state in which the phase shift between two neurons is Δ depends on the derivative of the function $\Gamma'(\Delta)$. This quantity depends on the derivative of the membrane potential $V(t)$. From (15.15), the membrane potential can be written as the sum of three contributions:

1. A subthreshold part, during which the membrane potential increases with time from a value V_r to a threshold value V_T.
2. A resetting, where potential is instantaneously brought back from V_T to V_r.
3. A spike, modeled by a δ-function of amplitude θ.

Differentiating (15.54) with respect to ϕ one finds

$$\Gamma'(\phi) = -\frac{g}{2\pi}\int_0^{2\pi} Z(u)\, V'(u-\phi)\mathrm{d}u. \tag{15.55}$$

After doing a change of variables from time to phase and expanding the membrane potential V in its components (spike, reset, and subthreshold component), we have

$$V'(u) = 2\pi f\,\theta\,\delta'(u) + (V_r - V_T)\,\delta(u) + v'(u), \tag{15.56}$$

where f is the frequency. This allows us to rewrite $\Gamma'(\Delta)$:

$$\Gamma'(\Delta) = -\frac{g}{2\pi}\left(2\pi f\,\theta\int_0^{2\pi} Z(u)\,\delta'(u-\Delta)\mathrm{d}u + (V_r - V_T)\int_0^{2\pi} Z(u)\,\delta(u-\Delta)\mathrm{d}u\right.$$
$$\left. + \int_0^{2\pi} Z(u)\,v'(u-\Delta)\,\mathrm{d}u\right). \tag{15.57}$$

Integrating by parts the first term and using the properties of the δ-function, we finally obtain:

$$\Gamma'(\Delta) = g(S_{sp} + S_r + S_{sub}), \tag{15.58}$$

where

$$S_{sp} = f\theta\frac{dZ}{d\phi}(\Delta), \tag{15.59}$$

$$S_r = \frac{1}{2\pi}(V_T - V_r)Z(\Delta), \tag{15.60}$$

$$S_{sub} = -\frac{1}{2\pi}\int_0^{2\pi} Z(\phi)\frac{\mathrm{d}v}{\mathrm{d}\phi}(\phi - \Delta)\mathrm{d}\phi. \tag{15.61}$$

The first term, S_{sp}, is the contribution of the presynaptic spikes. It is proportional to the firing rate f, which means that it dominates in the limit of high firing rates ($f \to \infty$) and is negligible in the limit of low firing rates ($f \to 0$). The second term, S_r, is the contribution of the instantaneous reset of the membrane potential. The last term, S_{sub}, corresponds to the effect of the coupling between the two neurons when the presynaptic neuron is subthreshold. By evaluating these expressions in $\Delta = 0$ or $\Delta = \pi$, we can compute $\Gamma'(0)$ and $\Gamma'(\pi)$ which determine the stability of the in-phase and antiphase states in absence of noise, as

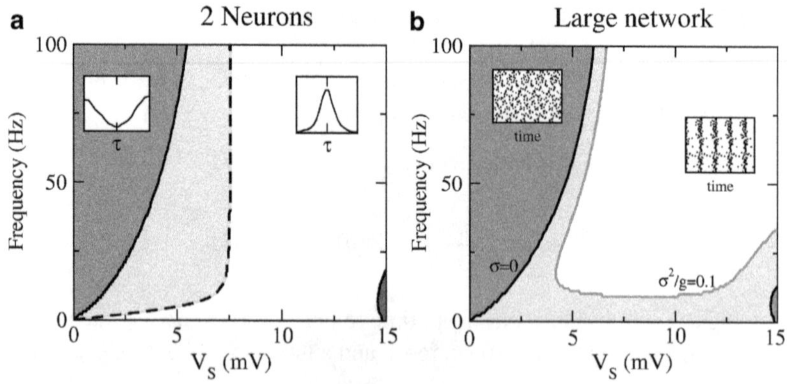

Fig. 15.3 (**a**), Phase diagram of the stability of the in-phase and antiphase states of a pair of identical QIF neurons coupled by electrical synapses in the weak coupling limit. The *solid and dashed boundary lines* are defined by the conditions $\Gamma'(\pi) = 0$ and $\Gamma'(0) = 0$, respectively (15.58). The *inset* shows the cross-correlation function $C(\tau)$. In the *dark region*, the antiphase state is stable and the maximum of $C(\tau)$ is at $\tau = \pi$. In the *white region*, the in-phase state is stable the maximum of $C(\tau)$ is located at 0. In the *gray intermediate region* the stable solution is neither in-phase nor antiphase. A similar phase diagram, but with V_r in the x-axis, is shown in Chap. 14. (**b**), The phase diagram for the stability of the asynchronous state in a large network of identical QIF neurons all-to-all coupled with electrical synapses in the weak coupling limit in presence of weak noise. The *inset* shows examples of raster plots. In the *dark region*, the AS state is stable even for 0 noise. The *solid lines* are the boundary between the domains in which the asynchronous state is stable and unstable for $\sigma^2/g = 0$ and $\sigma^2/g = 0.01$ defined by the condition $\mu_1 = 0$ (15.63). In the white region, the AS state is stable for $\sigma^2/g > 0.01$ Results are obtained using the QIF model with the time constant $\tau_0 = 10$ ms and the size of the spikes $\theta = 1$. The parameter V_s is varied but the reset depth is kept constant: $V_T - V_r = 15$ mV

well as the shape of CC distributions in presence of noise. A similar expression of the phase interaction function is also given in Chap. 14.

In Fig. 15.3a, we show the phase diagram for the stability of antiphase and in-phase locking as a function of V_s (that can be thought as the resting potential) and the firing frequencies. For $V_s < (V_T + V_r)/2$, $Z'(\pi)$ is negative (Fig. 15.2) such that S_{sp} contributes to stabilize the antiphase state for enough small value of V_s, and CC displays a peak at π. Conversely, the always negative value of S_{sub} for $\Delta = 0$ ensures that the in-phase state will be stable for V_s sufficiently large. Between these two regions, the two neurons are locked with a phase between 0 and π. Moreover, $S_r + S_{sub}$ is positive except for very small or very large values of V_s, and S_{sp} is negligible at low frequency. Therefore, for small firing frequency and large V_s, there is a narrow domain of stability of antiphase locking that coincides with a domain of stability of in-phase locking, which means that bistability occurs in the absence of noise and the CC displays two peaks at 0 and π in the presence of noise.

3.2 Synchrony in a Large Network of QIF Neurons Electrically Coupled

According to (15.51), the stability of the AS state depends on the Fourier components of the phase interaction function Γ. For the electrical coupling defined above, these components are given by:

$$\Gamma_n^{\mathrm{gap}} = gZ_n\,(v_{-n} + f\theta), \tag{15.62}$$

where Z_n, and v_n are the nth-Fourier components of the functions $Z(\phi)$ and $v(\phi)$ defined in Eqs. 15.23, 15.20. The variance of the noise, D, is given by $D = \sigma^2(Z^2)_0$ (Kuramoto 1984) [see also (15.28)]. This yields the stability condition for the asynchronous state

$$\mu_n = ngf\theta\; Im(Z_n) + ng\; Im(Z_n v_{-n}) - n^2\sigma^2(Z^2)_0 < 0. \tag{15.63}$$

For the asynchronous state to be stable, $\mu_n < 0$ for all n.

The first two terms in (15.63) correspond to the effect of the interaction. The first term represents the effect of the spikes and can be written as

$$\mu_n^{'} = -ngf\theta\; Im(Z_n) = -\frac{ngf\theta}{2\pi}\int_0^{2\pi} Z(\phi)\sin(n\phi)\mathrm{d}\phi. \tag{15.64}$$

The second term,

$$\mu_n^{''} = ng\; Im(Z_n v_{-n}) \tag{15.65}$$

corresponds to the combined effect of the reset of the membrane potential and its subsequent subthreshold evolution. The sign of the sum of these two terms depends on the parameters, V_s and on the firing frequency, f.

The last term in (15.63) corresponds to the effect of the noise. It is always positive. Therefore, as should be expected, noise increases the stability of the asynchronous state (because of the negative sign in front of this term in (15.63)). The stability of the asynchronous state depends on the competition between the first two terms and the last term. In particular, the sign of μ_n depends on the ratio between σ^2 and the coupling strength g. Note that because of the factor n^2 in the last term of (15.63), the stability of the modes increases rapidly with their order.

Note also that $(Z^2)_0$ decreases when the firing rate increases. Using Eqs. 15.28 and 15.24, we find that in the limit of large external current I_{ext} the leading term is given by

$$(Z^2)_0 \approx \frac{3\tau_0^2}{8I_{\mathrm{ext}}^2}. \tag{15.66}$$

As in the limit of large firing rate f is proportional to I_{ext} (see (15.22)), the noise is less efficient at stabilizing the asynchronous state at high firing rates. This effect can be seen in Fig. 15.3b. At large firing rate, the noise affects the stability line much less than at low firing rate.

The qualitative behavior of the quantities μ_1' and μ_1'' is similar to those of S_{sp} and $S_{\text{sub}} + S_{\text{r}}$, respectively, for a pair of neurons. In particular, μ_1'' is symmetric around $V_{\text{s}} = (V_{\text{T}} + V_{\text{r}})/2$. Moreover, μ_1' increases monotonically from negative values to positive values when V_{s} increases and it changes sign at $V_{\text{s}} = (V_{\text{T}} + V_{\text{r}})/2$. This is because for $V_{\text{s}} > (V_{\text{T}} + V_{\text{r}})/2$ the function $Z(\phi)$ is skewed toward $\phi < \pi$ (resp. $\phi > \pi$) where $\sin(\phi) > 0$ (resp. $\sin(\phi) < 0$) therefore, for $n = 1$, the integral in (15.64) is negative (resp. positive) and $\mu_1' < 0$ (resp. $\mu_1' > 0$). In particular, since for $V_{\text{s}} = (V_{\text{T}} + V_{\text{r}})/2$, $Z(\phi)$ is symmetric around $\phi = \pi$, the integral in (15.64) vanishes for $n = 1$ (because $\sin(\phi + \pi) = -\sin(\phi)$).

The phase diagram for the stability of the asynchronous state as a function of V_{s} and the firing frequency is plotted in Fig. 15.3b. It is very similar to the phase diagram for the stability of the antiphase state for a pair of neurons (Fig. 15.3a): asynchronous (vs antisynchronous) state is stable for (1) V_{s} sufficiently small and frequency sufficiently large or (2) V_{s} sufficiently large and frequency sufficiently small. The only qualitative difference between these two phase diagrams is associated with the presence of noise. For large enough firing rates, the instability lines for $\sigma^2/g = 0$ and $\sigma^2/g = 0.01$ are very close to each other. The distance between the two lines increases when the frequency decreases. At some value, $f = f_*$, the two lines separate completely and continue toward the right of the phase diagram (large V_{s}). In particular, for $\sigma^2/g = 0.01$ the asynchronous state is stable for small enough firing rate regardless the value of V_{s}. As a matter of fact, one can show analytically that the asynchronous state is always stable if the firing rate is smaller than a critical value, $f_*(\sigma^2/g)$, which vanishes with σ^2/g and increases with the noise level.

There are two noticeable regimes where the change of synchrony properties is nonmonotonic:

1. In some intermediate range of firing rates, the stability of the asynchronous state varies nonmonotonically with V_{s}: when V_{s} increases from 0, the asynchronous state is first stable, then unstable, and stable again for large V_{s}. The size of the intermediate region in which the asynchronous state is unstable decreases when the noise level increases.
2. If the noise is not too strong, the stability of the asynchronous state varies nonmonotonically in some intermediate range of $V_{\text{s}} < (V_{\text{T}} + V_{\text{r}})/2$. In this domain, the asynchronous state is stable at low firing rates, loses stability when the rate increases but becomes stable again at high firing rates.

3.3 *Synchrony in Networks of Conductance-Based Neurons Electrically Coupled*

In previous subsections, we have shown analytically how the shape of the iPRC, which depends on the QIF neuronal model parameter V_s and firing frequency, determines the stability of antisynchronous (for two neurons) and asynchronous states in networks of QIF neurons electrically coupled. These results provide a unified framework explaining how synchronization properties depend on detailed intrinsic neuronal properties of conductance-based neurons (models described in Appendix 1). We therefore simulate the dynamics of networks of two or a large number of conductance-based neurons, measure the synchronization properties and interpret the synchronization behavior in term of the shape of the iPRC. The quantities that measure synchrony in the case of two or large number of neurons are defined in the Appendix 2.

3.3.1 Synchrony via Electrical Synapses Depends on the Intrinsic Conductances of the Neurons

Our analysis of the QIF model suggests that more generally synchrony may be predicted from the firing frequency and the shape of the iPRC independently of the details of the neuronal dynamics. In particular, neurons coupled with electrical synapses will be more easily synchronized if their iPRC are skewed to the right than to the left. We check whether specific intrinsic neuronal properties that shape the iPRC also determine the synchronization properties of a pair of neurons or a large number of neurons coupled with electrical synapses according to this qualitative rule. In Fig. 15.4, we show the effect of specific currents frequently found in neurons both on neuronal synchronization and the shape of the iPRC.

1. Adding a slow potassium current increases the degree of synchrony between a pair of neurons (peak of CC) or in a large network (amplitude of the mean potential membrane). The presence of such current gives also rise to a diminution of the iPRC after the spike and to an increase in the value of $Z'(\pi)$ (Fig. 15.4b).
2. In presence of a persistent sodium current, the network activity becomes asynchronous and the two neurons are locked in antiphase. Incidentally, such current leads to an increase of the iPRC after the spike, such that the peak is shifted on the left (Fig. 15.4c).
3. Increasing the leak conductance (due for instance to synaptic conductance) enhances the synchrony level in large network or pair of neurons incorporating a persistent sodium current. This is because the shape of the iPRC is skewed to the right (Fig. 15.4d).

One can understand intuitively how potassium and sodium currents shape the iPRC. The current, I_K, increases the refractoriness of the neuron; therefore, it reduces its responsiveness to small depolarization after a spike. A similar effect,

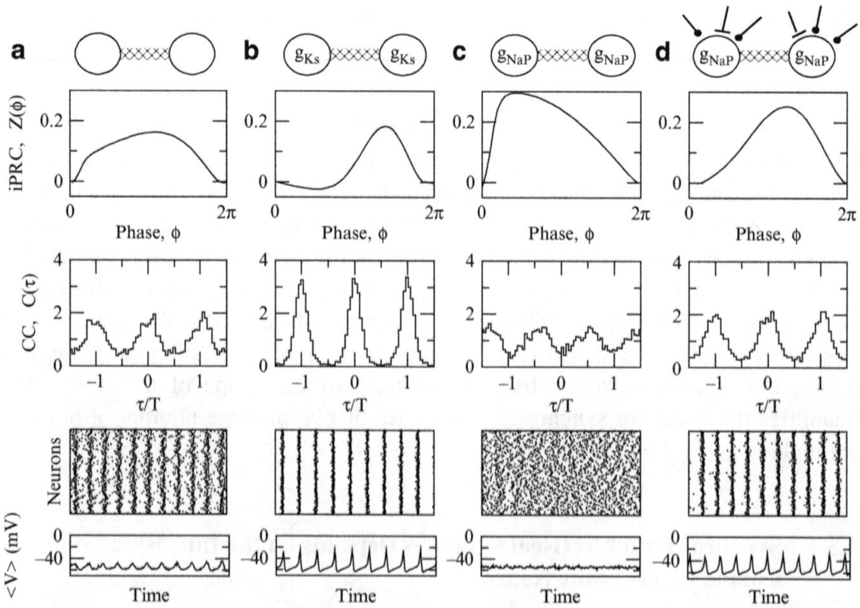

Fig. 15.4 Effect of various intrinsic neuronal conductances on synchronization properties of conductance-based neurons coupled with electrical synapses. Adding various intrinsic conductances to the original conductance-based model (see Appendix 1) modifies the shape of the iPRC (*top*), and the neuronal synchronization measured by a CC in the case of a pair of neurons (*middle*), or by a raster-plot and the mean potential in the case of a large network (*bottom*). Parameters used are $g_{\text{gap}} = 0.005\,\text{mS/cm}^2$, $f \approx 50\,\text{Hz}$, and $\sigma = 0.3\ \text{mV/ms}^{1/2}$. (**a**), In the original model, the iPRC has a right-side peak and electrical synapses tend to synchronize. (**b**), Adding a slow-potassium current ($g_{Ks} = 0.4\,\text{mS/cm}^2$) decreases the left-side of the iPRC and increases the synchrony level. (**c**), Adding a persistent sodium current ($g_{NaP} = 0.4\,\text{mS/cm}^2$) shifts the peak of the iPRC to the right and leads to antiphase or asynchronous firing. (**d**), Increasing the leak conductance in comparison with the model used in (**c**) shifts the peak of the iPRC to the right and promotes synchronization

but stronger and more lasting, occurs with I_{Ks}. This intuitively explains why we found that larger conductances of these currents skew the iPRC toward the second half of the firing period. Similar effects have been found by Ermentrout et al. (2001) for M and AHP potassium currents. The current I_{NaP} is an inward current. It is already activated near rest and depolarizing perturbations amplify this activation. Hence, I_{NaP} increases the responsiveness of the neuron after a spike and shifts the maximum of the response function toward the first half of the period as we have found here.

In all these examples, changes in intrinsic conductances of neurons affect the synchrony level of electrically coupled neurons, by skewing the iPRC toward the left or the right, similarly to changing V_s in the QIF model. Any other neuronal intrinsic currents that modify the shape of the iPRC may also have impact on synchronization properties, increasing or decreasing it.

We predict that potassium currents, like I_{K} and I_{Ks}, tend to promote synchrony of neurons coupled via electrical synapses and that, in contrast, I_{NaP} tends to oppose it. Our other predictions concern the dependency of the synchrony level with the firing rate. We expect that these conclusions do not depend on the particular models that we have chosen for the intrinsic currents.

4 Interplay Between Electrical and Chemical Inhibitory Synapses

Electrical synapses are frequently found to connect GABAergic interneurons in the central nervous system. Therefore, certain types of neurons (e.g., FS cells Gibson et al. 1999; Galarreta and Hestrin 2001 or multipolar bursting Blatow et al. 2003 in neocortex) often interact through both electrical and GABAergic synapses. This raises the question of the effect of their combination in neural synchrony.

4.1 Neuronal Synchronization with Inhibitory Interactions

First, we consider the case of inhibitory interactions alone, for which we have chosen a "current-based" model. If the presynaptic neuron fires a spike at time 0, it injects a current on the postsynaptic neurons equal to $g_{\mathrm{inh}} s(t)$ where

$$s(t) = \frac{\exp(-t/\tau_1) - \exp(-t/\tau_2)}{\tau_1 - \tau_2} \tag{15.67}$$

and τ_1 (resp. τ_2) is the decay (resp. rise) time of the synaptic interaction. If the presynaptic neuron is firing periodically with frequency f, the total current can be evaluated by adding up all the spikes. We obtain

$$s(\phi) = \frac{1}{\tau_1 - \tau_2}\left(\frac{\mathrm{e}^{-\phi/(2\pi f \tau_1)}}{1 - \mathrm{e}^{-1/(2\pi f \tau_1)}} - \frac{\mathrm{e}^{-\phi/(2\pi f \tau_2)}}{1 - \mathrm{e}^{-1/(2\pi f \tau_2)}}\right) \tag{15.68}$$

for $0 \leq \phi < 2\pi$ and $s(\phi) = s(\phi + 2\pi n)$ for any integer number n.

The phase interaction function is given by [see (15.27)]:

$$\Gamma^{\mathrm{inh}}(\phi) = \frac{g_{\mathrm{inh}}}{2\pi} \int_0^{2\pi} Z(u)\, s(u - \phi)\, \mathrm{d}u. \tag{15.69}$$

The stability of the asynchronous state is determined by the condition:

$$-Im(\Gamma_n^{\mathrm{inh}}) - n\sigma^2 < 0. \tag{15.70}$$

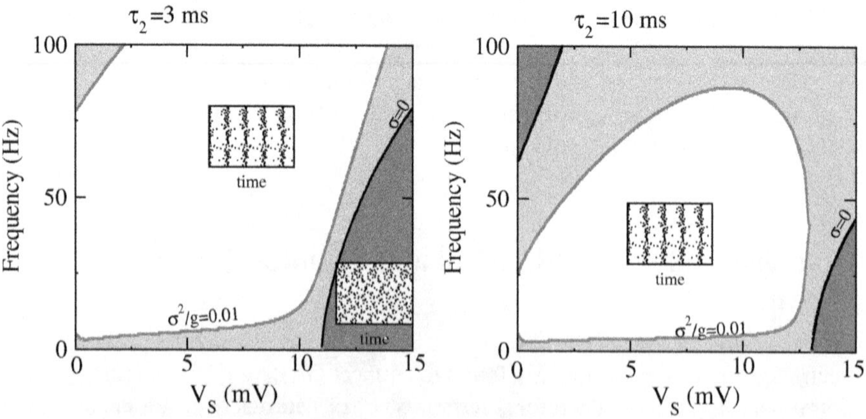

Fig. 15.5 The phase diagram for the stability of the asynchronous state in a large network of identical QIF neurons all-to-all coupled with inhibitory synapses in the weak coupling limit in presence of weak noise. The *solid lines* are the boundary between the domains in which the asynchronous state is stable and unstable for $\sigma^2/g = 0$ and $\sigma^2/g = 0.01$ defined by the condition $\Gamma_1 = 0$. Results are obtained using the QIF model with the time constant $\tau_0 = 10$ ms and synaptic rise time constant $\tau_1 = 1$ ms. *Left* and *right phase diagrams* are associated with two different synaptic decay time constants of $\tau_2 = 3$ ms and $\tau_2 = 10$ ms, respectively

Combining (15.69) and (15.68), $Im(\Gamma_n^{\text{inh}})$ writes:

$$Im\left(\Gamma_n^{\text{inh}}\right) = \frac{g_{\text{inh}}}{A}\left[\left(\frac{1}{\tau_1\,\tau_2} - n^2\right) Im(Z_n) - n\left(\frac{1}{\tau_1} + \frac{1}{\tau_2}\right) Re(Z_n)\right] \quad (15.71)$$

with:

$$A = 2\pi\tau_1\tau_2\left(\frac{1}{(\tau_1)^2} - n^2\right)\left(\frac{1}{(\tau_2)^2} - n^2\right). \quad (15.72)$$

Inhibitory interactions are characterized by significant synaptic time constants that influence their synchronization properties. The interplay between synaptic time constants and intrinsic properties of neurons is shown in Fig. 15.5, which displays the phase diagram for the stability of asynchronous state as a function of V_s, the firing frequencies and the decay time of inhibitory synapses τ_2. For rapidly decaying synaptic current, the asynchronous state is stable for low-firing frequencies and large values of V_s. A slower decay time constant tends to reduce such domain of stability of the asynchronous state, while it stabilizes the asynchronous state for small values of V_s and large firing rates.

4.2 Cooperation and Competition Between Electrical and Inhibitory Synapses

The approach involving weak coupling theory and iPRC can be used to study systems with combined interactions., i.e., chemical inhibitory and electrotonic. Here we study this situation. We analyze the interplay between these two types of interactions in the synchronization of large networks. A similar analysis can be done in the context of synchronization between pairs of neurons (Pfeuty et al. 2005). In the weak coupling case, any number of interactions can be combined additively. For the case of the combinations of electrical and inhibitory couplings, the effective phase interaction function for dual interaction is:

$$\Gamma^{\mathrm{dual}} = \Gamma^{\mathrm{inh}} + \Gamma^{\mathrm{gap}}. \tag{15.73}$$

From (15.69), (15.50), and (15.51), it is straightforward to see that in a network with electrical and inhibitory synapses, the stability of asynchronous state is determined by the condition:

$$Re(\lambda_n^{\mathrm{dual}}) = -Im(\Gamma_n^{\mathrm{gap}}) - Im(\Gamma_n^{\mathrm{inh}}) - n\sigma^2 < 0, \tag{15.74}$$

where $Im(\Gamma_n^{\mathrm{gap}})$ and $Im(\Gamma_n^{\mathrm{inh}})$ are evaluated using (15.63) and (15.71).

From the phase diagram shown in Figs. 15.3 and 15.5 and from (15.74), we can predict synchronization properties of combined interaction from those of each interaction separately. Specifically, distinct domains of parameters (such as f, V_s, τ_1, τ_2) for which the two interactions have opposite or similar effects in synchrony, would define cooperative and competitive regimes of synaptic interplay in synchronization. Figure 15.6 displays these different regimes in the case of a network of QIF neurons.

When V_s is sufficiently small and the frequency sufficiently large, the electrical and inhibitory synapses have opposite effects: electrical synapses reduce the synchronization induced by inhibitory synapses (region C1 of the Figs. 15.6a and the example in Fig. 15.6b). When V_s is sufficiently large and the frequency sufficiently small, the electrical and inhibitory synapses have also opposite effects: inhibitory synapses reduce the synchronization induced by electrical synapses (region C2 of Fig. 15.6a and the example in Fig. 15.6b). Finally for an intermediary region, the two types of synapses cooperate because each type promote synchronization induced by the other type (region C3 in Fig. 15.6a and the example in Fig. 15.6b).

In the cooperative regime, electrical and inhibitory synapses combine their effect to reinforce the robustness of synchrony. When electrical and inhibitory synapses both promote synchrony, this means that the asynchronous state becomes unstable below a given amount of noise σ_c such that $\lambda(\sigma_c) = 0$. From (15.74), we can write that $\sigma_c^{\mathrm{gap}} = \sqrt{-Im(\Gamma_n^{\mathrm{gap}})/n}$ and $\sigma_c^{\mathrm{inh}} = \sqrt{-Im(\Gamma_n^{\mathrm{inh}})/n}$. When both synapses are present, the robustness (defined as the maximal amount of noise compatible with a totally or partially synchronous state) is given by σ_c^{dual}. Figure 15.7 shows

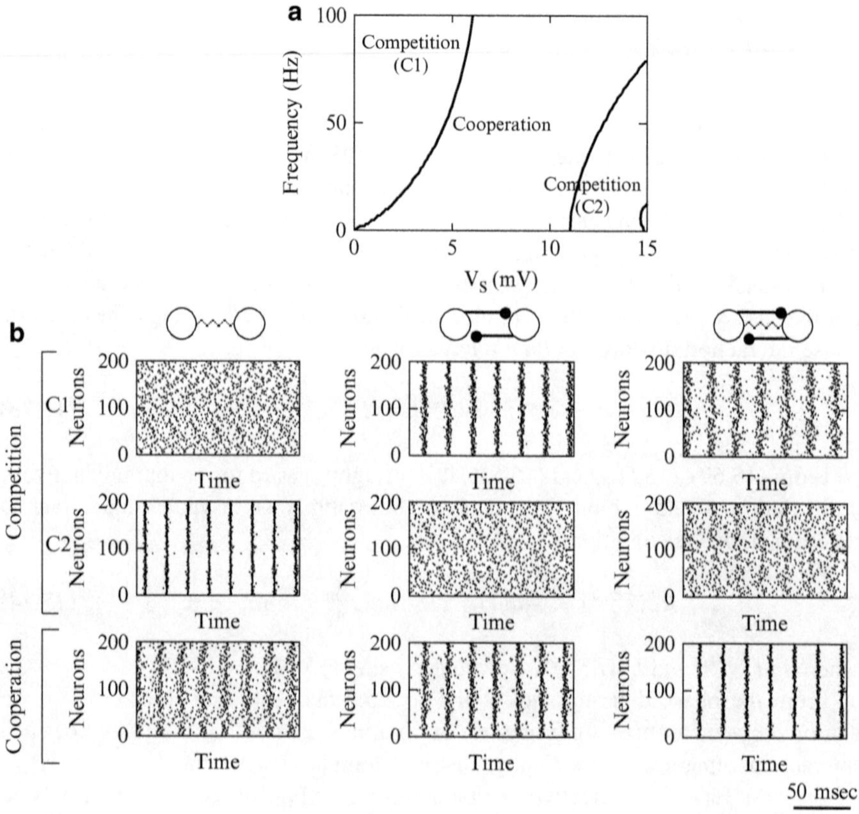

Fig. 15.6 Interplay between electrical and inhibitory synapses in the dynamics of network of weakly coupled QIF neurons. Activating and inactivating time associated with inhibitory currents are $\tau_1 = 1$ ms et $\tau_2 = 3$ ms. (**a**), phase diagram showing three domains of synaptic interplay: In C_1 and C_2 regions, electrical and inhibitory synapses have competitive effect in synchronization, one promoting it and the other destroying it. Between these regions, both types of synapses cooperate to promote synchronization. (**b**), Raster-plots obtained from numerical simulation of large network of QIF neurons in these cooperative and competitive regimes

the respective values of σ_c^{dual}, σ_c^{gap} and σ_c^{inh} as a function of V_s. For intermediate values of V_s for which both interactions cooperate to promote synchronization, the robustness σ_c increases compared to each interaction alone. In contrast, for extreme values of V_s for which both interactions compete, the robustness decreases. These results obtained with QIF model predict that these cooperative and competitive regimes and their robustness properties also arise in network of conductance-based neurons depending on the frequency, properties of intrinsic currents, and synaptic characteristics.

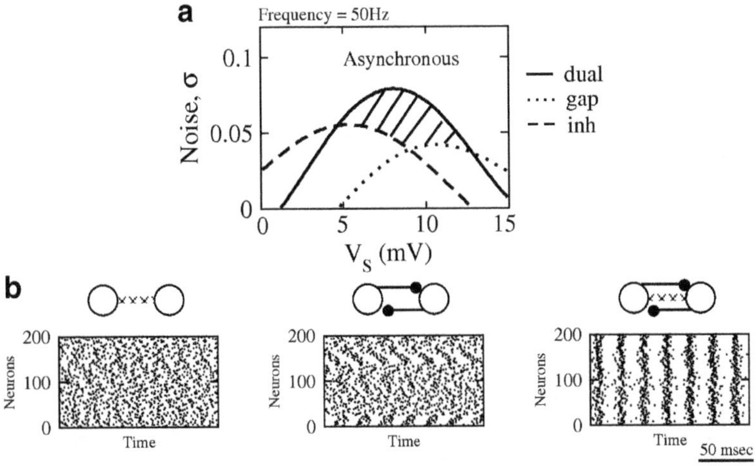

Fig. 15.7 Cooperation and competition regimes between electrical and inhibitory synapses promoting the robustness of synchronization of weakly coupled QIF neurons. (**a**), Phase diagram displaying the effect of noise on the stability of the asynchronous state. For intermediate value of V_s, synchrony is more robust to noise in presence of two interactions. (**b**), Raster plots displaying an example where combining the two types of synapses compensate the desynchronizing effect of noise (hatched domain)

4.3 Beyond Weak Coupling: Extension of the Cooperative Regime

Beyond the weak coupling regime, (15.74) cannot predict anymore the synchronization properties of small or large numbers of neurons. This is the case for each kind of synaptic interaction alone, but especially when the two types of coupling are combined. This can be shown in both networks of interacting QIF neurons or networks of interacting conductance-based neurons.

Figure 15.8 shows the case of a network of conductance-based neurons that incorporate a persistent sodium current. The electrical synapses oppose the synchronization induced by inhibitory synapses in agreement with the prediction of weak coupling theory (Fig. 15.8a). However, electrical synapses, when combined with sufficiently strong inhibitory synapses, favor synchronization (Fig. 15.8b). The amplitude of this synergetic effect depends on the strength of electrical synapses (Fig. 15.8c). Such synergetic cooperation between electrical synapses and strong enough inhibitory synapses contributes to reinforce robustness of synchronized states in the face of different types of noise: (1) the noise in external current, (2) the heterogeneity of external current, (3) the random connectivity (Fig. 15.9). Figure 15.9a shows that the noise in external current reduces synchronization and stabilizes the asynchronous state for a critical value $\sigma_c = 0.4\text{mV}/\text{ms}^{1/2}$ without electrical synapses and $\sigma_c = 0.8\text{mV}/\text{ms}^{1/2}$ in presence of electrical synapses. Similarly, the presence of electrical synapses increases the robustness to

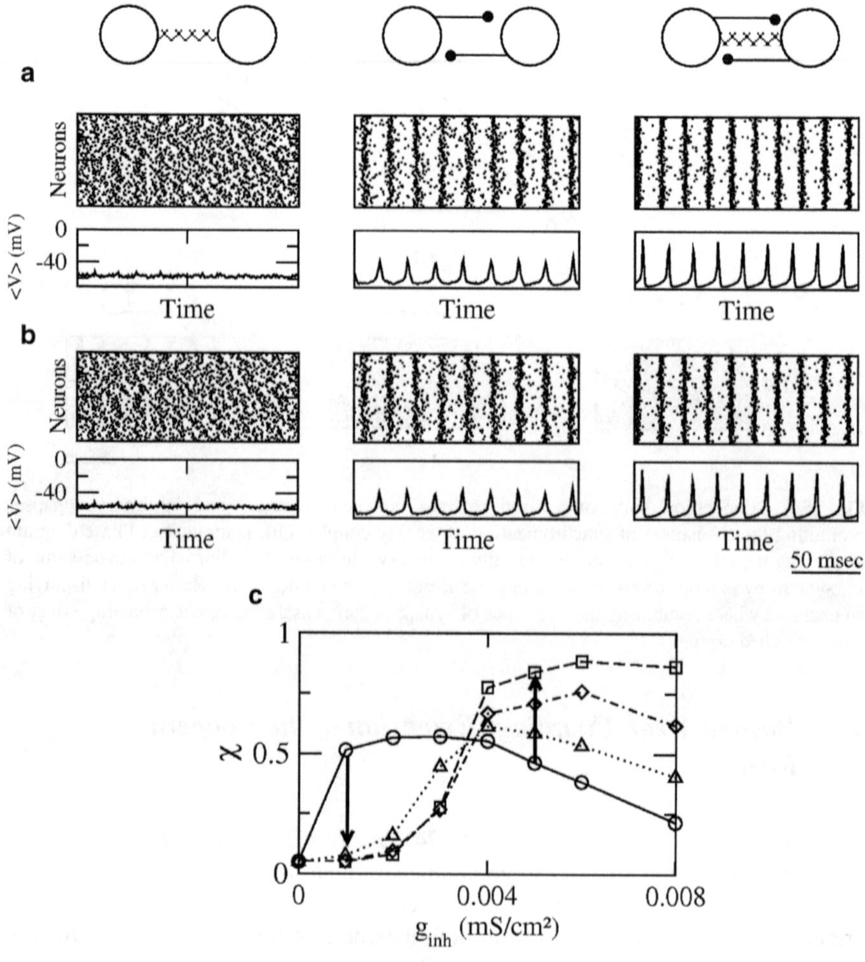

Fig. 15.8 Synergetic cooperation between electrical and inhibitory synapses beyond the regime of weak coupling. Simulations of conductance-based models ($g_{NaP} = 0.4\,\mathrm{mS/cm^2}$, $f \approx 50\,\mathrm{Hz}$, $\sigma = 0.1\,\mathrm{mV/ms^{1/2}}$). (**a**), In large networks, weak electrical synapses ($g_{\mathrm{gap}} = 0.001\,\mathrm{mS/cm^2}$) that stabilize the asynchronous state decrease the synchrony induced by weak inhibitory synapses($g_{\mathrm{inh}} = 0.001\,\mathrm{mS/cm^2}$). (**b**), Stronger electrical synapses ($g_{\mathrm{gap}} = 0.004\,\mathrm{mS/cm^2}$), although stabilizing the asynchronous state when alone, increase the synchronization induced by strong inhibitory synapses ($g_{\mathrm{inh}} = 0.005\,\mathrm{mS/cm^2}$). (**c**), Synchrony measure χ as function of g_{gap} and g_{inh}. The synergetic effect requires sufficiently strong inhibitory synapses ($g_{\mathrm{inh}} > 0.003\,\mathrm{mS/cm^2}$) while its intensity depends on the strength of electrical coupling (circles: $g_{\mathrm{gap}} = 0$; triangles: $g_{\mathrm{gap}} = 0.002\,\mathrm{mS/cm^2}$; diamonds: $g_{\mathrm{gap}} = 0.005\,\mathrm{mS/cm^2}$; squares: $g_{\mathrm{gap}} = 0.01\,\mathrm{mS/cm^2}$)

heterogeneity or spatial fluctuations in connectivity as shown in Fig. 15.9b,c. For the sake of argument, we have shown an instance of synergetic cooperation in the case where electrical synapses alone are desynchronizing; however, such synergetic effect still occurs when electrical synapses alone are synchronizing.

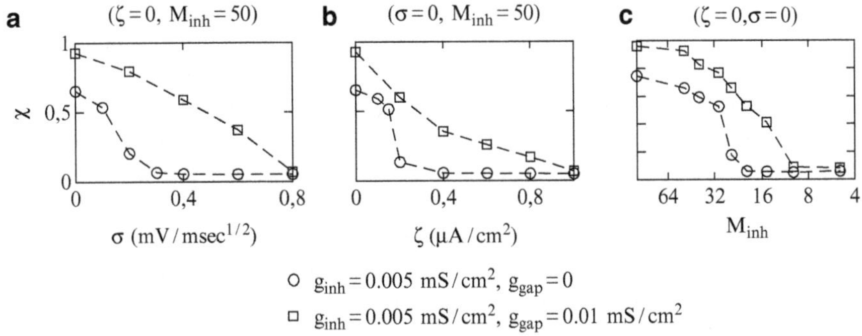

Fig. 15.9 Effect of synergetic synaptic cooperation on synchronization robustness. Simulations of conductance-based models with a persistent sodium current ($g_{NaP} = 0.4\,\text{mS/cm}^2$, $f \approx 50\,\text{Hz}$). The synchrony measure χ is compared between networks coupled with inhibitory synapses ($g_{\text{inh}} = 0.005\,\text{mS/cm}^2$) without (*circles*) or with (*squares*) electrical synapses ($g_{\text{gap}} = 0.01\,\text{mS/cm}^2$) as function of the noise, σ (**a**), the heterogeneity ζ (**b**), and the mean number of inhibitory synapses M_{inh} (**c**)

5 Discussion

5.1 Role of Intrinsic Neuronal Properties in Synchrony Mediated by Electrical Synapses

The dynamics of networks of neurons interacting via chemical synapses have been extensively studied (see e.g., Golomb et al. 2001 and references therein). However, much less theoretical studies have addressed the dynamics of networks in which neurons are coupled by electrical synapses. Previous modeling studies have shown that a pair of electrically coupled neurons may fire in synchrony or in antisynchrony depending on the model properties or on the firing rate (Sherman and Rinzel 1992; Han et al. 1995; Chow and Kopell 2000; Lewis and Rinzel 2003; Cymbalyuk et al. 1994; Alvarez et al. 2002; Saraga et al. 2006). This chapter provides a general framework for understanding this diversity of synchronization behaviors.

We have shown that the synchronization of neurons interacting via electrical synapses can be predicted if one knows the shape of their iPRC. Right-skewed iPRC promotes synchronization while left-skewed iPRC impedes it. In the case of the QIF model, the shape of the iPRC mainly depends on the parameter V_s and the time course of the potential. The effects of V_r and V_T can be studied in a similar way. In conductance-based models as in real neurons, ionic channels and morphology determine the iPRC (Crook et al. 1998; Ermentrout et al. 2001; Oprisan and Canavier 2002; Pfeuty et al. 2003). We have focused here on the effect of I_{NaP} which shifts the maximum of the iPRC toward the left. Let us note that in Chap. 14, the effect of a passive dendritic tree (modeled as two compartments) is also analyzed and it is found that it also shifts the peak of the iPRC to the left. This effect was also found in Pfeuty et al. (2007) for dendritic trees with one compartment. Other inward currents, such as calcium currents, have the same effects on the iPRC and may affect synchrony in a similar way. Potassium currents, in contrast, skew the iPRC to the right (Ermentrout et al. 2001; Pfeuty et al. 2003). Therefore, we predict that potassium and sodium currents have mirror effects on synchrony (Pfeuty et al. 2003). We have verified these predictions in simulations of conductance-based models.

Electrical synapses located on the dendritic compartment of the conductance-based neuron are less synchronizing than if they are located on the soma. One contribution to this effect is the additional delay induced by the dendritic compartment which is like shifting the iPRC to the left. One can similarly predict the effect of the synaptic delays in the inhibitory interactions. Although we did not include these delays in our model, our approach can be straightforwardly extended to take them into account. For given cellular properties this would amount to shifting the iPRC of the neurons to the left. Therefore, sufficiently large delays with respect to the firing period have the same effect as increasing g_{NaP} without delays.

Phase locking of neurons connected by electrical synapses has been also investigated analytically in Chow and Kopell (2000) and Lewis and Rinzel (2003). Both studies found stable antiphase locking at low firing rates and destabilization for large enough firing rates. This can be understood as follows. For a passive integrator, $Z(\phi) = 1/[I_{\mathrm{ext}}T]\ \exp(\phi T/2\pi\tau_{\mathrm{m}})$ where T is the period, I_{ext} the external current and τ_{m}, the neuronal membrane time constant (Kuramoto 1984). The value of Z diverges exponentially with the period at all ϕ and so do S_{sub} and S_{r} [(15.61) and (15.60)]. Detailed analysis reveals that the most divergent of these terms is $S_{\mathrm{sub}} < 0$. Hence, for low enough firing rates $\Gamma'(\pi) < 0$. The term S_{sp} is positive ($Z'(\phi) > 0$) and decreases with T, whereas S_{sub} and S_{r} increase. Hence, for a large enough frequency, $\Gamma'(\pi)$ changes sign. Therefore, for passive integrators antiphase locking is stable at low firing rates and loses stability as the firing rate increases. In contrast, we have found that the QIF model behaves differently than the LIF except in the limit of very large V_{s}. Therefore, our results show that the predictions of Chow and Kopell (2000) and Lewis and Rinzel (2003) may be relevant only for neurons with strong potassium currents.

5.2 *Validity of the Weak Coupling Approximation*

The stability of the asynchronous state of a large network of all-to-all connected QIF neurons can be calculated analytically in the absence of noise for any coupling strength using the population density method (Abbott and van Vreeeswijk 1999; van Vreeswijk 1996; Hansel and Mato 2001, 2003). An alternative approach is the phase reduction method. Although it is only mathematically justified for weak interactions, it leads to nontrivial results, which remain valid in a reasonable range of interaction strengths. Moreover, it is straightforward to study the effect of noise in this framework. These were the motivations to use the phase reduction method in the framework of which all the analytical results of this paper were derived.

In order to assess the validity of the approximation, it is necessary to perform numerical simulations for different coupling strengths. For instance, in Hansel et al. (1993) the synchronization properties of a system of two chemically coupled HH neurons were analyzed numerically. It was found that the coupling strength could be strong enough to change the firing rate of the coupled system in 30% with respect to the uncoupled neurons, and still the predictions of the phase models analysis were qualitatively correct.

For gap junctions, numerical simulations in this chapter were performed for the coupling conductance $g = 0.005\ \mathrm{mS/cm^2}$. This corresponds to coupling coefficients (Amitai et al. 2002) $\approx 5\%$ and to a spikelet size of about 0.5 mV. We have checked that similar results are found for conductances four times larger (data not shown). Therefore, the conclusions of the present work are relevant for electrical synapse conductances in the physiological range (Amitai et al. 2002; Galarreta and Hestrin 2001, 2002).

However, one must keep in mind that electrical synapses, in the limit of strong coupling, are always synchronizing regardless of intrinsic properties of neurons or firing rates. At finite coupling strength, an additional instability of the asynchronous state appears at low firing rates and large V_s. It corresponds to the impossibility of controlling low firing rates when the coupling is too strong and the AHP is too small. This is because for small AHP, the recurrent electrical synapses interactions tend to depolarize the neurons on the average and this increases their firing rates. This instability is similar to the rate instability in networks of excitatory neurons (Hansel and Mato 2003).

5.3 Combining Electrical and Inhibitory Synapses: In and Beyond the Weak Coupling Regime

In the weak coupling regime, the interplay between electrical and inhibitory synapses gives rise to a straightforward outcome (Lewis and Rinzel 2003; Pfeuty et al. 2005). The synchronizing or desynchronizing effects of each interaction alone, which depends on synaptic and neuronal parameters, add when interactions are combined, such as to define cooperative and competitive regimes. In the cooperative regime, electrical synapses can reinforce the robustness of synchrony induced by inhibitory synapses. In the competitive regime, electrical synapses can destroy the neural synchrony induced by inhibitory synapses rather than promoting it.

However, in the case of the combined interaction, weak coupling approximation is no longer valid for realistic coupling strength. In particular, electrical synapses become less versatile when the neurons also interact via inhibition, as they always promote synchrony (Pfeuty et al. 2007). This is because of a modulatory effect of inhibition which "renormalizes" the synchronization properties of electrical synapses. This effect was depicted here in the case of electrical synapses which are desynchronizing because of the intrinsic properties of neurons. We found that this is significant for spike-to-spike synchrony as well as for stochastic synchrony. Similar modulatory effects occur when the electrical synapses are desynchronizing because of their distal position (Pfeuty et al. 2007) on dendrites or because of their too sparse connectivity. In all these cases, sufficiently strong inhibition can switch the electrical synapses from being desynchronizing to synchronizing. It should be noted that inhibition also potentiates the synchronizing effect of electrical synapses which, when alone, are already synchronizing. In this case, the function $\sigma_c(g_{gap})$ increases more rapidly with g_{gap} when the inhibition is strong than when it is weak (Pfeuty et al. 2007).

5.4 Physiological Relevance of the Cooperative Interplay Between Electrical and Inhibitory Synapses

Experimental studies in neocortical and hippocampal preparations have revealed that both electrical and inhibitory synapses are often required to generate synchrony of firing. This has been shown in large ensemble of neurons in vitro (Hormuzdi et al. 2001; Traub et al. 2001; LeBeau et al. 2002; Blatow et al. 2003) as well as in pairs of neurons (Tamas et al. 2000; Szabadics et al. 2001). Our results showing that electrical synapses combined with inhibition improve the robustness of synchrony in a way which, to a large extent, does not depend on their location, is in line with these experimental results.

It has been proposed that fast synchronous rhythmic episodes in the γ range (30–100 Hz) observed in cortex and hippocampus, in-vivo or in vitro, are generated within local networks of GABAergic inhibitory interneurons (Whittington et al. 1995; Wang and Buzsáki 1996; Bartos et al. 2007). Although in γ oscillations found in slice preparations, interneurons are likely to fire about one action potential per cycle of the rhythm (Cunningham et al. 2003), recent experiments indicate that in γ oscillations observed in-vivo during specific behavioral states, GABAergic interneurons fire in a more sparse and irregular way, with a probability of firing per cycle which is between 0.1 and 0.6 (Csicsvari et al. 1999; Tukker et al. 2007). These two modes of synchrony are reminiscent of the two regimes described above, namely spike-to-spike synchrony and stochastic synchrony.

Acknowledgements This work was supported in part by a NATO PST-CLG (reference 977683), the PICS-CNRS (no. 837), the ACI "Neurosciences integratives et computationnelles" (Ministère de la Recherche, France), project SECyT-ECOS A04B04. We gratefully acknowledge David Golomb for fruitful collaboration and discussions.

Appendix 1

A1.1 Conductance-Based Models

The conductance-based model has two compartments although only one somatic compartment is considered in most situations. The second compartment, which corresponds to the dendrite, is passive. For simplicity, we assume that the two compartments have equal areas.

The membrane potential, V, of the neuron follows the equation:

$$C\frac{\mathrm{d}V}{\mathrm{d}t} = -I_{\mathrm{L}} - \sum_{\mathrm{ion}} I_{\mathrm{ion}} + I_{\mathrm{ext}} + I_{\mathrm{noise}} - g_c(V - V_d), \tag{15.75}$$

$$C\frac{\mathrm{d}V_d}{\mathrm{d}t} = -I_{\mathrm{Ld}} - g_c(V_d - V), \tag{15.76}$$

where $I_{\mathrm{L}} = g_L(V - V_L)$ and $I_{\mathrm{Ld}} = g_{Ld}(V_d - V_{Ld})$ are leak currents and $\sum_{\mathrm{ion}} I_{\mathrm{ion}}$ is the sum of all voltage dependent ionic currents. The currents I_{ext} and I_{noise} are a constant external current and a Gaussian white noise with a zero mean and standard deviation (SD), σ, respectively.

Our neuron model incorporates an inactivating sodium current, I_{Na}, a delayed rectifier potassium current, I_{K}, a slow potassium current, I_{Ks} (Erisir et al. 2001) and a noninactivating (persistent) sodium current I_{NaP} (French et al. 1990).

$$I_{\mathrm{Na}} = g_{Na}\, m_{\infty}^3\, h\,(V - V_{Na}), \tag{15.77}$$

$$I_{\mathrm{K}} = g_K\, n^4\,(V - V_K), \tag{15.78}$$

$$I_{\mathrm{Ks}} = g_{Ks}\, s^4\,(V - V_K), \tag{15.79}$$

$$I_{\mathrm{NaP}} = g_{NaP}\, p_{\infty}\,(V - V_{Na}). \tag{15.80}$$

The kinetics of the gating variable h, n, s are given by:

$$\frac{\mathrm{d}x}{\mathrm{d}t} = \alpha_x(V)(1 - x) - \beta_x(V)x \tag{15.81}$$

with $x = h, n, s$ and $\alpha_h(V) = 0.21\,\mathrm{e}^{-(V+58)/20}$, $\beta_h(V) = 3./(1 + \mathrm{e}^{-(V+28)/10})$, $\alpha_n(V) = 0.03\,(V + 34)/(1 - \mathrm{e}^{-(V+34)/10})$, $\beta_n(V) = 0.375\,\mathrm{e}^{-(V+44)/80}$, $\alpha_s(V) = 0.07(44 + V)/(1 - \mathrm{e}^{-(V+44)/4.6})$, $\beta_s(V) = 0.008\,\mathrm{e}^{(V+44)/68}$. The activation functions, m_∞ and p_∞, are given by: $m_\infty(V) = \alpha_m(V)/(\alpha_m(V) + \beta_m(V))$, where $\alpha_m(V) = 0.1(V + 35)/(1 - \mathrm{e}^{-(V+35)/10})$, $\beta_m(V) = 4\mathrm{e}^{-(V+60)/18}$, and $p_\infty(V) = 1/(1 + \mathrm{e}^{-(V+40)/6})$.

Throughout this work, the parameters $g_{Na} = 35\,\mathrm{mS/cm^2}$, $V_{Na} = 55\,\mathrm{mV}$, $g_K = 9\,\mathrm{mS/cm^2}$, $V_K = -75\,\mathrm{mV}$, $g_L = 0.1\,\mathrm{mS/cm^2}$, $V_L = -65\,\mathrm{mV}$ and $C = 1\,\mu\mathrm{F/cm^2}$ are kept constant. $g_{Ks} = 0, g_{Nap} = 0$ in the standard WB model.

In our simulations of network of conductance-based neurons, the network size is $N = 1,600$ and the average connectivity is $M = 10$.

Appendix 2

A2.1 Synchrony Measure for Two Neurons

The synchronization between pairs of neurons is usually defined in terms of the CC function. If the membrane potentials $V_1(t)$ and $V_2(t)$ are known during a period of time T, we define

$$C_{12}(\tau) =< V_1(t+\tau)V_2(t) > - < V_1(t) >< V_2(t) >, \tag{15.82}$$

where the bracket $< \cdots >$ denotes a temporal average:

$$< x(t) >\equiv \frac{1}{T}\int_0^T x(t)\,\mathrm{d}t. \tag{15.83}$$

In phase synchrony will be indicated by a CC with a peak near $\tau = 0$, while systems in antiphase will have a peak at $\tau \approx \frac{1}{2f}$, where f is the frequency of the oscillation.

A2.2 Synchrony Measure for Large Networks

In a large network of neurons, the activity is asynchronous if at any time the number of action potentials fired in the network is the same up to some random fluctuations. This implies that in a large network, the asynchronous state is unique. When this state is unstable, the network activity is necessarily synchronous. In this case, neurons tend to fire preferentially in some windows of time.

We can characterize the degree of synchrony in a population of N neurons by measuring the temporal fluctuations of macroscopic observables, as, for example, the membrane potential averaged over the population (Hansel and Sompolinsky 1992, 1996; Golomb and Rinzel 1993, 1994; Ginzburg and Sompolinsky 1994). The quantity

$$V(t) = \frac{1}{N}\sum_{i=1}^{N} V_i(t) \tag{15.84}$$

is evaluated over time and the variance $\sigma_V^2 = \left\langle [V(t)]^2 \right\rangle_t - [\langle V(t) \rangle_t]^2$ of its temporal fluctuations is computed, where $\langle \ldots \rangle_t$ denotes time averaging, as defined in (15.83). After normalization of σ_V to the average over the population of the single cell membrane potentials, $\sigma_{V_i}^2 = \left\langle [V_i(t)]^2 \right\rangle_t - [\langle V_i(t) \rangle_t]^2$, we define

$$\chi(N) \equiv \sqrt{\frac{\sigma_V^2}{\frac{1}{N}\sum_{i=1}^{N}\sigma_{V_i}^2}}. \tag{15.85}$$

This quantity goes from 0 to 1. The central limit theorem implies that in the limit $N \to \infty$, $\chi(N)$ behaves as:

$$\chi(N) = \chi(\infty) + \frac{a}{\sqrt{N}} + O\left(\frac{1}{N}\right), \tag{15.86}$$

where $a > 0$ is a constant, and $O(1/N)$ means a term of order $1/N$. In particular, $\chi(N) = 1$, if the activity of the network is fully synchronized (i.e., $V_i(t) = V(t)$ for all i), and $\chi(N) = O(1/\sqrt{N})$ if the state of the network activity is asynchronous. In the asynchronous state, $\chi(\infty) = 0$. More generally, the larger $\chi(\infty)$ the more synchronized the population. Note that this measure of synchrony is sensitive not only to the correlations in the spike timing but also to the correlations in the time course of the membrane potentials in the subthreshold range.

References

Abbott, L.F. & van Vreeswijk, C. (1993). Asynchronous states in networks of pulse-coupled oscillators. *Phys Rev E 48*, 1483–1490.

Alvarez, L.F., Chow, C., van Bockstaele, E.J. & Williams, J.T. (2002) Frequency-dependent synchrony in locus coeruleus: Role of electrotonic coupling. *P Natl Acad Sci USA 99*, 4032–4036.

Amitai, Y., Gibson, J.R., Patrick, A., Ho, B., Connors, B.W. & Golomb, D. (2002) Spatial organization of electrically coupled network of interneurons in neocortex. *J Neurosci 22*, 4142–4152.

Bartos, M,, Vida, I. & Jonas, P. (2007). Synaptic mechanisms of synchronized gamma oscillations in inhibitory interneuron networks. *Nat Rev Neurosci 8*, 45–56.

Beierlein, M., Gibson, J.R. & Connors, B.W. (2000). A network of electrically coupled interneurons drives synchronized inhibition in neocortex. *Nat Neurosci 3*, 904–910.

Berger, H. (1929) Uber das Elektrenkephalogramm des Menschen. *Archiv für Psychiatrie und Nervenkrankheiten 87*, 527–570.

Blatow, M., Rozov, A., Katona, I., Hormuzdi, S.G., Meyer, A.H., Whittington, M.A., Caputi, A. & Monyer, H. (2003). A novel network of multipolar bursting interneurons generates theta frequency oscillations in neocortex. *Neuron 38*, 805–817.

Bou-Flores, C. & Berger, A.J. (2001). Gap junctions and inhibitory synapses modulate inspiratory motoneuron synchronization. *J Neurophysiol 85*, 1543–1551.

Brown, E., Moehlis, J. & Holmes, P. (2004). On the phase reduction and response dynamics of neural oscillator populations. *Neural Comput 16*, 673–715.

Chow, C.C. & Kopell, N. (2000). Dynamics of spiking neurons with electrical coupling. *Neural Comput 12*, 1643–1678.

Crook, S.M., Ermentrout, G.B. & Bower, J.M. (1998). Spike frequency adaptation affects the synchronization properties of networks of cortical oscillators. *Neural Comput 10*, 837–854.

Csicsvari, J,, Hirase, H., Czurko, A., Mamiya, A. & Buzsáki, G. (1999). Fast network oscillations in the hippocampal CA1 region of the behaving rat. *J Neurosci 19:RC20*, 1–4.

Cunningham, M.O., Davies, C.H., Buhl, E.H., Kopell, N. & Whittington, M.A. (2003). Gamma oscillations induced by kainate receptor activation in the entorhinal cortex in vitro. *J Neurosci 23*, 9761–9769.

Cymbalyuk, G.S., Nikolaev, E.V. & Borisyuk, R.M. (1994). In-phase and antiphase self-oscillations in a model of two electrically coupled pacemakers. *Biol Cybern 71*, 153–160.

Deans, M.R., Gibson, J.R., Sellitto, C., Connors, B.W. & Paul, D.L. (2001). Synchronous activity of inhibitory networks in neocortex requires electrical synapses containing connexin36. *Neuron 31*, 477–485.

Erisir, A., Lau, D., Rudy, B. & Leonard, C.S. (2001). Function of specific K(+) channels in sustained high-frequency firing of fast-spiking neocortical interneurons. *J Neurophysiol 82*, 2476–2489.

Kopell, N. (1988), Toward a theory of modeling central pattern generators. In: *Neural control of rhythmic movements in vertebrates*. Cohen, A. (ed.). 369–413. New York: Wiley.

Ermentrout, G.B. (1996). Type I membranes, phase resetting curves and synchrony. *Neural Comput 8*, 979–1001.

Ermentrout, G.B. (2002). *Simulating, analyzing, and animating dynamical systems: a guide to Xppaut for researchers and students (software, environments, tools)*. Philadelphia: SIAM.

Ermentrout, G.B. & Kopell, N. (1986). Parabolic bursting in an excitable system coupled with a slow oscillation. *SIAM J Appl Math 46*, 233–253.

Ermentrout, G.B. & Kopell, N. (1991). Multiple pulse interactions and averaging in systems of coupled neural oscillators, *J Math Biol 29*, 195–217.

Ermentrout, B., Pascal, M. & Gutkin, B. (2001). The effect of spike frequency adaptation and negative feedback on the synchronization of neural oscillators. *Neural Comput 13*,1285–1310.

French, C.R., Sah, P., Buckett, K.J. & Gage, P.W. (1990). A voltage-dependent persistent sodium current in mammalian hippocampal neurons. *J Gen Physiol 95*, 1139–1157.

Friedman, D. & Strowbridge, B.W. (2003). Both electrical and chemical synapses mediate fast network oscillations in the olfactory bulb. *J Neurophysiol 89*, 2601–2610.

Fukuda, T. & Kosaka, T. (2000). Gap-junction coupling linking the dendritic network of GABAergic neurons in the hippocampus. *J Neurosci 20*, 1519–1528.

Furshpan, E.J. & Potter, D.D. (1959). Transmission at the giant motor synapses of the crayfish. *J Physiol 145*, 289–325.

Galarreta, M. & Hestrin, S. (1999). A network of fast spiking cells in the neocortex connected by electrical synapses. *Nature 402*, 72–75.

Galarreta, M. & Hestrin, S. (2001). Electrical synapses between GABA-releasing neurons. *Nature Neurosci 2*, 425–433.

Galarreta, M. & Hestrin, S. (2002). Electrical and chemical synapses among parvalbumin fast-spiking GABAergic interneurons in adult mouse neocortex. *P Natl Acad Sci USA 19*, 12438–12443.

Gibson, J.R., Beierlein, M. & Connors, B. (1999). Two networks of inhibitory neurons electrically coupled. *Nature 402*, 75–79.

Ginzburg, I. & Sompolinsky, H. (1994). Theory of correlations in stochastic neuronal networks. *Phys Rev E 50*, 3171–3191.

Golomb, D., Hansel, D. & Mato, G. (2001). Theory of synchrony of neuronal activity. In: *Handbook of Biological Physics, Vol 4: Neuro-informatics and neural modeling*. Gielen, S. & Moss, M. (eds.). 887–968. Amsterdam: Elsevier Science.

Golomb, D. & Rinzel, J. (1993). Dynamics of globally coupled inhibitory neurons with heterogeneity. *Phys Rev E 48*, 4810–4814.

Golomb, D. & Rinzel, J. (1994). Clustering in globally coupled inhibitory neurons. *Physica D 72*, 259–282.

Han, S.K., Kurrer, C. & Kuramoto, Y. (1995). Dephasing and bursting in coupled neural oscillators. *Phys Rev Lett 75*, 3190–3193.

Hansel, D. & Mato, G. (2001). Existence and stability of persistent states in large neuronal networks. *Phys Rev Lett 86*, 4175–4178.

Hansel, D. & Mato, G. (2003). Asynchronous states and the Emergence of Synchrony in Large Networks of Interacting Excitatory and Inhibitory Neurons. *Neural Comput 15*, 1–56.

Hansel, D., Mato, G. & Meunier, C. (1993). Phase dynamics for weakly coupled Hodgkin-Huxley neurons. *Europhys Lett 23*, 367–372.

Hansel, D., Mato. G. & Meunier, C. (1995). Synchrony in excitatory neural networks. *Neural Comput 7*, 307–337.

Hansel, D. & Sompolinsky, H. (1992). Synchrony and computation in a chaotic neural network. *Phys Rev Lett 68*, 718–721.

Hansel, D. & Sompolinsky, H. (1996). Chaos and synchrony in a model of a hypercolumn in visual cortex. *J Comp Neurosci 3*, 7–34.

Hodgkin, A.L. & Huxley, A.F. (1952). A quantitative description of membrane current and application to conductance in excitation nerve. *J Physiol-London 117*, 500–544.

Hormuzdi, S.G., Pais, I., Lebeau, F.E.N., Towers, S.K., Rozov, A., Buhl, E., Whittington, M.A. & Monyer, H. (2001). Impaired electrical signaling disrupts gamma frequency oscillations in connexin 36-Deficient Mice. *Neuron 31*, 487–495.

Kita, H., Kosaka, T. & Heizmann, C.W. (1990). Parvalbumin-immunoreactive neurons in the rat neostriatum: a light and electron microscopic study. *Brain Res 536*, 1–15.

Kopell, N. & Ermentrout, G.B. (2002). Mechanisms of phase-locking and frequency control in pairs of coupled neural oscillators. In: Fiedler B (ed) Handbook of Dynamical Systems 2:5–55. Amsterdam: Elsevier.

Kuramoto, Y. (1984). *Chemical Oscillations, Waves and Turbulence*. New York: Springer.

Kuramoto, Y. (1991) Collective synchronization of pulse-coupled oscillators and excitable units. *Physica D 50*, 15–30.

Landisman, C,E,, Long, M.A., Beierlein, M., Deans, M.R., Paul. D.L. & Connors, B.W. (2002) Electrical synapses in the thalamic reticular nucleus. *J Neurosci 22*, 1002–1009.

LeBeau, F.E.N., Towers, S., Traub, R.D., Whittington, M.A. & Buhl, E. (2002). Fast oscillations networks induced by potassium transient in the rat hippocampus in vivo. *J Physiol 542*, 167–179.

Lewis, T. & Rinzel, J. (2003). Dynamics of spiking neurons connected by both inhibitory and electrical coupling. *J Comput Neurosci 14*, 283–309.

Mann-Metzer, P. & Yarom, Y. (1999). Electrical coupling interacts with intrinsic properties to generate synchronized activity in cerebellar networks of inhibitory interneurons. *J Neurosci 19*, 3298–3306.

Oprisan, S.A. & Canavier, C.C. (2002). The influence of limit cycle topology on the phase resetting curve. *Neural Comput 14*, 1027-1057.

Perez-Velazquez, J.L., Valiente, T.A. & Carlen, P.L. (1994). Modulation of gap junctional mechanisms during calcium-free induced field burst activity: a possible role for electrotonic coupling in epileptogenesis. *J Neurosci 14*, 4308–4317.

Pfeuty, B., Mato, G., Golomb, D. & Hansel, D. (2003). Electrical synapses and synchrony: the role of intrinsic currents. *J Neurosci 23*, 6280–6294.

Pfeuty, B., Mato, G., Golomb, D. & Hansel, D. (2005). The combined effect of electrical synapses in synchrony. *Neural Comput 17*, 633–670

Pfeuty, B., Golomb, D., Mato, G. & Hansel, D. (2007). Inhibition potentiates the synchronizing action of electrical synapses. *Front Comput Neurosci* 1, art. 8.

Saraga, F., Ng, L. & Skinner, F.K. (2006). Distal gap junctions and active dendrites can tune network dynamics. *J Neurophysiol 95*, 1669–1682.

Sherman, A. & Rinzel, J. (1992). Rhythmogenic effects of weak electrotonic coupling in neuronal models. *P Natl Acad Sci USA 89*, 2471–2474.

Szabadics, J., Lorincz, A. & Tamas, G. (2001). Beta and gamma frequency synchronization by dendritic gabaergic synapses and gap junctions in a network of cortical interneurons. *J Neurosci 21*, 5824–5831.

Tamas, G., Buhl, E.H., Lorincz, A. & Somogyi, P. (2000). Proximally targeted GABAergic synapses and gap-junctions precisely synchronize cortical interneurons. *Nature Neurosci 3*, 366–371.

Traub, R.D., Kopell, N., Bibbig, A., Buhl, E., Lebeau, F.E.N. & Whittington, M. (2001). Gap junction between interneuron dendrites can enhance synchrony of gamma oscillations in distributed networks. *J Neurosci 21*, 9478–9486.

Tukker, J.J., Fuentealba, P., Hartwich, K., Somogyi, P. & Klausberger, T. (2007). Cell type-specific tuning of hippocampal interneuron firing during gamma oscillations in vivo. *J Neurosci 27*, 8184–8289.

van Kampen, N.G. (1981). *Stochastic processes in physics and chemistry*. Amsterdam: Elsevier.

van Vreeswijk, C., Abbott, L.F. & Ermentrout, G.B. (1994). When inhibition not excitation synchronizes neural firing. *J Comput Neurosci 1*, 313–321.

van Vreeswijk, C. (1996). Partial synchronization in populations of pulse-coupled oscillators. *Phys Rev E 54*, 5522–5537.

Venance, L., Rozov, A., Blatow, M., Burnashev, N., Feldmeyer, D. & Monyer, H. (2000). Connexin expression in electrically coupled postnatal rat brain neurons. *P Natl Acad Sci USA 97*, 10260–10265.

Wang, X.-J. & Buzsáki, J. (1996). Gamma oscillation by synaptic inhibition in a hippocampal interneuronal network model. *J Neurosci 16*, 6402–6413.

Watanabe, A. (1958) The interaction of electrical activity among neurons of lobster cardiac ganglion. *Jpn J Physiol 8*, 305–318.

Whittington, M.A., Traub, R.D. & Jefferys, J.G.R. (1995). Synchronized oscillations in interneuron networks driven by metabotropic glutamate receptor activation. *Nature 373*, 612–615.

Chapter 16
A PRC Description of How Inhibitory Feedback Promotes Oscillation Stability

Farzan Nadim, Shunbing Zhao, and Amitabha Bose

Abstract Using methods of geometric dynamical systems modeling, we demonstrate the mechanism through which inhibitory feedback synapses to oscillatory neurons stabilize the oscillation, resulting in a flattened phase-resetting curve. In particular, we use the concept of a synaptic phase-resetting curve to demonstrate that periodic inhibitory feedback to an oscillatory neuron locks at a stable phase where it has no impact on cycle period and yet it acts to counter the effects of extrinsic perturbations. These results are supported by data from the stable bursting oscillations in the crustacean pyloric central pattern generator.

1 Introduction

Oscillations in the nervous system often originate from neurons that have intrinsic pacemaker properties or from populations of neurons that are coupled with excitatory connections or gap junctions that result in synchronous patterned activity (Grillner, Markram, De Schutter, Silberberg, & LeBeau 2005). While purely inhibitory circuits can give rise to oscillations in the absence of excitation (Bartos

F. Nadim (✉)
Department of Mathematical Sciences, New Jersey Institute of Technology, Newark, NJ, USA

Department of Biological Sciences, Rutgers University, Newark, NJ, USA
e-mail: farzan@njit.edu

S. Zhao
Department of Biological Sciences, Rutgers University, Newark, NJ, USA
e-mail: shunbing@gmail.com

A. Bose
Department of Mathematical Sciences, New Jersey Institute of Technology, Newark, NJ, USA

School of Physical Sciences, Jawaharlal Nehru University, New Delhi, India
e-mail: bose@njit.edu

N.W. Schultheiss et al. (eds.), *Phase Response Curves in Neuroscience: Theory, Experiment, and Analysis*, Springer Series in Computational Neuroscience 6,
DOI 10.1007/978-1-4614-0739-3_16,

et al. 2007; Friesen 1994; Manor et al. 1999), feedback inhibition in pacemaker or excitatory oscillatory networks has been proposed to be important in determining the oscillation frequency (Borgers & Kopell 2003; Wang & Buzsaki 1996; Whittington et al. 2000). Alternatively, feedback inhibition may be prevalent but have little effect on network oscillation frequency. In such cases inhibition has been proposed to promote stability in the oscillatory network (Mamiya & Nadim 2004; Thirumalai, Prinz, Johnson, & Marder 2006).

We examine the idea of feedback inhibition as a promoter of oscillation stability, motivated by our recent data in the pyloric central pattern generator (CPG) of crustacean decapods in which the absence of feedback inhibition leads to a larger variability in the relatively stereotyped triphasic oscillations (freq ~1 Hz). We provide experimental evidence that oscillations in this CPG have less variability in response to perturbations and the effect of factors that inherently increase variability, such as intercircuit interactions, is buffered by the inhibitory synapse. Our results include a demonstration that the experimentally measured phase-resetting curve (PRC) of the pyloric pacemaker neurons lies closer to zero in the presence of feedback inhibition, indicating less sensitivity to extrinsic inputs.

We use a generic two-variable neuronal oscillator model to demonstrate how feedback inhibition can act to increase stability. This model has a typical cubic nullcline and is a simplified model built to mimic the activity of the pyloric pacemaker neurons. Our presentation, however, uses the geometric structure and not the details of the model. As such, the argument is quite general and can be expanded to different oscillator types.

2 Model

We use a single cell oscillator(O) described by two equations of the form

$$\varepsilon \frac{\mathrm{d}v}{\mathrm{d}t} = F(v,h) - I_{\mathrm{syn}} - I_{\mathrm{pert}},$$
$$\frac{\mathrm{d}h}{\mathrm{d}t} = G(v,h), \tag{16.1}$$

where v is the membrane potential, h is a recovery variable, I_{syn} denotes the feedback synaptic current, I_{pert} represents a perturbation of the system and ε is a positive parameter. We denote by P_0 the oscillation period in the absence of synaptic input and perturbations. The synaptic current is given as

$$I_{\mathrm{syn}} = g_{\mathrm{syn}} s(t)\left(v - V_{\mathrm{syn}}\right), \tag{16.2}$$

where V_{syn} is the synaptic reversal potential. $s(t)$ is a periodic function with period P_0 and is equal to 1 for a time duration of D and is equal to 0 at all other times. In our simulations below, we use a two-variable oscillator based on an inactivating calcium current with instantaneous activation and inactivation variable h based on a simplified model of the pyloric CPG pacemaker AB neuron (Kintos, Nusbaum, & Nadim 2008)[1]. In this model, we have

$$F(v,h) = I_{\mathrm{ext}} - g_{\mathrm{L}}\,(v - E_{\mathrm{L}}) - g_{\mathrm{Ca}} m_{\infty}{}^{3}(v) h\,(v - E_{\mathrm{Ca}}),$$

$$G(v,h) = (h_{\infty}(v) - h)/\tau_h(v).$$

We assume that, in the presence of the inhibitory feedback, the oscillator neuron O will oscillate with period P_0 and will lock to a certain phase relationship with the synaptic input (to be explained later). This assumption implies that the inhibitory feedback does not affect the oscillation frequency. The neuron presynaptic to O does not need to be explicitly modeled but will be referred to as F. The feedback synapse arrives periodically and, when active, has a constant conductance. Thus, we can assume that, if O fires at phase 0, then the synaptic feedback has an onset and an offset phase. The onset phase is also referred to as the synaptic phase φ_{syn}. The synaptic duty cycle (s_{DC}) is defined as the ratio between the active duration D of the synapse and the period of O. Thus the offset phase is $\varphi_{\mathrm{syn}} + s_{\mathrm{DC}}$. We will refer to the oscillation of O in the presence of feedback inhibition as the control case and in the absence of inhibition ($g_{\mathrm{syn}} \equiv 0$) as the uninhibited case.

Much of the analysis that we will conduct will be in the v–h phase plane (Fig. 16.1a) where we will track the behavior of the O trajectory in response to both synaptic inputs and perturbations. The $v-$ nullcline is defined as the set of points $\{(v,h)\colon F(v,h) = 0\}$ and the h-nullcline is the set of points $\{(v,h) : G(v,h) = 0\}$. The former is a cubic-shaped curve that has at lower voltage a local maximum $\mathrm{LK}_0 = (v_{\mathrm{LK0}}, h_{\mathrm{LK0}})$ and at higher voltage a local minimum $\mathrm{RK}_0 = (v_{\mathrm{RK0}}, h_{\mathrm{RK0}})$. The left and right branches of the cubic are denoted as LB_0 and RB_0, respectively. The $h-$nullcline is a decreasing sigmoidal shaped curve that, depending on parameters, intersects the $v-$nullcline on the middle or left branch. For the uninhibited case, parameters are chosen such that the intersection of the two nullclines occurs on the middle branch of the v-nullcline, allowing O to exhibit a stable limit cycle (Fig. 16.1a). When the O trajectory is subjected to inhibition from F (for the duration D), $s(t) = 1$. The effect of the synapse is to raise the $v-$nullcline in the v–h phase plane resulting in a new local maximum $\mathrm{LK}_1 = (v_{\mathrm{LK1}}, h_{\mathrm{LK1}})$, local minimum $\mathrm{RK}_1 = (v_{\mathrm{RK1}}, h_{\mathrm{RK1}})$, and left and right branches denoted LB_1 and RB_1(Fig. 16.1b). If the strength of the synapse, which is governed by g_{syn}, is sufficiently large when $s(t) = 1$ the intersection of the $v-$ and $h-$nullcline occurs along LB_1 (Fig. 16.1b, top left black circle). In a neighborhood of this intersection point, the O trajectory will slow its rate of evolution as this point represents a stable fixed point in the $v - -h$ phase space where both $\mathrm{d}v/\mathrm{d}t$ and $\mathrm{d}h/\mathrm{d}t$ equal 0. At other points along LB_1, because of the shape of the h-nullcline—namely that it is

---

[1] $\varepsilon v' = (i_{\mathrm{ext}} - i_L - i_{\mathrm{Ca}} - i_{\mathrm{syn}})/c_m$; $m_8(v) = 1/(1 + \exp(-(v+61)/4.2))$; $h_8(v) = 1/(1 + \exp((v+88)/8.6))$; $\tau_h(v) = (270/(1+\exp((v+84)/7.3)))(\exp((v+162)/30.0)) + 54$; $i_{\mathrm{ext}} = -0.45$, $g_L = 0.3142$, $g_{\mathrm{Ca}} = 1.2567$, $g_{\mathrm{syn}} = 0.0235$(active for 219.4 ms starting at $t_0 = 302$ ms in each cycle), $E_L = -62.5$, $E_{\mathrm{Ca}} = 120$, $E_{\mathrm{syn}} = -80$, $c_m = 7$, $\varepsilon = 1$.

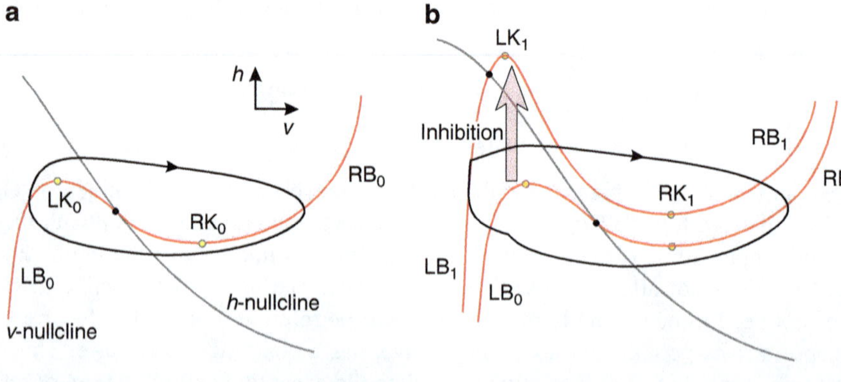

Fig. 16.1 The phase plane of the oscillator neuron.(**a**) In the uninhibited case, the h-nullcline intersects the cubic v-nullcline in the middle branch resulting in an unstable fixed point (filled circle) and a stable limit cycle (*black curve*). (**b**) Inhibition results in a shift of the cubic v-nullcline so that it intersects the h-nullcline in the left branch resulting in a stable fixed point (*filled circle*). Periodic inhibition results in transient movement of the limit cycle trajectory (*black curve*) to the left branch (LB_1) of the inhibited v-nullcline for the duration D of the inhibition. LK/RK denote the local mimima/maxima of the respective v-nullclines. Refer to text for detailed explanation of abbreviations

decreasing and has nonzero slope near LB_0 and LB_1—the rate at which trajectories evolve near LB_1 is greater than near LB_0. The opposite is true for trajectories near RB_1 and RB_0.

For parts of the analysis, we will assume that the system is singularly perturbed (Mishchenko & Rozov 1997). By this, we mean that the parameter ε in (16.1) is small and the O trajectory can by analyzed by separating a slow time scale (t) from a fast one ($\tau = t/\varepsilon$). The fast time-scale governs the rapid increase and decrease of voltage v corresponding to fast depolarization during a spike and repolarization at the end of the spike. The slow time-scale governs the evolution of the slow inactivation variable h during times when the O neuron is in its silent or active state. Mathematically, this separation of time-scales is achieved by setting $\varepsilon = 0$ in (16.1) to obtain the slow subsystem

$$\begin{aligned} 0 &= F(v,h) - I_{\text{syn}} - I_{\text{pert}} \\ \frac{\mathrm{d}h}{\mathrm{d}t} &= G(v,h) \end{aligned} \tag{16.3}$$

or by replacing t with τ in the same equation and then setting $\varepsilon = 0$ to obtain the fast subsystem

$$\begin{aligned} \frac{\mathrm{d}v}{\mathrm{d}\tau} &= F(v,h) - I_{\text{syn}} - I_{\text{pert}}, \\ \frac{\mathrm{d}h}{\mathrm{d}\tau} &= 0. \end{aligned} \tag{16.4}$$

In the singularly perturbed system, the O trajectory lies in a neighborhood of the v−nullcline at all moments of time, except when it makes the very fast transitions between the left and right branches of the nullcline. When this reduction is made, the uninhibited O trajectory can be tracked by following the evolution of the slow h-variable along either LB_0 or RB_0 because the evolution of the v-variable can be obtained by solving the constraint $F(v,h) = 0$. The transition between LB_0 and RB_0 occurs from either LK_0 or RK_0 and is assumed to be arbitrarily fast (i.e., taking no time) with respect to the slower h-evolution. This implies that the period of the uninhibited O orbit can be computed by understanding how much time the trajectory spends near LB_0 and RB_0. For the Control case when inhibition is present, whenever $s(t) = 1$, the O trajectory lies close to either LB_1 or RB_1. Thus when $s(t)$ switches from 0 to 1 or vice versa, the trajectory chooses which branch to lie close to. For example, if the trajectory is near LB_0 with $s(t) = 0$, then at the switch to $s(t) = 1$, the trajectory transitions quickly to LB_1. If it were near RB_0 and above the point RK_1 then it transitions to RB_1, but if it were below RK_1, then it would transition to LB_1. Similarly if with $s(t) = 1$ the trajectory is near LB_1 and above LK_0 and $s(t)$ switches to 0, the trajectory jumps to RB_0. Otherwise, it goes to LB_0. Thus, as the onset location and duration of the inhibition changes, the O trajectory will lie in different places in phase space. We will show how this is related to the PRC of the O neuron.

3 Results

3.1 *Experimental*

We examined the effect of inhibitory feedback on oscillation activity in the crab pyloric network, a pacemaker-driven oscillatory network with an extremely regular oscillation frequency. Pyloric pacemaker (AB and PD) neurons receive their sole inhibitory feedback though a synapse (LP to PD) from the follower neuron LP that bursts out of phase with the pacemakers. Perturbations to the pyloric rhythm arise from many sources, such as the intrinsic noise in pyloric neurons and excitatory inputs from descending projections. We examined whether the LP to PD synapse has an effect on the pyloric cycle period or on how the cycle period is affected by perturbations. After recording the ongoing oscillations (Fig. 16.2; control), the LP to PD synapse was removed by hyperpolarizing the LP neuron (uninhibited). Forty cycle periods were measured in each condition and the cycles immediately following the LP neuron hyperpolarization were not included in this analysis. On average, cycle period was not affected by the removal of the feedback inhibitory synapse; however, the coefficient of variation was significantly smaller in control than uninhibited (Fig. 16.2b). These results suggest that, during the normal ongoing pyloric activity, the LP to PD synapse does not affect the mean network cycle period but it significantly reduces its variability.

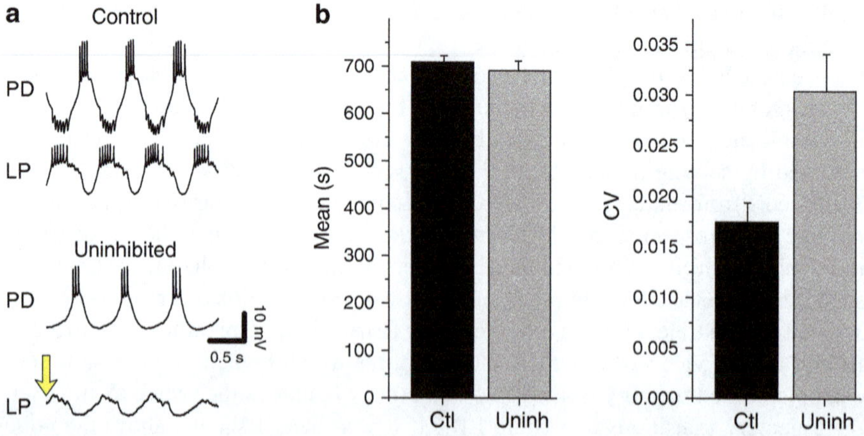

Fig. 16.2 (Experimental) Pyloric oscillation variability is reduced in the presence of the inhibitory feedback synapse but period is not affected. (**a**) Intracellular voltage traces from the pacemaker group neuron PD and the follower LP indicate the activity of pyloric circuit neurons. Recordings were done under control or uninhibited (LP hyperpolarized by injection of -5 nA DC current, arrow, to remove the LP to PD synapse) conditions. (**b**) There was no significant change in the mean pyloric period between control and uninhibited conditions but the coefficient of variation was larger in uninhibited compared to control ($P < 0.05$)

The effect of extrinsic perturbations on the activity of a neural oscillator is often measured by examining the phase-resetting curve (PRC) (Achuthan & Canavier 2009; Oprisan, Thirumalai, & Canavier 2003; Pinsker 1977). To examine the role of the LP to PD feedback inhibitory synapse on the effect of perturbations to the pyloric oscillations, we constructed and compared the PRCs measured by perturbing the PD neuron in control and uninhibited conditions. To construct the PRCs, we injected brief positive current pulses (2 nA, 50 ms) into the PD neuron. Using specialized software (phase response: http://stg.rutgers.edu/software), we injected the current pulses at phases 0.1–0.9 of the cycle to be able to average the response across different trials and preparations. Figure 16.3a shows an example of one such injection under control conditions. Figure 16.3b shows the PRC measured when the reset phase $\Delta\varphi_{\mathrm{PD}} = (P_0 - P)/P_0$ was plotted against the phase of the perturbation $\varphi_{\mathrm{pert}} = \Delta t / P_0$. Here, Δt was measured as the time of the perturbation onset after the first spike in PD burst, P_0 was the free run period, and P was defined as the perturbed period (Fig. 16.3a).

With positive current perturbation the period was prolonged at early perturbation phases and shortened at late phases in both control and uninhibited conditions. However, in the uninhibited case, the perturbations had a stronger effect on the cycle period which was seen in the significant shift of the PRC away from zero compared with the control case. Negative-current perturbations had the opposite effect on cycle period: the period was shortened at early perturbation phases but prolonged at late phases in both conditions (not shown). Yet again, in the control case the PRC was

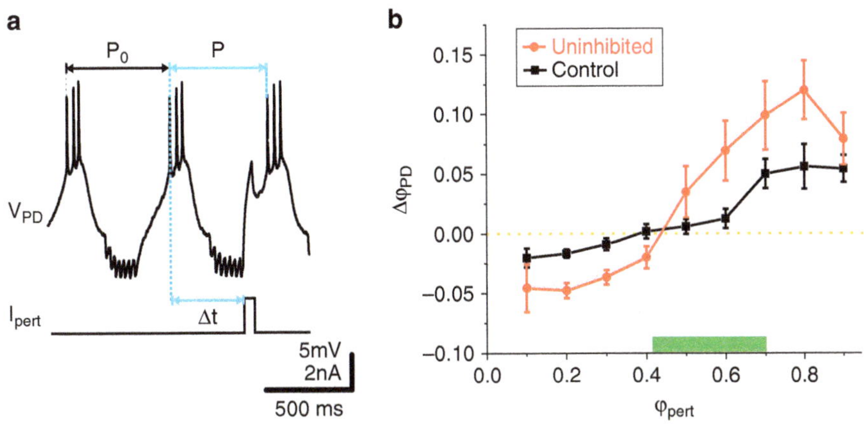

Fig. 16.3 (Experimental) The inhibitory feedback synapse attenuated the effect of extrinsic perturbations as measured through the PRC. (**a**) Intracellular voltage traces from the PD neuron are used to monitor pyloric oscillatory activity. A brief current pulses (2 nA, 50 ms) was injected into the PD neuron at different phases in the presence or absence (not shown) of the LP to PD synapse. (**b**) The PRC in response to the perturbation was measured as the reset phase ($\Delta\varphi_{PD} = ((P_0 - P)/P_0)$) and plotted against the perturbation phase φ_{pert}. In control, the PRC lies closer to zero compared to uninhibited ($N = 4$). The *green bar* indicates the duration of the synapse activity

closer to zero compared to the uninhibited case. These results show that the LP to PD synapse significantly reduces the effect of perturbations by "flattening" the overall PRC. In particular, there was significant flattening of the PRC even at phases were the synapse was typically not active.

4 Model Description of the PRC Effects

We now turn to our basic mathematical model to explain the observations regarding the PRCs for the two cases. Recall from Fig. 16.2 that the PD neuron cycle period has some natural variability and that this variability is reduced in the presence of the LP to PD feedback inhibitory synapse, as reflected in the cycle period coefficient of variation and more descriptively in the PRC (Fig. 16.3). First, we show that a simplified model as described by (16.1) can reproduce these experimental results. For this, we use the two-variable model of the AB neuron defined in the Model section.

Let $\Lambda(t)$ be a limit cycle of period T of the differential equation $\mathrm{d}X/\mathrm{d}t = f(X)$, $X \in \mathbb{R}^n$. Then the PRC of Λ, for an infinitesimal perturbation vector y can be calculated as the inner product $Z(\varphi) \cdot y$ where $Z(\varphi)$ satisfies the adjoint equation to the linearization of the original differential equation, given by $\mathrm{d}Z/\mathrm{d}t = -A(t)^T Z$, where $A(t) = \mathrm{d}f_{\Lambda(t)}$ (Ermentrout & Terman 2010). The close relationship between

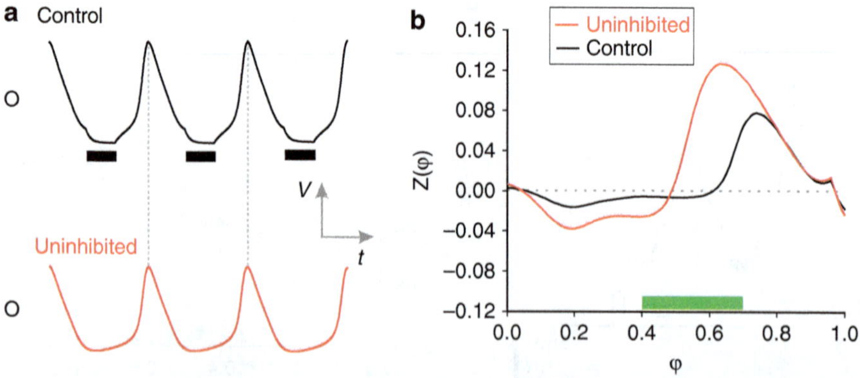

Fig. 16.4 The model PRC lies closer to zero in the presence of feedback inhibition. (**a**) Voltage traces from the model neuron O in the presence (control) and absence(uninhibited) of synaptic inhibition (*black bars*). Both cases have the same cycle period. (**b**) The numerically calculated solutions $Z(\varphi)$ to the adjoint equation calculated at the limit cycle describe the PRC in the control and uninhibited cases. φ represents the phase of the limit cycle. Model as described for the AB neuron in Sect. 2.2.1 and Table 3 of (Kintos et al. 2008). The synaptic input was added from phase 0.3 to 0.7 (*green bar*) of oscillation with $g_{max} = 0.0235$ nS and $E_{syn} = -80$ mV

the PRC and the adjoint solution has allowed some researchers to refer to these terms interchangeably. Although adjoint solutions are not always easy to solve for analytically, they can be calculated numerically once a limit cycle has been found (Ermentrout 2002) and graphed to show the shape of the PRC.

Figure 16.4 shows the adjoint solutions $Z(\varphi)$ at the limit cycle, generated for both the control model and when inhibition is removed (uninhibited).The curves $Z(\varphi)$ were generated by numerically calculating the solutions to the adjoint equation of (16.1) at the limit cycle, for one cycle of oscillation, and represent the PRC for an infinitesimal perturbation (Brown et al. 2004). (A similar PRC can be generated by subjecting the trajectories in this model to perturbations of duration 20 ms and amplitude 0.125 nA.) Note that the curves $Z(\varphi)$ shown in Fig. 16.4b qualitatively match the experimental PRCs obtained in Fig. 16.3b.

To explain why the control PRC is generally smaller in amplitude at any phase of the perturbation, i.e., why the oscillation is more robust to perturbation in the presence of the inhibitory feedback synapse, we first introduce the concept of the synaptic phase resetting curve (sPRC). The sPRC documents the change in the O phase as a result of F inhibition and depends primarily on two factors: the synaptic (onset) phase φ_{syn} defined as the phase in the O cycle that the inhibition arrives, and the synaptic duty cycle s_{DC}, defined as the fraction of the cycle that the synapse remains active. To conduct this analysis, we will assume that ε is small and use separation of time scales as described in the Model section above.

Divide the uninhibited O trajectory by phase (φ between 0 and 1 in one cycle), as shown in Fig. 16.5a, such that the point with the maximum voltage along RB_O is labeled $\varphi_0 = 0$. Label the value φ_1 as the point along RB_O that has the same h-value

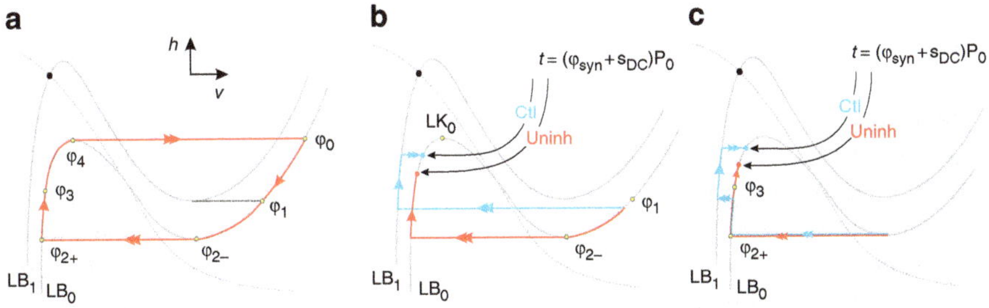

Fig. 16.5 The oscillation phases in the singularly-perturbed case where $\varepsilon = 0$. (**a**) The phase points defined on the uninhibited trajectory. (**b**) Phase relationships for φ_{syn} in $(\varphi_1, \varphi_{2-})$. (**c**) Phase relationships for φ_{syn} in (φ_2, φ_3). Blue trajectory denotes the Control case and the red trajectory is the uninhibited case. The trajectories shown in B and C start at the same time and terminate at the time $t = (\varphi_{\text{syn}} + s_{\text{DC}})P_0$

as RK_1, φ_{2-} as the point RK_0, φ_{2+} as the point on LB_0 with $h = h_{\text{RK0}}$, φ_3 as the point on LB_0 that is $D = s_{\text{DC}} \times P_0$ time away from the point LK_0 but where the evolution is considered using the dynamics along LB_1, and φ_4 as the point LK_0. We will derive the shape of the sPRC using the different phase points defined above. In particular, we will show that the sPRC is a monotone decreasing function of φ_{syn} on a certain interval of phase values. Within this interval, there exists a unique phase φ^* at which the synaptic input from F does not change the O phase. Moreover, we will show that φ^* is a stable attracting phase in the dynamics so that O locks to the synaptic input such that this input always arrives at $\varphi = \varphi^*$. Recall that the synaptic input is periodic with period P_0. This input to O will, in general, change the period of O to a new value we call P. We define $\text{sPRC}(\varphi_{\text{syn}}) = (P_0 - P)/P_0$ and call this the change in phase.

Consider the phase interval $\varphi_{\text{syn}} \in (\varphi_1, \varphi_{2-})$ as shown in Fig. 16.5b. In this case, the effect of the synapse is to shorten P relative to P_0. This happens for two different reasons. First, at the moment the inhibition begins, the trajectory immediately jumps down to LB_1. Thus the phase of the trajectory is advanced because it leaves RB_0 prematurely and does not need to spend time reaching the local minimum point RK_0. Second, for the duration of the inhibition, the trajectory lies on LB_1. Since the speed at which the trajectory evolves (vertically) on LB_1 is greater than on LB_0, the trajectory moves further up in the h-direction on LB_1 than it would have on LB_0. In addition it starts from a higher h-value on LB_1 as shown in Fig. 16.5a. Thus at $t = (\varphi_{\text{syn}} + s_{\text{DC}})P_0$, when the trajectory returns to LB_0, it lies closer to the local maximum point LK_0 than it would have in the absence of the inhibition. Therefore it takes less time to reach LK_0 and as a result $P < P_0$. Therefore $\text{sPRC}(\varphi_{\text{syn}})$ is positive on the interval $(\varphi_1, \varphi_{2-})$.

Next consider the case where φ_{syn} is in the interval $(\varphi_{2+}, \varphi_3)$ as shown in Fig. 16.5c. By the definition of φ_3, for any φ_{syn} in this interval, the trajectory will spend the duration of the inhibition on LB_1 and when the inhibition ends it will return to LB_0 at a point lower than LK_0. Because the evolution rate on LB_1 is larger than on LB_0 the point to which the trajectory returns to LB_0 will be higher than it would have been in the absence of inhibition. Thus, the remaining time to LK_0 will be less and as above $P < P_0$ which implies that sPRC(φ_{syn}) > 0. In this interval of synaptic phases, in the singular limit, sPRC(φ_{syn}) is constant. The reason for this has to do with the fact that there is no time compression between trajectories that receive synaptic inputs during this interval. Let us elaborate by considering two values of φ_{syn} called $\varphi_{synA} < \varphi_{synB}$. The synaptic input corresponding to φ_{synA} arrives when the trajectory is at some h value called h_A. Similarly φ_{synB} arrives at $h = h_B$ with $h_A < h_B$. At this moment in time, there is a certain time distance, say t_d, between the points h_A and h_B that is determined by the rate of evolution along LB_0. After the inhibition starts, both trajectories now evolve along LB_1. There is a new time distance, say t_e, between the h values of the trajectories. But this time remains invariant as long as the trajectories evolve on LB_1. The trajectories spend the same amount of time D on LB_1, before returning to LB_0 where there time distance again becomes t_d. Thus neither trajectory has advanced in phase relative to one another since after the inhibition their time distance apart remains the same as before. This implies that sPRC(φ_{syn}) is constant on this interval.

The above result is a generic feature of many systems that are decomposed into fast-slow pieces. It follows from the fact that if two trajectories follow the same one-dimensional path for the same amount of time, then, while the Euclidean distance between the trajectories may change, the time distance between them does not. In fact, it is just an application of the group property of flows of a dynamical system (Guckenheimer & Holmes 1997). Finally we note that when we relax, the separation constraint on timescales, sPRC(φ_{syn}) need not be constant on this interval. Indeed from our numerical simulations, shown later, sPRC may in fact be nonmonotonic. It is not so obvious how to explain why this is the case other than noting that trajectories are no longer constrained to exactly LB_0 and LB_1 and therefore may take slightly different times to reach different locations in phase space. Nonetheless, the fact that sPRC(φ_{syn}) > 0 for this interval of φ_{syn} follows for the same reasons as above.

If $\varphi_{syn} = \varphi_3$, when the inhibition ends, the trajectory has $h = h_{LK0}$ and immediately jumps back to the RB_0 (Fig. 16.6a). The uninhibited trajectory will lie below $h = h_{LK0}$ still on LB_0 when the control trajectory jumps to RB_0. Now consider the interval (φ_3, φ_4). Since we have already established that sPRC(φ_{syn}) > 0 at $\varphi_{syn} = \varphi_3$, let us determine what occurs at the opposite end of the interval when $\varphi_{syn} = \varphi_4$ (Fig. 16.6b). Here, the inhibition starts just before the trajectory would have jumped from LK_0 to RB_0. The trajectory, instead, must now spend an amount of time $s_{DC} \times P_0$ on LB_1 until the inhibition ends. Moreover, once the trajectory returns to RB_0, it will have a higher h-value than h_{LK0}. Thus there will be an additional amount of time needed for the trajectory to reach the location of the phase point φ_0. Therefore, $P > P_0$ and sPRC(φ_{syn}) < 0. By continuity with respect to φ_{syn}, there exists a value $\varphi^* \in (\varphi_3, \varphi_4)$ such that if $\varphi_{syn} = \varphi^*$, then $P = P_0$ and sPRC(φ_{syn}) = 0. Further, note that the value of φ^* is such that $\varphi^* + s_{DC} \approx 1$.

The synaptic locking phase adjusts itself according to the duration of the synaptic input. To prove uniqueness of φ^*, let us show that $\Delta\varphi$ is monotone decreasing on

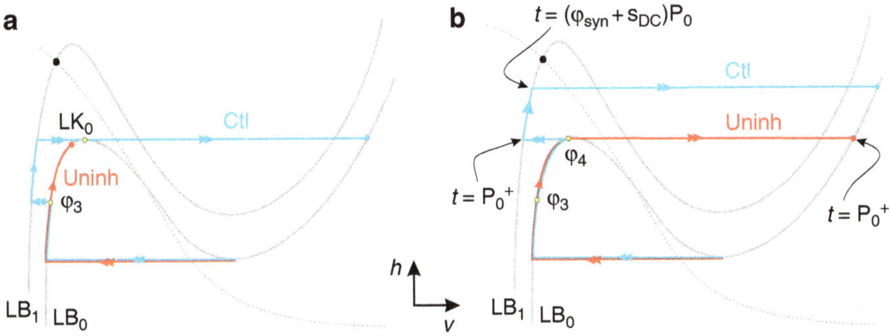

Fig. 16.6 The oscillation phases in the singularly-perturbed case where $\varepsilon = 0$.(**a**) Phase relationships for $\varphi_{\text{syn}} = \varphi_3$. (**b**) Phase relationships for φ_{syn} in (φ_3, φ_4). *Blue* trajectory denotes the control case and the red trajectory is the uninhibited case. The trajectories in both panels start at the same time and terminate at the time $t = (\varphi_{\text{syn}} + s_{\text{DC}})P_0$

this interval. Consider two different values of φ_{syn} denoted φ_{C} and φ_{E}, both in the interval (φ_3, φ_4) such that $\varphi_{\text{C}} < \varphi_{\text{E}}$. When $\varphi_{\text{syn}} = \varphi_{\text{C}}$, the trajectory moves along LB_1 and rises above LK_0 before the inhibition turns off. Say it reaches a point with h-value $h = h_{\text{C}}$. Similarly, when $\varphi_{\text{syn}} = \varphi_{\text{E}}$, the trajectory reaches a point with $h = h_{\text{E}}$ when the inhibition turns off. It is clear that $h_{\text{C}} < h_{\text{E}}$ by continuity with respect to initial conditions since the two trajectories spend the same amount of time on LB_1. Therefore, when the trajectories return to RB_0, the one associated with h_{C} lies below that associated with h_{E}. Now observe that the position of the uninhibited trajectories retain their original orientation. That is the uninhibited trajectory E reaches RB_0 first and begins evolving down, followed at some later time by the uninhibited trajectory C. Define $h(\varphi_{\text{syn}}) = h_{\text{RK}_0} e^{(1-\varphi_{\text{syn}}-s_{\text{DC}})P_0}$ as the value of h along the uninhibited trajectory at the moment its *inhibited* counterpart returns to RB_0. Then, $h(\varphi_{\text{C}}) > h(\varphi_{\text{E}})$ and the following ordering of points holds $h_{\text{E}} > h_{\text{C}} > h(\varphi_{\text{C}}) > h(\varphi_{\text{E}})$. Thus, the change in phase is more for trajectory E than for trajectory C, again implying monotonicity of sPRC on this interval. Therefore, sPRC is monotone decreasing on the interval (φ_3, φ_4) and the value φ^* is unique. To summarize, we have now shown that on the larger interval (φ_1, φ_4), sPRC contains a unique value $\varphi_{\text{syn}} = \varphi^*$ at which the $\text{sPRC}(\varphi_{\text{syn}}) = 0$.

Let us now establish that in the O–F control network, the locked phase φ^* at which $P = P_0$ is a stable, attracting phase of the dynamics (implying that F phase-locks to O activity at phase φ^*).To this end, we define a map f(modulo 1) which records the onset phase φ_{syn} of the synapse at each cycle of inhibition. This map is defined on the interval $[\varphi_1,\varphi_4]$ as the sum of the current synaptic phase and the sPRC at that phase.

$$\varphi_{\text{syn}}^{n+1} = f\left(\varphi_{\text{syn}}^{n}\right)$$
$$f(\varphi) = \varphi + sPRC(\varphi)(\text{mod } 1)$$

In other words, it takes the value of φ_{syn} at one cycle and computes what this value will be at the next cycle. The range of this map is the entire interval [0,1]. This new value of φ_{syn} obviously depends on how the synapse affected the period in the previous cycle and how this changed the phase relationship between O and the F inhibition. From the above analysis, it is easy to establish that $f(\varphi)$ is a continuous function on [0,1], that it has a unique fixed point φ^* and is such that $f(\varphi_1) > \varphi_1$ while $f(\varphi_4) < \varphi_4$.

Proving the stability of φ^* is an exercise in applying the concepts of Fast Threshold Modulation due to Somers & Kopell (1993). Namely, consider two synaptic phases $\varphi_{\text{syn1}} < \varphi_{\text{syn2}}$ in a neighborhood of φ^*. These phases correspond to different h-values $h_i(0)$ at which the trajectory receives the inhibition. There exists a time difference Δt_0 between these two initial conditions which is computed by flowing the trajectory at $h_1(0)$ forward along LB_0 for the time Δt_0 until it reaches the position of $h_2(0)$. When inhibition arrives to these two trajectories, they both move to LB_1 and there exists a new time Δt_1 between them. Since evolution is faster on LB_1 than LB_0, $\Delta t_1 < \Delta t_0$.Both trajectories now flow along LB_1 for exactly the same amount of time $D = s_{\text{DC}} \times P_0$ and the time distance between them remains invariant. After time D, the trajectories are released from inhibition and transition to RB_0, yielding, once there, a new time distance Δt_2. In (Somers & Kopell 1993) conditions are given so that $\Delta t_2 < \Delta t_1$. The main conditions involve a compression factor which is directly related to the rate of evolution on LB_1 immediately before the jump to the rate on RB_0 immediately after the jump. In our case, these rates are strongly governed by the vertical distance of the trajectory to the h-nullcline at these two moments and the time constants governing evolution on those branches. Provided that these time constants are of the same order—a reasonable assumption for bursting neurons—it follows that $\Delta t_2 < \Delta t_1 < \Delta t_0$. This compression in time implies compression in phase. Thus, locally, any two values of φ_{syn} lying in a neighborhood of φ^* are brought closer together by the map f after one iterate. This implies stability of the point φ^*. It must be noted that when φ_{syn1} and φ_{syn2} lie in (φ_2, φ_3), the respective trajectories also make jumps between LB_0, LB_1, and RB_0, as discussed earlier. However, the difference in these phases is that the trajectories return to LB_0 before jumping to RB_0 through the point RK_0. The fact that when φ_{syn1} and φ_{syn2} lie in (φ_3, φ_4) that the trajectories jump from LB_1 directly to RB_0 accounts for the compression.

The stability of φ^* has an immediate consequence. It means that any perturbation applied to the control trajectory will immediately begin to decay due to the fact that $\varphi = \varphi^*$ is the attracting phase and that any change in phase induced by the perturbation will not persist. Later, we will use this observation to show why the PRC for the control case lies closer to 0 than the PRC for the uninhibited case.

Next let us see how the sPRC is affected by the synaptic duty cycle s_{DC}. In Fig. 16.7, we plot numerically-generated examples of the sPRC for three different values of s_{DC}. In general, when s_{DC} decreases, part of the sPRC shifts up in the φ_{syn}–sPRC plane while another portion of it shifts down. Consider two distinct synaptic duty cycles $s_{\text{DC2}} < s_{\text{DC1}}$. To understand which part of sPRC shifts up, first let $\varphi_{\text{syn}} < \varphi_3$, where φ_3 is defined as above but now applied to the s_{DC1} case. The trajectories

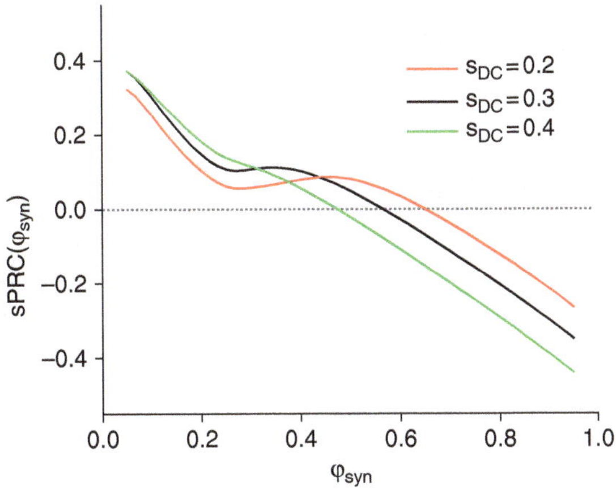

Fig. 16.7 Numerically generated sPRC for different values of s_{DC}. The sPRC was generated for three different values of s_{DC}, 0.2, 0.3 and 0.4, using the model described in Fig. 16.4 with $\varepsilon = 0.14$ to approximate the singular limit. The value φ^* is indicated by the intersection of the sPRC with 0

for both of the s_{DC}'s jump to LB_1 at φ_{syn} and return to LB_0 after the time interval $s_{\mathrm{DCj}} \times P_0(j = 1, 2)$ before leaving the silent state through LK_0. Since $s_{\mathrm{DC2}} < s_{\mathrm{DC1}}$, the trajectory associated with s_{DC2} spends less time on LB_1 than the other trajectory. Therefore, it advances less in phase since the speed of evolution on LB_1 is greater than on LB_0. Thus, the trajectory with longer duty cycle has a greater phase advance for small values of φ_{syn} resulting in the sPRC shifting up as shown in Fig. 16.7 (green curve above black and red). Alternatively, now consider the case where φ_{syn} is larger, say closer to φ_4. Trajectories in both cases will now move along LB_1 to points above LK_0 when the inhibition turns off. The one associated with s_{DC2} will have a smaller $h-$value and will thus need to spend less time on RB_0 to return to the $\varphi = 0$ location. While both trajectories will be delayed in phase relative to the uninhibited trajectory, the s_{DC2} trajectory will be ahead of its counterpart in phase. As a result, the SPRC curve for s_{DC2} will lie above that for s_{DC1} for larger values of φ_{syn} as shown in Fig. 16.7 (red curve above black and green). Note that larger φ_{syn} values have a larger effect on the s_{DC} than smaller φ_{syn} values. At larger φ_{syn}, there are two sources of phase delay. One is the fact that trajectory may still be on LB_1 under the influence of inhibition at the time that the uninhibited trajectory would have returned to RB_0. This means that the trajectory spends extra time on the LB. This occurs for all $\varphi_{\mathrm{syn}} \in (\varphi_3, \varphi_4)$. Second, when this happens, the trajectory must also spend extra time on RB_0 while returning to the $\varphi = 0$ starting point. In contrast, changes in s_{DC} have less of an effect for lower values of φ_{syn} as here the different duty cycles simply change the time spent on LB_1, but do not change the fact that these trajectories still leave LB_0 through LK_0 to return to the right branch and the $\varphi = 0$ starting point.

To understand how the sPRC counteracts the effect of perturbations, we first check to see whether the perturbation arrives before, during, or after the inhibition from F. The phase of the perturbation, φ_{pert} is defined with respect to the $\varphi = 0$ reference point on the uninhibited or control trajectory. If $\varphi_{\mathrm{pert}} < \varphi^*$, then the uninhibited and control trajectory are roughly following identical paths from $\varphi = 0$

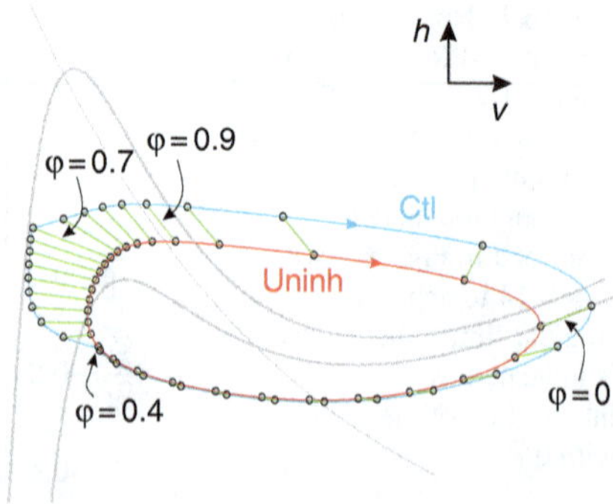

Fig. 16.8 Corresponding phases of the control and uninhibited trajectories. Line segments connect the corresponding phases in the two limit cycles of the model described in Fig. 16.4

to φ_{pert}. For this interval of phases, the sPRC can be composed with the PRC to derive the overall change $\Delta\varphi$ in the phase of O. Specifically, consider an inhibitory perturbation to the control trajectory. This perturbation will change the phase of O by an amount denoted $\Delta\varphi_{\text{prc}}$ and result in a new period P for that cycle. In turn, this causes a change in the value of φ_{syn}. This new value of $\varphi_{\text{syn}} = \varphi^*{}_{\text{syn}} P_0/P$. Since $P = P_0(1-\Delta\varphi_{\text{prc}})$, it follows that $\varphi_{\text{syn}} = \varphi^*{}_{\text{syn}}/(1-\Delta\varphi_{\text{prc}})$. For example, if $\Delta\varphi_{\text{prc}} > 0$, then $\varphi_{\text{syn}} > \varphi^*{}_{\text{syn}}$, which, referring to Fig. 16.7, shows that the O phase is delayed. Further, s_{DC} also changes to a new value given by $s_{\text{DC}} = s_{\text{DC}}{}^*/(1-\Delta\varphi_{\text{prc}})$, where $s_{\text{DC}}{}^*$ refers to the synaptic duty cycle of the unperturbed control case. Denote the effect of sPRC on the O phase by sPRC(φ_{syn}, s_{DC}). The effects of both the perturbation and sPRC on the change in O phase can be calculated as $\Delta\varphi =$ sPRC($\varphi^*{}_{\text{syn}}/(1-\Delta\varphi_{\text{prc}})$, $s_{\text{DC}}{}^*/(1-\Delta\varphi_{\text{prc}})$). In general, if the perturbation increases O phase, then the sPRC will decrease the O phase and vice versa. This is to be contrasted with the uninhibited trajectory. In that case, any change in phase induced by the perturbation is never counteracted and persists indefinitely.

Let us explore how the sPRC counteracts the perturbation a bit more closely by considering the singular case in which we again impose timescale separation. From before, we know that φ^* satisfies $\varphi^* + s_{\text{DC}} \approx 1$. That is, φ^* adjusts itself so that the time the inhibition turns off corresponds to the time that the control trajectory jumps to RB_0. Thus the above argument about φ_{pert} arriving before the inhibition holds for all $\varphi_{\text{pert}} < \varphi^* = 1 - s_{\text{DC}}$. If $\varphi_{\text{pert}} \in (\varphi^*, 1)$, then the perturbation will change the current value of the O phase, denoted φ_{O}, to some new value, $\varphi_{\text{O}}{}^{\text{new}}$. Note that $\varphi_{\text{O}}{}^{\text{new}}$ is not determined by the PRC of the uninhibited trajectory since the control trajectory does not lie near the uninhibited one for these phases. The perturbed trajectory will lie in a neighborhood of LB_1 until the moment in time that the inhibition ends, which occurs at $t \approx P_0$. Provided that $\varphi_{\text{O}}{}^{\text{new}} > \varphi_3$, the perturbed trajectory will lie above LK_0 at this moment, independent of the actual value of its phase, and at this time, will jump up to LB_0, and spend a small amount of time reaching the $\varphi = 0$ starting point. This argument suggests that, in the singular limit, $\Delta\varphi \approx 0$ for all $\varphi_{\text{pert}} \in (\varphi^*, 1)$.

Next, let us see how the argument extends to the nonsingular case. Now the trajectory spends a nonzero amount of time making the transition from the silent to the active state. In Fig. 16.8, we show the control and uninhibited trajectories where we have broken up each trajectory by phase and drawn short line segments to indicate places of equal phase along both. Note for this set of parameters, $\varphi^* = 0.4$ and $s_{\mathrm{DC}} = 0.3$. The control trajectory stays in a neighborhood of LB_1 from $\varphi = 0.4$ to $\varphi = 0.7$, while from $\varphi = 0.7$ to $\varphi = 1$ it returns to the active state. Note the position of the control and uninhibited trajectories at both $\varphi = 0.4$ and $\varphi = 1$. The control trajectory is longer (in Euclidean distance), but both traverse this range of phases in the same total amount of time. This means that the average speed with which the control trajectory moves in this range of phases is larger than for the uninhibited one. This is most clearly seen for phases in the range (0.4, 0.7) and (0.9, 1) when the control trajectory "catches and passes" the uninhibited trajectory (first in the vertical direction, then in the horizontal direction) in phase space. This observation is not unique to this particular model or parameters we have chosen. Indeed any time two periodic orbits have the same time length, but different total arc lengths over one period, there must be regions in phase space over which the trajectories evolve at different speeds. In our particular case, the main factors governing the speed near LB_0 and LB_1 are the h-dynamics since near these branches $\mathrm{d}v/\mathrm{d}t$ is close to 0. Across the jump to the active state, the primary factor controlling speed is the vertical distance to the v-nullcline. This can be seen from the fact that both trajectories are relatively flat in the region $(\varphi^*, 1)$ since $\mathrm{d}v/\mathrm{d}t$ dominates $\mathrm{d}h/\mathrm{d}t$ across this jump.

Now consider an inhibitory perturbation with phase φ_{pert} between φ^* and $\varphi^* + s_{\mathrm{DC}}$. Assume that the perturbation affects them in the same way by shifting them to the left by some amount Δv. For both trajectories, this causes a change in the phase. For the uninhibited trajectory, this change in phase persists for the remainder of the cycle resulting in a new value for the phase as documented by the uninhibited PRC. The situation for the control trajectory is different. Independent of the current phase of the perturbed control trajectory, the synaptic inhibition remains on until at $t = (\varphi^* + s_{\mathrm{DC}})P_0$. Note that the trajectory is constrained by the synaptic inhibition to lie close to the inhibited v-nullcline during this time. In particular, at $t = (\varphi^* + s_{\mathrm{DC}})P_0$. The perturbed control trajectory will lie in a neighborhood of the $\varphi = \varphi^* + s_{\mathrm{DC}}$ point of the control trajectory. This is in fact the key point. Namely, the inhibition places a time constraint on when the perturbed control trajectory can leave LB_1. Thus, because of this constraining effect any phase change the perturbation may have supplied is largely wiped out by the inhibition. For example, if the perturbation had advanced O's phase, the inhibition constrains the trajectory until $t = (\varphi^* + s_{\mathrm{DC}})P_0$, counteracting the advance by delaying the trajectory from leaving for the active state until that time. For $t > (\varphi^* + s_{\mathrm{DC}})P_0$, the perturbed trajectory lies in a neighborhood of the control trajectory in the v–h phase plane. Thus, any change in phase between the control and perturbed control trajectory is due solely to the minor difference in speed across this jump. In summary for this subcase, the synaptic inhibition largely wipes out any phase resetting supplied by the perturbation. In contrast, in the absence of F inhibition, the phase resetting persists for the entire cycle. The above argument suggests that, as in the singular case, the control PRC should be fairly insensitive to perturbations arriving in the interval

$(\varphi^*, \varphi^* + s_{DC})$ and should thus be quite flat since the only increase in phase is due to slightly different speeds across the jump up to the active state. This is consistent with the PRCs that we computed as shown in Fig. 16.4.

The biggest changes in the PRC occur when φ_{pert} lies between $\varphi^* + s_{DC}$ and 1. Consider an inhibitory perturbation coming during this interval that causes a change $\Delta v < 0$ in both the inhibited and uninhibited trajectories. Now the only difference that the inhibited trajectory has with its uninhibited counterpart is that it lies higher in phase space and thus has a larger dv/dt value over this portion of its orbit. Thus, it can get to the active state more quickly and therefore result in a smaller overall change in phase than the uninhibited trajectory does. Here, there is no direct constraining effect of the synapse. Instead, it is the inhibition that places the control trajectory in a particular part of phase space where it can utilize its advantage in speed to minimize phase changes.

5 Discussion

An important property of many CPG networks is that they exhibit very robust rhythmic oscillations. In many cases, this property is critically important for the proper function of the organism. For example, CPG networks have been shown to control heartbeat, feeding, ventilation, and locomotion (Dickinson 2006; Marder & Calabrese 1996). In all of these cases, stable periodic solutions are necessary for the survival of the animal.

Oscillations often arise in a CPG either through a reciprocally inhibitory pair of neurons or as a result of an endogenous bursting pacemaker neuron(s). In the former case, oscillations are driven by the antiphase interactions of the two neurons. The stability of the oscillations created by this oscillatory pair naturally makes the CPG relatively insensitive to perturbations. In the latter case, however, where oscillations are due to a pacemaker, there is no apparent mechanism that would promote robustness of the rhythm. In this paper, we proposed that a feedback inhibitory synapse that, in general, has no role in creating oscillations instead plays the role of minimizing the pacemaker's sensitivity to perturbations. To demonstrate this capability of the inhibitory neuron, we computed the pacemaker's PRC in the presence and absence of inhibitory feedback. The PRC of a neuron documents the amount that the neuron changes in phase (or period) in response to perturbations applied at different parts of the rhythmic cycle. We showed that the control PRC lies closer to 0 than the uninhibited PRC. This implies that the control pacemaker is less sensitive to perturbations than its uninhibited counterpart.

From a mathematical viewpoint, the effect of the feedback inhibition in making the oscillator less sensitive to perturbations can be explained simply through the notions of neutral and asymptotic stability. The uninhibited oscillator in this model is a neutrally stable limit cycle. While it attracts all trajectories that lie in a neighborhood of itself, it does not assign a phase preference to the attraction. In other words, the perturbed uninhibited trajectory has no means to compensate the change in phase. This can easily be seen by considering a perturbation of the uninhibited trajectory that lands the perturbed trajectory exactly on the limit cycle itself. Here, the phase is immediately changed, and since the perturbed trajectory is itself on the limit cycle, there will be no additional change in phase. In contrast, the control trajectory is an asymptotically stable limit cycle both in the sense of attracting nearby trajectories and also in the sense of assigning a phase preference. This can be seen through the stability of the fixed point φ^* of the map $f(\varphi_{\text{syn}}) = \varphi_{\text{syn}} + \text{sPRC}(\varphi_{\text{syn}})$. The fixed point φ^* forces the oscillator to lock at a specific phase of the inhibitory feedback. Thus, any perturbation to the control trajectory immediately begins to decay back to the $\varphi = \varphi^*$ fixed point. The rate of decay is governed by the magnitude of the derivative (eigenvalue) obtained through the linearization of the map. The key point is that this decay begins immediately from the moment of the perturbation. This is the reason why the feedback inhibition can have an effect on the period even when the perturbation occurs after the inhibition in that particular cycle has already ended (e.g. for $\varphi_{\text{pert}} > \varphi_{\text{syn}} + s_{\text{DC}}$ in the non-singular cases shown in Fig. 16.4b).

Several other studies have documented how neurons are affected by various synaptic or perturbing inputs. Prinz et al. (2003) found that the PRC of a neuron was more sensitive to the duration of synaptic input than to the strength of the input. This was established by showing that the PRC saturates at biologically plausible synaptic conductances, but continues to change over all biologically realizable synaptic durations (at least in the case of PD neurons). Oprisan et al. (2003) found similar sensitivity to and importance of the duration of short perturbative or longer synaptic inputs in their modeling study in which they reconstructed the PRC of a bursting neuron from time series data. Our work builds on the results of these earlier studies and asks not just how the duration or phase of an input can affect the PRC, but also how two different inputs interact with one another to determine the PRC. In particular, we explored how the perturbation to a pacemaker neuron is buffered by the existence of a specific synaptic component of the network, a component that seemingly has very little effect on the existence or frequency of oscillations in the pacemaker.

Results from our laboratory (Figs. 16.2 and 16.3) and previous experimental studies of the stomatogastric pyloric network show that the inhibitory feedback LP to PD synapse to the pyloric pacemaker neurons has no effect on the average pyloric cycle period in control conditions (Mamiya & Nadim 2004; Zhou, LoMauro, & Nadim, (2006)) even if the synapse is drastically strengthened in the presence of neuromodulators (Thirumalai et al. 2006). We and others had previously proposed that such feedback inhibition may in fact act to stabilize the pyloric rhythm cycle period in response to perturbing inputs (Mamiya & Nadim 2004; Thirumalai et al. 2006). Our current work now demonstrates that the stabilizing effect of feedback inhibition is in fact quite generic and applies to counteract both longlasting perturbations(such as through neuromodulation) that tend to alter network frequency and fast perturbations typically modeled as noise.

Acknowledgments We thank Dr. Lian Zhou for earlier ideas on this work. Supported by NIH MH-60605 (FN), NSF DMS0615168 (AB) and a Fulbright–Nehru Fellowship (AB).

References

Achuthan, S., & Canavier, C. C. (2009). Phase-resetting curves determine synchronization, phase locking, and clustering in networks of neural oscillators. *J Neurosci, 29*(16), 5218–5233. doi: 29/16/5218 [pii] 10.1523/JNEUROSCI.0426–09.2009

Bartos, M., Vida, I., & Jonas, P.(2007). Synaptic mechanisms of synchronized gamma oscillations in inhibitory interneuron networks. *Nat Rev Neurosci, 8*(1), 45–56. doi: nrn2044 [pii] 10.1038/nrn2044

Borgers, C., & Kopell, N.(2003). Synchronization in networks of excitatory and inhibitory neurons with sparse, random connectivity. *Neural Comput, 15*(3), 509–538. doi: 10.1162/089976603321192059

Brown, E., Moehlis, J., Holmes, P., Clayton, E., Rajkowski, J., & Aston-Jones, G.(2004). The influence of spike rate and stimulus duration on noradrenergic neurons. *J Comput Neurosci, 17*(1), 13–29. doi: 10.1023/B:JCNS.0000023867.25863.a4 5273291 [pii]

Dickinson, P. S.(2006). Neuromodulation of central pattern generators in invertebrates and vertebrates. *Curr Opin Neurobiol, 16*(6), 604–614. doi: S0959–4388(06)00148–6 [pii] 10.1016/j.conb.2006.10.007

Ermentrout, Bard.(2002). *Simulating, analyzing, and animating dynamical systems: a guide to XPPAUT for researchers and students*. Philadelphia: Society for Industrial and Applied Mathematics.

Ermentrout, Bard, & Terman, David H.(2010). *Mathematical foundations of neuroscience*. New York: Springer.

Friesen, W. O.(1994). Reciprocal inhibition: a mechanism underlying oscillatory animal movements. *Neurosci Biobehav Rev, 18*(4), 547–553.

Grillner, S., Markram, H., De Schutter, E., Silberberg, G., & LeBeau, F. E.(2005). Microcircuits in action–from CPGs to neocortex. *Trends Neurosci, 28*(10), 525–533. doi: S0166–2236(05)00211–0 [pii] 10.1016/j.tins.2005.08.003

Guckenheimer, John, & Holmes, Philip.(1997). *Nonlinear oscillations, dynamical systems, and bifurcations of vector fields*(Corr. 5th print. ed.). New York: Springer.

Kintos, N., Nusbaum, M. P., & Nadim, F.(2008). A modeling comparison of projection neuron- and neuromodulator-elicited oscillations in a central pattern generating network. *J Comput Neurosci, 24*(3), 374–397. doi: 10.1007/s10827–007–0061–7

Mamiya, A., & Nadim, F.(2004). Dynamic interaction of oscillatory neurons coupled with reciprocally inhibitory synapses acts to stabilize the rhythm period. *J Neurosci, 24*(22), 5140–5150.

Manor, Y., Nadim, F., Epstein, S., Ritt, J., Marder, E., & Kopell, N. (1999). Network oscillations generated by balancing graded asymmetric reciprocal inhibition in passive neurons. *J Neurosci, 19*(7), 2765–2779.

Marder, E., & Calabrese, R. L.(1996). Principles of rhythmic motor pattern generation. *Physiol Rev, 76*(3), 687–717.

Mishchenko, E., & Rozov, N.(1997). *Differential Equations with Small Parameters and Relaxation Oscillations*. New York: Plenum Press.

Oprisan, S. A., Thirumalai, V., & Canavier, C. C.(2003). Dynamics from a time series: can we extract the phase resetting curve from a time series? *Biophys J, 84*(5), 2919–2928. doi: S0006–3495(03)70019–8 [pii] 10.1016/S0006–3495(03)70019–8

Pinsker, H. M.(1977). Aplysia bursting neurons as endogenous oscillators. I. Phase-response curves for pulsed inhibitory synaptic input. *J Neurophysiol, 40*(3), 527–543.

Prinz, A. A., Thirumalai, V., & Marder, E.(2003). The functional consequences of changes in the strength and duration of synaptic inputs to oscillatory neurons. *J Neurosci, 23*(3), 943–954.

Somers, D., & Kopell, N.(1993). Rapid synchronization through fast threshold modulation. *Biol Cybern, 68*(5), 393–407.

Thirumalai, V., Prinz, A. A., Johnson, C. D., & Marder, E.(2006). Red pigment concentrating hormone strongly enhances the strength of the feedback to the pyloric rhythm oscillator but has little effect on pyloric rhythm period. *J Neurophysiol, 95*(3), 1762–1770.

Wang, X. J., & Buzsaki, G.(1996). Gamma oscillation by synaptic inhibition in a hippocampal interneuronal network model. *J Neurosci, 16*(20), 6402–6413.

Whittington, M. A., Traub, R. D., Kopell, N., Ermentrout, B., & Buhl, E. H.(2000). Inhibition-based rhythms: experimental and mathematical observations on network dynamics. *Int J Psychophysiol, 38*(3), 315–336. doi: S0167876000001732 [pii]

Zhou, L., LoMauro, R. and Nadim, F.(2006). The interaction between facilitation and depression of two release mechanisms in a single synapse. *Neurocomputing, 69*, 1001–1005.

Chapter 17
Existence and Stability Criteria for Phase-Locked Modes in Ring Networks Using Phase-Resetting Curves and Spike Time Resetting Curves

Sorinel Adrian Oprisan

Abstract Both phase resetting curves (PRCs) and spike time resetting curves (STRCs) are tools used for predicting the phase locked modes in neural networks and investigating their stability. The PRC tabulates the relative change in the duration of the interspike intervals due to a presynaptic input. In networks of neurons that generate brief action potentials (spiking neurons) with pulsatile couplings, the first order PRC, which accounts only for the change in the duration of the interspike interval that includes the perturbation itself, accurately predicts the phase-locked modes. Although the STRC-based method is equivalent to the first order PRC, it has the advantages of unambiguously defining the temporal intervals in networks of dissimilar neurons and a straightforward geometric interpretation. When using the PRC to predict the phase-locked modes it is generally assumed that (1) the open loop stimulus is identical to the recursive inputs received by the same neuron in closed loop setup, (2) the isolated neural oscillator returns to its unperturbed limit cycle before receiving the next input, and (3) each neuron receives only one presynaptic input and sends its output to only one other neuron in the network. These assumptions limit the applicability of the PRC and the STRC methods to 1:1 phase-locked modes to ring networks. For spiking neurons, the PRC/STRC-based theoretical predictions regarding the phase-locked modes and their stability were in very good agreement with the closed loop numerical simulations. Since neurons that generate sustained and closely spaced trains of action potentials (bursts) maintain their postsynaptic coupling over a significant duration, their influence on the postsynaptic neurons extend beyond the first order PRC. For bursting neurons, the first two assumptions were successfully relaxed by considering higher order PRCs and the burst resetting curves (BRCs) in predicting the phase-locked modes in hybrid networks controlled via dynamic clamp.

S.A. Oprisan (✉)
Department of Physics and Astronomy, College of Charleston, 66 George Street, Charleston, SC 29424, USA
e-mail: oprisans@cofc.edu

N.W. Schultheiss et al. (eds.), *Phase Response Curves in Neuroscience: Theory, Experiment, and Analysis*, Springer Series in Computational Neuroscience 6,
DOI 10.1007/978-1-4614-0739-3_17, © Springer Science+Business Media, LLC 2012

1 Introduction

Both the phase resetting curves (PRCs) and the spike time resetting curves (STRCs) reduce the complexity of a neuron to a black box whose output is a train of regularly spaced and brief action potentials (spikes) sensitive to timing, or phase, of the external perturbations, such as injected currents, presynaptic inputs, neuromodulators, etc. (Canavier et al. 1997; Canavier et al. 1999; Dror et al. 1999; Oprisan and Boutan, 2008; Oprisan, 2010). The generality of both the PRC and the STRC methods stems from the fact that they can be applied to heterogeneous networks and are not limited to small amplitude and short duration (pulsatile, or instantaneous) couplings among neurons. Moreover, the PRC and the STRC methods can be used to predict phase-locked modes in neural networks even when the explicit mathematical models of component neurons are not available. In such cases, the PRC simply tabulates the experimentally measured transient changes in the firing rate of an isolated neuron (*open loop*) subject to an external stimulus identical to that received by the same neuron as part of a fully-connected network (*closed loop*) (Oprisan and Boutan, 2008; Oprisan, 2010). Depending on the strength and the type of coupling (inhibitory/excitatory) among neural oscillators, as well as the topology of the neural network (ring, all-to-all, etc.), neural oscillators can develop stationary phasic relationships throughout the entire network, which are called phase-locked modes. Memory storage (Haenschel et al. 2010; Rutishauser et al. 2010) and information processing (Zacksenhouse and Ahissar, 2006; Cheron et al. 2007; Wingerden et al. 2010) are important examples of phase-locking processes that take place in neural networks.

One of the main assumptions made when using both the PRC and the STRC methods to predict the phase-locked firing patterns of neural networks is that the external stimulus injected into an isolated neuron, for the purpose of determining its PRC or STRC in *open loop*, is identical to the input received by the same neuron under the recurrent inputs of a phase-locked mode when it is part of the network in *closed loop*.

An equally important assumption is that the isolated neural oscillator must very rapidly return to its unperturbed limit cycle following any external perturbation, so that its response to a given input is always the same for a given phase of the stimulation. This assumption can be relaxed by taking into consideration higher order PRCs (see Sect. 4). However, the reduced system becomes harder to analyze in this case.

Another important assumption, which can be relaxed based on our geometric method of predicting the PRCs in the case of multiple inputs per cycle (see Chap. 6 on geometric interpretation of the PRC), is that each neuron receives only one presynaptic input and sends its output to only one neuron in the network. Ring networks satisfy such constraints and were extensively studied as possible neural circuits capable of mimicking legged locomotion (Golubitsky et al. 1998, 1999; Buono, 2001; Nakanishi et al. 2004; Proctor and Holmes, 2010; Proctor et al. 2010). Furthermore, stimuli-induced resetting was considered the main mechanism that allowed ring networks to switch between multiple firing patterns

(Canavier et al. 1997, 1999; Golubitsky et al. 1998, 1999; Buono 2000; Hooper and DiCaprio, 2004; Marder and Bucher, 2007; Ijspeert, 2008; Xu, 2008).

Although the PRCs and the STRCs can be used to study arbitrarily large ring networks, we start with a very detailed study of two-neuron networks, or half centers, due to their biological relevance. *Half centers* are reciprocally inhibitory functional antagonists (Brown, 1914). Half centers can rely on tonic external stimuli to trigger oscillations or to set the appropriate frequency of the rhythm (Calabrese, 1995; Aiken et al. 2003), or rely on exogenous rhythmic input, as it is the case with the pyloric network of the lobster stomatogastric nervous system (Miller and Selverston, 1982). Other well-studied two-neuron reciprocally inhibitory network controls the heartbeat of the leech (Calabrese et al. 1995; Efimov, 2011) and the stomatogastric nervous system of the crab (Bartos et al. 1999; Thuma et al. 2003) and lobster (Harris-Warrick et al. 1992; Marder and Bucher, 2001; Hooper and DiCaprio, 2004; Marder and Bucher, 2007). In addition, larger networks often can be divided into loosely connected half centers. Therefore, predicting firing patterns of half centers based on the intrinsic properties of individual oscillators and their connections is biologically relevant.

For pedagogical reasons, and without reducing the generality of the results regarding the prediction of the phase-locked modes based on the PRC or the STRC, a generic two-dimensional neural oscillator was used both in open and closed loops. Throughout this chapter, we used a Morris–Lecar (ML) model oscillator, originally formulated to model action potentials in barnacle muscle (Morris and Lecar, 1981; Ermentrout, 1996). ML model neuron can switching between type 1 and type 2 excitability by changing the values of only three control parameters: V_3, V_4, and ϕ (see Appendix for a detailed mathematical model). Neural excitability refers to the capability of a neuron to fire repetitively in response to an externally applied current. In their seminal work on giant squid axon, Hodgkin and Huxley (Hodgkin and Huxley, 1952) experimentally identified three classes, or types, of axonal excitability: type 1, where the repetitive firing is controlled by the intensity of an external stimulus; type 2, where the firing frequency is almost independent of stimulus, intensity; and type 3, where there are no endogenous bursters regardless the intensity or the duration of the stimulus. For type 1 excitability, the onset of repetitive firing occurs at arbitrarily low frequencies, whereas for type 2 the repetitive firing occurs at nonzero frequencies. The type of membrane excitability also reflects the bifurcation structure leading to stable oscillations. Type 1 oscillations arise via an unstable saddle-node fix point, whereas type 2 oscillations are determined by a Hopf bifurcation of an unstable fix point (Hoppensteadt and Izhikevich, 1997; Izhikevich, 2007). The PRC of a neuron depends upon the type of excitability it exhibits (Ermentrout, 1996). In general, the voltage trace of a type 1 neuron suggests a spiking behavior and its PRC exhibits either only delays or only advances in response to an external perturbation (Fig. 17.1a). For the same perturbation, the voltage trace of a type 2 neuron shows a long lasting depolarization and its PRC exhibits both phase advances and delays, depending upon the stimulus phase. The long-lasting depolarization of a type 2 neuron mimics the envelope of the membrane voltage for a generic bursting neuron (see Fig. 17.8a).

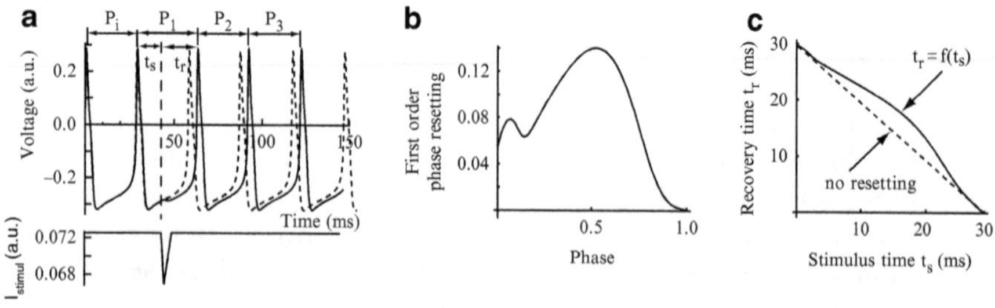

Fig. 17.1 (**a**) An external stimulus (lower panel trace) injected at a stimulus time t_s measured with respect to the most recent spike induces a transient change in the duration of the current cycle P_1 and possible long-lasting effects in the subsequent cycles P_2 and P_3. The dashed line represents the unperturbed firing pattern, whereas the continuous line is the perturbed pattern. (**b**) A typical first order PRC shows a delay (positive resetting) due to current cycle lengthening effect of the inhibitory pulse shown in panel **a**. (**c**) Similar information can be conveyed using a STRC that tabulates the recovery time t_r versus stimulus time t_s. The line of no resetting (*dashed line*) is given by $t_s + t_r = P_i$

The intrinsic, or free running, period P_i of a neural oscillator is the time interval between two successive crossings of an arbitrary membrane voltage threshold with the same slope in the absence of any external perturbation (Fig. 17.1a). Externally injected currents, chemical, or electrical synaptic inputs, neuromodulators, etc. induce transient changes in the firing period that could be captured by the PRCs (Ermentrout, 1996; Canavier et al. 1999; Dror et al. 1999; Galan et al. 2005; Govaerts and Sautois, 2006; Oprisan and Boutan, 2008; Oprisan, 2010), or the STRCs (Kopell and Ermentrout, 1988; Oprisan and Boutan, 2008; Oprisan, 2010). Usually, the most significant change in the period of oscillation occurs during the current cycle, P_1, which is the cycle that includes the perturbation, and is tabulated by the first order PRC (Fig. 17.1a). Long lasting transient changes of the intrinsic firing period P_i can be captured by tabulating the interspike intervals for the subsequent cycles P_2, P_3, etc. For limit cycle oscillators, a stimulus phase variable could be unambiguously and conveniently defined as the normalized stimulus time with respect to the intrinsic period of oscillation $\varphi = t_s/P_i$. The reference for the stimulus time (zero phase) is arbitrary and it can be selected as the most depolarized/hyperpolarized membrane potential, a crossing of an arbitrary voltage threshold with a positive/negative slope, etc. In all numerical simulations, the stimulus time reference was set by a crossing of $V_{thr} = 0\,\mathrm{mV}$ voltage threshold with a positive slope (Fig. 17.1a). The PRC measures the relative change in the interspike interval due to an external stimulus. For example, the first order transient PRC is defined by (Fig. 17.1b):

$$F^1(\varphi) = P_1/P_i - 1, \tag{17.1}$$

where P_1 is the transiently modified firing period of the current cycle, and $F^1(\varphi)$ measures the relative advance ($F^1(\varphi) < 0$) or delay ($F^1(\varphi) > 0$) of the subsequent spike due to a perturbation applied at phase $\varphi = t_s/P_i$ during the current cycle. Long lasting transients are quantified by higher order PRCs.

The STRC tabulates the recovery time t_r, which is the time it takes for a neuron to fire again after the most recent input, versus the stimulus time t_s (Fig. 17.1c) (Kopell and Ermentrout, 1988; Galan et al. 2005; Oprisan and Boutan, 2008; Oprisan, 2010):

$$t_r = f(t_s). \tag{17.2}$$

Based on (17.1), the transiently modified firing period is $P_1 = (1 + F^1(\varphi))P_i$, which is also equal to the sum of the stimulus and the recovery times $P_1 = t_s + t_r = t_s + f(t_s)$. Therefore, the first order PRC, $F^1(\varphi)$, and the STRC, $f(t)$, are related through:

$$F^1(\varphi) = (t_s + f(t_s))/P_i - 1. \tag{17.3}$$

From a mathematical point of view, all the existence and stability criteria derived by using the first order PRC could be rewritten in terms of the STRC. From a computational and geometric point of view, it is sometimes more efficient and intuitive to use the STRC instead of the PRC. However, the equivalence between the PRC- and STRC-based methods breaks down when stimuli affect multiple subsequent cycles. In such cases, higher order PRC-based methods must be used in order to correctly predict the phase-locked modes of the network (see Sect. 4).

In order to maintain the necessary consistency with the mathematical notations used in other chapters, an upper numerical index always refers to the order of the PRC, e.g., F^1 is the first order PRC. A lower numerical index identifies the neural oscillator in a network, e.g., t_{s1} refers to the stimulus time t_s of the input received by neuron "1" and P_{12} refers to the transiently modified period of the current cycle P_1 for the neuron labeled "2". Open brackets () are reserved for functional arguments, as in $F(\varphi)$ or $f(t)$, whereas closed brackets [] label the firing cycle counted from an arbitrary reference. The notation becomes particularly cumbersome when considering higher order corrections to the PRC (see Sect. 4).

Most of the work in this chapter arises from (Oprisan and Canavier, 2000, 2001, 2002, 2003, 2005, 2006; Oprisan et al. 2003, 2004; Oprisan and Boutan, 2008; Oprisan, 2009, 2010).

2 Entrainment of Spiking Neural Oscillators

In order to investigate the usefulness of the PRC and the STRC methods to predict the phase-locked modes in a ring networks, we first investigate the entrainment in which one neuron drives a postsynaptic neuron without a feedback loop. In this case, the two neurons maintain a constant phase difference between them and, therefore, generate a 1:1 phase-locked firing pattern, which was called 1:1 entrainment in order to emphasize that the feedback loop is missing. The two circles symbolically represent the open loop network with unidirectional coupling through a synapse marked by a small solid circle (Fig. 17.2a). The spikes are represented by thick vertical lines along a horizontal temporal axis running from left to right. The input sent by the driving neuron "2" to the driven neuron "1" is marked by vertical arrows.

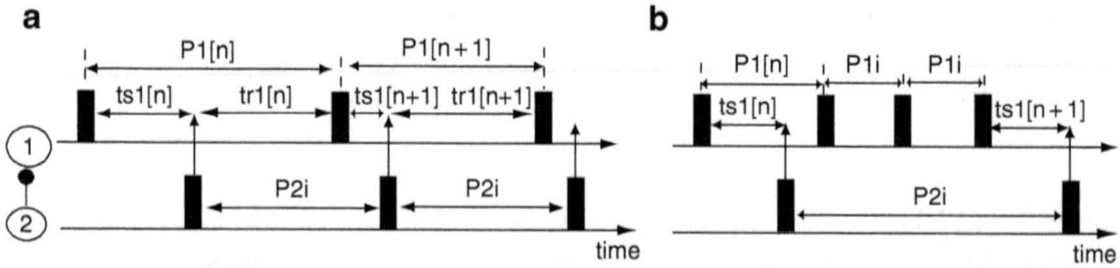

Fig. 17.2 (**a**) A schematic representation of two neural cells (*empty circles* labeled "1" and, respectively, "2") with a synapse (*closed circle*) from the driving neuron "2" to the driven neuron "1". To the right, the temporal evolution of the spikes generated by each neuron shows a possible 1:1 entrainment. The vertical thick lines on each temporal axis mark a spike of the corresponding neuron shown at the left side. Every synaptic input from neuron "2" to neuron "1" induces a phase resetting and is marked by a vertical arrow pointing upward (from the driving to the driven neuron). (**b**) A possible 3:1 entrainment mode for the same two-neuron network occurs when the two free-running periods are very dissimilar

2.1 *PRC-based Existence and Stability Criteria for 1:1 Entrainment of a Spiking Neuron*

Based on Fig. 17.2a, the transient interspike interval P_1 of the driven neuron is:

$$t_{s1}[n] + (P_{2i} - t_{s1}[n+1]) = P_1[n], \tag{17.4}$$

which states that the sum of the stimulus time $t_{s1}[n]$ and the recovery time $t_{r1}[n] = P_{2i} - t_{s1}[n+1]$ must add up to the firing period $P_1[n]$ of the driven neuron and n is cycle's index. Using the definition (17.3), the recursion (17.4) becomes:

$$t_{s1}[n+1] = t_{s1}[n] + P_{2i} - P_1[u] = t_{s1}[n] + P_{2i} - \left(1 + F^1{}_1(t_{s1}[n]/P_{1i})\right) P_{1i}, \tag{17.5}$$

which gives the PRC-based *recursive existence criterion* for 1:1 entrainment:

$$\varphi_1[n+1] = \varphi_1[n] - F^1{}_1(\varphi_1[n]) + P_{2i}/P_{1i} - 1. \tag{17.6}$$

For identical neurons, $P_{1i} = P_{2i} = P_i$, (17.6) reduces to:

$$\varphi_1[n+1] = \varphi_1[n] - F^1{}_1(\varphi_1[n]). \tag{17.7}$$

Every state of a neural network must obey the *causality condition*: the response due to a perturbation cannot precede the perturbation itself, which for entrainment patterns means $P_1[n] \geq t_{s1}[n]$. Since $(1 + F^1{}_1(\varphi_1[n]))P_{1i} \geq t_{s1}[n]$, the causality condition in terms of oscillator's phases is $F^1{}_1(\varphi_1[n]) \geq \varphi_1[n] - 1$ (dashed line at 45° in Fig. 17.3a).

Any 1:1 phase-locked mode is a fixed point φ^* of the *recursive existence criterion* (17.6):

$$\varphi_1{}^* = \varphi_1{}^* - F^1{}_1(\varphi_1{}^*) + P_{2i}/P_{1i} - 1 \Rightarrow F^1{}_1(\varphi_1{}^*) = P_{2i}/P_{1i} - 1. \tag{17.8}$$

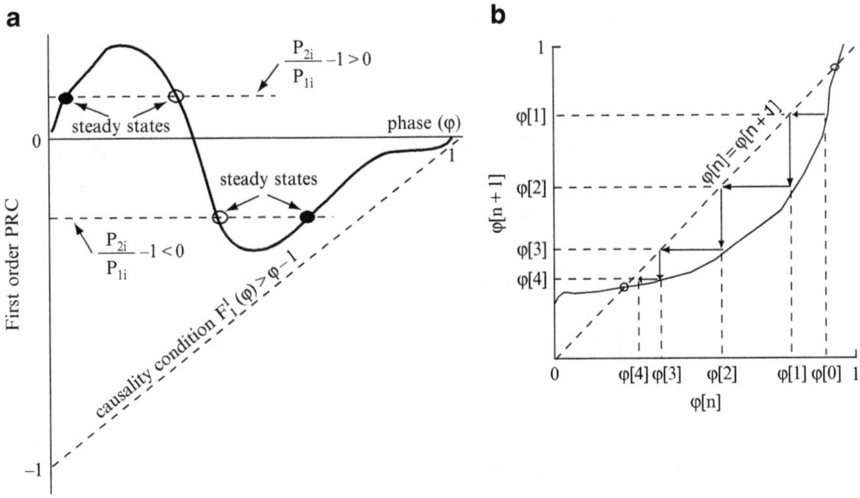

Fig. 17.3 (**a**) A typical PRC for a type 2 neuron has regions where the current cycle is both lengthened (positive resetting) and shortened (negative resetting) depending on the phase of the perturbation. For a given synaptic strength, a steady 1:1 entrainment can occur only if $F_{\min} < P_{2i}/P_{1i}-1 < F_{\max}$, where $F_{\min}$ and $F_{\max}$ are the minimum and maximum values of the PRC, respectively. Generally, there are multiple possible 1:1 entrainment solutions: a stable (*solid circle*) entrainment at positive slopes of the PRC and an unstable entrainment (*open circle*) at negative slopes. (**b**) The solution of the existence criterion could also be determined using a cobwebbing technique between $\varphi - F^1(\varphi_1)$ (*solid line*) and the bisectrix (dashed line at 45°)

Geometrically, the fixed points, or steady-states φ^*, determined by (17.8) are at the crossing of the PRC with the horizontal dashed lines $P_{2i}/P_{1i}-1$ in Fig. 17.3a. In particular, for identical neurons, the 1:1 entrainment phase is the solution of $F^1{}_1(\varphi_1{}^*) = 0$. During a 1:1 stable entrainment, both neurons fire with the same entrained period, which is the free-running period P_{2i} of the driving neuron (Oprisan and Boutan, 2008; Oprisan, 2010).

According to the existence criterion (17.8), it is possible to have multiple intersections of the line $P_{2i}/P_{1i}-1$ with the PRC of the entrained neuron (Fig. 17.3a). In such a case, the question is: *which entrainment is stable to small perturbations*? In order to determine if a steady entrainment at a certain phase $\varphi^* = t_{s1}{}^*/P_{\mathrm{i}}$ is stable, it is assumed that a small perturbation in the entrainment time occurred due to varying factors (biological noise, peripheral input, etc.) and slightly perturbed the stimulus time $t_{s1}[n] = t_{s1}{}^* + \delta t_{s1}[n]$, with $\delta t_{s1}[n] \ll t^*_{s1}$. Assuming that the PRC is a smooth function, a standard linear stability analysis of the fixed points in 1-D maps can be performed by using a Taylor series expansion of the PRC around the steady 1:1 entrainment solution $t_{s1}{}^*$, which gives

$$F^1{}_1(t_{s1}[n]/P_{1i}) = F^1{}_1\left(\left(t_{s1}{}^* + \delta t_{s1}[n]\right)/P_{1i}\right) \approx$$
$$F^1{}_1\left(t_{s1}{}^*/P_{1i}\right) + m^1{}_1\delta t_{s1}[n]/P_i + O\left(\left(\delta t_{s1}[n]/P_{1i}\right)^2\right), \qquad (17.9)$$

where $m^1{}_1$ is the slope of the first order PRC of the driven neuron at the steady entrainment phase $\varphi_1{}^* = t_1^*/P_{1i}$ and $O((\delta t_{s1}[n]/P_{1i})^2)$ collects all nonlinear terms in a Taylor series expansion. By substituting (17.9) into (17.7), one obtain the following recursion for the stimulus time perturbation δt_{s1}, which is the *recursive stability criterion* (Oprisan and Boutan, 2008; Oprisan, 2010):

$$\delta t_{s1}[n+1] = \left(1 - m^1{}_1\right)\delta t_{s1}[n] = \lambda_{PRC}\delta t_{s1}[n], \qquad (17.10)$$

with the solution $\delta t_{s1}[n] = (1 - m^1{}_1)^n \delta t_{s1}[0] = \lambda^n_{PRC}\delta t_{s1}[0]$. As a result, the initial perturbation $\delta t_{s1}[0]$ dies out and the steady entrainment $\varphi_1{}^*$ is stable if the stability index $\lambda_{PRC} = |1 - m^1{}_1| \le 1$. Otherwise, the steady-state $\varphi_1{}^*$ is unstable and any perturbation grows exponentially (in linear approximation) over time. The *stability criterion* gives the stability index values, λ_{PRC}, by explicitly solving the recursive stability criterion (17.10). For 1:1 entrainment, the stability criterion is $\lambda_{PRC} = |1 - m^1{}_1| \le 1$, and it breaks down to two important subcases: (1) $0 \le m^1{}_1 \le 1$, which means that the perturbation $\delta t_{s1}[n]$ monotonically dies out and all perturbations have the same signs, and (2) $1 \le m^1{}_1 \le 2$, which means that the perturbation δt_{s1} still dies out but with alternating signs. The latter case corresponds to a stable entrainment with alternating leading neurons that flip their firing order from one cycle to the next and determine a combination of 2:1 and 1:2 entrainments. Stable fixed points are marked by solid circles in Fig. 17.3a. Based on the existence criterion (17.8), the steady-states are found at the intersections of the line $P_{2i}/P_{1i} - 1$ and the PRC of driven neuron. The PRC of the entrained neuron has a finite amplitude $F_{min} < F^1{}_1(\varphi) < F_{max}$, that depends on the strength of the synaptic coupling. As a result, the relative interspike intervals of the two neurons must obey $(1 + F_{min}) \le P_{2i}/P_{1i} \le (1 + F_{max})$ in order to lead to a steady 1:1 entrainment for a given synaptic coupling. Therefore, the neural networks with very dissimilar intrinsic firing periods could still lead to 1:1 entrainment by increasing the synaptic strength among them.

We leave it as an exercise for our readers to show that, based on the relationship (17.3) between the first order PRC and the STRC, and the relationship between the phase and the stimulus time $\varphi = t_{s1}/P_i$, (17.4) becomes the STRC-based *recursive existence criterion*:

$$t_{s1}[n+1] + f_1(t_{s1}[n]) = P_{2i}. \qquad (17.11)$$

In a similar manner, we leave to our readers to show that, based on (17.3), the slopes, m^1, of the first order PRC, $F^1(\varphi)$, and the slope, n, of the STRC, $f(t_s)$, are related through $m^1 = 1 + n$. As a result, based on the already derived PRC-based stability criterion $\lambda_{PRC} = |1 - m^1{}_1| \le 1$, it follows that the STRC-based stability criterion for 1:1 entrainment is $\lambda_{STRC} = |-n_1| \le 1$.

2.2 PRC-based Geometric Interpretation of the Existence and the Stability Criteria for 1:1 Entrainment of a Spiking Neuron

In addition to the geometric visualization of the PRC with its steady-states determined by (17.8) as shown in Fig. 17.3a, the entrainment phase $\varphi_1{}^*$ could be also determined by a cobwebbing method based on the recursive existence criterion (17.6). For this purpose, a new PRC-based function, equal to the right hand side of (17.6) was defined:

$$g(\varphi_1) = \varphi_1 - F^1{}_1(\varphi_1) + P_{2i}/P_{1i} - 1. \tag{17.12}$$

Successive iterations of an arbitrary initial condition $\varphi_1[0]$ determine the next stimulus phase $\varphi_1[1]$ in agreement with the recursive existence criterion (17.6). The newly determined stimulus phase $\varphi_1[1]$ along the vertical axis must become the next initial condition for the map (17.6), and this is accomplished geometrically with the aid of the horizontal arrow between the map (17.6) and the dashed line $\varphi_1[n+1] = \varphi_1[n]$, which is the bisectrix of the plane. The vertical arrow between the bisectrix and the map (17.6) determines the subsequent firing phase $\varphi_1[2]$ and so forth. The two intersections of (17.12) and the bisectrix determine the possible 1:1 entrained phases (steady-states, or fixed points) of the network. The generic PRC shown in Fig. 17.3 was used to numerically compute (17.12), which is represented by a continuous line in Fig. 17.3b. At the crossing between $g(\varphi_1)$ and the bisectrix there are two steady-states: (1) a stable fixed point that attracts all nearby iterations marked by a solid circle, and (2) an unstable fixed point that repels all nearby iterations marked by an empty circle. In the above examples, the two steady-states are global attractors, respectively, repellers of successive iterations. For neurons with a more complex PRCs, a higher number of steady-states are possible and each steady-state has a local basin of attraction with multiple coexisting firing patterns (Canavier et al. 1997, 1999). Neural networks with multiple stable steady-states were found in CPGs involved in respiration, locomotion, etc. Due to external stimuli, such as visual, auditory, olfactory, etc. cues, the same neural network could switch between walking, trotting, jumping, running, etc. (Eldridge, 1991; Taiga et al. 2003; Xu, 2008)

2.3 Numerical Verification of the Existence and the Stability Criteria for 1:1 Entrainment of a Spiking Neuron

The validity of the existence and the stability criteria derived above was tested numerically with a two-neuron network of type 1 ML model neurons (see Appendix for detailed equations and implementation). We considered instantaneous one way, i.e., rectifying, electrical coupling described by a current $I_{syn} = g_{syn}(V_{driver} - V_{syn})$,

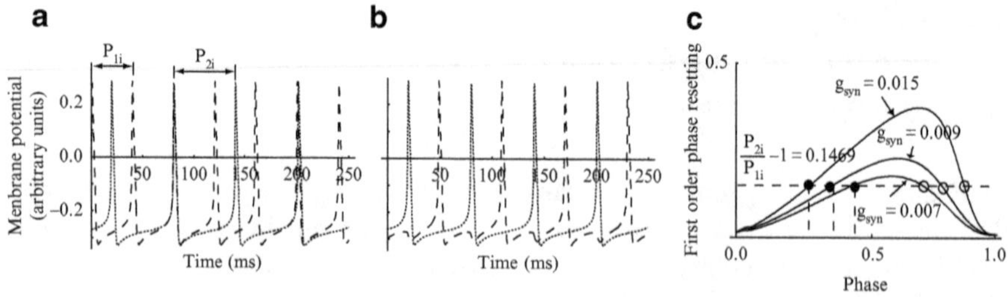

Fig. 17.4 (**a**) Membrane potential traces from two uncoupled type 1 ML model neurons running with different intrinsic periods P_{1i}, respectively, P_{2i}. (**b**) After coupling, the entrained neuron fires with the same period as the driving neuron and maintains a fix phase difference with respect to the driving neuron. (**c**) The amplitude of the PRC is proportional to the synaptic coupling. The predicted 1:1 entrainment phase, which is at the crossing between the PRC and the horizontal dashed line P_{2i}/P_{1i}, decreases for stronger couplings. Solid and open circles indicate stable and unstable steady 1:1 entrainments, respectively

where V_{driver} is the membrane potential of the driving neuron and V_{syn} is the synaptic voltage (Oprisan and Boutan, 2008; Oprisan, 2010). The two uncoupled neurons have different intrinsic periods (Fig. 17.4a) and spike durations. By turning on the synaptic coupling, the driven neuron adjusts its firing pattern such that its new driven firing period equals the intrinsic period of the driving neuron and they fire out of phase (Fig. 17.4b).

The open loop PRC was numerically computed by applying a single current pulse similar to the synaptic current the same neuron would receive in a close loop setup. The predicted phase difference between neurons was numerically computed from the PRCs (Fig. 17.4c). According to the graphical method highlighted in the preceding sections, the steady 1:1 entrainment, if it exists, is determined by the intersection between the first order PRC, $F^1{}_1(\varphi)$, and the horizontal line $P_{2i}/P_{1i}-1$, which for the particular case shown in Fig. 17.4c is $60.37/52.6349-1 = 0.1456$. For strong enough couplings, the driving neuron effectively entrains the driven neuron. According to Fig. 17.4c, the minimum synaptic strength $g_{syn,min}$ that produces 1:1 entrainment is determined by the condition $P_{2i}/P_{1i} - 1 = F^1{}_{1\,max}$ where the maximum amplitude of the PRC depends upon the synaptic strength. In general, a larger ratio of the two intrinsic periods P_{2i}/P_{1i} requires a stronger synaptic coupling in order to force the two neural oscillators to maintain a steady firing pattern. If the synaptic coupling is above the minimum value required for an effective 1:1 entrainment, then there are two intersections between the PRC and the horizontal line $P_{2i}/P_{1i} - 1$ (see Fig. 17.4c). If the slope of the PRC at the entrainment point is $0 < m^1{}_1 < 2$ then the entrainment pattern is stable, otherwise it is unstable. Figure 17.4c also predicts that by increasing the synaptic coupling, the steady phase difference between the two neurons decreases. The error between the theoretically predicted 1:1 entrainment phase based on the PRC method and the actual entrainment phase increases with the strength of the synaptic coupling for g_{syn} in the range (0, 0.02) a.u., but the error remains under 5% (data not shown). The explanation is that higher order PRCs become more significant while increasing the coupling strength. The coupled differential equations for the two neurons in closed loop were numerically integrated in order to determine the actual time difference between spikes during the 1:1 entrainment.

2.4 Generalized PRC-based Existence and Stability Criteria for 1:p Entrainment of a Spiking Neuron

For unidirectional coupling, it is possible to derive general existence and the stability criteria for the firing patterns consisting of one spike of the driving neuron followed by p spikes of the driven neuron (Oprisan and Boutan, 2008; Oprisan, 2010). Assuming that the driven neuron returns to its unperturbed limit cycle before receiving the next input, a $1:p$ entrainment means that the driving neuron must have an intrinsic period $(p-1)P_{1i} \leq P_{2i} \leq p\,P_{1i}$ such that it perturbs the driven neuron exactly once in very p periods (see Fig. 17.2b). Based on Fig. 17.2b, which shows a 1:3 entrainment, the *recursive existence criterion* for a $1:p$ entrainment writes:

$$t_{r1}[n] + (p-1)P_{1i} + t_{s1}[n+1] = P_{2i}. \tag{17.13}$$

Using the PRC-based approach, the recursive existence criterion (17.13) generalizes (17.6) to:

$$\varphi_1[n+1] = \varphi_1[n] - F^1{}_1(\varphi_1[n]) + P_{2i}/P_{1i} - p, \tag{17.14}$$

which leads to the *1:p existence criterion* that gives the steady-state solution $\varphi_1{}^*$:

$$F^1{}_1(\varphi_1{}^*) = P_{2i}/P_{1i} - p. \tag{17.15}$$

The $1:p$ *stability criterion* is identical to (17.10) for 1:1 entrainment and states that the stability index is $\lambda_{PRC} = |1 - m^1{}_1| \leq 1$. However, because the steady-states are at the crossing between the line $P_{2i}/P_{1i} - p$ and the PRC of the driven neuron, the relative periods of the two neurons must obey $(p + F_{min}) \leq P_{2i}/P_{1i} \leq (p + F_{max})$ in order to lead to a steady 1:1 entrainment. In conclusion, for the same strength of the synaptic coupling, i.e., the same range (F_{min}, F_{max}) of the first order PRC, the network selects the appropriate $1:p$ entrainment based on the ratio of the intrinsic firing periods P_{2i}/P_{1i}.

3 Existence and Stability Criteria for 1:1 Phase-Locked Modes in Ring Networks of Spiking Neurons Using the STRC

Coupling spiking neurons with fast synapses produces brief synaptic currents that induce soft resetting. Such brief synaptic inputs change the firing period of the postsynaptic neuron and their effect can be captured entirely by the first order PRC

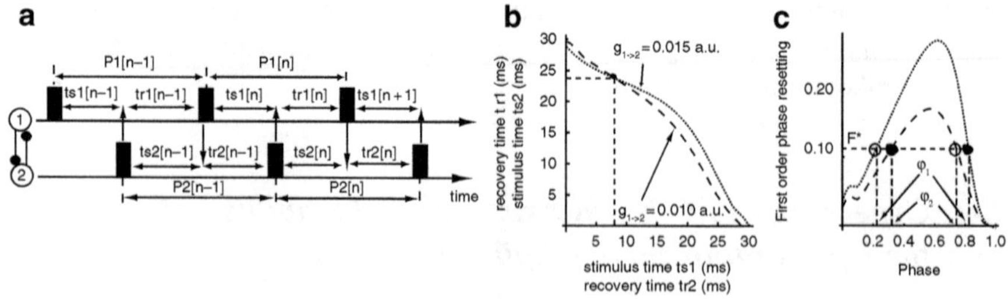

Fig. 17.5 (**a**) A schematic representation of the recurrent inhibition of the postsynaptic neuron in a two-neuron reciprocally coupled network. Solid vertical arrows indicate the moment a neuron generates an action potential and sends an inhibition to its postsynaptic neuron. The recovery interval for the postsynaptic neuron t_r is the same as the stimulus time t_s of the presynaptic neuron. (**b**) The STRC of one neuron must be overlapped with the STRC of the other neuron with the axes flipped in order to satisfy the existence criterion. The predicted 1:1 phase-locked mode is at the intersection of the two STRCs. (**c**) The PRCs for the asymmetric network of two neurons gives four crossing points with any horizontal dashed line representing a possible stable steady-state resetting F^*. After checking both the existence and stability criteria, only one combination (*solid circles*) gives a stable 1:1 phase-locked mode

or, equivalently, by the STRC. The advantage of the STRC-based prediction of phase-locked modes over the first order PRC method is its straightforward geometric interpretation. The previous section started with a PRC-based approach and left some STRC-based derivations as exercises for our readers. In this section, we will exploit the straightforward geometric interpretation of the STRC and leave some PRC-based derivations as exercises.

3.1 *STRC-based Existence and Stability Criteria for 1:1 Phase-Locked Modes in a Two-neuron Network of Spiking Neurons*

In a two-neuron network of reciprocally coupled spiking neurons, the recovery time of one neuron is the same as the stimulus time of the other neuron (Fig. 17.5a):

$$t_{s1}[n] = t_{r2}[n] \quad \text{and} \quad t_{s2}[n] = t_{r1}[n-1]. \tag{17.16}$$

Using the STRC definitions $t_{r2}[n] = f_2(t_{s2}[n])$ and $t_{r1}[n] = f_1(t_{s1}[n])$, (17.16) becomes:

$$t_{s1}[n] = f_2(t_{s2}[n]) \quad \text{and} \quad t_{s2}[n] = f_1(t_{s1}[n-1]). \tag{17.17}$$

By substituting the second equation of (17.17) into the first, one finds the *recursive existence criterion*:

$$t_{s1}[n] = f_2(f_1(t_{s1}[n-1])), \tag{17.18}$$

which determines the nth cycles stimulus time based on a known stimulus time for the previous cycle. The *STRC-based existence criterion* gives the fixed points of the recursive existence criterion (17.18):

$$t_{s1}{}^* = f_2\left(f_1\left(t_{s1}{}^*\right)\right). \tag{17.19}$$

The STRC-based existence criterion (17.19) has a straightforward geometric interpretation: the 1:1 phase-locked mode is found at the intersection of the two STRCs in which one of them has the axes flipped in order to satisfy the existence criterion (17.16) (Fig. 17.5b) (Oprisan and Boutan, 2008; Oprisan, 2010). For example, in Fig. 17.5b we plotted the STRC of neuron "1" with the recovery time on the vertical axis and the stimulus time along the horizontal axis. For the second neuron, we plotted its STRC with the axes flipped, i.e., the recovery time along the horizontal axis and the stimulus time along the vertical. In our example, the only predicted phase-locked mode occurs for $t_{s1}{}^* = 8.5$ ms and $t_{s2}{}^* = 23.5$ ms and corresponds to an asymmetric coupling of two identical type 1 ML model neurons with synaptic couplings $g_{2\to1} = 0.010$ a.u., $g_{1\to2} = 0.015$ a.u. (Fig. 17.5b).

According to the STRC-based existence criterion (17.19), it is possible to have multiple intersections of the two STRCs. In this case, the question is: *which phase-locked mode is stable to small perturbations?* In order to determine if a phase-locked mode is stable to small perturbations due to varying factors, such as biological noise, peripheral input, etc., a linear stability analysis of all fixed points of (17.18) was performed. Let the perturbed stimulus time during the nth cycle be $t_{s1}[n] = t_{s1}{}^* + \delta t_{s1}[n]$, where the perturbation $\delta t_{s1}[n]$ from the steady-state values $t_{s1}{}^*$ is assumed to be very small ($|\delta t_{s1}[n]| \ll t_{s1}{}^*$). Assuming that the STRCs are smooth functions, Taylor's expansion around $t_{s1}{}^*$ gives:

$$\begin{aligned} f_1(t_{s1}[n-1]) &= f_1\left(t_{s1}{}^* + \delta t_{s1}[\mathrm{n}-1]\right) = f_1\left(t_{s1}{}^*\right) + (\partial f_1/\partial t_{s1})_{t_{s1}*}\delta t_{s1}[n-1] \\ &+ O((\delta t_{s1}[n-1])^2) \approx t_{s2}{}^* + n_1\,\delta t_{s1}[n-1], \end{aligned} \tag{17.20}$$

where the last relationship is a consequence of (17.16), and n_1 is the slope of the STRC of the first neuron at steady-state. By substituting the above relationship into (17.18) one finds:

$$\begin{aligned} t_{s1}[n] &= f_2(f_1(t_{s1}[n-1])) \approx f_2\left(t_{s2}{}^* + n_1\,\delta t_{s1}[n-1]\right) \\ &= f_2(t_{s2}{}^*) + (\partial f_2/\partial t_{s2})_{t_{s2}}{}^* n_1 \delta t_{s1}[n-1] \\ &+ O\left(\delta t_{s1}{}^* + n_2 n_1\right)\delta t_{s1}[n-1], \end{aligned} \tag{17.21}$$

which leads to the *STRC-based recursive stability criterion:*

$$\delta t_{s1}[n] = n_2 n_1 \delta t_{s1}[n-1] = \lambda_{\mathrm{STRC}}\delta t_{s1}[n-1]. \tag{17.22}$$

The solution of (17.22) is $\delta t_{s1}[n] = \lambda_{\mathrm{STRC}}{}^n\,\delta t_{s1}[0]$ and describes the evolution of a small perturbation $\delta t_{s1}[0]$ from the steady-state $t_{s1}{}^*$, which is the phase-locked mode, over n cycles. If the stability index $|\lambda_{\mathrm{STRC}}| = |n_2 n_1| < 1$, then the initial perturbation decays over time and the 1:1 phase-locked mode is stable, otherwise the mode is unstable.

As in the previous section, we leave it as an exercise for our readers to show that, based on the relationship (17.3) between the first-order PRC and the STRC, and the relationship between the phase and the stimulus time $\varphi = t_{s1}/P_i$, the PRC-based *recursive existence criterion* (17.16) rewrites as:

$$\varphi_1[n]P_{1i} + \varphi_2[n+1]P_{2i} = \left(1 + F^1{}_1(\varphi_1[n])\right) P_{1i},$$
$$\varphi_2[n]P_{2i} + \varphi_1[n]P_{1i} = \left(1 + F^1{}_2(\varphi_2[n])\right) P_{2i}. \qquad (17.23)$$

In a similar manner, we leave to our readers to show that, based on the relationship $m^1 = 1 + n$ between the slopes m^1 of the first order PRC, $F^1(\varphi)$, and the slope, n, of the STRC, $f(t_s)$, the PRC-based *recursive stability criterion* (17.22) rewrites:

$$\delta\varphi_2[n+1] = \left(1 - m^1{}_1\right)\left(1 - m^1{}_2\right)\delta\varphi_2[n] = \lambda_{PRC}\delta\varphi_2[n], \qquad (17.24)$$

where the stability index is $\lambda_{PRC} = (1 - m^1{}_1)(1 - m^1{}_2)$ and $m^1{}_k$ is the slope of the first order PRC for neuron k. As before, the solution of (17.24) is $\delta\varphi_2[n] = \lambda^n_{PRC}\delta\varphi_2[0]$, and it means that the phase-locked mode is stable to small perturbations if the PRC-based stability index $|\lambda_{PRC}| = |(1 - m^1{}_1)(1 - m^1{}_2)| < 1$, and unstable otherwise. For identical neurons, the stability criterion (17.28) reduces to $(1-m)^2 < 1$, which leads to $0 < m < 2$.

The PRC-based method has also a geometric interpretation, but it is not as intuitive as the STRC-based approach and requires two steps: (1) determine all pairs (φ_1, φ_2) that satisfy $F^1{}_1(\varphi_1) = F^1{}_2(\varphi_2) = F^*$ (the horizontal dashed line in Fig. 17.5c), and (2) of all solutions (φ_1, φ_2), retain only the pairs that also satisfy $\varphi_1 + \varphi_2 = 1 + F^*$. For example, the asymmetrically coupled two-neuron network with synaptic couplings $g_{2\to1} = 0.010$ a.u., $g_{1\to2} = 0.015$ a.u. gives the steady-state phases $\varphi_1{}^* = 0.82$, which corresponds to $t_{s1}{}^* = \varphi_1{}^*\ P_{i1} = 0.82 \times 28.8$ ms $=$ 23.6 ms, and $\varphi_2{}^* = 0.27$, which corresponds to $t_{s2}{}^* = \varphi_{12}\ P_{i2} = 0.27 \times 28.8$ ms $=$ 7.8 ms. The actual phase-locked mode obtained by effectively integrating the ML model equations gave $t_1 = 21$ ms, and respectively, $t_2 = 12$ ms.

3.2 *Numerical Verifications of the Existence and the Stability Criteria for 1:1 Phase-Locked Modes in a Two-neuron Network of Spiking Neurons*

In order to check the validity of both the existence and the stability criteria, we compared the predicted phase-locked modes obtained from open loop PRCs/STRCs against the results of numerical simulations for a closed loop asymmetric two-neuron network made of two identical type 1 ML model neurons. The current injected into the driven neuron during the open loop was $I_{syn} = g_{syn}(V_{pre} - E_{syn})$, where V_{pre} is the membrane potential of the presynaptic neuron E_{syn} is the reversal potential of the synapse (Oprisan and Boutan 2008; Oprisan, 2010). Since the firing period of any type 1 neuron is sensitive to the bias current, we tested our predictions

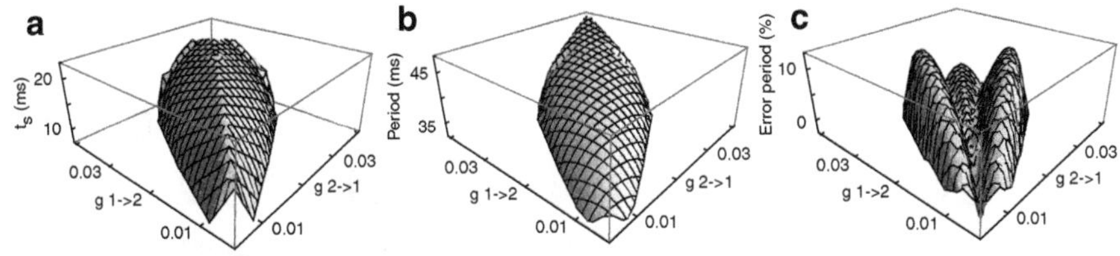

Fig. 17.6 Exploration of a wide range of synaptic couplings for asymmetric two-neuron networks of type 1 ML model neurons. Phase-locked mode predictions both for stimulus time (**a**) and firing period (**b**) gave a relatively small error ($< 10\%$) between the PRC-based prediction and the actual firing period of the coupled neurons (**c**). For strongly asymmetric coupling higher order correction to the PRC method are required to reduce the prediction errors

against closed loop results for multiple bias currents. All predictions were in very good agreement with the closed loop numerical simulations and, therefore, only the results for a bias current of $I_0 = 0.0725$ a.u. are shown in Fig. 17.6. As expected, all graphs are symmetric with respect to the permutation of the two neurons. For symmetric coupling, the steady stimulus time of the first neuron, which is the same as the recovery time for the second neuron, increases by increasing the synaptic conductance (Fig. 17.6a). Asymmetric combinations with one very strong synapse and the other very weak do not give a stable 1:1 firing pattern. The explanation is that, for any asymmetric synaptic couplings, the phase φ_1 at which one neuron is entrained via the weak synapse must have a very weak resetting $F^1{}_1(\varphi_1) = F^* \approx 0$, and, at the same time, the phase φ_2 at which the other neuron is entrained by the very strong synapse must have a large resetting $F^1{}_2(\varphi_2) = F^*$, which cannot be both satisfied at the same time. As expected, the recovery from a stronger inhibition takes longer and it impacts the phase-locked period, which also increases with the synaptic conductance (Fig. 17.6b). The prediction error for the phase-locked period below 10% (Fig. 17.6c).

3.3 STRC-based Generalization of the Existence and the Stability Criteria for 1:1 Phase-Locked Modes in a Ring Network of N Spiking Neurons

Another advantage of the STRC-based method, in addition to a simple geometric interpretation of the phase-locked modes for two-neuron network, is that the generalization of the existence and stability criteria for 1:1 phase-locked modes in a ring network of N neurons is straightforward.

For this purpose, we first stated the rules used for labeling the firing cycles: we selected an arbitrary "leading" neuron and labeled it "1" (see Fig. 17.7a for a three-neuron network) and chose to attribute the same cycle index to the other neurons that spike after the leading neuron. For example, in a three-neuron network with leading neuron "1" and the geometry shown in Fig. 17.7a, during the nth cycle, the neuron "1" spikes earlier than "2", which spikes earlier than "3".

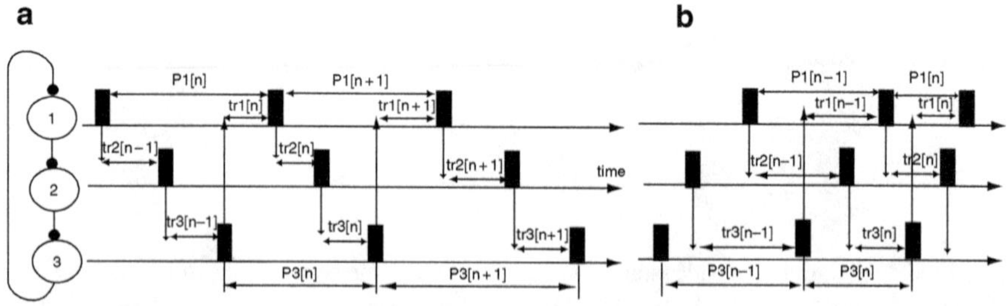

Fig. 17.7 A schematic representation of the three-neuron ring network. In such a network, two possible firing patterns can emerge: (**a**) $1 \to 2 \to 3$, respectively, (**b**) $3 \to 2 \to 1$

Another important observation that helps writing general recursive existence criteria for networks of N neurons is that the recursive stimulus time of a neuron during the current cycle is the sum of the recovery times of all the other neurons during appropriately labeled cycles determined according to the above cycle labeling rule. For example, for a three-neuron network, the symmetry of the network only allows for two phase-locked modes: $1 \to 2 \to 3$ (Fig. 17.7a) and $3 \to 2 \to 1$ (Fig. 17.7b).

Based on the above observations, the phase-locked mode $1 \to 2 \to 3$ (Fig. 17.7a) is described by the following *recursive existence criterion*:

$$t_{s1}[n] = t_{r2}[n-1] + t_{r3}[n-1];\; t_{s2}[n] = t_{r1}[n] + t_{r3}[n-1];\; t_{s3}[n] = t_{r1}[n] + t_{r2}[n], \tag{17.25}$$

which gives all possible steady-state solutions $(t_{s1}{}^*, t_{s2}{}^*, t_{s3}{}^*)$ by solving the *existence criterion*:

$$t_{s1}{}^* = f_2(t_{s2}{}^*) + f_3(t_{s3}{}^*);\, t_{s2}{}^* = f_1(t_{s1}{}^*) + f_3(t_{s3}{}^*);\, t_{s3}{}^* = f_1(t_{s1}{}^*) + f_2(t_{s2}{}^*). \tag{17.26}$$

The *recursive stability criterion* results from the recursive existence criterion (17.25) by linearization around the steady states $(t_{s1}{}^*, t_{s2}{}^*, t_{s3}{}^*)$:

$$\begin{cases} \delta t_{s1}[n+1] = n_2 \delta t_{s2}[n] + n_3 \delta t_{s3}[n], \\ \delta t_{s2}[n+1] = n_1 \delta t_{s1}[n+1] + n_3 \delta t_{s3}[n], \\ \delta t_{s3}[n+1] = n_1 \delta t_{s1}[n+1] + n_2 \delta t_{s2}[n+1], \end{cases} \tag{17.27}$$

which could be rewritten in matrix form:

$$\begin{pmatrix} 1 & 0 & 0 \\ -n_1 & 1 & 0 \\ -n_1 & -n_2 & 1 \end{pmatrix} \begin{pmatrix} \delta t_{s1}[n+1] \\ \delta t_{s2}[n+1] \\ \delta t_{s3}[n+1] \end{pmatrix} = \begin{pmatrix} 0 & n_2 & n_3 \\ 0 & 0 & n_3 \\ 0 & 0 & 0 \end{pmatrix} \begin{pmatrix} \delta t_{s1}[n] \\ \delta t_{s2}[n] \\ \delta t_{s3}[n] \end{pmatrix}. \tag{17.28}$$

The recursive stability criterion (17.28) has the general form $A\delta t[n+1] = B\delta t[n]$, or $\delta t[n+1] = A^{-1} B\delta t[n] = C\delta t[n]$, with the eigenvalues (stability indices) of matrix C given by $\lambda_{\mathrm{STRC}} = (\alpha \pm \sqrt{4n_1n_2n_3} + \alpha^2)/2)$, where $\alpha = n_1n_2 + n_1n_3 + n_2n_3 + n_1n_2n_3$. As expected, the above eigenvalues (stability indices) are symmetric with respect to cyclic permutations of the neurons.

Similarly, the phase-locked mode $1 \to 2 \to 3 \to \cdots \to N$ is described by the following *recursive existence criterion*:

$$t_{\mathrm{s1}[n]} = t_{\mathrm{r2}}[n-1] + t_{\mathrm{r3}}[n-1] + \cdots + t_{\mathrm{r}N-2}[n-1] + t_{\mathrm{r}N-2}[n-1] + t_{\mathrm{r}N}[n-1];$$

$$t_{\mathrm{s2}[n]} = t_{\mathrm{r1}}[n] + t_{\mathrm{r3}}[n-1] + \cdots + t_{\mathrm{r}N-2}[n-1] + t_{\mathrm{r}N-1}[n-1] + t_{\mathrm{r}N}[n-1];$$

$$\ldots$$

$$t_{\mathrm{s}N-1}[n] = t_{\mathrm{r1}}[n] = t_{\mathrm{r2}}[n] + t_{\mathrm{r3}} + \cdots + t_{\mathrm{r}N-2}[n] + t_{\mathrm{r}N[n-1]}$$

$$t_{\mathrm{s}N}[n] = t_{\mathrm{r1}}[n] + t_{\mathrm{r2}}[n] + t_{\mathrm{r3}}[n] + \cdots + t_{\mathrm{r}N-2}[n] + t_{\mathrm{r}N-1}[n] \tag{17.29}$$

We leave it as an exercise for the readers to complete the derivation for three-neuron ring and determine the fixed points of (17.29) and prove that the linear stability analysis leads again to $A\delta t[n+1] = B\delta t[n]$, where the matrix A is a lower triangular matrix with 1s along the main diagonal, whereas B is an upper diagonal matrix with 0s along the main diagonal similar to (17.28).

Similarly, using the same labeling rule for the firing cycles and the observation that the only other possible firing pattern in a three-neuron network is $1 \to 3 \to 2$ (Fig. 17.7b) the readers can arrive to the conclusion that, in this case, the recursive recovery time of a neuron during the current cycle is the sum of the stimulus times of the other neurons during appropriately labeled cycles. We also leave it as an exercise to show that the corresponding *recursive existence criterion* writes:

$$t_{\mathrm{r1}}[n-1] = t_{\mathrm{s2}}[n] + t_{\mathrm{s3}}[n];\ t_{\mathrm{r2}}[n-1] = t_{\mathrm{s1}}[n-1] + t_{\mathrm{s3}}[n];\ t_{\mathrm{r3}}[n] = t_{\mathrm{s1}}[n] + t_{\mathrm{s2}}[n], \tag{17.30}$$

which leads to the stability indices $\lambda_{\mathrm{STRC}} = (-\beta \pm \sqrt{4n_1n_2n_3 + \beta^2})/2$, where $\beta = 1 + n_1 + n_2 + n_3$. The generalization to N neurons is also straightforward.

4 Higher Order Corrections for the PRC-based Method

In the previous sections, we derived existence and stability criteria for 1:1 entrainment and phase-locked modes based on the STRC and its equivalent counterpart, the first order PRC, assuming that a presynaptic perturbation only influences the current cycle. In other words, we assumed that the first order PRC defined by (17.1) is the only quantity required to predict the network's dynamics.

The algorithm we used to successfully predict 1:1 phase-locked modes and their stability was as follows: (1) assume a fix firing order, (2) tabulate the algebraic relationships between the transiently perturbed cycle periods P_1 and either the stimulus times t_{s} or the recovery intervals t_{r} (Fig. 17.1a), (3) solve for the fixed points of the recursions determined at step 2 (*existence criterion*), and (4) linearize the PRC or the STRC about the fixed points found at step 3 and solve the characteristic equation of the linearized system (*stability criterion*).

Although the most significant transient change occurs during the current firing period P_1 (Fig. 17.1), the transient changes during the subsequent firing periods (P_2, $P_3, \ldots$) cannot always be neglected.

4.1 Second Order Correction for the PRC-based Method

We generalized the existence and stability criteria ((Dror et al. 1999) and references therein) by considering higher order PRCs (Oprisan and Boutan, 2008; Oprisan, 2010):

$$F^k(\varphi) = P_k/P_\mathrm{i} - 1, \tag{17.31}$$

where P_i is the intrinsic firing period of the isolated neurons, P_k with $k = 1, 2, \ldots$ is the order of the transiently modified firing period due to a presynaptic input at a stimulus phase $\varphi = t_\mathrm{s}/P_\mathrm{i}$ (Fig. 17.1a). For any 1:1 phase-locked modes in a ring network, the sum of the stimulus times t_s must add up to the common entrained period, or a multiple thereof (the same is true for recovery times t_r). The entrained period is defined as the sum of all phase resettings:

$$P_k[n] = (1 + F^1(\varphi[n]) + F^2(\varphi[n-1]) + \cdots)P_\mathrm{i}, \tag{17.32}$$

where the index n refers to the appropriate firing cycle shown in Fig. 17.5a. In the absence of higher order corrections for the PRC, the entrained stimulus time is simply $t_\mathrm{s}[n] = \varphi[n]P_\mathrm{i}$. Assuming that the second order resetting due to a stimulus applied during the previous $(n-1)$th cycle propagates to the current nth cycle and, moreover, that its effect manifests entirely before the current stimulus occurs, the required correction of the stimulus time is:

$$t_\mathrm{s}[n] = (\varphi[n] + F^2(\varphi[n-1]))P_\mathrm{i}. \tag{17.33}$$

The recovery time is defined as:

$$t_\mathrm{r}[n] = (1 - \varphi[n] + F^1(\varphi[n]))P_\mathrm{i}. \tag{17.34}$$

Therefore, the *existence criterion* states that:

$$\sum_{i=1}^{N}(\varphi_k{}^* + F^2(\varphi_k{}^*))P_{ki} = j(1 + F^1(\varphi_m{}^*) + F^2(\varphi_m{}^*))P_{mi}, \tag{17.35}$$

where N is the number of oscillators in the ring network, $m = 1, 2, \ldots, N$ is the index of an arbitrary oscillator and j is an integer number. The quantity on the left-hand side of (17.35) is the sum of the stimulus intervals and the right-hand side is an integer multiple j of the common phase-locked period of oscillation. Based on Fig. 17.5a and the derivation of the stability criteria from Sect. 3, the existence criteria in terms of the recovery times writes:

$$t_{r1}[n] + t_{r2}[n-1] = P_1[n] \text{ and } t_{r1}[n] + t_{r2}[n] = P_2[n]. \tag{17.36}$$

By substituting (17.32) and (17.34) into (17.36) one obtains the *recursive existence criterion*:

$$\begin{aligned}
&(1 - \varphi_1[n] + F^1{}_1(\varphi_1[n]))P_{1i} + (1 - \varphi_2[n-1] + F^1{}_2(\varphi_2[n-1]))P_{2i}\\
&\quad = (1 + F^1{}_1(\varphi_1[n]) + F^2{}_1(\varphi_1[n-1]))P_{1i},\\
&(1 - \varphi_1[n] + F^1{}_1(\varphi_1[n]))P_{1i} + (1 - \varphi_2[n] + F^1{}_2(\varphi_2[n]))P_{2i}\\
&\quad = (1 + F^1{}_2(\varphi_2[n]) + F^2{}_2(\varphi_2[n-1]))P_{2i}.
\end{aligned} \tag{17.37}$$

The system (17.37) could be solved for the steady states $(\varphi_1{}^*, \varphi_2{}^*)$:

$$\begin{aligned}
&(1 - \varphi_1{}^* + F^1{}_1(\varphi_1{}^*))P_{1i} + (1 - \varphi_2{}^* + F^1{}_2(\varphi_2{}^*))P_{2i}\\
&\quad = (1 + F^1{}_1(\varphi_1{}^*) + F^2{}_1(\varphi_1{}^*))P_{1i},\\
&(1 - \varphi_1{}^* + F^1{}_1(\varphi_1{}^*))P_{1i} + (1 - \varphi_2{}^* + F^1{}_2(\varphi_2{}^*))P_{2i}\\
&\quad = (1 + F^1{}_2(\varphi_2{}^*) + F^2{}_2(\varphi_2{}^*))P_{2i}.
\end{aligned} \tag{17.38}$$

As usual, the *linear stability* of the steady states $(\varphi_1{}^*, \varphi_2{}^*)$ can be assessed by using Taylor series expansions for $\varphi_1[n] = \varphi_1{}^* + \delta\varphi_1[n]$ and $\varphi_2[n] = \varphi_2{}^* + \delta\varphi_2[n]$ in (17.37):

$$\begin{aligned}
&(1 - \varphi_1{}^* - \delta\varphi_1[n] + F^1{}_1(\varphi_1{}^* + \delta\varphi_1[n]))P_{1i} + (1 - \varphi_2{}^* - \delta\varphi_2[n-1]\\
&\quad + F^1{}_2(\varphi_2{}^* + \delta\varphi_2[n-1]))P_{2i} = (1 + F^1{}_1(\varphi_1{}^* + \delta\varphi_1[n])\\
&\quad + F^2{}_1(\varphi_1{}^* + \delta\varphi_1[n-1]))P_{1i},\\
&(1 - \varphi_1{}^* - \delta\varphi_1[n] + F^1{}_1(\varphi_1{}^* + \delta\varphi_1[n]))P_{1i} + (1 - \varphi_2{}^* - \delta\varphi_2[n]\\
&\quad + F^1{}_2(\varphi_2{}^* + \delta\varphi_2[n]))P_{2i} = (1 + F^1{}_2(\varphi_2{}^* + \delta\varphi_2[n])\\
&\quad + F^2{}_2(\varphi_2{}^* + \delta\varphi_2[n-1]))P_{2i}.
\end{aligned} \tag{17.39}$$

The *recursive stability criterion* (17.39) can be written in matrix form as:

$$\begin{pmatrix} P_{1i} & 0 \\ (1 - m_1^1)\, P_{1i} & P_{2i} \end{pmatrix} \begin{pmatrix} \delta\varphi_1[n] \\ \delta\varphi_2[n] \end{pmatrix} + \begin{pmatrix} P_{1i,}m_2^1 & (1 - m_1^2)\, P_{2i} \\ 0 & P_{2i}m_2^2 \end{pmatrix} \begin{pmatrix} \delta\varphi_1[n-1] \\ \delta\varphi_2[n-1] \end{pmatrix} = 0, \tag{17.40}$$

which has the characteristic equation

$$\lambda_{PRC}{}^2 - \left(\left(m^1{}_1 - 1\right)\left(m^1{}_2 - 1\right) - m^2{}_1 - m^2{}_2\right)\lambda_{PRC} + m^2{}_1 m^2{}_2 = 0. \tag{17.41}$$

Equation (17.41) is the *PRC-based stability criterion* for a two-neuron ring network with a second-order correction to the PRC and, as expected, it is symmetric with respect to neurons' permutation. A phase-locked mode $(\varphi_1{}^*, \varphi_2{}^*)$ is stable if the stability indices $|\lambda_{\mathrm{PRC}}| < 1$. Moreover, our results can be generalized to synchrony by setting the steady state to $(\varphi_1{}^* = 0, \varphi_2{}^* = 1)$. However, since we assumed a fix firing order in our derivations, (17.41) gives only the stability condition for monotonic convergence to a steady state $(\varphi_1{}^*, \varphi_2{}^*)$. For pairs of identical oscillators, oscillating convergence to a stable phase-locked mode $(\varphi_1{}^*, \varphi_2{}^*)$ cannot be achieved because the eigenvalues given by (17.41) are, therefore, leading to monotonic convergence.

Based on the detailed derivation of the PRC-based existence and stability criteria for 1:1 phase locked modes in two-neuron networks presented above and the generalization to ring networks composed of N neurons presented in the previous section, we leave the details of generalization of the second order PRC method to the reader. For example, it can be shown that the *recursive existence criterion* for three-neuron network becomes:

$$\begin{aligned}
(\varphi_1[n] + F_1{}^2(\varphi_1[n-1]))P_{1\mathrm{i}} &= (1 - \varphi_2[n-1] + F_2{}^1(\varphi_2[n-1]))P_{2\mathrm{i}} \\
&\quad + (1 - \varphi_3[n-1] + F_3{}^1(\varphi_3[n-1]))P_{3\mathrm{i}}, \\
(\varphi_2[n] + F_2{}^2(\varphi_2[n-1]))P_{2\mathrm{i}} &= (1 - \varphi_1[n] + F_1{}^1(\varphi_1[n]))P_{1\mathrm{i}} + (1 - \varphi_3[n-1] \\
&\quad + F_3{}^1(\varphi_3[n-1]))P_{3\mathrm{i}}, \\
(\varphi_3[n] + F_3{}^2(\varphi_3[n-1]))P_{3\mathrm{i}} &= (1 - \varphi_1[n] + F_1{}^1(\varphi_1[n]))P_{1\mathrm{i}} + (1 - \varphi_2[n] \\
&\quad + F_2{}^1(\varphi_2[n]))P_{2\mathrm{i}}, \qquad (17.42)
\end{aligned}$$

which gives the steady-state solutions $(\varphi_1{}^*, \varphi_2{}^*, \varphi_3{}^*)$:

$$\begin{aligned}
(\varphi_1{}^* + F_1{}^2(\varphi_1{}^*))P_{1\mathrm{i}} &= (1 - \varphi_2{}^* + F_2{}^1(\varphi_2{}^*))P_{2\mathrm{i}} + (1 - \varphi_3{}^* + F_3{}^1(\varphi_3{}^*))P_{3\mathrm{i}}, \\
(\varphi_2{}^* + F_2{}^2(\varphi_2{}^*))P_{2\mathrm{i}} &= (1 - \varphi_1{}^* + F_1{}^1(\varphi_1{}^*))P_{1\mathrm{i}} + (1 - \varphi_3{}^* + F_3{}^1(\varphi_3{}^*))P_{3\mathrm{i}}, \\
(\varphi_3{}^* + F_3{}^2(\varphi_3{}^*))P_{3\mathrm{i}} &= (1 - \varphi_1{}^* + F_1{}^1(\varphi_1{}^*))P_{1\mathrm{i}} + (1 - \varphi_2{}^* + F_2{}^1(\varphi_2{}^*))P_{2\mathrm{i}},
\end{aligned} \qquad (17.43)$$

The *recursive stability criterion* results from (17.42) by linearization around the steady-states (17.43):

$$\begin{cases}
\delta\varphi_1[n] + m_1^2\delta\varphi_1[n-1] = \left(-1 + m_2^1\right)\delta\varphi_2[n-1] + \left(-1 + m_3^1\right)\delta\varphi_3[n-1], \\
\delta\varphi_2[n] + m_2^2\delta\varphi_2[n-1] = \left(-1 + m_1^1\right)\delta\varphi_1[n] + \left(-1 + m_3^1\right)\delta\varphi_3[n-1], \\
\delta\varphi_3[n] + m_3^2\delta\varphi_3[n-1] = \left(-1 + m_1^1\right)\delta\varphi_1[n] + \left(-1 + m_2^1\right)\delta\varphi_2[n],
\end{cases} \qquad (17.44)$$

which could be rewritten in a matrix form as:

$$\begin{pmatrix} 1 & 0 & 0 \\ 1-m_1^1 & 1 & 0 \\ 1-m_1^1 & 1-m_2^1 & 1 \end{pmatrix} \begin{pmatrix} \delta\varphi_1[n] \\ \delta\varphi_2[n] \\ \delta\varphi_3[n] \end{pmatrix} = \begin{pmatrix} -m_1^2 & -1+m_2^1 & -1+m_3^1 \\ 0 & -m_2^2 & -1+m_3^1 \\ 0 & 0 & -m_3^2 \end{pmatrix} \begin{pmatrix} \delta\varphi_1[n-1] \\ \delta\varphi_2[n-1] \\ \delta\varphi_3[n-1] \end{pmatrix}. \tag{17.45}$$

The characteristic equation of (17.45) is:

$$\begin{aligned} &\lambda_{\mathrm{PRC}}^3 + (2 - m_1{}^1 - m_1{}^2 - m_2{}^1 - m_2{}^2 - m_3{}^1 - m_3{}^2 + m_1{}^1 m_2{}^1 m_3{}^1)\lambda_{\mathrm{PRC}}^2 \\ &\quad + (-1 + m_1{}^1 + m_1{}^2 + m_2{}^1 - m_1{}^1 m_2{}^1 - m_1{}^2 m_2{}^1 + m_2{}^2 - m_1{}^1 m_2{}^2 \\ &\qquad - m_1{}^2 m_2{}^2 + m_3{}^1 - m_1{}^1 m_3{}^1 - m_1{}^2 m_3{}^1 - m_2{}^1 m_3{}^1 + m_1{}^1 m_2{}^1 m_3{}^1 \\ &\qquad + m_1{}^2 m_2{}^1 m_3{}^1 - m_2{}^2 m_3{}^1 + m_1{}^1 m_2{}^2 m_3{}^1 + m_3{}^2 - m_1{}^1 m_3{}^2 - m_1{}^2 m_3{}^2 \\ &\qquad - m_2{}^1 m_3{}^2 + m_1{}^1 m_2{}^1 m_3{}^2 - m_2{}^2 m_3{}^2)\lambda_{\mathrm{PRC}} - m_1{}^2 m_2{}^2 m_3{}^2 = 0, \end{aligned} \tag{17.46}$$

and it reduces to:

$$\begin{aligned} \lambda_{\mathrm{PRC}}^2 &+ (2 - m_1{}^1 - m_2{}^1 - m_3{}^1 + m_1{}^1 m_2{}^1 m_3{}^1)\lambda_{\mathrm{PRC}} + (1 + m_1{}^1 + m_2{}^1 \\ &+ m_3{}^1 - m_1{}^1 m_2{}^1 - m_1{}^1 m_3{}^1 - m_2{}^1 m_3{}^1 + m_1{}^1 m_2{}^1 m_3{}^1) = 0, \end{aligned} \tag{17.47}$$

if only the first order PRCs are considered. The solution of (17.47) is similar to the STRC-based stability criterion derived in Sect. 3, although somewhat more involved. Moreover, both characteristic equations (17.46) and (17.47) are symmetric with respect to neurons' permutations, as expected.

4.2 *Numerical and Experimental Verification of the Existence and the Stability Criteria Using Higher Order PRC Corrections*

In addition to the traditional numerical modeling that we developed and implemented for all previous sections of this chapter in order to test the validity of the existence and stability criteria, dynamic clamp experiments were conducted in order to check the predicted phase-locked modes in two-neuron hybrid networks (Oprisan et al. 2003). We predicted the firing patterns of 164 distinct hybrid circuits created by utilizing a bursting pyloric dilator (PD) neuron from the stomatogastric ganglion of *Homarus americanus* reciprocally coupled to different model neurons at varying values of the synaptic conductances in each direction. The accuracy of the results was satisfactory for 161 of the 164 circuits and in almost every case was within the 10% variability in the period of the biological neurons. The three failures signify that a complex mode was observed despite a prediction of 1:1 phase-locking. In each of

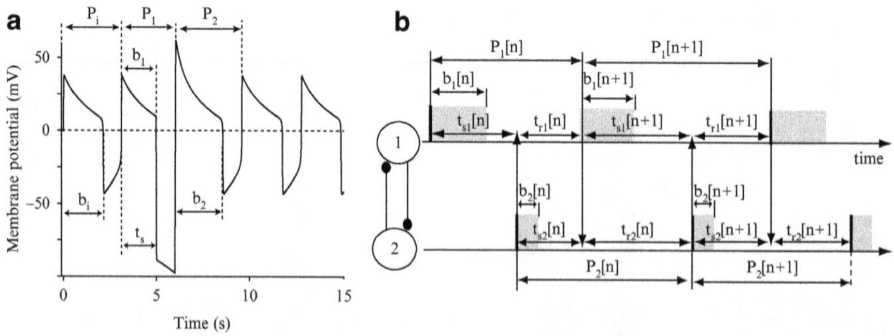

Fig. 17.8 (**a**) Long lasting perturbations significantly change not only the current P_1 and subsequent P_2, etc. firing periods, but also the burst durations b_1, b_2, etc. A type 2 ML model neuron was used to simulate the envelope of a bursting neuron with an intrinsic period P_i and burst duration b_i. (**b**) A schematic representation of a two-neuron network of bursting neurons with a 1:1 firing pattern that shows both changing period and burst durations

these three cases, at least one synaptic conductance was less than or equal to 2 nS, so we conclude that the coupling was too weak to actually affect the firing pattern of the biological neuron. This observation is in agreement with our previous findings (Sect. 2) stating that in a hybrid network a 1:1 entrainment can only occur if the free running periods $P_{2i}/P_{1i}-1$ (horizontal dashed line in Fig. 17.8) crosses the PRC of the driven neuron. In other words the condition $F^1{}_{\min} < P_{2i}/P_{1i}-1 < F^1{}_{\max}$ must hold for the given value of the synaptic conductance. However, if the two intrinsic firing periods are very dissimilar and a synapse is too weak (meaning that the amplitude of the PRC is very small) it is possible that 1:1 entrainment never occurs. Out of the 161 hybrid circuits with firing patterns predicted correctly, 21 circuits required a second order PRC correction either for the existence and/or the stability criterion in order to match the experimental findings.

4.3 *Prediction of 1:1 Phase-Locked Modes Using the PRCs and the Burst Resetting Curves (BRCs)*

Spiking neurons represent a significant class of neural oscillators and the first three sections of this chapter focused on predicting the 1:1 phase-locked modes in networks of spiking neurons using the first order PRC and its equivalent STRC counterpart. One reason the PRC/STRC methods are so successful in predicting the phase-locked modes for spiking neurons is because each postsynaptic neuron is perturbed for a brief time, which leads to soft perturbations of the limit cycle oscillator, i.e., small amplitude disturbances of the unperturbed limit cycle of the free-running neuron. However, bursting neurons maintain a significantly longer depolarized state and a corresponding longer synaptic coupling with other neurons. As a result, bursting neurons often induce significant (hard) perturbations of the limit cycle oscillators that alter not only the duration but also the waveform and the duty cycle of the membrane potential. Endogenous bursters play a central role in numerous CPGs, including the pyloric circuit (Hartline, 1979; Miller and Selverston, 1982; Bal et al. 1988; Huerta et al. 2000; Hooper and DiCaprio, 2004), the heartbeat of crustaceans (Tazaki and Cooke, 1990; Calabrese et al. 1995), the gastric CPGs of the crustacean stomatogastric system (Harris-Warrick et al. 1992; Marder and Bucher, 2001, 2007), and the feeding CPG in mollusks (Arshavsky et al. 1989; Arshavsky et al. 1997).

So far we only considered networks of spiking neurons, which fire brief action potentials lasting a few milliseconds that only represent a tiny fraction of the intrinsic firing period P_{i}. In other words, we only considered spiking neurons because their duty cycle, which is the ratio between the duration of the depolarized state normalized by the intrinsic period, is very small and does not induce long lasting postsynaptic effects. Therefore, the synaptic current injected by such a spiking neuron into its postsynaptic neuron during the depolarized phase can be approximated as very brief, or "infinitesimal." Although the most obvious effect of such brief synaptic currents is a measurable shortening or lengthening of the current cycle of the postsynaptic neuron, it also alters all ionic mechanisms responsible for generating an action potential. In particular, a synaptic input, however brief, also alters the neurons' depolarization and repolarization mechanisms. Since in spiking neurons these mechanisms are responsible for the very brief action potential lasting a few milliseconds, it is reasonable to assume that the action potential duration (APD) and its shape also change. Such changes in the duty cycle are more significant and easier to quantify in bursting neurons, which are characterized by a significantly larger duty cycle than the spiking neurons. As a result, a very strong presynaptic perturbation of a significant duration applied during the current cycle changes both the firing period from P_{i} to P_1 and the burst duration from b_{i} to b_1 of the postsynaptic neuron (Fig. 17.8a).

There are similarities between the derivation of the existence and the stability criteria for the 1:1 phase-locked modes in two-neuron networks of spiking and bursting neurons. First, the same set of basic assumption should be observed: (1) no synaptic delays, (2) each neuron receives one perturbation (synaptic input) per cycle, (3) the perturbation takes the same form in the closed loop circuit as in an open loop circuit, and (4) each neuron returns to its unperturbed limit cycle before the next input is received. However, there are two major differences: (1) strong and long-lasting presynaptic inputs have significant impact not only on the current cycle P_1 but also on many subsequent cycles P_2, P_3, etc. (Fig. 17.8a), and (2) a presynaptic input changes not only the firing period of the postsynaptic neuron but also its burst duration from its intrinsic value b_{i} to b_1, b_2, etc. (Oprisan and Boutan, 2008; Oprisan, 2010).

In addition to the higher order PRCs defined by (17.37), we defined the kth ($k = 1, 2, \ldots$) order effects of the presynaptic perturbation on the oscillator's burst duration, which we called the burst resetting curves (BRCs):

$$G^k(\varphi) = b_k/b_{\mathrm{i}} - 1. \tag{17.48}$$

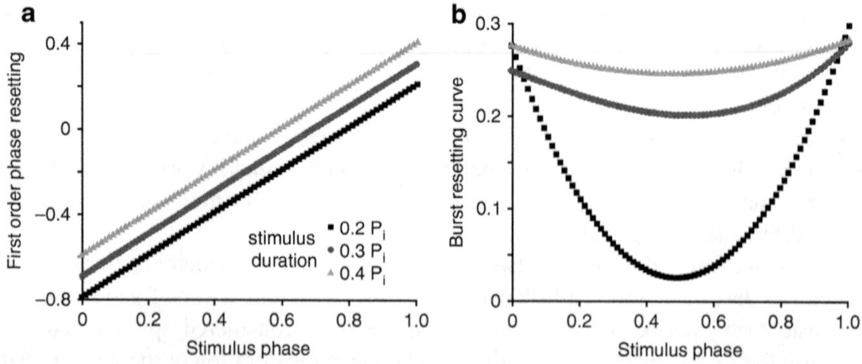

Fig. 17.9 (**a**) PRCs for a type 2 ML model neuron subject to a long lasting perturbation covering 20% (*squares*), 30% (*circles*) and, respectively 40% (*triangles*) of the intrinsic period P_i. (**b**) The BRCs show significant resetting at the beginning of the burst ($\varphi = 0$) and close to the end of the cycle ($\varphi = 1$) due to the burst truncation

The PRCs and the BRCs shown in Fig. 17.9 were obtained for a type 2 ML model neuron, which mimics the envelope of a true bursting neuron, subject to a synaptic current

$$I_{\mathrm{syn}} = g_{\mathrm{syn}} m(V) \cdot (V - E_{\mathrm{syn}}), \tag{17.49}$$

where g_{syn} is the maximum conductance of the synapse, m is the synaptic activation function, V is the membrane potential of the postsynaptic neuron, and E_{syn} is the synaptic reversal potential, which was set to −90 mV for inhibitory synapses. The synaptic activation function was chosen such that the synapse is fully activated during the burst and deactivated between bursts. Since the burst duration can change significantly compared to its intrinsic value b_i (Fig. 17.9), in order to correctly predict the existence and the stability of the 1:1 phase-locked modes, we developed new existence and stability criteria that take into account both PRCs and BRCs (Oprisan and Boutan, 2008; Oprisan, 2010).

The idea that the duration of the quiescent interval, during which a cell is inhibited by a presynaptic cell, serves as an effective mechanism of controlling the firing rate of a network is extensively used in predicting the phase-locked modes in other excitable cells. It was first introduced by Nolasco and Dahlen (1968) and called APD restitution curve in order to explain the "alternans," steady oscillations that periodically switch between long APD followed by a short quiescent interval, which in turn is followed by a short APD, and then followed by a long quiescent interval. They derived a so-called APD restitution curve, which plotted the APD versus the duration of the quiescent interval (Tolkacheva et al. 2003; Kalb et al. 2004; Cherry and Fenton, 2007). The APD restitution curve is successfully used to predict the phase-locked modes in networks of cardiac cells (Tolkacheva et al. 2003). Obviously, by replacing the APD with the relative change in APD, which is the

PRC, we arrive at what we called BRC. Moreover, the dynamic adaptation of the quiescent interval duration and its effect on the overall firing rate in cardiac cells is similar to the phase-locking in neural networks.

The derivations for the existence and the stability criteria follow the procedure highlighted in the preceding sections. However, the PRC, $F_{\text{post}}()$, of the postsynaptic neuron strongly depends on both the phase φ_{post} of the stimulus and its duration τ_{pre}. Therefore, in general, a 1:1 phase-locked firing pattern is no longer only characterized by the set of relative firing phases $(\varphi_1{}^*, \varphi_2{}^*, \ldots, \varphi_N{}^*)$, but also by a set of self-adjusted burst durations $(\tau_1{}^*, \tau_2{}^*, \ldots, \tau_N{}^*)$ that must be determined.

In agreement with the definitions (17.33) and (17.34), the stimulus intervals are:

$$\begin{aligned} t_{\text{s}1}[n] &= (\varphi_1[n] + F_1^2(\varphi_1[n-1], \tau_2[n-1]))P_{1\text{i}}, \\ t_{\text{s}2}[n] &= (\varphi_2[n] + F_2^2(\varphi_2[n1], \tau_1[n]))P_{2\text{i}}, \end{aligned} \tag{17.50}$$

and the recovery times are:

$$\begin{aligned} t_{\text{r}1}[n] &= (1 - \varphi_1[n] + F_1^1(\varphi_1[n], \tau_2[n]))P_{1\text{i}}, \\ t_{\text{r}2}[n] &= (1 - \varphi_2[n] + F_2^1(\varphi_2[n], \tau_1[n+1]))P_{2\text{i}}, \end{aligned} \tag{17.51}$$

where the PRC, $F(\varphi, \tau)$, depends both on the temporal phase $\varphi = t_{\text{s}}/P_{\text{i}}$ and the normalized duration $\tau = b/b_{\text{i}}$ of the presynaptic perturbation. Based on Fig. 17.8b, the *recursive existence criterion* for phase-locked modes in a two-neuron network of bursting neurons is:

$$t_{\text{r}1}[n+1] + t_{\text{r}2}[n] = P_1[n+1]t_{\text{r}1}[n] + t_{\text{r}2}[n] = P_2[n]. \tag{17.52}$$

By substituting (17.50) and (17.51) into (17.52), we obtain:

$$\begin{aligned} &(1 - \varphi_1[n+1] + F^1{}_1(\varphi_1[n+1], \tau_2[n+1]))P_{1\text{i}} + (1 - \varphi_2[n] \\ &\quad + F^1{}_2(\varphi_2[n], \tau_1[n+1]))P_{2\text{i}} = 1 + F^1{}_1(\varphi_1[n+1], \tau_2[n+1]) \\ &\quad + F^2{}_1(\varphi_1[n], \tau_2[n])P_{1\text{i}}, \\ &(1 - \varphi_1[n] + F^1{}_1(\varphi_1[n], \tau_2[n]))P_{1\text{i}} + (1 - \varphi_2[n] + F^1{}_2(\varphi_2[n], \tau_1[n+1]))P_{2\text{i}} \\ &\quad = (1 + F^1{}_2(\varphi_2[n], \tau_1[n+1]) + F^2{}_2(\varphi_2[n-1], \tau_1[n]))P_{2\text{i}}. \end{aligned} \tag{17.53}$$

Let $(\varphi_1{}^*, \varphi_2{}^*, \tau_1{}^*, \tau_2{}^*)$ be the steady 1:1 phase-locked mode determined by the recursive existence criterion (17.53) with $\delta[n] = (\delta\varphi_1[n], \delta\varphi_2[n], \delta\tau_1[n], \delta\tau_2[n])^{\text{T}}$ the infinitesimal phase and burst fluctuations, where the superscript T means transposed. In order to determine the stability of the 1:1 phase-locked mode, if it exists, (17.59) is linearized around the steady state $(\varphi_1{}^*, \varphi_2{}^*, \tau_1{}^*, \tau_2{}^*)$. By direct substitution of the current cycle phase, $\varphi_i[n] = \varphi_i{}^* + \delta\varphi_i[n]$, and burst durations, $\tau_i[n] = \tau_i{}^* + \delta\tau_i[n]$

into (17.53), and considering a Taylor series expansions around the steady 1:1 phase-locked mode, we get a system of two equations with four unknowns, which is part of the *recursive stability criterion*:

$$\begin{aligned}
&P_{1i}(-\delta\varphi_1[n] + m^1{}_1\delta\varphi_1[n] + h_1{}^1\delta\tau_2[n]) + P_{2i}(-\delta\varphi_2[n-1] + m^1{}_2\delta\varphi_2[n-1]\\
&\quad + h^1{}_2\Delta\delta\tau_1[n-1]) \cong P_{10}(m^1{}_1\delta\varphi_1[n] + h^1{}_1\delta\tau_2[n] + m^2{}_1\delta\varphi_1[n-1]\\
&\quad + h^2{}_1\Delta\delta\tau_2[n-1]),\\
&P_{1i}(-\delta\varphi_1[n] + m^1{}_1\delta\varphi_1[n] + h^1{}_1\delta\tau_2[n]) + P_{2i}(-\delta\varphi_2[n] + m^1{}_2\delta\varphi_2[n] + h^1{}_2\delta\tau_1[n])\\
&\quad \cong P_{2i}(m^1{}_2\delta\varphi_2[n] + h^1{}_2\delta\tau_1[n] + m^1{}_2\delta\varphi_2[n-1] + h^2{}_2\delta\varphi_1[n-1]),
\end{aligned} \tag{17.54}$$

where $m^i{}_j = (\partial F^i{}_j/\partial\varphi)_\tau$ is the slope of the ith order PRC for neuron j when the phase is changing at a constant burst duration in the presynaptic neuron, and $h^i{}_j = (\partial F^i{}_j/\partial\tau)_\varphi$ is the slope of the ith order PRC for neuron j with respect to changes in the presynaptic neuron's burst duration at a given (constant) stimulus phase. The recursive stability criterion (17.54) cannot be explicitly solved unless the recursive BRCs are considered as part of the existence criterion (17.53):

$$\begin{aligned}
b_1[n] &= b_{1i}(1 + G^1{}_1(\varphi_1[n], \tau_2[n]) + G^2{}_1(\varphi_1[n-1], \tau_2[n-1])),\\
b_2[n] &= b_{2i}(1 + G^1{}_2(\varphi_2[n], \tau_1[n]) + G^2{}_2(\varphi_2[n-1], \tau_1[n-1])),
\end{aligned} \tag{17.55}$$

where $G^i{}_k = b_i/b_{ki} - 1$ is the ith order ($i = 1, 2$, etc.) relative change in the burst duration for the neuron labeled k. By linearizing the BRCs around the phase-locked solution ($\varphi_1{}^*, \varphi_2{}^*, \tau_1{}^*, \tau_2{}^*$) we get the second half of the *recursive stability criterion*:

$$\begin{aligned}
\Delta\tau_1[n] &\cong p^1{}_1\Delta\varphi_1[n] + q^1{}_1\Delta\tau_2[n] + p^2{}_1\Delta\varphi_1[n-1] + q^2{}_1\Delta\tau_2[n-1],\\
\Delta\tau_2[n] &\cong p^1{}_2\Delta\varphi_2[n] + q^1{}_2\Delta\tau_1[n] + p^2{}_2\Delta\varphi_2[n-1] + q^2{}_2\Delta\tau_1[n-1],
\end{aligned} \tag{17.56}$$

where $p^i{}_j = (\partial G^i{}_j/\partial\varphi)_\tau$ is the slope of the ith order BRC for neuron j when the phase is changing at a constant burst duration in the presynaptic neuron, and $q^i{}_j = (\partial G^i{}_j/\partial\tau)_\varphi$, is the slope of the ith order BRC for neuron j with respect to changes in the presynaptic neuron's burst duration at a given (constant) stimulus phase. By combining both the PRCs and the BRCs, the matrix form of the recursive stability criterion writes:

$$\begin{pmatrix} P_{1i} & 0 & -P_{2i}h^1{}_2 & 0\\ P_{1i}(1-m^1{}_1) & P_{2i} & P_{2i}h^2{}_2 & -P_{1i}h^1{}_1\\ p^1{}_1 & 0 & -1 & q^1{}_1\\ 0 & p^1{}_2 & q^1{}_2 & -1 \end{pmatrix}\delta[n] + \begin{pmatrix} P_{1i}m^1{}_2 & P_{2i}(1-m^2{}_1) & 0 & P_{1i}h^2{}_1\\ 0 & P_{2i}m^2{}_2 & 0 & 0\\ p^2{}_1 & 0 & 0 & q^2{}_2\\ 0 & p^2{}_2 & q^2{}_2 & 0 \end{pmatrix}\delta[n-1] = 0. \tag{17.57}$$

The characteristic polynomial of (17.57) is of fourth degree, but it can be reduced to a second-degree polynomial if all the BRC contributions are neglected ($p^i{}_j = 0$ and

$q^i{}_j = 0$), and the PRC is not sensitive to changes in the burst duration ($h^i{}_j = 0$). In this particular case, the characteristic equation associated to the recursive stability criterion (17.57) becomes:

$$\lambda_{\text{PRC}}{}^2 - ((1 - m^1{}_1)(1 - m^1{}_2) - m^2{}_1 - m^2{}_2)\lambda_{\text{PRC}} + m^2{}_1 m^2{}_2 = 0,$$

which coincides with the stability criterion for networks made of spiking neurons.

Another peculiar case derived from the general stability criterion (17.57) refers to the PRC that are not sensitive to stimulus duration (all $h^i_j = 0$):

$$[\lambda_{\text{PRC}}{}^2 - \lambda_{\text{PRC}}(1 - m^1{}_1)(1 - m^1{}_2) - m^2{}_1 - m^2{}_2 + m^2{}_1 m^2{}_2] \times [\lambda_{\text{PRC}}{}^2(-1 + q^1{}_1 q^1{}_2) + \lambda_{\text{PRC}}(q^1{}_1 q^2{}_2 + q^2{}_1 q^1{}_2) + q^2{}_1 q^2{}_2] = 0, \quad (17.58)$$

which shows that, for the stability of the steady 1:1 phase-locked mode, the relationship between the postsynaptic burst duration and the presynaptic stimulus phase is not relevant ($p^i{}_j$ do not affect the stability criterion). In this particular case, the characteristic polynomial is the product of two second-degree polynomials: one determines stability of the phase-locked mode ($\varphi_1{}^*, \varphi_2{}^*$) and the other characteristic polynomial determines the stability of the steady-state bursts ($\tau_1{}^*, \tau_2{}^*$).

5 Conclusions

The PRCs are extensively used to predict phase relationships among neural oscillators (existence criterion) and to investigate the stability of the predicted phase-locked modes (stability criterion). In this chapter, we reviewed some analytical methods used to predict the 1:1 phase-locked modes in ring networks and presented both numerical and experimental results to support the predictions. The PRCs in this chapter were generated using a Morris–Lecar (ML) model neuron, which was tuned to act either as a type 1 (spiking) or as type 2 excitable cell, whose membrane potential envelope mimic the activity of a bursting neuron with a significant duty cycle. The first order PRC, which tabulates the relative change in the duration of the current cycle due to a presynaptic input, is mathematically equivalent to the STRC approach. However, the STRC method has an intuitive geometric interpretation and is the natural choice for networks made of dissimilar neurons.

The limitations of the PRCs method come from the assumptions made.

1. The external stimulus injected into an isolated neuron, for the purpose of determining its PRCs or STRCs in open loop, is identical to the recursive inputs received by the same neuron in closed loop. For neural networks of spiking neurons with pulsatile couplings, this requirement is often met since the action potentials of a presynaptic spiking neuron last a few milliseconds and cannot significantly change the ionic mechanisms of the postsynaptic cell. Therefore, the action potential of an isolated (open loop) spiking neuron is generally close

to its shape and duration in a closed loop setup. This assumption is much harder to fulfill for bursting neurons that maintain a synaptic coupling with their postsynaptic neurons over a much longer duration of their depolarized state. Therefore, when predicting the phase-locked modes in networks of bursting neurons, higher order correction to both the PRCs and the BRCs should be considered.

2. An equally important assumption is that the isolated neural oscillator returns to its unperturbed limit cycle between inputs, so that its response to any input is always the same for a given phase of stimulation. Due to this constraint, this chapter focused mainly on 1:1 phase-locked modes and always compared the peak-to-peak amplitude of the second (and higher) order PRCs against the first order PRC to make sure that the neuron returns to its unperturbed limit cycle by the time it receives another synaptic input. However, by considering higher order PRCs it is possible to improve the accuracy of the phase-locked predictions. Such higher order PRC corrections have two drawbacks: (1) they are analytically and numerically much more involved than the first order PRC and its equivalent STRC counterpart, and (2) they require additional assumptions. For example, in order to apply the second order PRC, it is assumed that this entire correction occurs before the subsequent input.
3. It was also assumed that each neuron receives only one presynaptic input and sends its output to only one other neuron in the network. This assumption could limit the applicability of the PRCs method to ring networks. However, in Chap. 6 on the geometrical interpretation of the PRCs, we showed that the PRCs were successfully expanded to more than one input per cycle, therefore, allowing more general network topologies. The price for such a flexibility is a steep increase in both the analytical and the numerical complexity of the problem.

The general procedure followed throughout this chapter in order to derive analytically the existence and the stability criteria was as follows. (1) Assume a fix firing order. In this chapter, we assume most of the time a 1:1 phase-locked mode. (2) Explicitly state the relationships between the transiently perturbed cycle period(s) and either the stimulus times or the recovery intervals. Such a relationship is called a recursive existence criterion since it involves quantities spanning multiple firing cycles. It could be as simple as stating that in a reciprocally coupled two-neuron network the stimulus time of one neuron equals the recovery time of the other (see Sect. 2), or as involved as stating that the stimulus time of one neuron is the sum of recovery times of all the other neurons in a ring network of N neurons (see Sect. 3). (3) Numerically solve for the fixed points of the recursive existence criterion derived at step 2 using either the PRC or the STRC generated in open loop with the appropriate input. (4) Linearize the PRCs or the STRCs about the fixed points found at step 3 and numerically solve the characteristic equation of the linearized system in order to determine the stability index of each predicted phase-locked mode. The linearized existence criterion around the fixed points is called the recursive stability criterion. (5) Check the predicted phases for stable modes against numerical or experimental results obtained in closed loop. We also provided intuitive geometric interpretation for the existence criterion.

Section 2 is a review of the analytical methods used to predict 1:1 phase-locked modes in two-neuron networks with unidirectional synaptic coupling between spiking neurons in which one neuron drives the other (no feedback loop). This type of phase-locking was called entrainment to distinguish it from the phase-locked modes in fully connected networks that allow each neuron both to influence and be influences by other neurons. The prediction is that, based on the first order PRC, for a given synaptic strength, the ratio of the intrinsic firing periods determines the relative phase of the entrainment. Moreover, by increasing the strength of the coupling the phase difference between neurons should decrease. The stability of a given 1:1 entrainment mode is solely determined by the slope of the first order PRC of the driven neuron at the entrained relative phase. All these predictions were confirmed by close loop numerical simulations carried out with type 1 ML model neurons. A generalization to $1{:}p$ entrainment was also briefly introduced.

Section 3 is a review of the analytical methods used to predict 1:1 phase-locked modes in ring networks of spiking neurons based on the STRCs method. For two reciprocally coupled neurons, the recursive existence criterion simply states that the stimulus time of one neuron equals the recovery time of the other. Since the STRCs tabulate the recovery time versus the stimulus time, prediction the phase-locked mode in a two-neuron network is as simple as determining the crossing between the STRCs of the two neurons. The stability index is the product of the slopes of the two STRCs. Numerical simulations using type 1 ML model neurons confirmed the validity of the theoretical predictions. A generalization of the 1:1 phase-locked existence and stability criteria was also reviewed for a ring network of N neurons.

Section 4 is a review of analytical tools and experimental results regarding higher order PRCs when some of the original assumptions are violated. Both the existence and the stability criteria that include second order corrections to the PRC were analytically derived. Such corrections are necessary especially when the network is made of bursting neurons since, in addition to the phase, the duration of the long-lasting burst of a presynaptic neuron significantly changes the duration of the current and subsequent cycles. We used dynamic clamp to reciprocally couple a bursting AB/PD pair of the stomatogastric ganglion of lobster with a computer-generated bursting neuron using artificial synapses. We found that in 21 out of 161 hybrid circuits our prediction matched the experimental results only after applying the second order PRC correction. In all cases, the second order correction improved the accuracy of predictions. Another very important experimental observation was that in almost all hybrid circuits the burst duration of both neurons in closed loop was significantly different (more than 10%) from the open loop value. Therefore, we were led to considering a generalized PRC that depends both on the stimulus phase and the stimulus duration. Although the analytical procedure for deriving the existence and the stability conditions for the generalized PRC is similar to the previous cases it requires additional information, which we provided by measuring the effect of the stimuli with different durations on the length of a burst – called the BRCs. The BRCs were extensively used in investigations of cardiac arrhythmia and were called APD restitution curves.

Acknowledgments I would like to thank my former postdoc advisor, Dr. Carmen C. Canavier, for introducing me to the fascinating field of phase resetting, and for the freedom she has allowed me in pursuing my own ideas and interests. I deeply appreciate the guidance, support, and inspiration she has given me.

I am especially grateful to my wife, Dr. Ana Oprisan, for standing beside me throughout my career. *Her* patience and unwavering support made this work possible. I also thank my wonderful children: Andrei and Andra for their understanding and good humor. I dedicate this work to the memory of mother-in-law and both my parents.

I gratefully acknowledge the helpful comments and feedback I received from reviewers while preparing this chapter.

This work was partly supported by the National Science Foundation CAREER award IOS – 1054914 to SAO.

Appendix

Our simulations were performed using a type 1, single compartment, neural oscillators described by a standard conductance-based, or Hodgkin–Huxley (HH), mathematical model. The rate of change of the membrane potential is proportional to the sum of the ionic plus externally injected currents flowing through the cell's membrane:

$$\begin{aligned} dV/dt &= -I_{Ca} - I_K - I_{Leak} + I_0 \\ &= -\bar{g}_{Ca} m_\infty(V)(V - V_{Ca}) - \bar{g}_K w(V - V_K) - \bar{g}_L(V - V_L) + I_0, \end{aligned}$$

where V is the membrane potential, $\bar{g}_{ch}$ and V_{ch} are the maximum conductance and, respectively, the reversal potential for the ionic channel *ch* (only calcium, potassium, and leak were considered), w is the normalized potassium concentration and stands for the instantaneous probability that a potassium channel is open, and I_0 is an externally applied bias current. Each ionic current is the product of a voltage-dependent conductance and a driving force $I_{ch} = g_{ch}(V)(V - V_{ch})$ where $g_{ch}(V)$ is the product between the maximum conductance for that channel and a voltage-dependent gating variable. The Morris and Lecar (ML) model has two noninactivating voltage-sensitive gating variables: one instantaneous, voltage-dependent, calcium activation $m_\infty(V)$ and a delayed voltage-dependent potassium w given by a first order differential equation (Morris and Lecar, 1981):

$$dw/dt = \phi(w_\infty(V) - w)/\tau(V),$$

where ϕ is a temperature-dependent parameter and the voltage-dependent relaxation time constant is defined by $\tau(V) = Cosh^{-1}((V - V_{w\,0.5}/2V_{wslope}))$. All open-state probability functions, or steady-state gating variables x_∞, have a sigmoidal form (Morris and Lecar 1981):

$$x_\infty(V) = (1 + Tanh((V - V_{x\,0.5})/V_{x\,slope}))/2,$$

where $V_{x\ 0.5}$ is the half-activation voltage and $V_{x\ \mathrm{slope}}$ is the slope factor for the gating variable x. The ML model is widely used in computational neuroscience because it captures relevant biological processes and, at the same time, by changing only a small subset of its parameters it can behave either as a type 1 or a type 2 neural oscillator. The dimensionless parameters for a type 1 ML neuron are: $V_{m\ 0.5} = -0.01$; $V_{m\ \mathrm{slope}} = 0.15$; $V_{w\ 0.5} = 0.1$; $V_{w\ \mathrm{slope}} = 0.145$; $V_{\mathrm{K}} = -0.7$; $V_{\mathrm{L}} = -0.5$; $V_{\mathrm{Ca}} = 1.0$; $g_{\mathrm{Ca}} = 1.33$; $g_{\mathrm{K}} = 2.0$; $g_{\mathrm{L}} = 0.5$; $I_0 = 0.070$; $\phi = 0.6$ (Ermentrout 1996). The model's equations and its parameters are in dimensionless form with all voltages divided by the reversal potential of calcium $V_{\mathrm{Ca}}{}^0 = 120\,\mathrm{mV}$, all conductances were divided by $g_{\mathrm{Ca}}{}^0 = 4\,\mathrm{mS/cm^2}$, and all currents normalized by $V_{\mathrm{Ca}}{}^0 g_{\mathrm{Ca}}{}^0 = 480\,\mu\mathrm{A/cm^2}$ (Ermentrout 1996). For example, a dimensionless reversal potential for the leak current of $V_{\mathrm{L}} = -0.5$ means $V_{\mathrm{L}} = -0.5 V_{\mathrm{Ca}}{}^0 = -0.5 \times 120\,\mathrm{mV} = -60\,\mathrm{mV}$.

References

Aiken SP, Kuenzi FM, Dale N (2003) Xenopus embryonic spinal neurons recorded in situ with patch-clamp electrodes – conditional oscillators after all? European Journal of Neuroscience 18:333–343.

Arshavsky Y, Deliagina TG, Orlovsky GN, Yu V. P (1989) Control of feeding movements in the pteropod mollusc, Clione limacina. Exp Brain Res 78:387–397.

Arshavsky YI, Deliagina TG, Orlovsky GN (1997) Pattern generation. Current Opinion in Neurobiology 7:781–789.

Bal T, Nagy F, Moulins M (1988) The pyloric central pattern generator in Crustacea: a set of conditional neuronal oscillators. Journal of Comparative Physiology A: Neuroethology, Sensory, Neural, and Behavioral Physiology 163:715–727.

Bartos M, Manor Y, Nadim F, Marder E, M/P. N (1999) Coordination of fast and slow rhythmic neuronal circuits. J Neurosci 19:6650–6660.

Brown TG (1914) On the nature of the fundamental activity of the nervous centres; together with an analysis of the conditioning of rhythmic activity in progression, and a theory of the evolution of function in the nervous system. J Physiol 48:18–46.

Buono P-L (2001) Models of central pattern generators for quadruped locomotion II. Secondary gaits. Journal of Mathematical Biology 42:327–346.

Buono PL, Golubitsky M, Palacios A (2000) Heteroclinic cycles in rings of coupled cells. Physica D 143:74–108.

Calabrese RL (1995) Oscillation in motor pattern-generating networks. Current Opinion in Neurobiology 5:816–823.

Calabrese RL, Nadim F, Olsen OH (1995) Heartbeat control in the medicinal leech: a model system for understanding the origin, coordination, and modulation of rhythmic motor patterns. J Neurobiol 27:390–402.

Canavier CC, Baxter DA, Clark JW, Byrne JH (1999) Control of multistability in ring circuits of oscillators. Biological Cybernetics 80:87–102.

Canavier CC, Butera RJ, Dror RO, Baxter DA, Clark JW, Byrne JH (1997) Phase response characteristics of model neurons determine which patterns are expressed in a ring circuit model of gait generation. Biol Cybern 77:367–380.

Cheron G, Cebolla AM, Saedeleer CD, Bengoetxea A, Leurs F, Leroy A, Dan B (2007) Pure phase-locking of beta/gamma oscillation contributes to the N30 frontal component of somatosensory evoked potentials. BMC Neuroscience 8.

Cherry EM, Fenton FH (2007) A tale of two dogs: analyzing two models of canine ventricular electrophysiology. Am J Physiol 292:H43–H55.

Dror RO, Canavier CC, Butera RJ, Clark JW, Byrne JH (1999) A mathematical criterion based on phase response curves for stability in a ring of coupled oscillators. Biological Cybernetics 80:11–23.

Efimov DV (2011) Phase resetting control based on direct phase response curve. Journal of Mathematical Biology 63(5):855–879.

Eldridge FL (1991) Phase resetting of respiratory rhythm – experiments in animals and models. In: Springer Series in Synergetics, Rhythms in Physiological Systems, pp 165–175. Berlin: Springer-Verlag.

Ermentrout B (1996) Type I Membranes, Phase Resetting Curves, and Synchrony. Neural Computation 8:979–1001.

Galan RF, Ermentrout GB, Urban NN (2005) Efficient Estimation of Phase-Resetting Curves in Real Neurons and its Significance for Neural-Network Modeling. Physical Review Letters 94:158101.

Golubitsky M, Stewart I, Buono P-L, Collins JJ (1998) A modular network for legged locomotion. Physica D: Nonlinear Phenomena 115:56–72.

Golubitsky M, Stewart I, Buono P-L, Collins JJ (1999) Symmetry in locomotor central pattern generators and animal gaits. Nature 401.

Govaerts W, Sautois B (2006) Computation of the Phase Response Curve: A Direct Numerical Approach. Neural Computation 18:817–847.

Haenschel C, Linden DE, Bittner RA, Singer W, Hanslmayr S (2010) Alpha phase locking predicts residual working memory performance in schizophrenia. Biol Psychiatry 68:595–598.

Harris-Warrick RM, Marder E, Selverston AI, Moulins M (1992) Dynamic Biological Networks. The Stomatogastric Nervous System. Cambridge: MIT Press

Hartline D (1979) Pattern generation in the lobster (Panulirus) stomatogastric ganglion. II. Pyloric network simulation. Biol Cybern 33:223–236.

Hodgkin AL, Huxley AF (1952) A quantitative description of membrane current and its application to conduction and excitation in nerve. J Physiol 117:500–544.

Hooper SL, DiCaprio RA (2004) Crustacean motor pattern generator networks. Neurosignals 13:50–69.

Hoppensteadt FC, Izhikevich EM (1997) Weakly connected neural networks. New York: Springer-Verlag.

Huerta R, Sánchez-Montañés MA, Corbacho F, Sigüenza JA (2000) A Central pattern generator to control a pyloric-based system. Biological Cybernetics 82:85–94.

Ijspeert AJ (2008) Central pattern generators for locomotion control in animals and robots: A review. Neural Networks 21:642–653.

Izhikevich EM (2007) Dynamical Systems in Neuroscience: The Geometry of Excitability and Bursting: MIT Press, Cambridge, Massachusetts & London, England.

Kalb SS, Dobrovolny HM, Tolkacheva EG, Idriss SF, Krassowska W, Gauthier DJ (2004) The restitution portrait: a new method for investigating rate-dependent restitution. J Cardiovasc Electrophysiol 15:698–709.

Kopell N, Ermentrout B (1988) Coupled oscillators and the design of central pattern generators. Mathematical Biology 90:87–109.

Marder E, Bucher D (2001) Central pattern generators and the control of rhythmic movements. Current Biology 11:R986–R996.

Marder E, Bucher D (2007) Understanding circuit dynamics using the stomatogastric nervous system of lobsters and crabs. Annu Rev Physiol 69:291–316.

Miller JP, Selverston AI (1982) Mechanisms underlying pattern generation in lobster stomatogastric ganglion as determined by selective inactivation of identified neurons. II. Oscillatory properties of pyloric neurons J Physiol 48:1378–1391.

Morris C, Lecar H (1981) Voltage oscillations in the barnacle giant muscle fiber. Biophysical Journal 35:193–213.

Nakanishi J, Morimoto J, Endo G, Cheng G, Schaal S, Kawato M (2004) Learning from demonstration and adaptation of biped locomotion. Robotics and Autonomous Systems 47: 79–91.

Oprisan S. A, Canavier CC (2005) Stability criterion for a two-neuron reciprocally coupled network based on the phase and burst resetting curves. Neurocomputing 65:733–739.

Oprisan SA (2009) Stability of Synchronous Oscillations in a Periodic Network. International Journal of Neuroscience 119:482–491.

Oprisan SA (2010) Existence and stability criteria for phase-locked modes in ring neural networks based on the spike time resetting curve method. Journal of Theoretical Biology 262:232–244.

Oprisan SA, Canavier CC (2000) Phase response curve via multiple time scale analysis of limit cycle behavior of type I and type II excitability. Biophysical Journal 78:218–230.

Oprisan SA, Canavier CC (2001) Stability Analysis of Rings of Pulse-Coupled Oscillators: The Effect of Phase Resetting in the Second Cycle After the Pulse Is Important at Synchrony and For Long Pulses. Differential Equations and Dynamical Systems 9:243–258.

Oprisan SA, Canavier CC (2002) The Influence of Limit Cycle Topology on the Phase Resetting Curve. Neural Computation 14:1027–1057.

Oprisan SA, Canavier CC (2003) Stability analysis of entrainment by two periodic inputs with a fixed delay. Neurocomputing 52–54:59–63.

Oprisan SA, Canavier CC (2006) Technique for eliminating nonessential components in the refinement of a model of dopamine neurons. Neurocomputing 69:1030–1034.

Oprisan SA, Boutan C (2008) Prediction of Entrainment And 1:1 Phase-Locked Modes in Two-Neuron Networks Based on the Phase Resetting Curve Method. International Journal of Neuroscience 118:867–890.

Oprisan SA, Thirumalai V, Canavier CC (2003) Dynamics from a Time Series: Can We Extract the Phase Resetting Curve from a Time Series? Biophys J 84:2919–2928.

Oprisan SA, Prinz AA, Canavier CC (2004) Phase resetting and phase locking in hybrid circuits of one model and one biological neuron. Biophys J 87:2283–2298.

Proctor J, Holmes P (2010) Reflexes and preflexes: on the role of sensory feedback on rhythmic patterns in insect locomotion. Biological Cybernetics 102:513–531.

Proctor J, Kukillaya RP, Holmes P (2010) A phase-reduced neuro-mechanical model for insect locomotion: feed-forward stability and proprioceptive feedback. Phil Trans R Soc A 368:5087–5104

Rutishauser U, Ross IB, Mamelak AN, Schuman EM (2010) Human memory strength is predicted by theta-frequency phase-locking of single neurons. Nature 464:903–907.

Taiga Y, Taishin N, Shunsuke S (2003) Possible functional roles of phase resetting during walking. Biological Cybernetics 88:468–496.

Tazaki K, Cooke IM (1990) Characterization of Ca current underlying burst formation in lobster cardiac ganglion motorneurons. Journal of Neurophysiology 63:370–384.

Thuma JB, Morris LG, Weaver AL, Hooper SL (2003) Lobster (Panulirus interruptus) Pyloric Muscles Express the Motor Patterns of Three Neural Networks, Only One of Which Innervates the Muscles. The Journal of Neuroscience 23:8911–8920.

Tolkacheva EG, Schaeffer DG, Gauthier DJ, Krassowska W (2003) Condition for alternans and stability of the 1:1 response pattern in a "memory" model of paced cardiac dynamics. Phys Rev E 67(3 Pt 1):031904.

Wingerden MV, Vinck M, Lankelma JV, Pennartz CMA (2010) Learning-Associated Gamma-Band Phase-Locking of Action–Outcome Selective Neurons in Orbitofrontal Cortex. Journal of Neuroscience 30:10025–10038.

Xu X (2008) Complicated dynamics of a ring neural network with time delays. Journal of Physics A: Mathematical and Theoretical 41:035102.

Zacksenhouse M, Ahissar E (2006) Temporal decoding by phase-locked loops: unique features of circuit-level implementations and their significance for vibrissal information processing. Neural Computation 18:1611–1636.

Chapter 18
Phase Resetting Curve Analysis of Global Synchrony, the Splay Mode and Clustering in N Neuron all to all Pulse-Coupled Networks

Srisairam Achuthan, Lakshmi Chandrasekaran, and Carmen C. Canavier

Abstract We use phase resetting curve (PRC) theory to analyze phase-locked patterns in pulse-coupled all to all network of N neurons that receive multiple inputs per cycle. The basic principles are that the phase must be updated each time an input is received, and simultaneous inputs do not sum linearly for strong coupling, but the conductances do. Therefore, the dependence of the resetting on conductance must be known. We analyze a splay mode in which no neurons fire simultaneously, global synchrony in which all neurons fire together, and clustering modes in which the firing breaks up into a small number of clusters. The key idea is to identify the appropriate perturbation in order to determine the stability of a given mode. For the splay mode, jitter is introduced into all firing times. For synchrony, a single neuron is perturbed from the rest, and for the two cluster mode, a single neuron is perturbed from one cluster. Global synchrony can be destabilized by increasing the network size or the strength of the individual synapses. At most, a small number of M clusters form because the $M - 1$ locking points are more likely to sample destabilizing regions of the PRC as M increases. Between cluster interactions can enforce synchrony on subclusters that are incapable of synchronizing themselves. For the two cluster case, general results were obtained for clusters of any size. These results can be used to gain insights into the activity of networks of biological neurons whose PRCs can be measured.

S. Achuthan • L. Chandrasekaran • C.C. Canavier (✉)
Neuroscience Center of Excellence, LSU Health Sciences Center, New Orleans, LA 70112, USA
e-mail: sairam.achuthan@gmail.com; chandrasekaran.lakshmi@gmail.com; ccanav@lsuhsc.edu

N.W. Schultheiss et al. (eds.), *Phase Response Curves in Neuroscience: Theory, Experiment, and Analysis*, Springer Series in Computational Neuroscience 6,
DOI 10.1007/978-1-4614-0739-3_18,

1 Introduction

The spontaneous organization of individual oscillators, such as pendulum-based clocks and fireflies, into a coherent entity with a common frequency has been widely investigated (Winfree 1967; Peskin 1975; Kuramato 1984; Buck 1988; Glass and Mackey 1988; Mirollo and Strogatz 1990; Singer 1999; Winfree 2001; Strogatz 2003; de la Iglesia et al. 2004; Yuste et al. 2005; Buszaki 2006). Global synchronization is relevant to a number of important biological problems. Transiently synchronized assemblies of neurons are believed to underlie cognitive functions (Buzsaki 2006). For example, in humans, the presence of synchronization between distant brain regions predicts whether a word will be recalled (Fell et al. 2001) and whether an image is recognized as a face (Rodriguez et al. 1999). In photosensitive epilepsy, exposure to rhythmic bright lights can induce a paroxysm of synchronous brain activity resulting in a seizure. Pathological synchrony could also lead to tremor (Hammond et al. 2007), and there is consistent evidence for a reduction of synchronization in schizophrenia (Uhlhaas and Singer 2006). On the other hand, the sequential firing of oscillators, which we term a splay mode, also has biological relevance, particularly for central pattern generators (CPGs). CPGs are neural circuits that produce a repetitive pattern of motor activity (Stein et al. 1997), such as locomotion or respiration, even in the absence of sensory feedback or patterned input from higher brain centers. An influential theory (Llinas 2002; Yuste et al. 2005) suggests that cortical rhythms and circuits evolved from motor rhythms and circuits, and are therefore likely to share basic principles of organization and dynamic function. In addition, pathological rhythms such as absence epileptic seizures have been postulated to arise from phase-locking between thalamic and cortical sites (Perez Velasquez et al. 2007).

In addition to synchrony between dissimilar clusters like thalamic and cortical sites, networks of similar neurons have also been hypothesized to form clusters. Global synchrony results in a population rhythm with a frequency equal to the frequency of an individual neuron, but if the population were to break into two clusters, the population frequency would double (Foffani et al. 2007). For example, it has been suggested that a synchronized population of entorhinal stellate cells (Pervouchine et al. 2006) breaks up into two similar clusters, with each cluster firing at theta frequency in antiphase with the other cluster, resulting in a beta peak seen by Cunningham et al. (2004) in slices of entorhinal cortex. Similarly, Terman et al. (1996) proposed that the frequency of sleep spindles is twice that of the delta rhythm because the thalamocortical cells firing individually at delta frequencies split into two clusters firing in antiphase during sleep spindles. Along the same lines, Foffani et al. (2007) found that individual cells in hippocampal area CA3 could fire no faster than 300–400 Hz, and suggested the fast ripple activity at observed at up to 600 Hz results from the interaction between local generators of the ripple activity, perhaps from clusters in antiphase. It is difficult to get more than a few clusters in networks of similar neurons with similar connectivity (Golomb et al. 1994; Achuthan and Canavier 2009).

In this chapter, we apply phase resetting (PRC) theory (Glass and Mackey 1988; Kopell and Ermentrout 1988; van Vreeswijk et al. 1994; Hansel et al. 1995; Rinzel and Ermentrout 1998; Canavier et al. 1997, 1999; Goel and Ermentrout 2002; Kopell and Ermentrout 2002; Netoff et al. 2005a,b; Maran and Canavier 2008; Achuthan and Canavier 2009; Canavier et al. 2009) to the analysis of phase-locked patterns in a pulse-coupled network of oscillators in which each oscillator receives multiple inputs per firing cycle. The results obtained are robust to variability in the form of noise and heterogeneity in frequency between neural oscillators (Achuthan et al. 2011). Our PRC study is based on pulsatile coupling assumptions (see Canavier and Achuthan, this volume), namely that (1) each neuron in the network can be represented as a limit cycle oscillator, (2) that the trajectory of each neuron returns to its limit cycle between inputs rendering the coupling effectively pulsatile, and (3) that effect of each pulsatile interaction on the phase of the oscillator is known.

The simple, homogeneous model networks used to establish proof of principle in this chapter do not capture all the complexities of neural dynamics; the point of these examples is that if the appropriate PRCs can be obtained, no further information is required in order to understand network synchronization properties. The solutions for homogeneous networks approximate those of mildly heterogeneous networks, and we suggest that modulation of the PRC in such networks may be a viable strategy to promote or discourage synchronization in realistic biological systems.

2 Methods

2.1 Phase Resetting Curve

A PRC plots the normalized change in cycle period of an oscillator as a function of the phase at which a perturbation is applied. The normalized change in the length of the cycle containing the perturbation and the one following it are called the first order resetting, $f_1(\phi) = (P_1 - P_\mathrm{i})/P_\mathrm{i}$, and second order resetting, $f_2(\phi) = (P_2 - P_\mathrm{i})/P_\mathrm{i}$, respectively. P_1 is the length of the cycle containing the perturbation and P_2 is the length of the subsequent cycle (see Fig. 18.1 in this volume by Canavier and Achuthan). P_i is the intrinsic period of the oscillator. The PRC generated in this fashion should approximate the input that the neuron will receive in the circuit in order to use the PRC for network analysis. Circuits in which each oscillator receives a single input, such as a two neuron circuit or a unidirectional ring, are very amenable to PRC-based analysis of phase-locked network activity under the assumption of pulsatile coupling (Dror et al. 1999; Canavier et al. 1997, 1999; Maran and Canavier 2008; Canavier et al. 2009). For the all to all circuits addressed in this chapter, each oscillator receives $N - 1$ inputs. The possibility that some or all of these inputs arrive simultaneously must be considered, making the application of PRCs to these circuits under the assumption of pulsatile coupling more difficult.

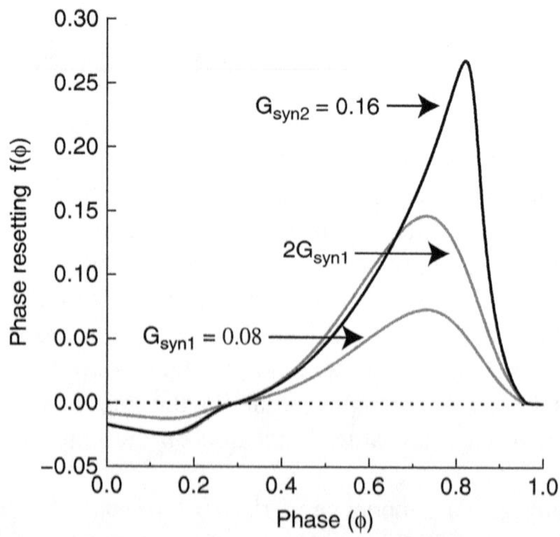

Fig. 18.1 PRCs are nonlinear with respect to conductance strengths. PRCs corresponding to $g_{syn1} = 0.08\,\text{mS/cm}^2$ (*gray color*) and $g_{syn2} = 0.16\,\text{mS/cm}^2$ (*black color*) for a type II excitable neuron with inhibitory input. For strong coupling, the resetting does not add linearly as evidenced by the changing shape as the conductance strength is increased to represent a single input ($g_{syn1} = 0.08\,\text{mS/cm}^2$), two simultaneous inputs ($g_{syn2} = 0.16\,\text{mS/cm}^2$). Note that $g_{syn2} \neq 2g_{syn1}$ (*gray color*). Here, conductance strength refers to the synaptic coupling strength of a neuron

For sufficiently weak coupling strength, it is appropriate to simply add the resetting due to simultaneous inputs. Our analysis does not assume weak coupling. In the strong coupling regime, the phase resetting due to simultaneous inputs does not sum linearly, because doubling the conductance does not double the resetting at each phase, as illustrated in Fig. 18.1. Therefore, the PRC needs to be known or measured for the conductance value that is the sum of any simultaneously active conductances. Even if the PRC of the summed inputs is not known precisely, if the trend is known, general insights into the collective synchronization behavior can be obtained.

3 Theoretical Results

3.1 Existence and Stability Criterion for the Synchronous Firing Mode

The synchronous mode exists due to symmetry in an network of all to all identical, identically connected neurons. In order to be stable, global synchrony must be robust to small perturbations. As described in Chap. 2, Goel and Ermentrout (2002) assumed a perturbation in which the firing of all neurons was separated such that they fire sequentially without changing order, then determined the PRC slopes that allowed convergence back to synchrony. In their scheme, all oscillators were absorbed into a synchronous cluster simultaneously, so the effects of sequential absorptions were not considered. On the other hand, we developed a criterion for robustness of global synchrony to the perturbation of a single neuron from global synchrony, resulting in a single neuron coupled to a cluster of $N-1$ neurons (see Fig. 18.2a). The stability criterion (Achuthan and Canavier 2009) is a straightforward extension of an existing proof for the case of two neurons (Dror et al. 1999; Oprisan et al. 2004). The within cluster effects of the $N-1$ neurons on each other are ignored, and the large cluster is treated as a single oscillator whose conductance is $N-1$ times that of the single neuron (Fig. 18.2a, left). Synchrony exists because in the fully synchronous state, each neuron receives simultaneous inputs from $N-1$ other neurons. The stability criterion for the synchronous state is derived as follows:

From Fig. 18.2a, we have:

$$\mathrm{ts}_{N-1}[k] = \mathrm{tr}_1[k-1],$$
$$\mathrm{tr}_{N-1}[k] = \mathrm{ts}_1[k].$$

The recovery interval tr_i is defined for $i = 1(N-1)$ as the time elapsed between when neuron 1 (cluster $N-1$) receives an input from the $N-1$ cluster (neuron 1) and when it next fires. The stimulus interval ts_i for $i = 1(N-1)$ is the time elapsed between when neuron 1 (or cluster $N-1$) fires and when it next receives an input. The intervals can be rewritten under the assumption of pulsatile coupling in terms of the phase at which an input is received. The stimulus interval is equal to the intrinsic period (P_i) times the phase at which in input is received, plus a correction for the second order phase resetting in the previous cycle, so $\mathrm{ts}_\mathrm{i}[k] = P_\mathrm{i}\phi_i[k] + P_\mathrm{i} f_{2,i}(\phi_i[k-1])$. The recovery interval is the time remaining in the cycle when the input is received plus the correction due to first order resetting, so $\mathrm{tr}_\mathrm{i}[k] = P_\mathrm{i} - P_\mathrm{i}\phi_i[k] + P_\mathrm{i} f_{1,i}(\phi_i[k])$. Here, $\phi_i[k]$ stands for the phase of the ith neuron (group of neurons) in the kth cycle and $f_{j,i}(..)$ stands for the jth order resetting ($j = 1$ or 2) produced by the activation of i neurons simultaneously. Substituting for these intervals in the above two equations and eliminating the common intrinsic period P_i, we get:

$$\phi_{N-1}[k] + f_{2,N-1}(\phi_{N-1}[k-1]) = 1 - \phi_1[k-1] + f_{1,1}(\phi_1[k-1]),$$
$$1 - f_{N-1}[k] + f_{1,N-1}(\phi_{N-1}[k]) = \phi_1[k] + f_{2,1}(\phi_1[k-1]).$$

Rewriting, the above equations we have:

$$\phi_{N-1}[k] = 1 - \phi_1[k-1] + f_{1,1}(\phi_1[k-1]) - f_{2,N-1}(\phi_{N-1}[k-1]),$$
$$\phi_1[k] = 1 - \phi_{N-1}[k] + f_{1,N-1}(\phi_{N-1}[k]) - f_{2,1}(\phi_1[k-1]).$$

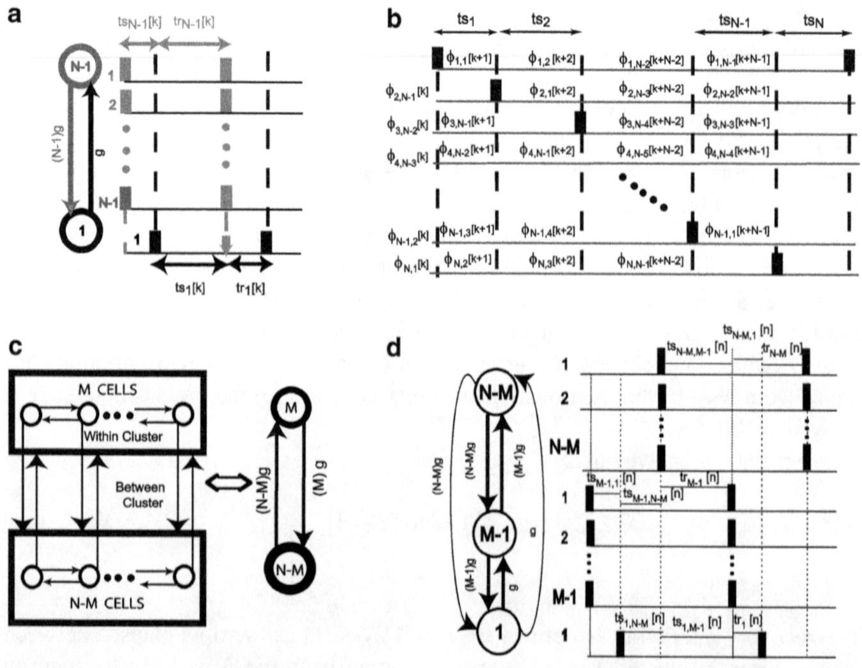

Fig. 18.2 Presumed firing patterns for stability proofs. The stimulus (ts) and response intervals (tr) in **a**, **b**, and **d** can be written in terms of the phases under the assumption of pulsatile coupling, allowing the construction of discrete maps of the successive phases. (**a**) One cluster of N neurons. A perturbation of a single neuron away from the synchronous firing was assumed. (**b**) Splay firing pattern in a network of N neurons. ϕ_{ij}[k] indicate the phase of the ith nonfiring oscillator just before the kth spike in the reference oscillator which produces the jth input to the other $N-1$ oscillators. (**c**) Two clusters of M and $N-M$ neurons, respectively. Cluster interactions were analyzed by collapsing the neurons within each cluster to a single oscillator, and by determining the existence and stability of a splay mode between two such oscillators using M times and $N-M$ times the conductance for a single synapse, respectively. This method generalized to more clusters and to larger clusters. (**d**) A perturbation map of the antiphase mode between M neurons in one cluster and $N-M$ neurons in the second cluster. A perturbation of a single neuron away from the synchronous firing of M neurons was assumed. ts$_{ij}$[k] and tr$_{ij}$[k] refer to the stimulus and recovery intervals (respectively) of the ith neuron or cluster due to an input from the jth neuron or cluster. The dashed vertical lines in (**a**), (**b**), and (**d**) indicate the firing times with an assumed sequential firing order. The strength of the synapses is indicated across the vertical arrows in (**a**), (**c**) and (**d**). The synaptic conductance strength of each neuron in (**b**) was assumed to correspond to that of a single neuron (not shown)

In the above two equations, we first assume that there is a fixed point $(\phi_1{}^*, \phi_{N-1}{}^*)$ corresponding to a one to one phased locked oscillation between the two circuit components. We then assume a perturbation $\Delta\phi_i[k]$ about this fixed point $\phi_i{}^*$ so that $\phi_i[k] = \phi_i{}^* + \Delta\phi_i[k]$, $i = 1$ or $N-1$. After linearizing the PRCs such that

$f(\phi) = f(\phi^*) + f'(\phi^*)\Delta\phi$ and cancelling the steady state terms from both sides of the equations, we obtain:

$$\Delta\phi_{N-1}[k] = (f'_{1,1}(\phi^*{}_1) - 1)\Delta\phi_1[k-1] - f'_{2,\mathrm{N}-1}(\phi^*{}_{N-1})\Delta\phi_{N-1}[k-1],$$
$$\Delta\phi_1[k] = ((f'_{1,N-1}(\phi^*{}_{N-1}) - 1)(f'_{1,1}(\phi^*{}_1) - 1) - f'_{2,1}(\phi^*{}_1))\Delta\phi_1[k-1]$$
$$+(1 - f_{1,\mathrm{N}-1}(\phi^*{}_{\mathrm{N}-1}))f'_{2,N-1}(\phi^*{}_{N-1})\Delta\phi_{N-1}[k-1].$$

The two equations above constitute a discrete linear system of two coupled difference equations.

$\Delta[k] = M\Delta[k-1]$ where Δ is a vector $\Delta[k] = [\Delta\phi_1[k], \Delta\phi_{N-1}[k]]^T$ and M is a matrix with two rows and two columns. For synchrony, $\phi^*{}_1 = 0^+$ and $\phi^*{}_{N-1} = 1^-$ or $\phi^*{}_1 = 1^-$ and $\phi^*{}_{N-1} = 0^+$. These two sets of limits correspond to approaching synchrony with neuron 1 leading or lagging the larger cluster, respectively (see Fig 18.2a). The characteristic polynomials corresponding to the matrix M evaluated at the two sets of limits approaching the synchronous fixed point are:

$$\lambda^2 - \lambda[(1 - f'_{1,1}(0^+))(1 - f'_{1,N-1}(1^-)) - f'_{2,1}(0^+) - f'_{2,N-1}(1^-)]$$
$$+f'_{2,1}(0^+)f'_{2,N-1}(1^-) = 0,$$
$$\lambda^2 - \lambda[(1 - f'_{1,N-1}(0^+))(1 - f'_{1,1}(1^-)) - f'_{2,N-1}(0^+) - f'_{2,1}(1^-)]$$
$$+f'_{2,N-1}(0^+)f'_{2,1}(1^-) = 0,$$

where λ refers to the eigenvalues corresponding to the matrix M. For stable synchrony absolute value of λ must be less than 1. In some cases (see Chap. 2), the above expression can be simplified to $\lambda = 1 - f'_{1,1}(0^+) - f'_{1,N-1}(0^+)$, from which it is easy to see that negative slopes are destabilizing, as are large positive slopes.

3.2 *Existence and Stability Criterion for the Splay Firing Mode*

The proof of Goel and Ermentrout (2002) for synchrony was based on the map given in Fig. 18.2b. Achuthan and Canavier (2009) used the same expressions in a different context, generalizing the proof to a splay firing pattern in which all neurons in the same order fire once per cycle, but no neurons fire simultaneously. Second order resetting is not considered in this analysis. In order to determine stability of a splay mode, we need to know how the phases at which an input is received at the time of any given spike depend on the phases at which an input is received at the time the previous spike was fired. Since only $N-1$ neurons receive an input at the time of each spike (no neuron sends an input to itself), we need to keep track of only $N-1$ phases per spike. The phases at which the other $N-1$ neurons receive inputs from the kth spike are $\phi_{ij}[k]$ where $j = 1$ to $N-1$. The subscript i does not change on each cycle but corresponds to the ith neuron for each of N neurons. The subscript

j is 1 for the neuron that spiked most recently and j is $N-1$ for the neuron that will spike next. The neuron that emitted the kth spike is the reference neuron for that cycle. Thus, reference neuron switches on each cycle as do the index of the phase in a given neuron (as shown in Fig. 18.2b). The switching after each spike is predictable because the firing order is presumed to be constant. We assume that a 1:1 phase locked exists in which the neurons fire in a constant sequential order. We discuss below how to determine the actual values of the phases in a one to one locked mode, but for now we assume they are $\phi^* = ([\phi_{N-1}{}^*, \phi_{N-2}{}^*, \ldots, \phi_1{}^*])$, ignoring the ith subscript and tracking only the jth. Assuming pulsatile coupling, we can write recurrence relations that give the phase prior to the $(k+1)$th spike in terms of the phases prior to the kth spike in terms of the phase resetting due to each input (Achuthan and Canavier 2009; Canavier and Achuthan 2010). If the stimulus intervals and the intrinsic periods for all neurons are equal, a nonlinear map from a spike in any neuron to the next spike in the neuron can be constructed by calculating the phase resetting that will be incurred on each cycle. Linearizing the phase resetting curves about a perturbation from the steady values allows us to write a set of coupled difference equations or a linear map $\phi[k+1] = M\phi[k]$. If all neurons are identical and if the intervals between the firing of each neuron are identical, a map for a single spike is sufficient due to symmetry. However, if the symmetry is broken, N nonlinear recurrence relations need to be combined to obtain a map from the firing of one neuron to the next firing in that same neuron. We then obtain from the recurrence relations N different resultant linearized maps M_{m}, where the m subscript refers to a map for the spike emitted by each neuron m within a cycle. The subscript m refers to a specific neuron and does not change on each cycle, since $\phi[k](= [\phi_{N-1}[k], \phi_{N-2}[k], \ldots, \phi_1[k]]^T)$ denotes the $N-1$ phases of the nonspiking neuron when the original reference neuron spikes, then $\phi[k+N]$ is the phase of the nonspiking neurons when that same neuron becomes the reference neuron again on its next spike, completing a cycle. The complete return map that is used to determine the stability of the splay mode is $\Delta\phi[k+N] = M_N M_{N-1} \ldots M_2 M_1 \Delta\phi[k]$.

For each spike, the PRCs that come into play are the ones that describe the effect of the reference neuron on each of the other neurons, thus each matrix M_{m} (given below) contains the slopes of the PRCs $(f'_{1,j})$ where j is modulo N and refers to the actual neuron (and not the firing order within the cycle) receiving an input from the reference neuron.

$$\begin{pmatrix} \alpha\left(\frac{P_{i,m+1}}{P_{i,m+2}}\right) & \left(1 - f'_{1,m+2}\left(\varphi^*_{m+2,N-2}\right)\right) & 0 & \cdots & 0 \\ \alpha\left(\frac{P_{i,m+1}}{P_{i,m+3}}\right) & 0 & \left(1 - f'_{1,m+3}\left(\varphi^*_{m+3,N-3}\right)\right) & \cdots & 0 \\ \vdots & \vdots & \vdots & \cdots & \vdots \\ \alpha\left(\frac{P_{i,m+1}}{P_{i,m+N-1}}\right) & \vdots & \vdots & \cdots & \left(1 - f'_{1,m+N-1}\left(\varphi^*_{m+N-1,1}\right)\right) \\ \alpha\left(\frac{P_{i,m+1}}{P_{i,m+N}}\right) & 0 & 0 & \cdots & 0 \end{pmatrix}$$

where $\alpha = f_{1,m+1'}(\phi_{m+1^*,N-1}) - 1$ The slopes of the first order PRC $(f'_{1m}(\phi_{m,j}))$ at the locking points of the jth input to the mth neuron determine stability of the splay mode. $P_{i,m}$ indicates the intrinsic period of the mth neuron. If all neurons are identical and all intervals are identical so that each neuron receives inputs at the same phases, the mth subscript is not necessary and a single matrix suffices.

In the computational results, we looked only at all-to-all networks of identical neurons, so we considered only splay modes in identical neurons in which all firing intervals are equal. This greatly simplifies the task of identifying the phases that give one to one locked splay firing pattern, and allows us to use only one subscript, the jth subscript that switches on each spike to indicate the neurons in the order that they will next fire. The cycle of each neuron is divided into stimulus intervals (ts_q, $q = 1, \ldots, N$) shown for neuron 1 in Fig. 18.2b. The first stimulus interval is bounded by the spike that begins the cycle and the first input, the second is bounded by the first and second inputs, and so on until the final interval is bounded by the last input and the spike that ends the cycle. We define the steady values for the stimulus intervals for a given neuron between successive inputs in Fig. 18.2b as follows. The first stimulus interval in each cycle can be written in terms of a single phase. The stimulus intervals for q from 2 to $N - 1$ can be calculated by adding the first order resetting due to the input that marks the start of the interval to the difference between the phases at which the inputs bounding the interval were received. The stimulus interval, $\mathrm{ts}_N[\infty]$ is the recovery interval due to the last input in the cycle. Thus,

$$\begin{aligned}
\mathrm{ts}_1[\infty] &= P_{\mathrm{i}}\{\phi_1^*\},\\
\mathrm{ts}_q[\infty] &= P_{\mathrm{i}}\{\phi_q^* - \phi_{q-1}^* + f_1(\phi_{q-1}^*)\},\\
\mathrm{ts}_N[\infty] &= P_{\mathrm{i}}\{1 - f_{N-1}^* + f_1(\phi_{N-1}^*)\}.
\end{aligned}$$

We use an iterative method to find the fixed points corresponding to a splay mode with N equal stimulus intervals. We sampled $\phi^*{}_{N-1}$ between 0 and 1 to determine the values that produce the same value of $\mathrm{ts}_N[\infty]$ whether it is calculated (1) directly from the expression for $\mathrm{ts}_N[\infty]$ or (2) iteratively from the expression for $\mathrm{ts}_q[\infty]$ starting from the value of $\phi^*{}_1$ that produces the same value of $\mathrm{ts}_N[\infty]$ as the direct method. We then applied the stability criterion to determine whether the splay mode was stable.

3.3 Existence and Stability Criterion for the Cluster Firing Mode

Achuthan and Canavier (2009) originally analyzed clusters by treating within and between cluster interactions as separate. The stability of synchrony within each subcluster in isolation was determined, and then each cluster was treated as a single oscillator (with a conductance strength scaled by the number of simultaneously active synapses—see Fig. 18.2c) in order to determine if any splay mode between clusters exists and if so, to determine the stability of the splay mode. The simplifying assumption was made that the synchronized subcluster has a phase response similar to that of a single neuron within the subcluster. It is possible to perform resetting experiments by giving each neuron in the intact subcluster the same input, but we did not find this necessary because the splay modes, including the antiphase mode, exhibited by the clusters closely matched those predicted using single neuron PRCs. On the other hand, treating between and within cluster interactions separately did not always produce correct stability results. Therefore, we developed a novel generic proof for determining the stability of two clusters of arbitrary size (M neurons in the first cluster and $N - M$ neurons in the second cluster) (Chandrasekaran et al. 2011). The fixed point corresponding to the antiphase mode is determined by considering each cluster as a single oscillator. The main idea for the proof was to consider perturbation of a single neuron within either cluster; keeping the rest of the cluster configuration intact (see Fig. 18.2d). For simplicity of notation, we define $g_j(\phi) = 1 - f_{1j}{}'(\phi)$, where $f_{1j}{}'$ is the slope of the first order PRC at the locking point ϕ produced by the activation of j synapses simultaneously. If all oscillators are presumed identical the eigenvalues that determine the stability of the two cluster mode are given by:

$$\lambda_1 = g_1(\phi^*) \cdot g_{M-1}(\phi^*) \cdot g_{N-M}(\phi^*),$$

$$\lambda_2 = \{g_1(1) \cdot g_{M-1}(0) - f_{21}'(1)\} \cdot g_{N-M}(\phi^*).$$

The above expressions correspond to a perturbation in which a single neuron lags the rest of the cluster. The expression also needs to be evaluated with the jth indices switched for 1 and $M - 1$ to evaluate the perturbation in which the single neuron leads. The terms that are evaluated at 0 and 1 represent within cluster interactions, whereas the terms that are evaluated at ϕ^* represent the between cluster interactions. If you ignore second order resetting at a phase of zero, the eigenvalue λ_2 is exactly the eigenvalue that determines within cluster stability scaled by a between cluster term $g_{N-M}(\phi^*)$. It is this term that allows the between cluster antiphase mode to stabilize within cluster synchrony. For clusters to be stable both $|\lambda_1|$ and $|\lambda_2|$ must be less than one. This proof for two clustered solution is quite general and is applicable even if the two clusters are composed of dissimilar neurons. In this case, the expressions also need to be evaluated with each cluster taken as the perturbed cluster M in Fig. 18.2d. A rigorous proof for cluster modes when there are more than two clusters is an open question, but the two cluster case is most likely to be physiologically significant.

4 Computational Results for Four Neuron Networks

In order to test our analyses of the modes (see Fig. 18.3e, f, g) described above, we applied them to four networks of four identical, identically coupled neurons. The PRC shapes of type I or type II (see below) model neurons in response to different synaptic polarities (inhibition or excitation) are shown in Figs. 18.3a1, 18.3b1, 18.3c1, and 18.3d1. Two neural models were chosen as exemplars of different excitability types as originally described by Hodgkin (1948). Neurons with Hodgkin's type I excitability exhibit a gradual transition from quiescence to repetitive spiking as the magnitude of an applied depolarizing current step is increased. At the onset of spiking, the frequency is arbitrarily low. Oscillators with type I excitability act as integrators (Izhikevich 2007) and generally have only advances due to brief excitations and delays due to brief inhibitions (Ermentrout 1996). Neurons with Hodgkin's type II excitability exhibit an abrupt transition from quiescence to repetitive spiking as the magnitude of an applied depolarizing current step is increased. Spiking cannot be sustained below a threshold frequency. Oscillators with type II excitability act as resonators (Izhikevich 2007) and can exhibit both advances or delays in response to either excitation or inhibition depending upon the timing of the input (Ermentrout 1996). The Wang and Buzsaki 1996 (WB) model as commonly parameterized is type I (see Appendix). Although the Morris and Lecar (1981) (ML) model can be parameterized in either regime, here we have used type II parameters (see Appendix).

In Fig. 18.3, the intervals between firing events were plotted on the y-axis, as measured during the integration of the differential equations that describe the network of model neurons (second column) or as predicted from analytical methods (third column). Synchronous firing of a group of neurons was considered as a single event. The analytical methods (Figs. 18.3a3, 18.3b3, 18.3c3, and 18.3d3) correctly predicted what parameter values would support global synchrony, the splay mode, and the two cluster mode (Figs. 18.3a2, 18.3b2, 18.3c2, and 18.3d2) in all four networks. The antiphase mode between two clusters was correctly predicted to exist and be stable at parameter values at which isolated clusters of two neurons cannot synchronize. In this case, the antiphase mode between clusters enforces within cluster synchrony, as follows. For $f_{1,N-M}(\phi^*)$ between 0 and 1, as is the case for this example, the factor $g_{N-M}(\phi^*)$ is also between 0 and 1. Therefore, scaling the eigenvalue that determines whether cluster stability by this factor decreases its magnitude so that the absolute value is less than one for the range of conductances in which this mode is observed. As the conductance is increased, the eigenvalue becomes too large for the scale factor to stabilize.

5 Computational Results for a Twelve Neuron Network

Although the rigorous result for two clusters that includes within and between cluster interactions is specific to two clusters, treating within and between cluster interactions separately works reasonably well for more clusters in large networks. As an illustrative example, we use a twelve neuron network that is a scaled up version of the four neuron network with inhibitory inputs to the type I neurons illustrated in Fig. 18.3a1. In this network, we predicted and observed multistability between the fully synchronous mode (not shown), two clusters of six (Fig. 18.4a2),

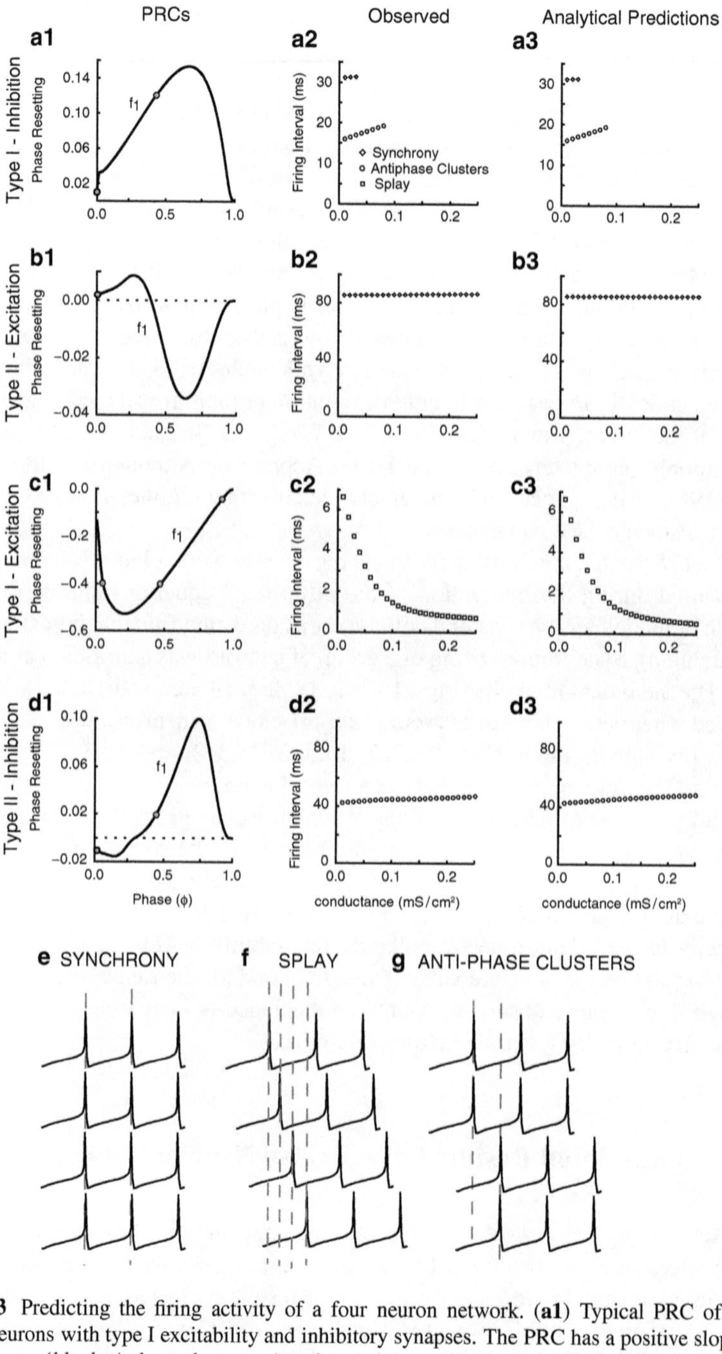

Fig. 18.3 Predicting the firing activity of a four neuron network. (**a1**) Typical PRC of a WB model neurons with type I excitability and inhibitory synapses. The PRC has a positive slope both at synchrony (*black circle* at phase zero) and at antiphase (*black circle filled with gray* at a phase < 0.5). (**a2**) Bistability between global synchrony and antiphase clusters of two was observed in the simulations of the full system of differential equations, with global synchrony losing stability first.

and three clusters of four (Fig. 18.4b2). Modes with more than three clusters (Fig. 18.4c2) were found but were always unstable. We can explain this result intuitively as follows. Most PRCs are not monotonically increasing, but instead have one or more regions of destabilizing slope. If there are only two clusters, each cluster receives one input from another cluster. If the locking point for this input falls in a stabilizing region of slope on the PRC, in general the mode will be stable. However, each additional cluster increases by one, the number of locking points, and the locking points are spread out across the PRC. The more clusters, the greater the chance that some of the points will fall in highly destabilizing regions of steep negative slopes. The more points that fall in a region of de-stabilizing slope, the less likely the cluster solution is to be stable, so solutions with fewer clusters are much more likely to be observed in practice. This is in direct agreement with the finding by Golomb et al. (1994) that the number of clusters is usually small even for large networks of N neurons. Therefore, the two cluster case is probably the most relevant. The stability proof for two cluster solution is generalizable to any N and clusters of different sizes.

Figure 18.4a2 shows that the two clusters of six mode is analytically predicted to lose stability at $g_{\text{syn}} > 0.02\,\text{mS/cm}^2$ (vertical dashed line), which coincides with the point at which this solution ceases to be observed in the full system of differential equations (black dash-gap style horizontal line). On the other hand, the within cluster mode for three clusters of four (Fig. 18.4b2) is predicted to loose stability at a smaller value of conductance than the value at which it ceases to be observed in the full system. The lack of precise agreement is probably due to the stabilizing effects of the between cluster interaction that we do not take into account for clusters larger than two. These effects enable the cluster mode to persist beyond the conductance value at which isolated subclusters lose the ability to synchronize themselves. The between cluster splay mode remains stable for all conductances

Fig. 18.3 (continued) (**a3**) Corresponding predictions by the PRC-based method (Fig. 18.2**a–c**). (**b1**) Typical PRC of a ML model neurons with type II excitability and excitatory synapses. The PRC has a positive slope at synchrony (*black circle*) but not at antiphase. (**b2**) Global synchrony was observed to be stable across the parameter range in the full system. (**b3**) Corresponding predictions by the PRC-based method. (**c1**) Typical PRC of a WB model neurons with type I excitability and excitatory synapses. The PRC has a negative slope at phase zero and positive slope for the last input (*black circle* at a phase ≈ 0.8). (**c2**) A sequential order of firing, i.e., a splay mode was observed to be stable in the full system. (**c3**) Corresponding predictions by the PRC-based method. (**d1**) Typical PRC of a ML model neuron with type II excitability and inhibitory synapses. The PRC has a negative slope at synchrony (*black circle* at phase zero) and positive at antiphase (*black circle* at a phase < 0.5). (**d2**) The antiphase mode between two clusters was observed in the full system to be stable at the intervals predicted by the existence criteria despite the absence of within cluster synchrony. (**d3**) Corresponding prediction by the PRC-based method. The phase-locked patterns that we can predict are (**e**) the fully synchronous mode (one cluster of four neurons), (**f**) four clusters of one neuron each in the splay mode and (**g**) two clusters of two neurons in antiphase with each other. Other modes are possible but were not examined here. The *vertical dashed lines* indicate the firing intervals plotted in columns 2 and 3 above

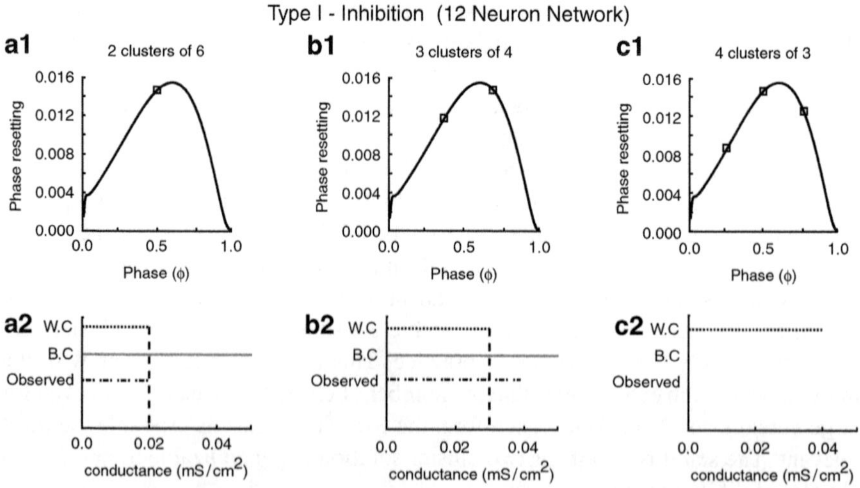

Fig. 18.4 Cluster predictions for three networks of twelve type I neurons with inhibitory inputs. The model neurons are type I WB neuron coupled by inhibition. PRCs are shown for g_{syn} = $0.01\,mS/cm^2$. The *black squares* on each PRC indicate the phases at which each cluster receives inputs from the other clusters (one, two, and three, respectively) in the network. The observed results are indicated by black dash-gap style horizontal lines. These results were predicted by an analytically derived method that considered separately the within cluster interactions (*black dotted horizontal lines*; denoted as W.C) and between cluster interactions (*gray horizontal lines*; denoted as B.C). The within cluster interactions were characterized by determining the stability of synchrony in an isolated M neuron network with g_{syn} set to the value for a single synapse. M in this case assumed a value of 3 (4 clusters of 3), 4 (3 clusters of 4) or 6 (2 clusters of 6). The between cluster interactions were characterized as before (see Fig. 18.2b). (**a**) Two clusters of six neurons. (**a1**) PRC. (**a2**) Two clusters of size six was observed to be stable up to $g_{syn} = 0.02\,mS/cm^2$. Within cluster synchrony was stable until $g_{syn} = 0.02\,mS/cm^2$. The splay mode captured the between cluster interaction and remained stable beyond $g_{syn} = 0.10\,mS/cm^2$. The dashed vertical line indicates that the two clusters of six neurons mode was predicted to be stable until g_{syn}= $0.02\,mS/cm^2$. (**b**) Three clusters of four neurons. (**b1**) PRC. (**b2**) Three clusters of size four was observed to be stable upto $g_{syn} = 0.04\,mS/cm^2$. Within cluster synchronous mode was stable until $g_{syn} = 0.03\,mS/cm^2$. The splay mode between clusters was predicted to stable for all conductance values shown. The *vertical line* indicates that the three clusters of four neurons mode was predicted to be stable until $g_{syn} = 0.03\,mS/cm^2$ in the regime where both within and between cluster modes were predicted to be stable. The stability of the cluster mode persists beyond the predicted region. (**c**) Four clusters of size three. (**c1**) PRC. (**c2**) Four clusters of size three were observed. The within cluster synchronous mode was stable until $g_{syn} = 0.04\,mS/cm^2$. The between cluster splay mode exists but was not stable for the given range of conductance values, which explains why it was not observed in the full system

examined. At large values of conductance, the inhibitory postsynaptic potential essentially clamps the membrane potential at the synaptic reversal potential so that the PRC does not change as synaptic strength is increased further. For four clusters of three (Fig. 18.4c2), the analytical method correctly predicted that this mode would not be observed in the full system of differential equations because the between cluster splay mode is predicted to be unstable at all conductance values. The black squares on each PRC (Fig. 18.4a1, 18.4b1 and 18.4c1) indicate the phases at which each cluster receives inputs from the other clusters (one, two, and three, respectively) in the network. Note that in both 4b1 and 4c1, one square falls in the destabilizing negative slope region, and that the slope is steeper and more destabilizing for Fig. 18.4c1. As the number of clusters increase, more slopes fall in the steeper negative regions, and the positive slopes cannot stabilize the negative one(s) as they do in Fig. 18.4b1.

6 Generalization to Heterogeneous Networks

PRCs accurately predict the phase-locked synchronous, splay, and cluster modes observed in networks of all to all identical and identically connected model neurons (Achuthan and Canavier 2009). However, in reality neuronal circuits are inherently heterogeneous and noisy (Mancilla et al. 2007). Nonetheless, the simplest cases in which we study synchronization frequently involve networks of identical, identically connected oscillators. When heterogeneity is introduced by allowing variability in the intrinsic frequency or coupling conductances, the solution structure of the homogeneous network is perturbed, such that exact synchrony within a cluster becomes near synchrony, and a symmetric splay mode between clusters, such as the antiphase mode, becomes a near antiphase mode. Figure 18.5 shows that the approximate solution structure is not preserved at very weak coupling strengths, but increasing the coupling strength above a threshold approximately restores the solution structure. The robustness of the solutions to heterogeneity can be predicted using the structural stability (Sieling et al. 2009) of the PRC-based prediction methods (Achuthan et al. 2011). Below the threshold, predictions made using model neurons with slightly different frequencies cause the fixed points corresponding to the approximate solutions to disappear.

7 Conclusions

Using phase resetting curves, three distinct types of neural firing modes were successfully predicted by analytically derived methods when applied to pulse-coupled N neuron networks with arbitrary connection strength. The analytically derived methods that we utilized for predictions required only the slopes of the PRCs at the locking points. Synchronous clusters were shown to be destabilized by increasing cluster size or increasing the individual synaptic conductances, because both of these manipulations increased the slope of the PRC. If a clustering mode was predicted to exist and be stable based on considering between and within cluster interactions it was always observed. However, considering the interactions between two clusters revealed additional cluster modes that were stabilized by the between

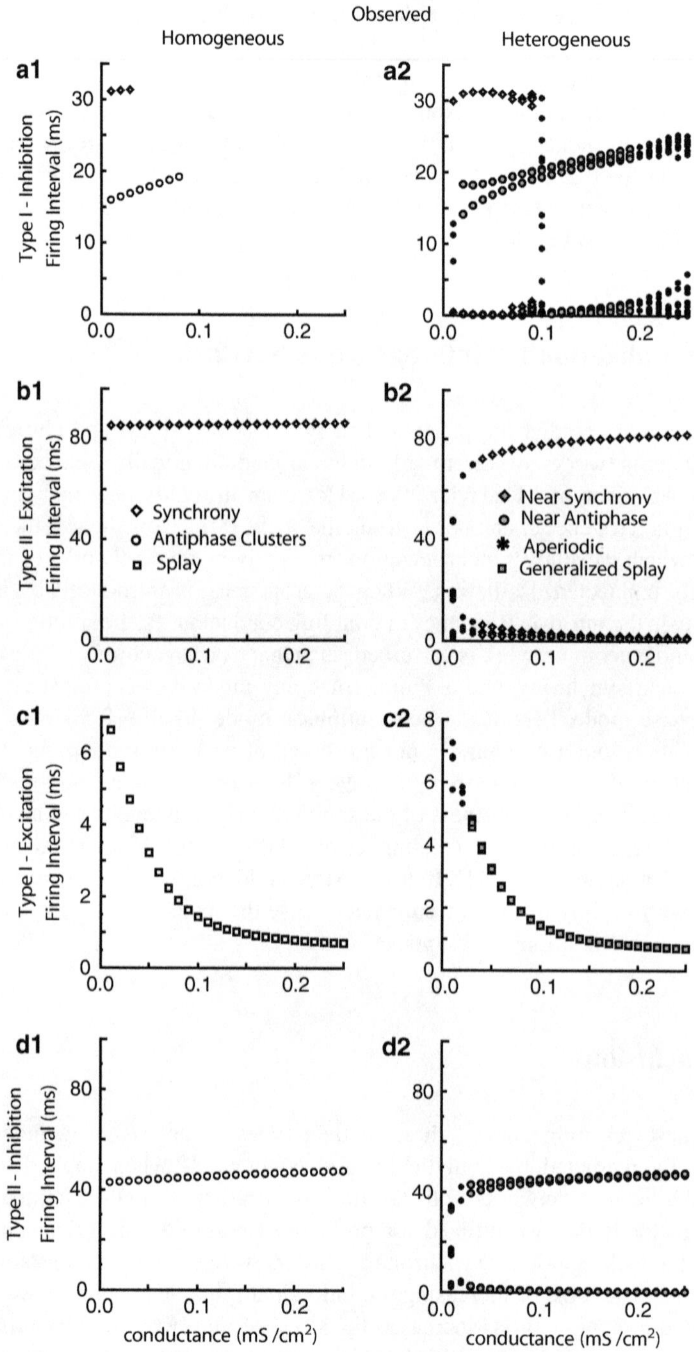

Fig. 18.5 Dynamics of homogenous networks are robust to small amounts of heterogeneity. Heterogeneous four neuron networks were produced by introducing 1% variability in the intrinsic frequencies of the four neurons. The observed homogeneous network dynamics resulting

cluster interactions. In large networks, it was shown to be progressively more difficult to find PRCs that support multiple clusters. When heterogeneity was introduced by allowing variability in the intrinsic frequency or coupling conductances, the solution structure of the homogeneous network was approximately preserved for strong but not weak coupling.

Acknowledgments This work was supported by NIH grants NS54281 and MH85387 under the CRCNS program.

Appendix

The differential equations for the Wang and Buzsaki (1996) and Morris and Lecar 1981 networks were simulated using a variable step size implicit fifth order Runge–Kutta method (Hairer 1991). The current balance equation for each WB model neuron is:

$$\mathrm{C}\mathrm{d}V/\mathrm{d}t = -I_{\mathrm{Na}} - I_{\mathrm{K}} - I_{\mathrm{L}} - I_{\mathrm{syn}} + I_{\mathrm{stim}},$$

where the capacitance $C = 1\ \mu\mathrm{F/cm}^2$, V is the cell membrane voltage in millivolts, and t is time in milliseconds. The leak current is given by $I_{\mathrm{L}} = g_{\mathrm{L}}(V - E_{\mathrm{L}})$. The sodium current is given by $I_{\mathrm{Na}} = g_{\mathrm{Na}} m_{\infty}{}^3 h(V - E_{\mathrm{Na}})$. The steady state activation $m_{\infty} = \alpha_{\mathrm{m}}/(\alpha_{\mathrm{m}} + \beta_{\mathrm{m}})$ where $\alpha_{\mathrm{m}}(V) = -0.1(V + 35)/\{\exp[-0.1(V + 35)] - 1\}$ and $\beta_{\mathrm{m}}(V) = 4\exp[-(V + 60)/18]$. The rate equation for the inactivation variable h in the expression for sodium current is:

$$\mathrm{d}h/\mathrm{d}t = \varphi\{\alpha_{\mathrm{h}}(V)(1 - h) - \beta_h(V)h\}$$

where $\varphi = 5$. (The symbol was changed from ϕ to φ to avoid confusion with the symbol for phase.) The rate constants for the inactivation variable h are given by $\alpha_h(V) = 0.07\exp[-(V + 58)/20]$ and $\beta_h(V) = 1/\{\exp[-0.1(V + 28)] + 1\}$. The potassium current is given by $I_{\mathrm{K}} = g_{\mathrm{K}} n^4(V - E_{\mathrm{K}})$, where the activation variable n satisfies the following equation:

$$\mathrm{d}n/\mathrm{d}t = \varphi\{\alpha_n(V)(1 - h) - \beta_n(V)n\},$$

Fig. 18.5 (continued) from integrating the full systems of differential equations were adapted from Fig. 18.3 (second column) and compared to those of the heterogeneous network dynamics. The qualitative correspondence is evident except at the weakest conductance strengths. Symbols that occur near the x-axis indicate small within cluster intervals that are observed during approximate but not exact synchrony where the rate constants for n are $\alpha_n(V) = -0.01(V+34)/\{\exp[-0.1(V+34)]-1\}$ and $\beta_n(V) = 0.125\exp[-(V+44)/80]$. The reversal potentials E_{Na}, E_{K}, and E_{L} were set to 55, −90, and −65 mV, respectively. The maximal sodium (g_{Na}), potassium (g_{K}), and leak (g_{L}) conductances were set to 35, 9 and 0.1 mS/cm^2, respectively. I_{stim} is the applied current and was set at 0.5 μA/cm^2. Unless otherwise stated, the values for the various parameters were equal to those given above.

The synaptic current is given by $I_{\mathrm{syn}} = g_{\mathrm{syn}}s(V - E_{\mathrm{syn}})$, where g_{syn} is the maximum synaptic conductance and E_{syn} is equal to −75 mV for inhibitory synaptic connectivity and equal to 0 mV for excitatory synaptic connectivity. The rate of change of the gating variable s is given by:

$$\mathrm{d}s/\mathrm{d}t = \alpha T(V_{\mathrm{pre}})(1-s) - s/\tau_{\mathrm{syn}},$$

with $T(V_{\mathrm{pre}}) = 1/[1+\exp(-V_{\mathrm{pre}}/2)]$, where V_{pre} is the voltage of the presynaptic cell, $\alpha 6.25\,\mathrm{ms}^{-1}$ is the rate constant of the synaptic activation (Bartos et al. 2001), and τ_{syn} is the synaptic decay time constant and was set to 1.0 ms.

The current balance equation for each ML model neuron is:

$$C\mathrm{d}V/\mathrm{d}t = -I_{\mathrm{Ca}} - I_{\mathrm{K}} - I_{\mathrm{L}} - I_{\mathrm{syn}} + I_{\mathrm{stim}},$$

where the capacitance $C = 20\,\mu\mathrm{F/cm}^2$, V is the cell membrane voltage in millivolts, and t is time in milliseconds. The calcium current is given by $I_{\mathrm{Ca}} = g_{\mathrm{Ca}}m_{\infty}(V)(V - E_{\mathrm{Ca}})$. The leak current is given by $I_{\mathrm{L}} = g_{\mathrm{L}}(V - E_{\mathrm{L}})$. The steady state activation is:

$$m_{\infty}(V) = 0.5\left[1 + \tan h\left\{(V - V_1)/V_2\right\}\right],$$

where $V_1 = -1.2$ mV and $V_2 = 18$ mV. The potassium current is given by $I_{\mathrm{K}} = g_{\mathrm{K}}w(V - E_{\mathrm{K}})$. The rate equation for the activation variable w in the expression for the potassium current is:

$$\mathrm{d}w/\mathrm{d}t = \varphi(w_{\infty}(V) - w)/\tau_{\mathrm{w}}(V),$$

where φ was set to 0.04 with the steady state activation amplitude as:

$$w_{\infty}(V) = 0.5\left[1 + \tan h\left\{(V - V_3)/V_4\right\}\right]$$

and activation rate as:

$$\tau_{\mathrm{w}}(V) = 1/\cos h\left\{(V - V_3)/2V_4\right\}$$

with $V_3 = 2$ mV and $V_4 = 30$ mV. The reversal potentials E_{Ca}, E_{K}, and E_{L} were set to 120, −84, and −60 mV, respectively. The maximal potassium (g_{K}) and leak (g_{L}) conductances were set to 8.0 and 2.0 mS/cm^2, respectively. For calcium, the maximal conductance maximal (g_{Ca}) was set to $4.4\,\mathrm{mS/cm^2}$ and I_{stim} was set at $100.0\,\mu\mathrm{A/cm^2}$. Unless otherwise stated, the values for the various parameters for the type II excitability regime were equal to those given above and were taken from Rinzel and Ermentrout (1998). The synaptic coupling current was the same as in the case of the WB model, except that the synaptic decay time constant τ_{syn} was set to 10.0 ms.

References

Achuthan S and Canavier CC (2009). Phase resetting curves determine synchronization, phase-locking, and clustering in networks of neural oscillators. *J. Neurosci.*, 29:5218–5233.

Achuthan S, Sieling FH, Prinz AA and Canavier CC (2011). Phase resetting curves in presence of heterogeneity and noise, in The dynamic brain an exploration of neuronal variability and its functional significance, M. Ding and D. Glanzman (Eds.), Oxford University Press, pp 104–118.

Bartos M, Vida I, Frotscher M, Geiger JRP and Jonas P (2001). Rapid signaling at inhibitory synapses in a dentate gyrus interneuron network. *J. Neurosci.*, 21:2687–2698.

Buck J (1988). Synchronous rhythmic flashing of fireflies II. *Quart. Rev. Biol.*, 63:265–289.

Buszaki G (2006). *Rhythms of the brain.* Oxford University Press Inc., New York.

Canavier CC, Gurel Kazanci F, and Prinz AA (2009). Phase resetting curves allow for simple and accurate prediction of robust N:1 phase locking for strongly coupled neural oscillators. *Biophys. J.*, 97:59–73.

Canavier CC, Butera RJ, Dror RO, Baxter DA, Clark JW and Byrne JH (1997). Phase response characteristics of model neurons determine which patterns are expressed in a ring circuit model of gait generation. *Biol. Cybern.*, 77:367–380.

Canavier CC, Baxter DA, Clark JW and Byrne JH (1999). Control of multistability in ring circuits of oscillators. *Biol. Cybern.*, 80:87–102.

Chandrasekaran L, Achuthan S and Canavier CC (2011). Stability of two clustered solutions in networks of neuronal oscillators using phase resetting curves. *J. Comput. Neurosci.*, 2:427–445.

Cunningham MO, Pervouchine DD, Kopell NJ and Whittington MA (2004). Cellular and network mechanisms of slow activity (<1 Hz) in the entorhinal cortex. *Soc. for Neurosci. Abstr.*, 638:9.

de la Iglesia HO, Cambras T, Schwartz WJ, and Diez-Noguera A (2004). Forced desynchronization of dual circadian oscillators within the rat suprachiasmatic nucleus. *Curr. Biol.*, 14:796–800.

Dror RO, Canavier CC, Butera RJ, Clark JW and Byrne JH (1999). A mathematical criterion based on phase response curves for stability in a ring of coupled oscillators. *Biol. Cybern.*, 80:11–23.

Ermentrout GB (1996). Type I membranes, phase resetting curves, and synchrony, *Neural Comput.*, 8:979–1002.

Fell J, Klaver P, Lehnertz K, Grunwald T, Schaller C, Elger CE, and Fernandez G (2001). Human memory formation is accompanied by rhinal-hippocampal coupling and decoupling. *Nat. Neurosci.*, 4:1259–1264.

Foffani G, Uzcategui YG, Gal B, and de la Prida LM (2007). Reduced spike-timing reliability correlates with the emergence of fast ripples in the rat epileptic hippocampus. *Neuron*, 55: 930–941.

Glass L and Mackey MC (1988). *From Clocks to Chaos: the Rhythms of Life*. Princeton, N.J. Princeton University Press.

Goel P and Ermentrout GB (2002). Synchrony, stability, and firing patterns in pulse-coupled oscillators. *Physica D*, 163(3– 4):191–216.

Golomb D, Wang XJ and Rinzel J (1994). Synchronization properties of spindle oscillations in a thalamic reticular nuclues model. *J. Neurophysiol.*, 72:1109–1126.

Hammond C, Bergman H and Brown P (2007). Pathological synchronization in Parkinson's disease: networks, models and treatments. *TINS*, 30:357–364.

Hansel D, Mato G and Meunier C (1995). Synchrony in excitatory neural networks. *Neural Comput.*, 7:307–337.

Hodgkin AL (1948). The local electric changes associated with repetitive action in a non-medullated axon. *J. Physiol.*, 107:165–181.

Izhikevich EM (2007). *Dynamical systems in neuroscience: the geometry of excitability and bursting*. Cambridge, MIT.

Kopell N and Ermentrout GB (1988). Symmetry and phase locking in chains of weakly coupled oscillators, *Commun. Pure Appl. Math.*, 90:87–109.

Kopell, N., & Ermentrout, G. B. (2002). Mechanisms of phase-locking and frequency control in pairs of coupled neural oscillators. In: B Fiedler, ed. *Handbook on dynamical systems*. Elsevier, New York, 3–54.

Kuramoto Y (1984). *Chemical Oscillations, Waves and Turbulence*. Springer, Berlin.

Llinas R. *I of the Vortex: From Neurons to Self.* MIT, Cambridge Massachusetts, 2002.

Luo C, Clark JW Jr, Canavier CC, Baxter DA and Byrne JH (2004) Multimodal behavior in a four neuron ring circuit: mode switching. *IEEE Trans. Biomed. Eng.* 51:205–218.

Mancilla JG, Lewis TJ, Pinto DJ, Rinzel J and Connors BW (2007). Synchronization of electrically coupled pairs of inhibitory interneurons in neocortex. *J. Neurosci.*, 27:2058–2073.

Maran SK, and Canavier CC (2008). Using phase resetting to predict 1:1 and 2:2 locking in two neuron networks in which firing order is not always preserved. *J. Comput. Neurosci.*, 24:37–55.

Mirollo RE and Strogatz SH (1990). Synchronization of pulse-coupled biological oscillators. *SIAM J. Appl. Math.*, 50:1645–1662.

Morris C and Lecar H (1981). Voltage oscillations in the barnacle giant muscle fiber. *Biophys. J.*, 35:193–213.

Netoff TI, Acker CD, Bettencourt JC and White JA (2005a). Beyond two-cell networks: experimental measurement of neuronal responses to multiple synaptic inputs. *J Comput. Neurosci.*, 18:287–295.

Netoff T, Banks M, Dorval A, Acker C, Haas J, Kopell N, and White J (2005b). Synchronization in hybrid neuronal networks of the hippocampal formation. *J. Neurophysiol.*, 93:1197–1208.

Oprisan, SA, Prinz, AA, Canavier, CC (2004). Phase resetting and phase locking in hybrid circuits of one model and one biological neuron, *Biophys. J.*, 87:2283–2298.

Perez Velasquez JL, Galan RF, Dominguez LG, Leshchenko Y, Lo S, Belkas J and Guevara Erra R (2007). Phase response curves in the characterization of epileptiform activity. *Phys. Rev. E*, 76:061912.

Peskin CS (1975). *Mathematical Aspects of Heart Physiology*, Courant Institute of Mathematical Sciences, New York University, New York, pp. 268–278.

Pervouchine DD, Netoff TI, Rostein HG, White JA, Cunningham MO, Whittington MA and Kopell NJ (2006). Low dimensional maps encoding dynamics in entorhinal cortex and hippocampus. *Neural Comput.*, 18:2617–2650.

Rinzel J and Ermentrout GB (1998). Analysis of neural excitability and oscillations. In C. Koch and I. Segev (Eds), *Methods in neuronal modeling from ions to networks*. Cambridge, MIT.

Rodriguez E, George N, Lachaux JP, Martinerie J, Renault B, and Varela F (1999). Perception's shadow: Long distance synchronization of human brain activity. *Nature*, 397:430–433.

Sieling FH, Canavier CC and Prinz AA (2009). Predictions of phase-locking in excitatory hybrid networks: Excitation does not promote phase-locking in pattern generating networks as reliably as inhibition.*J. Neurophysiol.*, 102:69–84.

Singer W (1999). Neural synchrony: a versatile code for definition of relations. *Neuron*, 24:49–65.

Stein SG, Grillner S, Selverston AI and Stuart DG, eds (1997). *Neurons, networks, and motor behavior*. Cambridge, MIT.

Strogatz SH (2003). Sync: *How order emerges from chaos in the universe, nature, and daily life*. New York: Hyperion.

Terman D, Bose A and Kopell NJ (1996). Functional reorganization in thalamocortical networks: Transition between spindling and delta sleep rhythms. *PNAS*, 93:15417–15422.

Uhlhaas PJ and Singer W (2006). Neural synchrony in brain disorders: relevance for cognitive dysfunctions and pathophysiology. *Neuron*, 52:155–168.
van Vreeswijk, C. A., Abbott, L. F., & Ermentrout, G. B. (1994). Inhibition, not excitation, synchronizes coupled neurons. *J. Comp. Neurosci.*, 1:303–313.
Wang XJ and Buzsaki G (1996). Gamma oscillation by synaptic inhibition in a hippocampal interneuronal network model. *J. Neurosci.*, 16:6402–6413.
Winfree AT (1967). Biological rhythms and the behavior of populations of coupled oscillators. *J. Theor. Biol.*, 16:15–42.
Winfree AT (2001). *The geometry of biological time*. Springer, New York.
Yuste R, MacLean JN, Smith J and Lansner A (2005). The cortex as a central pattern generator. *Nat. Rev. Neurosci.*, 6:477–483.

Chapter 19
Effects of the Frequency Dependence of Phase Response Curves on Network Synchronization

Christian G. Fink, Victoria Booth, and Michal Zochowski

Abstract Neuronal phase response curves (PRCs) generally fall into one of two classes. Type I PRCs exhibit exclusively phase advances and lead to decreased propensity for synchronization of excitatory networks, while Type II PRCs show regions of both phase delay and phase advance and better facilitate synchronization of excitatory networks. One little-investigated feature of neuronal PRCs is that they are attenuated as neuronal firing frequency increases. Interestingly, Type II PRCs often experience greater attenuation of their phase delay region compared to their phase advance region, while Type I PRCs typically show uniform attenuation of phase shifts. We simulate large-scale excitatory networks of Morris–Lecar neurons in order to investigate the effects of these phenomena upon network synchrony, and we show that they lead to Type I network synchrony *increasing* with increased frequency and Type II network synchrony *decreasing* with increased frequency.

1 Introduction

It is well known that individual neurons respond differently to external stimulation. Different response properties are primarily due to different excitation properties of

C.G. Fink (✉)
Department of Physics, University of Michigan Ann Arbor, MI 48109, USA
e-mail: tcgfink@umich.edu

V. Booth
Departments of Mathematics and Anesthesiology, University of Michigan Ann Arbor, MI 48109, USA
e-mail: vbooth@umich.edu

M. Zochowski
Department of Physics, University of Michigan Ann Arbor, MI 48109, USA

Biophysics Program, University of Michigan Ann Arbor, MI 48109, USA
e-mail: michalz@umich.edu

N.W. Schultheiss et al. (eds.), *Phase Response Curves in Neuroscience: Theory, Experiment, and Analysis*, Springer Series in Computational Neuroscience 6,
DOI 10.1007/978-1-4614-0739-3_19,

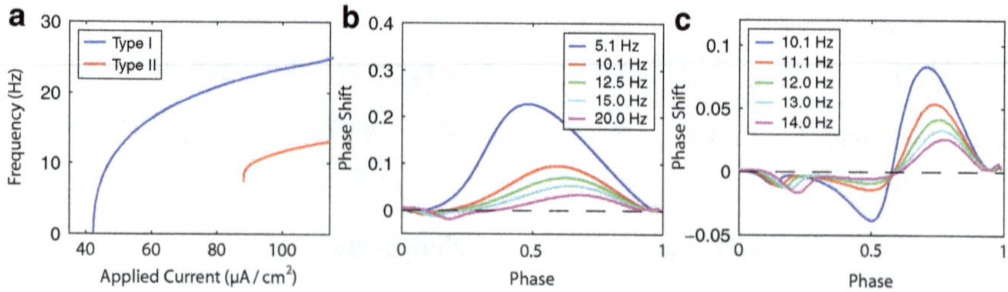

Fig. 19.1 (**a**) Frequency-current (f-I) curves for Type I and Type II Morris–Lecar neurons. The firing frequency of the Type I neuron goes to zero at firing threshold, while the Type II neuron exhibits a threshold frequency of about 8 Hz. (**b**) PRCs for various firing frequencies of the Type I Morris-Lecar neuron. PRCs were calculated using 0.5 ms excitatory current pulses of amplitude 200 μA/cm^2. (**c**) PRCs for various firing frequencies of the Type II Morris-Lecar neuron. PRCs were calculated using 0.5 ms excitatory current pulses of amplitude 200 μA/cm^2

the neuronal membrane, but differences in the anatomical structure of the neurites can also contribute (Schultheiss et al. 2010). Two basic measures characterizing general response properties of individual neurons are the frequency-current (f-I) curve and the phase response curve (PRC). The f-I curve measures neuronal firing frequency in response to continuous stimulation by external current. Typically, neurons can be divided into two groups based on their f-I curves, Type I and Type II (see Fig. 19.1a). The frequency of Type I neurons is strongly modulated by the magnitude of the external current and can take on very low values at firing threshold. The frequency of Type II neurons is much less responsive to changes in the current magnitude. These neurons start firing at threshold with a critical frequency and as current amplitude increases, firing frequency does not change as significantly.

The neuronal PRC is an experimentally obtainable measure characterizing the response of a periodically firing neuron to brief, weak external stimulation. The PRC is calculated as the time difference between the firing of the neuron when it is perturbed by the external pulse relative to its firing when it is unperturbed:

$$\Delta = \frac{T - T_{\text{new}}(\theta)}{T}, \tag{19.1}$$

where $\theta = t/T$ denotes the normalized phase of the neuron when the external pulse is applied (with $\theta \equiv 0$ at a voltage peak), t is the time of pulse application, T is the firing period of the unperturbed neuron and T_{new} is the period of a perturbed cycle. Thus, the value of Δ is negative if the perturbed period is longer than the unperturbed one (constituting a phase delay) and, conversely, it is positive if the perturbed period is shorter than the unperturbed one (constituting a phase advance). Neuronal PRCs have also been divided into two groups. Type I PRCs show phase advances for almost all stimulus phases, while Type II PRCs show a region of phase delay for early stimulus phases and a region of phase advance for late stimulus phases. Conveniently, neurons with Type I f-I curves typically have Type I PRCs and neurons with Type II f-I curves typically have Type II PRCs (Ermentrout 1996).

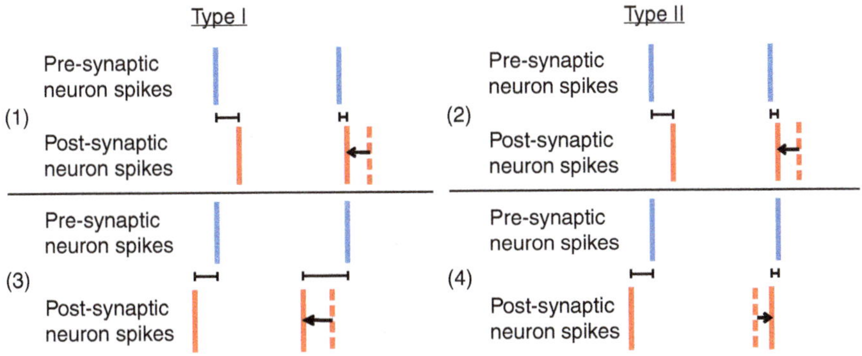

Fig. 19.2 Heuristic explanation of why excitatory neurons with Type II PRCs have a greater propensity for synchronization than excitatory neurons with Type I PRCs. Consider two neurons coupled with a unidirectional excitatory synapse and being driven to fire periodically at similar frequencies. For cases (1)–(4), *vertical ticks* represent spike times of the presynaptic (*blue*) and the postsynaptic (*red*) neuron during an unperturbed cycle (*dashed tick*) and a perturbed cycle (*solid tick, with black arrows* indicating the change in spike time). In cases (1) and (2), the presynaptic neuron fires near the end of the postsynaptic neuron's oscillatory cycle, which induces a phase advance in both Type I and Type II neurons. The phase advance results in the postsynaptic spike occurring closer in time to the next presynaptic spike, thereby enhancing synchrony. In cases (3) and (4), the presynaptic neuron fires near the beginning of the postsynaptic neuron's oscillatory cycle. This induces a phase advance in the Type I postsynaptic neuron but a *phase delay* in the Type II postsynaptic neuron. The phase advance of the Type I neuron increases the separation in time between spikes of the pre- and postsynaptic neurons, resulting in diminished synchrony. In contrast, the phase delay of the Type II neuron decreases the separation in time between spikes of the pre- and postsynaptic neurons, resulting in enhanced synchrony. Thus, input to Type II neurons at both early and late phases promotes synchrony, while input to Type I neurons only promotes synchrony when received at late phases

Properties of PRCs and their implications for neuronal responses have been studied extensively for single neurons as well as for small networks composed of a few cells (Hansel et al. 1995; Ermentrout 1996; Canavier et al. 1997; Dror et al. 1999; Maran and Canavier 2008; Luo et al. 2004). Theoretical results have shown that neurons with Type II PRCs tend to synchronize more readily when synaptically coupled compared to neurons with Type I PRCs (Hansel et al. 1995; Ermentrout 1996). This may generally be attributed to the absence of a phase delay region in the Type I PRC, which leaves Type I neurons with only one method of adapting to the signal of a coupled neuron. For a heuristic explanation of this phenomenon, see Fig. 19.2. Despite the progress that has been made in understanding the effects of PRCs upon network synchronization, it remains unclear how the shape of the PRC influences spatiotemporal activity in larger networks. Recently, we have shown using numerical simulations that the PRC of individual neurons is indeed a good predictor of synchronization in large networks (Bogaard et al. 2009). Specifically, we studied spatiotemporal activity patterns in large-scale excitatory networks composed of biophysically accurate model neurons exhibiting both types of f-I curves and PRCs. We showed that networks consisting of neurons with Type

II PRCs synchronize more easily for different network topologies and connectivity parameters. Furthermore, these neurons can drive synchronization in heterogenous networks composed primarily of neurons exhibiting Type I PRCs. While these results support the idea that the neuronal PRC can be a good predictor of synchrony in a large network, they only provide preliminary data indicating the relevance of the PRC to spatiotemporal patterning of neural activity in complex networks.

In this chapter, we first investigate links between the two measures which characterize neural response to external stimuli, then investigate their influence on network spatiotemporal activity patterns. Specifically, we measure changes in the PRC as a function of firing frequency of the neuron. We find that while the PRC of a given neuron depends on a number of factors, including the amplitude of the stimulating pulse and the firing frequency of the neuron at which the PRC is measured, for neurons with Type I PRCs these factors mainly influence the amplitude of the induced phase shifts but do not qualitatively change the profile of the PRC. However, for neurons with Type II PRCs, the amplitude of the phase delays evoked at early phases can decrease significantly as the intrinsic firing frequency increases, leading to a more Type I-like response. Thus, by modulating firing frequency, we can continuously monitor the transition from Type II to Type I-like PRCs and observe the effect of this transition upon spatiotemporal patterning in a large network. Specifically, we consider the propensity for synchronization as a function of network firing frequency in large-scale excitatory networks composed of either Type I or Type II Morris–Lecar model neurons.

2 Type I and Type II Dynamics of Morris–Lecar Model Neurons

In our model networks, individual neurons are modeled by the Morris–Lecar equations (Morris and Lecar 1981),

$$C\frac{\mathrm{d}V}{\mathrm{d}t} = -\overline{g}_{Ca}m_\infty(V)(V - V_{Ca}) - \overline{g}_K w(V - V_K) - \overline{g}_L(V - V_L) + I_{\text{app}}, \tag{19.2}$$

$$\frac{\mathrm{d}w}{dt} = \phi\frac{w_\infty(V) - w}{\tau_w(V)}, \tag{19.3}$$

$$m_\infty(V) = \frac{1}{2}\left[1 + \tanh\left(\frac{\mathrm{V} - \overline{\mathrm{V}}_1}{\overline{\mathrm{V}}_2}\right)\right], \tag{19.4}$$

$$w_\infty(V) = \frac{1}{2}\left[1 + \tanh\left(\frac{\mathrm{V} - \overline{\mathrm{V}}_3}{\overline{\mathrm{V}}_4}\right)\right], \tag{19.5}$$

$$\tau_w(V) = \frac{1}{\cosh\left(\frac{\mathrm{V}-\overline{\mathrm{V}}_3}{2\overline{\mathrm{V}}_4}\right)}, \tag{19.6}$$

Table 19.1 Parameters for the Morris-Lecar Type I and Type II model neurons used in our simulations, taken from Rinzel and Ermentrout (1998)

Parameter	Type I	Type II
C	20 μF/cm^2	20 μF/cm^2
g_{Ca}	4.0 mS/cm^2	4.4 mS/cm^2
g_K	8.0 mS/cm^2	8.0 mS/cm^2
g_L	2.0 mS/cm^2	2.0 mS/cm^2
V_{Ca}	120.0 mV	120.0 mV
V_K	−84.0 mV	−84.0 mV
V_L	−60.0 mV	−60.0 mV
$\overline{V}_1$	−1.2 mV	−1.2 mV
$\overline{V}_2$	18.0 mV	18.0 mV
$\overline{V}_3$	12.0 mV	2.0 mV
$\overline{V}_4$	17.4 mV	30.0 mV
ϕ	1/15	0.04
I_{app}	35.0 μA/cm^2	80.0 μA/cm^2

where V represents membrane voltage (in mV) and time is measured in ms. Action potentials are generated by an inward Ca^{2+}-mediated current and an outward K^+-mediated current whose dynamics are governed by the gating variable w. The functions $m_\infty(V)$ and $w_\infty(V)$ are the steady state activation functions of the Ca^{2+} and K^+ currents, and $\tau_w(V)$ is the time constant governing dynamics of the gating variable w. The term I_{app} represents an externally applied stimulus to the neuron.

Following Rinzel and Ermentrout (1998), we use the parameters listed in Table 19.1 to simulate Type I and Type II Morris–Lecar model neurons. Figure 19.1a depicts the f-I curves of both neuron types. Note that the frequency of the Type I neuron is capable of assuming arbitrarily low values at firing threshold, while the Type II neuron exhibits a critical frequency at threshold of about 8 Hz.

Figure 19.1b,c depicts the PRCs of Type I and Type II Morris–Lecar neurons in response to excitatory stimulation. Note that at low frequency, both types exhibit a phase delay when a current pulse is applied at values of θ slightly greater than 0. This is due to the fact that $\theta = 0$ is defined to occur at maximum membrane voltage, so that the voltage naturally decreases at the beginning of the cycle, when the neuron is repolarizing after an action potential. The excitatory applied current pulse therefore opposes this process and delays the natural oscillatory behavior. In the Type I neuron, this initial phase delay gives way to phase advance for stimulation at later phases in the firing cycle. The Type II neuron exhibits qualitatively different behavior. The initial phase delay region is followed by a further delay region, forming a second, larger "trough" before finally giving way to a phase advance region at late phases of stimulation.

2.1 Frequency Dependence of PRCs

In the Morris–Lecar model, PRC properties are strongly affected not only by the properties of membrane excitability but also by the induced firing frequency of

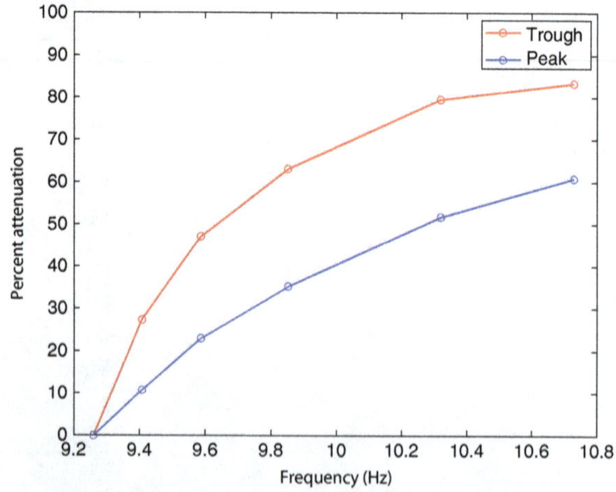

Fig. 19.3 Attenuation of both the peak and trough of the Type II PRC as a function of frequency, relative to the amplitudes of the peak and trough when the neuron is firing at approximately 9.3 Hz. PRCs were computed using 0.5 ms current pulses of amplitude 40 μA/cm^2

the unperturbed neuron. Figures 19.1b,c depict changes in the phase shifts for both types of neurons as the firing frequency of the neuron is increased by increasing the external current I_{app}. In both cases, there is an overall attenuation of the PRC as frequency increases. Additionally, the Type II PRC exhibits an extreme attenuation of the phase delay region relative to the phase advance region with increased frequency, rendering the neuron more Type I-like. This phenomenon is quantified in Fig. 19.3.

3 Network Model

To better understand the effects on network dynamics of frequency-dependent changes in the characteristics of the PRC, we simulated large-scale excitatory networks consisting of either Type I or Type II Morris–Lecar neurons for different average network firing frequencies. We varied synaptic weight and network connectivity and then quantified synchronization of the network spatiotemporal activity patterns.

3.1 Network Structure

In our networks of Morris–Lecar neurons, individual neurons are in an excitable state (I_{app} levels are set below firing threshold) and underlying network activity is driven by random, excitatory noise to each neuron. Thus, we modify (19.2) for the membrane voltage of the ιth neuron V_ι as follows:

$$C\frac{\mathrm{d}V_\iota}{\mathrm{d}t} = -\bar{g}_{Ca}m_\infty(V_\iota)(V_\iota - V_{Ca}) - \bar{g}_K w_\iota(V_\iota - V_K) - \bar{g}_L(V_\iota - V_L) + I_{app} + \sum_{\jmath\in\Gamma_\iota} I^{syn}_{\iota\jmath} + I^{noise}_\iota. \quad (19.7)$$

Here we have added two terms, one for the synaptic current exchanged between neurons, $\sum_{j \in \Gamma_i} I_{ij}^{\text{syn}}$, and another for random, excitatory inputs used to modulate average network firing frequency, I_i^{noise}. The synaptic current flowing from neuron j to i is defined to be non-zero if neuron j is directionally coupled to neuron i (i.e., $j \in \Gamma_i$) and the voltage of neuron j is greater than 0 mV. The magnitude of the synaptic current is given by

$$I_{ij}^{\text{syn}} = s \max(0, V_j). \tag{19.8}$$

The synaptic weight s is the same for all synapses, and its specific value is one of the parameters varied from simulation to simulation.

The term I^{noise} is used to control the average firing frequency of the network. Each neuron randomly receives square current pulses of amplitude 300 $\mu\text{A}/\text{cm}^2$ and duration 0.5 ms at a specified frequency. The noise frequency determines the average firing frequency of the network. Additionally, Type I and Type II neurons are both given subthreshold baseline currents, with $I_{\text{app}} = 35\ \mu\text{A}/\text{cm}^2$ for Type I and $I_{\text{app}} = 80\ \mu\text{A}/\text{cm}^2$ for Type II, to compensate for their differences in threshold current.

The neurons are connected in a one-dimensional network with periodic boundary conditions. The connectivity pattern is constructed using the Watts–Strogatz architecture for Small World Networks (Watts and Strogatz 1998). A radius of connectivity r is specified, and each neuron is directionally coupled to its $2r$ nearest neighbors. Then, every connection in the network is rewired with probability p to another neuron chosen at random. In this way, $p = 0$ results in a locally-connected network and $p = 1$ in a randomly connected network. The radius of connectivity r therefore determines the density of connections in the network, while the rewiring probability p determines the network connectivity structure. r is set to 5 for all simulations except those used to generate Fig. 19.6. All simulated networks have 250 neurons.

3.2 *Measuring Network Synchronization*

We monitor phase synchronization of firing in the network using the mean phase coherence, σ (Mormann et al. 2000). The mean phase coherence between a pair of neurons, $\sigma_{1,2}$, is defined by:

$$\sigma_{1,2} = \left| \frac{1}{N} \sum_{k=1}^{N} e^{i\phi_k} \right|, \tag{19.9}$$

$$\phi_k = 2\pi \left(\frac{t_{2,k} - t_{1,k}}{t_{1,k+1} - t_{1,k}} \right), \tag{19.10}$$

where $t_{2,k}$ is the time of the kth spike of neuron 2, $t_{1,k}$ is the time of the spike of neuron 1 that is largest while being less than $t_{2,k}$, $t_{1,k+1}$ is the time of the spike of neuron 1 that is smallest while being greater than or equal to $t_{2,k}$, N is the number of spikes of neuron 2, and $e^{i\phi_k} = \cos(\phi_k) + i\,\sin(\phi_k)$. The mean phase coherence of the entire network, σ, is calculated by averaging $\sigma_{i,j}$ over all pairs of neurons, excluding $\imath = \jmath$.

4 Effects of the PRC Frequency Dependence on Network Synchronization

We investigated the changes in network mean phase coherence as a function of network structure for different values of average frequency of network activity. Figures 19.4a–c summarize the dual effects of synaptic coupling and architecture on synchronization of Type I networks. For low coupling strengths s (vertical axis), the networks essentially fail to synchronize regardless of the value of the rewiring probability p (horizontal axis), indicated by the low values of mean phase coherence. Above $s \approx 0.3\,\text{mS/cm}^2$, however, the networks synchronize for network architectures with a large fraction of random connectivity. Figure 19.4c shows that the differences in mean phase coherence between networks driven at high frequencies and low frequencies, $\sigma_{\text{high}} - \sigma_{\text{low}}$, are positive, revealing that Type I networks synchronize better at higher frequency. This is hardly a surprising result, because higher network frequency implies more exchange of synaptic current, which leads to greater effective coupling among neurons. For $s \gtrsim 0.45\,\text{mS/cm}^2$, the differences in mean phase coherence, $\sigma_{\text{high}} - \sigma_{\text{low}}$, return to approximately 0, since σ saturates for both the high- and low-frequency networks at such high coupling.

Type II networks, on the other hand, exhibit quite different behavior. Results depicted in Figs. 19.4d–f demonstrate that appreciable network synchronization occurs at coupling strengths of $s \approx 0.2\,\text{mS/cm}^2$ and greater. At the same time, the differences in mean phase coherence between networks driven at high frequency and at low frequency, $\sigma_{\text{high}} - \sigma_{\text{low}}$, remain negative, implying that Type II networks synchronize better at *lower frequency* (see Fig. 19.4f, and note the difference in scale from Fig. 19.4c). These results differ from those obtained for Type I networks, in which synchrony increases with increasing mean network firing frequency. This contrast in synchronization behavior is even more clearly illustrated in Fig. 19.5, where the differences in mean phase coherence between high- and low-frequency networks are plotted as a function of synaptic weight for a number of different pairings of noise frequency. In all cases, the difference in synchronization between high- and low-frequency Type I networks starts at essentially zero, steeply rises to a large peak as synaptic weight increases, then returns to zero as both networks saturate. The presence of this pronounced region of positively-valued $\sigma_{\text{high}} - \sigma_{\text{low}}$ for all frequency pairings demonstrates the tendency of Type I networks to synchronize better at higher frequency. Type II networks, on the other hand, exhibit large regions of negative $\sigma_{\text{high}} - \sigma_{\text{low}}$ for all frequency pairings, indicating that Type II networks tend to synchronize better at lower frequency. Interestingly, Figs. 19.5c,d show a small region where $\sigma_{\text{high}} - \sigma_{\text{low}}$ is significantly positive for Type II networks. This feature is lacking in Figs. 19.5a,b. These contrasting results may be a result of the

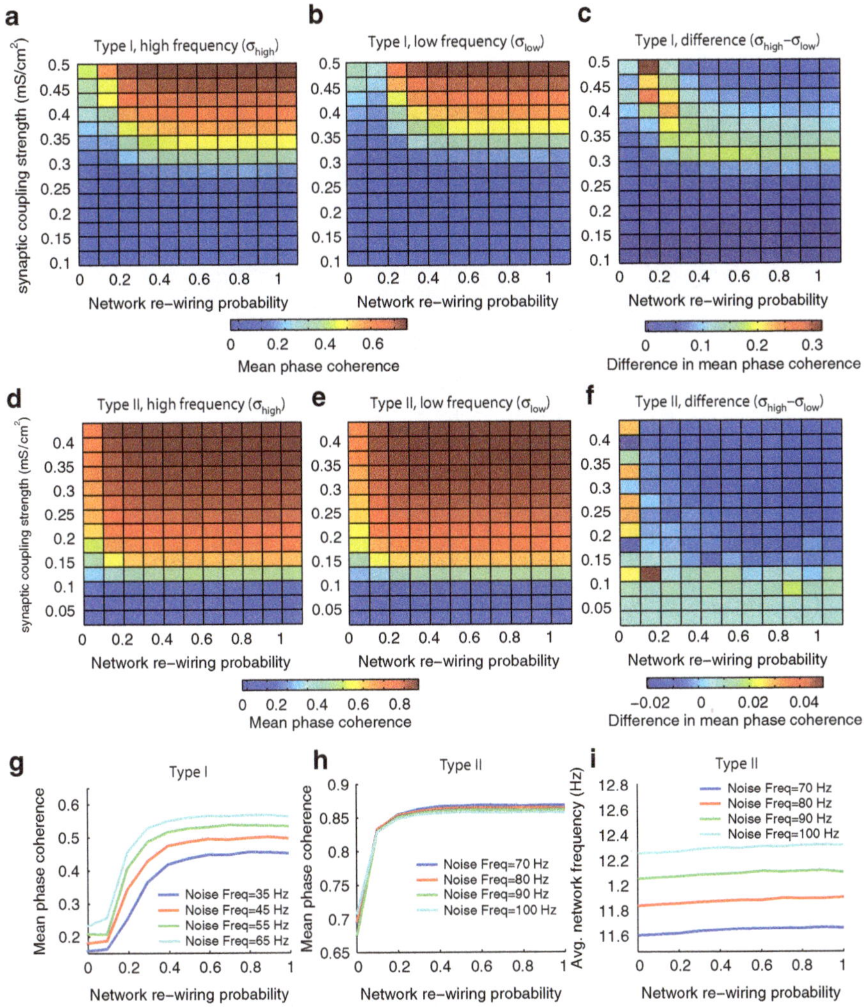

Fig. 19.4 (**a**,**b**) Mean phase coherence σ as a function of synaptic coupling strength s and rewiring probability p for networks of Type I neurons driven with a noise frequency of 65 Hz (σ_{high}, **a**) and of 35 Hz (σ_{low}, **b**). (**c**) Difference between the σ values for the high-frequency and low-frequency Type I networks, $\sigma_{\text{high}} - \sigma_{\text{low}}$, as a function of synaptic coupling strength s and rewiring probability p. (**d**,**e**) Mean phase coherence σ as a function of synaptic coupling s and rewiring probability p for networks of Type II neurons driven with a noise frequency of 100 Hz (σ_{high}, **d**) and of 70 Hz (σ_{low}, **e**). (**f**) Difference between the σ values for the high-frequency and low-frequency Type II networks, $\sigma_{\text{high}} - \sigma_{\text{low}}$, as a function of synaptic coupling strength s and rewiring probability p. Note the difference in scale between (**c**) and (**f**). (**g**) Cross sections of (**a**) and (**b**) with additional, intermediate frequency stimulation values and s fixed at $0.36\,\text{mS/cm}^2$. (**h**) Cross sections of (**d**) and (**e**) with additional, intermediate frequency stimulation values and s fixed at $0.36\,\text{mS/cm}^2$. (**i**) Average network firing frequency as a function of p for various values of the noise frequency in a Type II network, with $s = 0.36\,\text{mS/cm}^2$

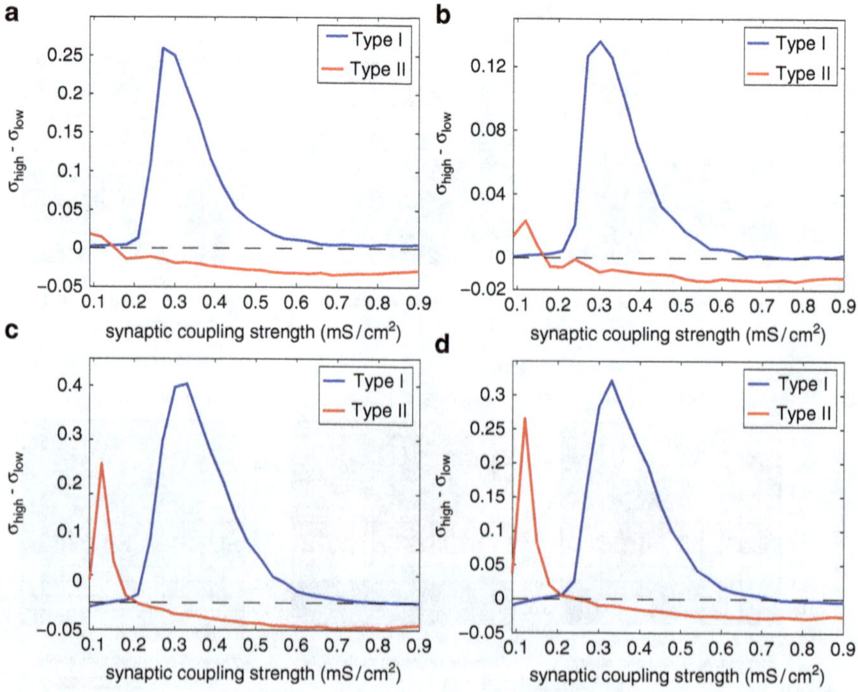

Fig. 19.5 Difference in mean phase coherence, $\sigma_{\mathrm{high}} - \sigma_{\mathrm{low}}$, as a function of synaptic coupling s for Type I (*blue*) and Type II (*red*) networks driven at high and low noise frequencies of (**a**) 130 Hz and 60 Hz, (**b**) 90 Hz and 60 Hz, (**c**) 130 Hz and 30 Hz, and (**d**) 90 Hz and 30 Hz, respectively. The network rewiring probability p is fixed at 0.3

competing influences of effective coupling and PRC phase delay. As frequency increases, more current is exchanged between neurons, leading to higher effective coupling and increased synchronization. At the same time, increasing frequency also leads to a diminishing phase delay region in the neuronal PRCs, which depresses synchronization. The interplay between these two influences is nontrivial, and we see that in Figs. 19.5a,b the depressive pressure of the neuronal PRC dominates for virtually all values of s, while in Figs. 19.5c,d the positive pressure of effective coupling dominates for small synaptic weight values, before giving way to the effects of the neuronal PRC at greater values of s.

We obtain similar results when considering cross sections of the color plots from Figs. 19.4a–f. In Figs. 19.4g,h the synaptic coupling is fixed and the mean phase coherence is plotted as a function of rewiring probability p for four different noise frequencies. The strength of synaptic coupling was chosen to lie in a transitional region, where Type I network synchrony takes on an intermediate value ($s = 0.36\,\mathrm{mS/cm^2}$). In both Type I and Type II networks, mean phase coherence increases with increasing rewiring probability, but increasing frequency affects the two types of networks in different ways. In the Type I network (Fig. 19.4g),

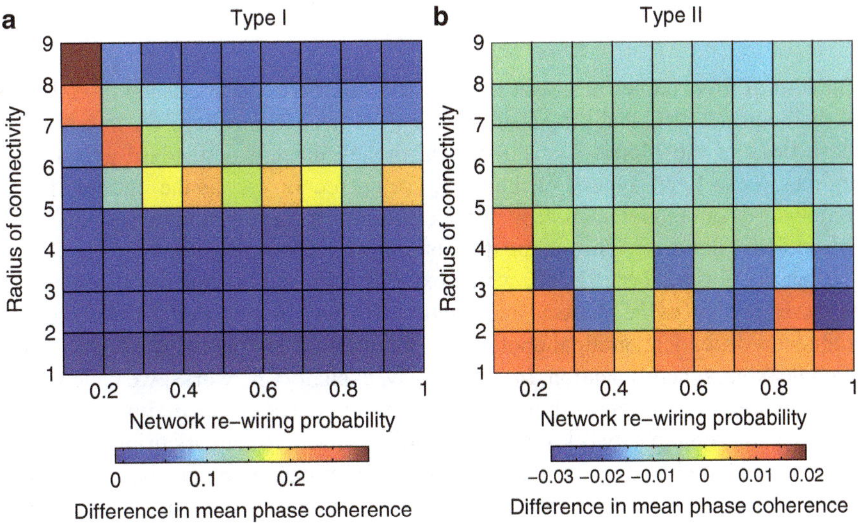

Fig. 19.6 (**a**) Difference in mean phase coherences, $\sigma_{\text{high}} - \sigma_{\text{low}}$, as a function of the radius of connectivity r and the rewiring probability p in a Type I network. s is fixed at $0.30\,\text{mS/cm}^2$, and the noise frequency is 65 Hz in the high frequency network and 35 Hz in the low frequency network. (**b**) $\sigma_{\text{high}} - \sigma_{\text{low}}$ as a function of r and p in a Type II network. s is fixed at $0.30\,\text{mS/cm}^2$, and the noise frequency is 100 Hz in the high frequency network and 70 Hz in the low frequency network

higher frequency implies greater σ values for all values of p, while in the Type II network (Fig. 19.4h) higher frequency yields slightly smaller σ values for virtually all values of p. This effect in the Type II network is not due to changes in network firing frequency with changes in p, since average network firing frequency remains virtually constant for all values of p (Fig. 19.4i). These results further indicate that for networks of Type II neurons, the increase in network propensity for synchronization due to higher mean firing frequency, and thus higher effective coupling, is offset by reduced cellular propensity for synchronization.

We also investigated how network connection density affects the emergence of network synchrony for the two network types. Figures 19.6a,b depict the differences in mean phase coherence for high- and low-frequency network firing, $\sigma_{\text{high}} - \sigma_{\text{low}}$, as a function of p and the radius of connectivity r, for Type I and Type II networks, respectively. Similar to our previous results in which Type I networks exhibited positive values of $\sigma_{\text{high}} - \sigma_{\text{low}}$ for a bounded range of coupling strength values (Fig. 19.4c), Type I networks also show greater synchrony at high frequency than at low frequency between $r \approx 5$ and $r \approx 7$. Type II networks synchronize slightly better at low frequency than at high frequency for $r \gtrsim 2$, which is analogous to the transition that occurs at $s \approx 0.2\,\text{mS/cm}^2$ in Fig. 19.4f. These results show that the networks behave in qualitatively the same manner for both increasing r and increasing s, which is not surprising, since the effect of both is a greater exchange of synaptic current within the network.

5 Conclusion

The aim of this chapter has been to show the link between frequency modulation of the neuronal PRC and propensity for network synchronization. We investigated these links in the Morris–Lecar neuronal model under two standard excitability regimes, Type I and Type II excitability. For each cell type, as the intrinsic firing frequency of our model neurons increases, phase shifts in spike timing (as measured by the PRC) decrease. This attenuation is uniform across all phases for the Type I cell but not for the Type II cell, which exhibits a greater attenuation of the phase delay region than of the phase advance region. We have investigated the effect of this frequency-dependent, gradual abolition of the phase delay region of Type II PRCs upon network synchronization. In large, noisy, excitatory networks, we have varied synaptic weight and connectivity architecture across parameter regimes that both prevent and promote network synchronization. As average network firing frequency is increased by changing the frequency of random excitatory inputs, we observe diminished synchronization (as measured by mean phase coherence) in networks composed of Type II neurons. This phenomenon is robust and can be attributed to the nonuniform, pronounced attenuation of phase delays in Type II neurons as frequency increases. In contrast, networks composed of Type I neurons, whose PRCs have no delay region and are uniformly attenuated as frequency increases, show an increase in propensity for synchronization as frequency increases.

These results paint a complex picture of interactions between network and cellular properties that may lead to nontrivial effects in terms of network dynamical patterns. For example, neurons having various intrinsic cellular properties may be differentially mixed to form networks having different capacities in terms of their pattern formation. As we have shown, Type I networks have a stronger propensity to synchronize at high compared to low frequencies, whereas Type II networks show the opposite tendency. It is therefore possible that Type I networks function as integrators, while Type II networks may act as low-pass filters in the temporal coding regime.

References

Bogaard, A., Parent, J., Zochowski, M., and Booth, V. (2009). Interaction of cellular and network mechanisms in spatiotemporal pattern formation in neuronal networks. *The Journal of Neuroscience*, 29:1677–1687.

Canavier, C., Butera, R., Dror, R., Baxter, D., Clark, J., and Byrne, J. (1997). Phase response characteristics of model neurons determine which patterns are expressed in a ring circuit model of gait generation. *Biological Cybernetics*, 77:367–380.

Dror, R., Canavier, C., Butera, R., Clark, J., and Byrne, J. (1999). A mathematical criterion based on phase response curves for stability in a ring of coupled oscillators. *Biological Cybernetics*, 80:11–23.

Ermentrout, B. (1996). Type I membranes, phase resetting curves, and synchrony. *Neural Computation*, 8:979–1001.

Hansel, D., Mato, G., and Meunier, C. (1995). Synchrony in excitatory neural networks. *Neural Computation*, 7:307–337.

Luo, C., Clark, J., Canavier, C., Baxter, D., and Byrne, J. (2004). Multimodal behavior in a four neuron ring circuit: Mode switching. *IEEE Transactions on Biomedical Engineering*, 51: 205–218.

Maran, S. and Canavier, C. (2008). Using phase resetting to predict 1:1 and 2:2 locking in two neuron networks in which firing order is not always preserved. *Journal of Computational Neuroscience*, 24:37–55.

Mormann, F., Lehnertz, K., David, P., and Elger, C. (2000). Mean phase coherence as a measure for phase synchronization and its application to the eeg of epilepsy patients. *Physica D-Nonlinear Phenomena*, 144:358–369.

Morris, C. and Lecar, H. (1981). Voltage oscillations in the barnacle giant muscle fiber. *Biophysical Journal*, 35:193–213.

Rinzel, J. and Ermentrout, B. (1998). Analysis of neural excitability and oscillations. In Koch, C. and Segev, I., editors, *Methods in Neuronal Modeling*. MIT Press.

Schultheiss, N., Edgerton, J., and Jaeger, D. (2010). Phase response curve analysis of a full morphological globus pallidus neuron model reveals distinct perisomatic and dendritic modes of synaptic integration. *The Journal of Neuroscience*, 7:2767–2782.

Watts, D. and Strogatz, D. (1998). Collective dynamics of “small-world” networks. *Nature*, 393:440–442.

Chapter 20
Phase-Resetting Analysis of Gamma-Frequency Synchronization of Cortical Fast-Spiking Interneurons Using Synaptic-like Conductance Injection

Hugo Zeberg, Nathan W. Gouwens, Kunichika Tsumoto, Takashi Tateno, Kazuyuki Aihara, and Hugh P.C. Robinson

Abstract Synchronous firing of cortical fast-spiking (FS) interneurons is important in generating gamma frequency (30–80 Hz) oscillations. These cells connect with each other, both through inhibitory chemical synapses and through gap junctions. To understand how these synaptic interactions control synchronization, we injected artificial synaptic conductances in FS cells, and measured the phase resetting produced by synaptic inputs, the synaptic phase-resetting function (SPRF), during gamma-frequency firing. We found that artificial $GABA_A$ receptor and gap junctional conductances made distinct contributions to the SPRF, which had a remarkably simple, piecewise linear form, with a sharp break between regions of phase delay and advance. Analysing the SPRF shows how the intrinsic properties

H. Zeberg
Department of Physiology, Development and Neuroscience, University of Cambridge, Cambridge, UK, CB2 3EG
e-mail: hugo.zeberg@gmail.com

Department of Neuroscience, Nobel Institute for Neurophysiology, Karolinska Institutet, S-171 77, Stockholm, Sweden

N.W. Gouwens • H.P.C. Robinson (✉)
Department of Physiology, Development and Neuroscience, University of Cambridge, Cambridge, UK, CB2 3EG
e-mail: ngouwens@gmail.com; hpcr@cam.ac.uk

K. Tsumoto • K. Aihara
Aihara Complexity Modelling Project, ERATO, Japan Science and Technology Agency (JST), 151–0065 Japan

Institute of Industrial Science, The University of Tokyo, 153–8505, Japan
e-mail: tsumoto@pharma2.med.osaka-u.ac.jp; aihara@sat.t.u-tokyo.ac.jp

T. Tateno
Graduate School of Engineering Science, Osaka University, 560–8531, Japan

PRESTO, Japan Science and Technology Agency, 4–1–8 Honcho Kawaguchi, Saitama, Japan
e-mail: tateno@fbs.osaka-u.ac.jp

N.W. Schultheiss et al. (eds.), *Phase Response Curves in Neuroscience: Theory, Experiment, and Analysis*, Springer Series in Computational Neuroscience 6,
DOI 10.1007/978-1-4614-0739-3_20,

of FS neurons and their synaptic connections allow synchronization of firing over a wide gamma frequency band, and upper and lower frequency limits of this band are shown to be controlled by gap junctions and GABAergic inhibition, respectively. This phase-resetting model gives insight into how FS cells synchronize so effectively at gamma oscillations, and can be a building block in large-scale simulations of the FS cell networks, in order to understand the onset and stability of patterns of gamma oscillation in the cortex.

1 Introduction

Synchronized oscillations in cortical neural activity at gamma frequencies (30–80 Hz) are thought to be linked to a variety of cognitive processes such as motor preparation and control, and sensory feature binding (Buszaki and Draguhn 2004; Singer 1999). Gamma oscillations can be generated locally in the neocortex, for example, by pharmacological (Fisahn et al. 1998; Whittington et al. 1995), electrical (Whittington et al. 1997) or optogenetic (Cardin et al. 2009) stimulation. It seems, then, that local neocortical circuits have an intrinsic capability for generating gamma oscillations, while sensory inputs and connections from other brain regions may shape complex spatial patterns of oscillatory interaction. Synchronized firing of cortical inhibitory interneurons has been implicated in the production of these rhythms in many experimental and modeling studies (Hasenstaub et al. 2005; Lytton and Sejnowski 1991; Morita et al. 2008; Vida et al. 2006). Fast-spiking (FS) inhibitory interneurons are interconnected by electrical synapses or gap junctions as well by $GABA_A$ receptor-mediated inhibitory synapses (Bacci and Huguenard 2006; Galarreta and Hestrin 1999; Gibson et al. 1999; Tamas et al. 2000). Electrical synapses alone (Galarreta and Hestrin 1999; Gibson et al. 1999) or together with GABAergic synapses (Tamas et al. 2000) can synchronize firing in pairs of these interneurons *in vitro*. In addition, the biophysical properties of FS neurons appear to be ideally suited to generating gamma rhythms: they have an abrupt, or hard ("type 2") onset of regular firing at about 30 Hz (Tateno et al. 2004), which means that they can be easily entrained at this frequency. They also show a strong intrinsic drive for spike generation at gamma frequencies when stimulated with broadbrand conductance noise (Tateno and Robinson 2008). Recently, selective optical stimulation of FS interneurons, but not of pyramidal neurons, was shown to cause gamma oscillations (Cardin et al. 2009). Electrical synapses among mutually inhibitory interneurons have been found to increase the precision of synchrony in simulation studies (Bartos 2002; Kopell and Ermentrout 2004; Traub 2001). However, the relative roles of chemical inhibition and gap-junctional coupling in shaping synchronous oscillations in the cortex are still unclear.

The theory of synchronization of coupled oscillators uses the concept of phase dynamics to evaluate the stability of the relative phase of coupled oscillators in time (Pikovsky et al. 2001; Rinzel and Ermentrout 1998). The key to this approach is to determine the effect of a very small perturbing input on the phase of oscillation ("phase resetting"), as a function of the point in the oscillation cycle at which it occurs. This is most often used, under the assumptions of weak coupling and linear summation of phase shifts, to account for how the relative phase of presynaptic and postsynaptic cells evolves from cycle to cycle.

However, as described above, FS cells in the cortex are actually coupled quite strongly to other FS neighbors, with large postsynaptic conductance changes caused by each presynaptic action potential (Galarreta and Hestrin 1999; Gibson et al. 1999). Here, rather than using "infinitesimal" pulse perturbations, we use synthetic conductance injection, or dynamic clamp, to directly measure the phase-resetting response to conductance inputs which closely mimic the effects of presynaptic action potentials, and systematically vary the relative strengths of electrical and GABAergic inhibitory conductances. The compound synaptic connections between FS neurons, together with the intrinsic spike-generating properties of FS neurons, give rise to a distinctively shaped phase-resetting relationship, or "synaptic phase-resetting function," which ensures rapid and precise synchronization over a large gamma-frequency range.

2 Results

2.1 Conductance Injection Reproducing Synaptic Input

Using whole-cell recording in young (postnatal day 13–19) rat somatosensory cortical slices, at a near-physiological temperature, and in the presence of blockers of synaptic conductances, we identified FS cells by their morphology, action potential shape and characteristic firing pattern in response to depolarizing current injection (Galarreta and Hestrin 1999; Gibson et al. 1999; Kawaguchi and Kubota 1993; 1997). FS cells fired high frequency, nonadapting trains of action potentials during depolarizing excitatory conductance steps, which are occasionally interrupted by pauses with subthreshold oscillations, particularly around threshold (Tateno et al. 2004). We used conductance injection/dynamic clamp (Robinson and Kawai 1993; Sharp et al. 1993) to reproduce the effects of electrical and chemical synapses (Fig. 20.1). For further details of methods and the design of the conductance injection system, see (Gouwens et al. 2010; Morita et al. 2008; Robinson 2008; 2009). In FS cells, both gap junctions and GABAergic synapses from neighboring cells are located perisomatically (Tamas et al. 2000), so that point conductance injection at the soma can reasonably reproduce the electrical effects of synaptic inputs. Gap junctions were implemented as a static conductance, g_e, between the recorded cell and a predetermined spike trajectory of "presynaptic" membrane potential, V_{pre}. At rest, this produced a small postsynaptic spikelet, very similar in size and shape to those observed with natural electrotonic coupling (Galarreta and Hestrin 1999; Gibson et al. 1999). We also measured coupling coefficients (the ratio of postsynaptic to presynaptic potential change) for gap-

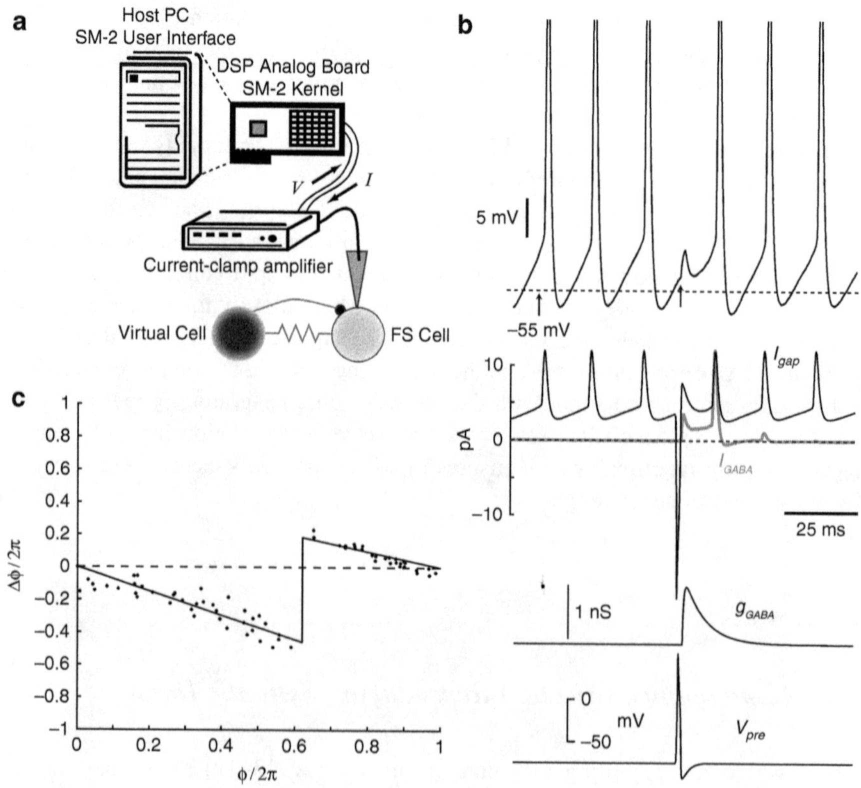

Fig. 20.1 Synaptic connections between FS cells simulated by conductance injection. (**a**) Schematic of experiment. Synthetic conductance reproducing the current flow from a spiking presynaptic cell, through both gap-junctional and GABAergic inhibitory synaptic connections, is injected by calculating it in real time based on feedback of the recorded membrane potential using a "SM2" conductance injection system (Robinson 2008), and injecting current through a whole-cell recording electrode. (**b**) The membrane potential trajectory (top, spike peaks truncated) and injected currents (bottom, gap-junctional current in *black*, current through GABAergic conductance in *gray*, outward current is represented *upward*) during application of a single compound conductance perturbation ($g_e = 0.2$ nS, $g_i = 1.4$ nS) starting at the time indicated by an *arrow*, in this case inducing a delay in the subsequent spike time. Membrane potential spikes are truncated in order to show the subthreshold potential at high resolution. (**c**) Relationship between the phase at which input is applied and the phase shift of the subsequent spike: the phase-resetting relationship, or synaptic phase-resetting function

junctional type conductance. These were similar to physiological values, and larger for step inputs (0.05–0.22) than for spike inputs (0.01–0.05), owing to low-pass filtering by the combined effects of gap junctional conductance and membrane resistance and capacitance (Galarreta and Hestrin 2001).

Many pairs of FS cells are connected by both GABAergic ($GABA_A$, chloride conductance) and electrical synapses (Galarreta and Hestrin 1999; Gibson et al. 1999; Tamas et al. 2000). We also simulated GABAergic synaptic input using conductance injection. The GABAergic conductance transient follows a difference of two exponentials: $g(t) = g_i[\exp(-t/\tau_2) - \exp(-t/\tau_1)]$, where g_i is the scaling amplitude of the inhibitory conductance, and τ_2 had values between 3 and 7 ms, and τ_1 was 0.5 ms. Transients were initiated 2 to 3 ms after the rising inflexion of the presynaptic spike, to simulate synaptic latency and presynaptic propagation delays. The GABA reversal potential (E_{GABA}) was set to −55 mV (Martina et al. 2001), such that from rest, the "IPSP" is a small depolarization lasting about 40 ms, again very similar to natural IPSPs in these cells At the resting potential, a stimulus with both electrical and GABAergic components produces a biphasic depolarizing response with the gap-junctional potential visible just before the larger GABAergic potential. Unlike the gap-junctional spikelet, though, the amplitude of the GABAergic potential is highly voltage-dependent, reversing around E_{GABA} as expected (Galarreta and Hestrin 1999).

2.2 *Perturbing Spike Timing: The Synaptic Phase-resetting Function and the Effect of Varying Electrical and Inhibitory Conductances*

To determine how this compound synaptic input shifts the timing of periodic firing in an FS cell, we applied conductance inputs during periodic firing elicited by a maintained excitatory stimulus, a step of excitatory conductance reversing at 0 mV. An example response to a compound "synaptic" perturbation is shown in Fig. 20.1b. In phase-resetting analysis of synchronization, the state of the neuron is characterized by a single quantity, the phase angle, $\phi(t)$, which—in the absence of any perturbations—increases linearly with time, and which is reset to zero whenever it reaches 2π, corresponding to the occurrence of a spike (Rinzel and Ermentrout 1998). The variability of interspike intervals can be represented by adding additional noise, due to stochastic gating of ion channels and other intracellular sources of variability, to the rate of change of $\phi(t)$. To measure the phase *resetting*, or shift in the phase, produced by synaptic-like conductance inputs, we applied isolated single inputs during long trains of periodic firing. From this, one can measure the relationship between the time at which an input (in this case a compound gap/GABA input) is applied (the time of its *onset*, although importantly the perturbation is far from being instantaneous, but rather plays out over a number of subsequent milliseconds), relative to the time of the preceding spike, and the time until the next spike occurs. This relationship has been termed a spike time response curve (STRC) (Netoff et al. 2005).

From the STRC, we can straightforwardly estimate the phase at the moment that each input is applied, and the amount of phase resetting $\Delta\phi$ produced by the input. The change in phase ($\Delta\phi$) caused by a stimulus was calculated as follows. Let ϕ be the phase reached at the moment of perturbation, ϕ' the phase immediately after, t_p the time after the *previous* spike at which the perturbation is applied, t_n

the time elapsed after the perturbation before the *next* spike, and T_0 the average interspike interval. Then $\varphi = 2\pi t_p/T_0$, $\varphi' = 2\pi(1 - t_n/T_0)$, and $\Delta\phi = \phi' - \phi$. A positive $\Delta\phi$ is a phase advance – the time to the next spike has been reduced, while a negative $\Delta\phi$ is a phase delay – the time to the next spike has been increased. Figure 20.1c shows an example of $\Delta\phi$ plotted as a function of ϕ. This relationship – the total phase-resetting effect of a synaptic input as a function of the phase at which it arrives – has been termed a synaptic interaction function (Ermentrout and Kopell 1991), to distinguish it from a classical phase response or phase-resetting curve, which normally describes responses to very small and brief inputs, whose effects can be considered to sum linearly. However, this term has referred to functions calculated from infinitesimal PRCs using the assumption of weak coupling. Therefore, we will refer to the strong-coupling relationship directly measured here distinctly, as a synaptic phase-resetting function (SPRF). We examined how the parameters of the synaptic input determine the shape of the SPRF, by varying the magnitude of gap junctional and GABAergic conductance, applied individually or together (Fig. 20.2a–f). These components vary physiologically, since FS cells' interconnections can be purely GABAergic (one-way or reciprocal), purely gap-junctional or both (Galarreta and Hestrin 1999; Gibson et al. 1999; Tamas et al. 2000). In addition, there is a wide range of electrical synaptic strengths (Galarreta and Hestrin 2001).

Purely GABA input produces a phase delay early in the cycle, which increases almost linearly during the cycle until an abrupt critical point, beyond which it had no effect (Fig. 20.2a). Introducing a small (250 pS) gap junction, causes a linear region of phase advance (Fig. 20.2b), as in Fig. 20.1d, which has an abrupt onset at a phase of about 1.2π. A sharp transition marks the boundary between this region and the first, phase delay part of the phase cycle. The slope of the phase advance region became more negative, and the boundary between the regions, termed the critical phase φ_c, shifted earlier in the cycle, as gap-junctional conductance increased (Fig. 20.2c, d, e). With no GABAergic input, a phase advance region produced by gap junctional input is seen in isolation (Fig. 20.2f).

Thus, GABAergic input retards, and gap-junctional input advances the phase. For the compound gap/GABA input, the early region of phase delay has a slope determined by the amplitude of inhibition, g_i, and switches abruptly in midcycle, to a region of decreasing phase advance, whose slope is determined by g_e, with no detectable sign of cancellation of the two regions at φ_c. The only clear interaction between the electrical and GABAergic components is that a larger gap junctional conductance shifts φ_c to earlier in the cycle.

A good approximation to the measured SPRF is given by a piecewise linear function:

$$\Delta\varphi(\varphi) = \begin{cases} -\alpha\varphi & (0 \le \varphi < \varphi_c) \\ \beta(2\pi - \varphi) & (\varphi_c \le \varphi < 2\pi) \end{cases}, \tag{20.1}$$

where $-\alpha$ is the slope in the phase delay section, $-\beta$ is the slope of the phase advance section, and φ_c is the breakpoint. SPRFs were fitted to experiments by least

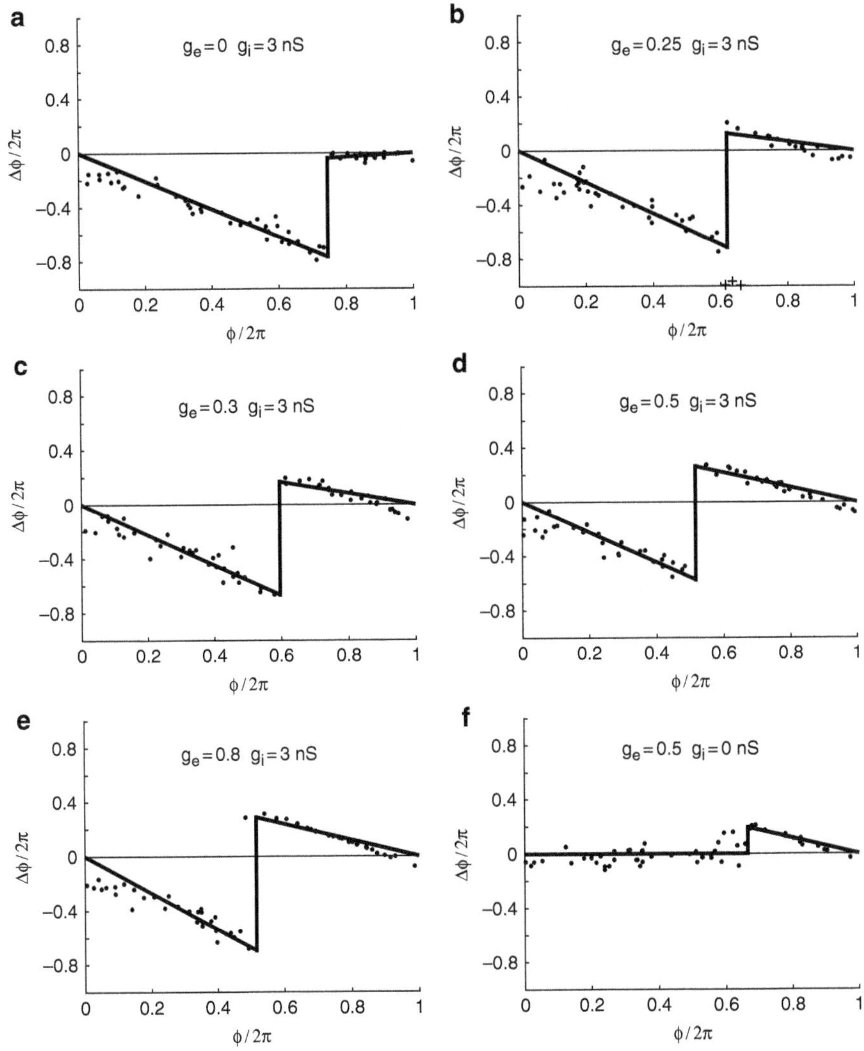

Fig. 20.2 SPRFs in one cell for different strengths of gap-junctional and inhibitory conductance. (**a**) Inhibition only. Phase delay increases linearly as the phase of onset of the synaptic perturbation increases, before an abrupt loss of sensitivity late in the cycle. (**b**) As gap-junctional conductance is introduced, phase delay switches to a region of linear phase advance late in the cycle "+" symbols indicate outliers excluded using Grubb's test (see text). (**c**))–**e**) As gap-junctional conductance is increased, the slope of the phase advance region becomes larger, and the point of switching is shifted progressively earlier in the cycle. (**f**) Switching off inhibition completely leaves only the late-phase advance (compare to (**d**))

squares, using Grubbs' test for outliers, to delete occasional outlying points near the point of maximum instability (in most cases none, but no more than three per SPRF) (Figs. 20.1c, 20.2), which otherwise reduce the quality of fit to all other points.

To quantify the goodness of fit of the piecewise linear SPRF, we performed a chi-square test of 130 phase-response curves (in total 6,111 data points, 10 cells). For each SPRF, variance of phase was estimated from an unperturbed spike train within the same experiment (median $\sigma^2 = 0.021\ (\mathrm{rad}/2\pi)^2$). 111 of 130 SPRFs contained no significant difference between the model fit and experimental result ($p < 0.05$). The average reduced chi-square value was 0.80, meaning that the overall fit of the model is extremely good, given the measured degree of variance in the phase. On the whole, the relatively simple piecewise linear model performs remarkably well.

The dependencies of the slopes and breakpoint on the strengths of g_i and g_e were also fitted by linear relationships (Fig. 20.3). The negative slope of the region of phase delay was proportional to inhibition ($\alpha = ag_i$, Fig. 20.3b), the negative slope of the phase advance region was proportional to excitation ($\beta = bg_e$, Fig. 20.3a), while φ_c was weakly sensitive to g_e ($\varphi_c = c - dg_e$, Fig. 20.3c). Average values of a and b of this piecewise linear model for the SPRF were $a = 0.16/\mathrm{nS}$ (n = 7 cells, 3 cells providing insufficient data for analyzing this dependency), $b = 0.69/\mathrm{nS}$ ($n =$ 10 cells). c and d were more variable from cell to cell, and the pooled data in fact showed little overall dependence on g_e (not shown). Nevertheless (e.g. Fig. 20.3c), the weak relationship is clear within individual cells.

2.3 *Entrainment by Synaptic Input*

Having established that conductances resembling the synaptic input of neighboring FS cells can consistently modify spike timing, we next tested the ability of FS cells to synchronize to, or to be entrained by this input. To visualize the time course of entrainment, we examined responses stroboscopically (Pikovsky et al. 2001), sampling the phase of the FS cell at the times of periodic stimuli. Figure 20.4a shows such an experiment. Before the conductance pulses are switched on (*open circles*), the phase changes in a "sawtooth" pattern, reflecting detuning—the continuously growing phase difference between two oscillators of different frequencies. After the conductance transients begin (Fig. 20.4a, *filled circles*), the phase quickly converges on a fixed value relative to the stimulus, at about 0.6π (dashed line), which matches the expected equilibrium phase difference from solving (20.2) (below) with parameters for this cell. Thus, the FS cell becomes phase locked and frequency locked to the stimulus train, with spikes occurring around 0.6π before, or equivalently 1.4π after each stimulus. After the end of the stimulation train, the phase reverts to the drifting detuned state.

Entrainment of spiking to periodic stimulation was simulated by the noisy map describing the evolution of the phase from stimulus n to stimulus $n+1$:

$$\varphi_{n+1} = G(\varphi_n) = \varphi_n + \Delta\varphi(\varphi_n) + 2\pi F/f + \xi_n \pmod{2\pi}, \qquad (20.2)$$

Fig. 20.3 SPRF parameters depend on the strength of synaptic perturbation in a simple manner. (**a**) Dependence of the phase advance slope ($-\beta$) on the gap-junctional conductance (for $g_i = 1.5$ nS). Data pooled from 120 measurements in 10 cells. (**b**) Dependence of the phase delay slope ($-\alpha$) on g_i. Data from 43 measurements in 7 cells. (**c**) Dependence of the critical phase at which delay switches to advance (φ_c) on the gap-junctional conductance g_e in one cell

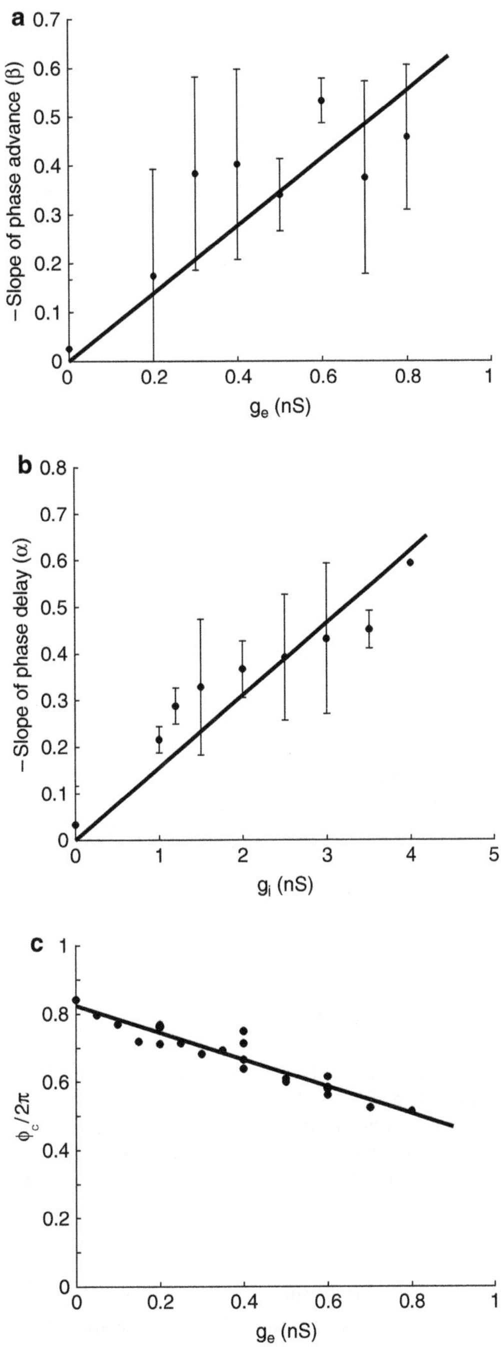

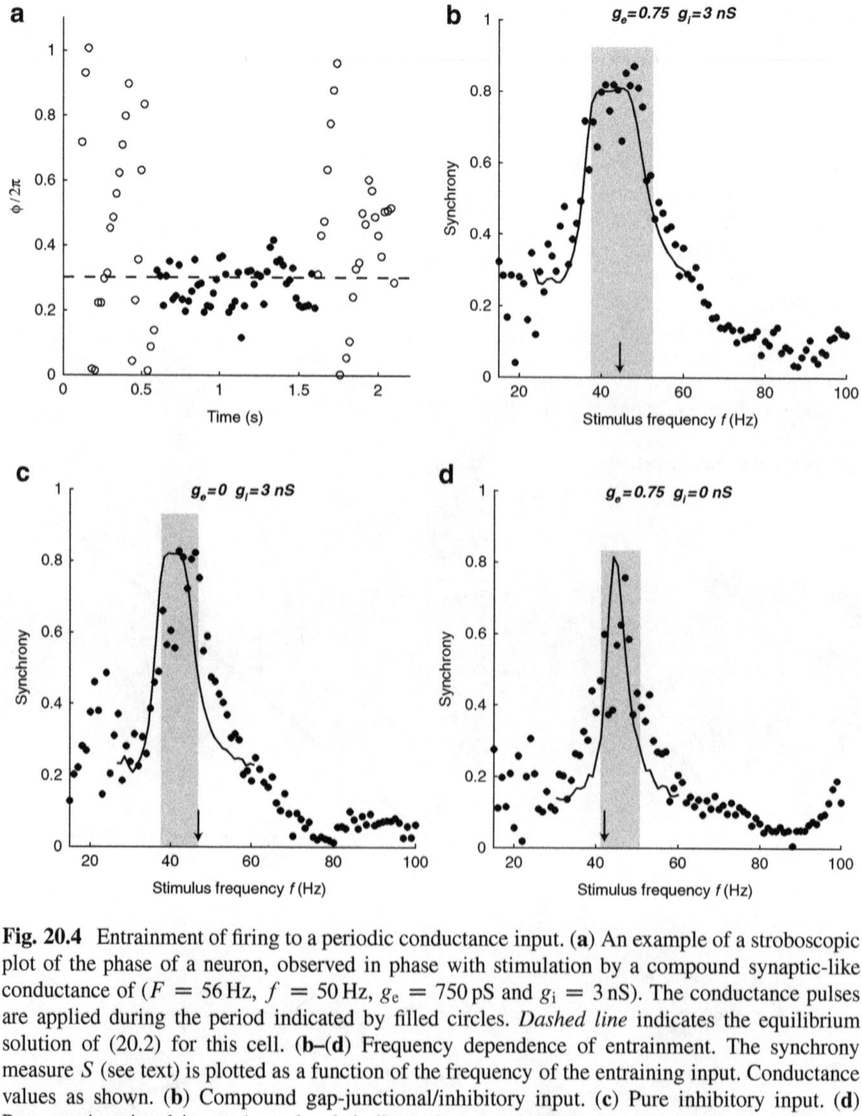

Fig. 20.4 Entrainment of firing to a periodic conductance input. (**a**) An example of a stroboscopic plot of the phase of a neuron, observed in phase with stimulation by a compound synaptic-like conductance of ($F = 56\,\text{Hz}$, $f = 50\,\text{Hz}$, $g_{\text{e}} = 750\,\text{pS}$ and $g_{\text{i}} = 3\,\text{nS}$). The conductance pulses are applied during the period indicated by filled circles. *Dashed line* indicates the equilibrium solution of (20.2) for this cell. (**b**–(**d**) Frequency dependence of entrainment. The synchrony measure S (see text) is plotted as a function of the frequency of the entraining input. Conductance values as shown. (**b**) Compound gap-junctional/inhibitory input. (**c**) Pure inhibitory input. (**d**) Pure gap-junctional input. *Arrowheads* indicate the natural firing frequency F in the absence of perturbations, and gray regions indicate the frequency bands of 1:1 synchronization predicted by the measured SPRF. *Solid curves* in (**b**–**d**) show the calculated steady-state synchrony of the fitted noisy SPRF model

where f is the stimulus angular frequency, F is the unperturbed (natural) angular frequency of the cell, and ξ_n is a Gaussian-distributed noise term, with variance σ^2.

The piecewise linear SPRF could also account for the frequency band over which synchronization was possible. Figure 20.4b–d shows an experiment in which an FS neuron firing at a steady frequency F was stimulated repeatedly with a periodic synaptic conductance input at frequency f, and an index of the synchrony of the cell with the input, S, was measured over a range of frequencies. This "phase order" parameter, or synchrony (also called circular statistic R^2) is the magnitude of the average phase vector over the ensemble $S = \sqrt{\langle\cos(\phi)\rangle^2 + \langle\sin(\phi)\rangle^2}$, and varies between ≈ 0 (when a very large number of phases are distributed uniformly between 0 and 2π) and 1 (phases all identical).

As seen in Fig. 20.4b–d, this changes from a low level when f is very different from F, to a high value approaching 1, when $f \approx F$. Because of the effects of noise in the neuron, there is no absolute phase locking ($S < 1$), and the change in synchrony with input frequency does not have abrupt boundaries, but falls away continuously as the difference between f and F grows. It is clear that the central region of high synchrony lies below the unperturbed or natural firing frequency F when only inhibition is applied (Fig. 20.4c), above F when only gap-junctional conductance is applied (Fig. 20.4d), or both above and below F when a compound input is applied (Fig. 20.4b). This observation was duplicated by the piecewise linear model of the SPRF, analysis of which predicted the 1:1 synchronized frequency bands shown in gray, for the deterministic (noise-free) case—in this neuron, these boundaries corresponds to a synchrony of about 0.7. The synchronized frequency band is much narrower for either gap-junctional stimulation alone (Fig. 20.4d) or GABAergic inhibition alone (Fig. 20.4c). Iterations of the noisy stroboscopic map derived from the fitted SPRF ((20.2)) showed that it could also reproduce the distribution of S adequately (black curves in Fig. 20.4b–d). Thus the piecewise linear model of the SPRF appears to account very well, both for the frequency range and the degree of synchronization in noise.

2.4 Frequency Bands of Deterministic and Stochastic Synchronization

We next used the SPRF to predict the frequency ranges of entrainment for different strengths of inhibition and electrical coupling (Fig. 20.5), by analyzing the bifurcations at the onset of synchrony in the stroboscopic map of the phase, i.e. (20.2). For the deterministic (zero noise) case, 1:1 entrainment corresponded to a stable fixed point of the map, labelled φ^* in the example shown in Fig. 20.5a. As the amount of detuning (difference between f and F) varies, the map shifts vertically, so that at certain stimulus frequencies, the fixed point disappears (at a "corner-collision" bifurcation (DiBernardo et al. 2001)). Thus, it is possible to plot the regions in which there is synchronization in the g_e, f plane (Fig. 20.5b) or the g_i, f plane (Fig. 20.5c,d). These form Arnol'd tongues (Pikovsky et al. 2001) in which the frequency range of entrainment shrinks as the synaptic strength is reduced.

This analysis shows a number of effects which are relevant to the physiological function of FS neurons. Increasing g_e strongly increased the upper frequency limit

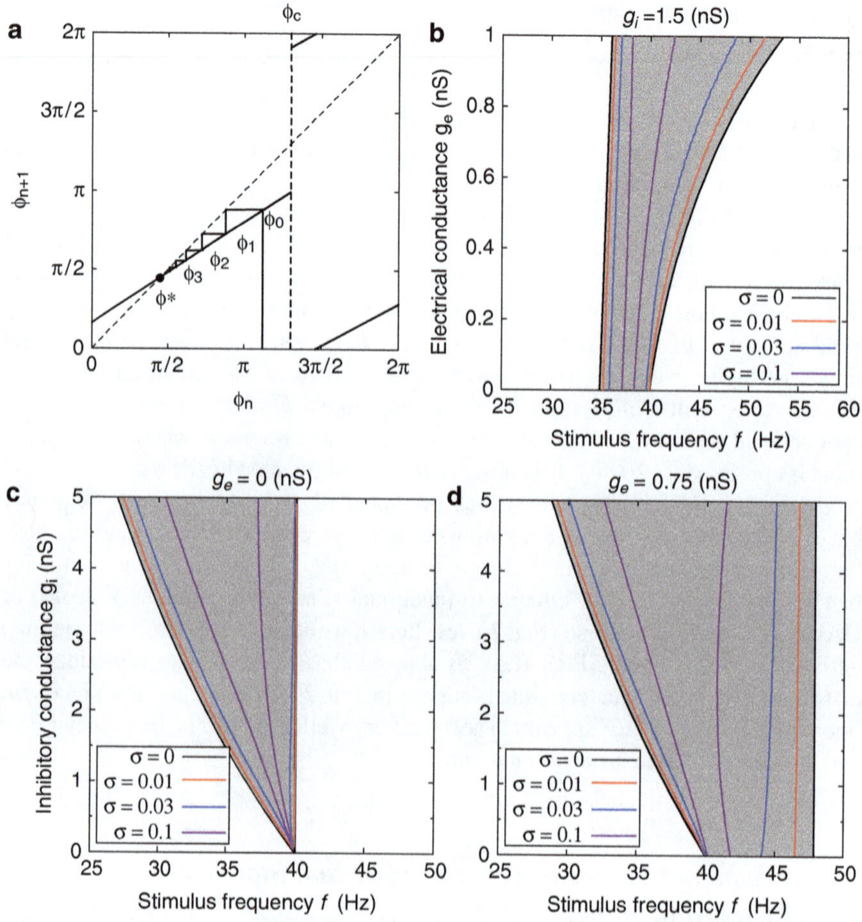

Fig. 20.5 Bifurcation analysis of frequency bands of synchronization. (**a**) Piecewise linear map between phase at stimulus n and phase at stimulus $n+1$. The point φ^* on the diagonal is a stable fixed point of the map, as illustrated by the converging orbit $\varphi_1, \varphi_2, \ldots$ showing that 1:1 entrainment occurs at this stimulus frequency. (**b**) Bifurcation points of 1:1 entrainment in the g_e, f plane, $g_i = 1.5\,\text{nS}$. 1:1 entrainment occurs in the gray regions. $\sigma = 0$ is the deterministic case. For $\sigma > 0$, stochastic bifurcation points with added Gaussian noise in the phase (see text). (**c**) Synchronization region in the g_i, f plane, with $g_e = 0$. (**d**) As in (**c**), with $g_e = 0.75\,\text{nS}$. Raising g_e strongly increases the upper frequency limit of entrainment, and weakly increases the lower limit. Noise shrinks the stochastic synchronization region. Parameters describing sensitivity of the SPRF to g_e and g_i (see p. 13): $a = 0.12/\text{nS}$, $b = 0.625/\text{nS}$, $c = 0.8 \times 2\pi\,\text{rad}$, $d = 0.2 \times 2\pi\,\text{rad/nS}$

of entrainment and weakly increased the lower limit (Fig. 20.5b). When $g_i = 0$ it is impossible to entrain firing with $f < F$. Conversely, with $g_e = 0$, it is impossible to entrain for $f > F$, and increasing g_i strongly reduces the lower frequency limit of entrainment (Fig. 20.5c,d).

Since physiologically, entrainment must occur in the face of considerable noise, we also investigated the effect of adding noise to the phase map. It is possible to define stochastic bifurcation points of the map defined by (20.2), at which there is a qualitative change in the nature of the stochastic dynamics (Gouwens et al. 2010; Inoue et al. 2001). The stochastic map of the phase between successive stimuli is represented by a Markov operator which is approximated by a stochastic transition matrix, and the frequency of onset of stochastic entrainment is defined by the point where the second eigenvalue of this matrix changes from real to complex. This coincides with the deterministic bifurcation frequency, as the noise level approaches zero. We examined the frequency extents of this kind of stochastic entrainment at different noise levels (Fig. 20.5b–d). In all cases, increasing the noise in the phase shrinks the region of entrainment. For $\sigma = 0.1\,\mathrm{rad}/2\pi$, which was a typical noise level in these cells in vitro, the area of stochastic entrainment shrank to a third or less of the noise-free case. This noise-induced distortion is not symmetrical in the frequency axis. For example, Fig. 20.5c shows that in the absence of electrical coupling, the lower frequency limit of entrainment was highly susceptible to noise while the upper limit was not. The greater the level of electrical coupling (g_e), the more the upper limit was reduced by noise.

The SPRF makes several predictions. First, FS cells receiving purely electrical synaptic input will synchronize effectively when driven at frequencies higher than F. Higher frequencies can be followed with stronger electrical input. Second, cells will synchronize to purely inhibitory input at frequencies lower than F, and stronger inhibition allows lower frequencies to be followed. Third, combined electrical and inhibitory input allows cells to synchronize to frequencies both above and below their unperturbed frequency. Although noise diminishes the frequency band of synchronization, sometimes asymmetrically, these conclusions remain valid in the presence of noise. For typical strengths of combined electrical-inhibitory synaptic connections, 1–2 Hz bandwidths of stochastic synchronization persist even in quite high levels of noise ($\sigma = 0.1$).

3 Discussion

3.1 Measuring the Effect of Synaptic Conductance on Phase of Periodic Firing

A number of previous theoretical and experimental studies have examined the phase-resetting properties of cortical neurons. (Reyes and Fetz 1993) stimulated synaptic inputs to regularly firing pyramidal neurons to measure the phase resetting produced by EPSPs, while (Stoop et al. 2000) used similar measurements to predict input frequency regions for entrainment and chaos. (Netoff et al. 2005) used dynamic-clamp to measure phase resetting (or spike-time response curves) by artificial excitatory or inhibitory conductances in excitatory stellate cells of medial entorhinal cortex, and oriens-lacunosum-molecular interneurons in the CA1 region of hippocampus, and were able to demonstrate synchronization in pairs of neurons connected by artificial conductances mimicking synaptic connections, or between biological neurons and simulated neurons. In fast-spiking inhibitory cells, Mancilla et al. (2007) measured phase-resetting relationships for small current pulses (weak coupling) and showed that they could account quite well for synchronization of pairs of gap-junction coupled FS cells, both experimentally and in a biophysical model of FS neurons (Mancilla et al. 2007). In this study, we have gone further, by using conductance injection (dynamic clamp) to reproduce the *combined* effect of gap-junctional and strong-synaptic connections, and using this to predict the resulting synchronized frequency bands, and their dependence on synaptic strength, including the effect of noise in the synaptic phase-resetting function on synchronization.

The conductance pulses which we have used are based on the physiological properties of the synaptic connections between FS neurons. In FS neurons of a basket morphology, APs initiate in the axon (Hu et al. 2009) arising usually from a proximal dendrite, (Martina et al. 2000) and receive many of their inhibitory connections and gap junctions from other fast-spiking interneurons perisomatically (Tamas et al. 2000). Thus, this should provide a reasonably realistic simulation of the natural gap-junctional and fast inhibitory input.

In order to carry out this analysis, we have made the approximation that the presynaptic voltage was held at a resting potential of $-70\,\mathrm{mV}$, instead of continuously depolarizing, between spikes. In other words, we have focused on the effect of current flow associated with the discrete event of the presynaptic spike rather than of the continuous phase difference of the presynaptic membrane potential in the rest of the interspike interval, which will contribute a maintained pull on the phase difference (along with the gap junction conductance to the syncytium of other, presumably inactive, connected FS cells in the network (Amitai et al. 2002)). We have also ignored the two-way nature of coupling between cell pairs. In other words, we model entrainment of one cell by another, rather than synchronization of a symmetrical coupled pair. Although both electrical and inhibitory coupling can often be asymmetrical (Galarreta and Hestrin 2002; Gibson et al. 1999), they may also be quite symmetrical. However, the entrainment studied here models the situation where the presynaptic cell is already imperturbably driven as part of a strong synchronously-firing assembly of FS neurons, so that the phase and frequency of its firing will be clamped to that of its predominant input. Thus, the SPRF should be an effective model for describing recruitment of new cells to such a synchronous assembly.

It is expected that the preferred firing frequency F of the postsynaptic cell may also affect the form of the SPRF, since the timing of intrinsic ion channel kinetics will shift relative to the phase as the cycle length changes. In a few experiments where we were able to address this issue, we indeed found evidence of a change in the parameters of the SPRF model. a, the dependence of α on g_{i} increased quite strongly as firing frequency increased, and φ_{c} shifted earlier in the cycle as firing frequency increased. The dependence of b and d on firing frequency was not marked.

3.2 The Synaptic Phase-resetting Function

The synaptic phase-resetting function for compound input was distinguished by the following features: an extremely abrupt midcycle switch from phase delay to phase advance, which shifted weakly toward the early part of the cycle as the strength of electrical coupling was increased; amplification of the slope in the phase delay region by increasing inhibition; and amplification of the slope in the phase advance region by increasing gap-junctional coupling. We found that these qualitative features were also present in a biophysical model of firing in fast-spiking cells (Erisir et al. 1999; Gouwens et al. 2010). However, there were also major differences. In experiments, phase advance was produced exclusively by gap-junctional conductance and phase delay exclusively by inhibition, while in the Erisir et al. model, gap-junctional input did affect phase delay strongly early in the cycle—this was never observed experimentally. This deficiency of the biophysical model suggests that additional conductances expressed in FS neurons somehow help to confer a complete immunity to gap-junctional stimulation in the early, phase-delay part of the cycle. We surmise that the voltage-gated potassium conductance in this part of the cycle may actually be much higher than in the model, and that this may allow phase delay and advance to be regulated completely independently. Also, because of their relative timing, the effect of inhibition will outlast that of the gap-junctional current transient—thus phase delays caused by inhibition starting early in the cycle may in fact be caused more by their persistence until later in the cycle. In addition, the model shows a pronounced curvature in the phase delay region of the SPRF which was not noticeable in any experimental recordings. This might reflect the presence of other voltage-dependent conductances in real FS cells which effectively linearize this part of the relationship.

The sharp discontinuity between phase delay and advance which emerges at high synaptic strengths is a result of the particular intrinsic biophysical properties and the nature of the synaptic perturbation. It appears to be related to the "type 2" nature of the FS neuron threshold (Tateno et al. 2004), and may be sensitively determined by the potassium conductance densities and kinetics (Ermentrout and Wechselberger 2009; Gutkin et al. 2005). We do not see it, for example, in simulations of a type 1 excitable Morris–Lecar model. The discontinuity is a critical decision point, or threshold, in the progression of the membrane potential toward spike initiation, at which hyperpolarization and depolarization both exert their maximal influence. The effect of this shape of SPRF is to ensure very rapid synchronization of the cell. Maximal phase shift occurs in the middle of the cycle when the phase difference is high—the postsynaptic cell either advances or delays its phase to achieve nearly immediate in-phase firing when detuning between pre- and postsynaptic cell is small. This extremely sharp midcycle transition is not observed in conventional phase-resetting relationships to weak brief inputs in these cells (Mancilla et al. 2007; Tateno and Robinson 2007): it is a consequence of the integration of the strong compound input.

The piecewise nature of the SPRF, with the phase advance contributed exclusively by gap-junctional input, and the delay component contributed exclusively by chemical inhibition, mean that these two types of connection have complementary roles in synchronization: gap junctions are necessary to entrain the firing of the postsynaptic cell to a frequency higher than its preferred frequency, while inhibitory synapses are necessary to entrain firing to a frequency lower than the preferred frequency (as seen in Fig. 20.5). That this must be so can be demonstrated by examining the map of the phase difference between the presynaptic cell (the stimulus) and the postsynaptic cell firing, as follows. Let H be the phase difference between postsynaptic and presynaptic cells ($\varphi_{\mathrm{post}} - \varphi_{\mathrm{pre}}$). The change in H over one period of the input, i.e., from input i to input $i+1$, is: $H_{i+1} - H_i = 2\pi\,(F/f - 1) + \Delta\varphi\,(H_i)$, which incorporates the change due to detuning (difference in frequencies of pre- and postsynaptic cells) and the phase resetting caused by each input. Therefore, when entrainment is achieved, $H_{i+1} = H_i = H_\infty$, and so if $F > f$, then $\Delta\phi\,(H_\infty) > 0$, and if $F < f$, then $\Delta\phi\,(H_\infty) < 0$.

3.3 *Using the SPRF to Predict Entrainment*

Using the SPRF to model entrainment assumes that the effect of each stimulus in the train is the same as if it was applied in isolation. The success of the SPRF in predicting entrainment shown here demonstrates that it is at least a good approximation for this purpose. This assumption implies that the entire dynamical state of the neuron may be represented by just a single number at any time, the phase, which assumes that its dynamical state always lies on a limit cycle, along which it is kicked instantaneously forward and backward by the synaptic inputs. The complex dynamics of a real neuron containing a large number of different voltage-dependent conductances distributed in a complex morphology, and the strong and non-instantaneous nature of the perturbation mean that this has to be a considerable simplification of the reality. An indication of whether the phase approximation is reasonably valid, is to test whether there is any higher-order phase resetting, i.e. changes in the interspike interval *following* that during which the input is applied, or in subsequent intervals. When we analyzed second order shifts, we found that they were sometimes detectable, but very small in relation to the first-order SPRF, consistent with the rather short memory of FS cells for input conductance fluctuations (Tateno and Robinson 2008).

3.4 *Physiological Consequences of the Synaptic Phase-resetting Function*

FS cell firing is suspected to be directly and primarily responsible for producing gamma oscillations in the neocortex (Cardin et al. 2009; Hasenstaub et al. 2005; Morita et al. 2008). Small subnetworks of pyramidal cells in layers 2 or 3 driven by specific layer 4 inputs, appear to interact locally with each other via the inhibitory interneuron network (Yoshimura and Callaway 2005), which is driven nonspecifically by layer 4. Synchronization of FS cells, therefore, may be essential for linking responses of pyramidal cells very rapidly to specific features of the synaptic input, as hypothesized to occur in sensory "binding" (Singer 1999). We have shown that the effect of conductance inputs which realistically mimic single synaptic connections on the phase of FS firing is very powerful, and is capable of entraining the postsynaptic cell even against strong noise. The strikingly sharp discontinuity between phase delay and advance in the SPRF causes a very rapid jump to nearly in-phase firing.

The relative strengths of electrical and inhibitory components can vary greatly from connection to connection (Galarreta and Hestrin 1999; Gibson et al. 1999), and some pairs of FS cells connected by gap junctions can synchronize their firing, while others cannot (Tamas et al. 2000). The strengths of these components will also vary dynamically. Electrical synapses can exhibit plasticity through G protein-coupled receptor activation, intracellular calcium, and phosphorylation (Bennett and Zukin 2004), and the GABAergic connections show strong short-term depression (Galarreta and Hestrin 1999; Gibson et al. 1999; Tamas et al. 2000). These effects presumably help to shape the spatiotemporal dynamics of synchronous firing. The model that we present here could easily accommodate independent plasticity rules for inhibition and gap junctions, by incorporating additional rules for modifying the slopes of the corresponding regions of the SPRF. In addition to such modulation, the $GABA_A$ receptor is also the target of many important neuroactive drugs, such as benzodiazepines, barbiturates, and ethanol. These will be expected to influence the shape of the SPRF, and the synchronization behavior of FS cells in the gamma frequency range. The SPRF, therefore, may be a useful tool for characterizing the action of such compounds on pathological network states treated by these drugs.

Firing is considerably more variable in vivo than in vitro (Holt et al. 1996), and it is important to consider the consequences of the SPRF in strong noisy fluctuations. The stochastic bifurcation analysis that we carried out (Fig. 20.5) delineated a well-defined boundary between entraining and nonentraining frequencies, based on a qualitative change in the nature of the motion of the phase (Inoue et al. 2001). The stronger the noise, the smaller the frequency region of stochastic entrainment—in line with intuition, noise acts to break down synchronization. The strength of the noise effect in controlling the boundary of the synchronized region is not symmetrical around F—thus noise can effectively shift, as well as shrink the synchronized frequency band.

When assemblies of noisy periodic oscillators obeying these phase-resetting rules are locally connected in a lattice (Fig. 20.6a), in a semblance of the FS cell network in the cortex, some interesting effects emerge. Plotting the global synchrony (the order parameter, as defined above) of a network containing 1000 cells versus time (Fig. 20.6b–d), we see that the synchrony grows and diminishes during each cycle, as each cascade of spiking events temporarily pulls the phases back together. Each simulation starts from an initially random assignment of phases. With inhibition only (Fig. 20.6b, blue), there is a transient synchrony which decays over several

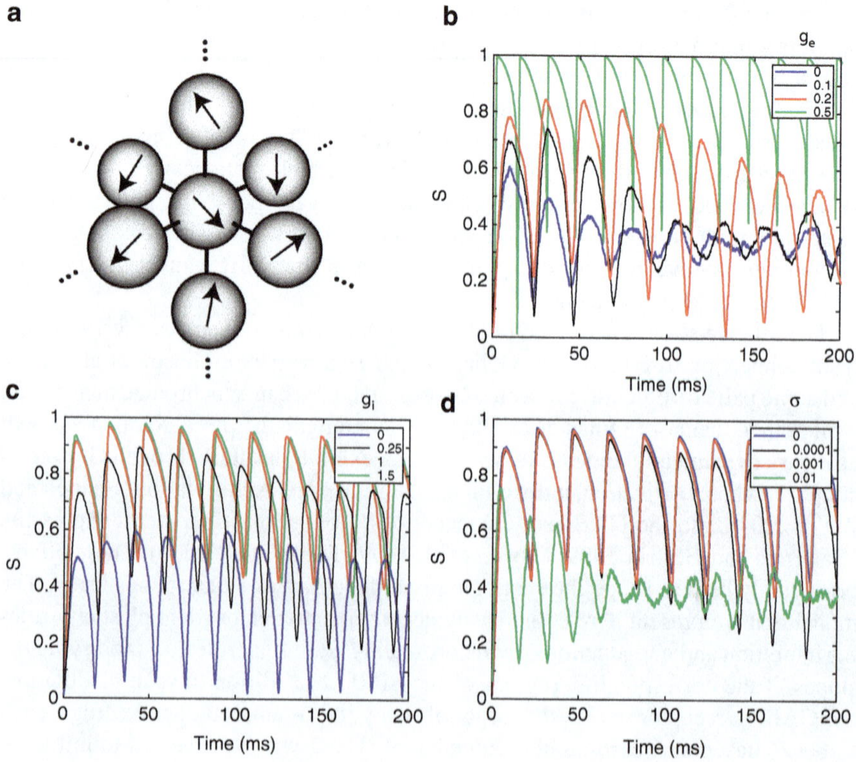

Fig. 20.6 Simulations of a lattice of FS cell phase oscillators. (**a**) Schematic of the local connections, with each of 1,000 oscillators connected to its six neighbors, in a 3D torus arrangement. The steady drift rate of each noisy oscillator is normally distributed around 40 Hz with a standard deviation of 8 Hz. Additionally an independent noise term is added to each phase every cycle. Firing corresponds to passing through phase zero, at which time the phases of connected neighbors are adjusted according to the SPRF. Synchrony versus time plots for (**b**) different levels of g_e as indicated, with $\sigma = 0.001$ rad$/2\pi$ and g_i fixed at 0.75 nS. (**c**) Different levels of g_i, as indicated, with g_e fixed at 0.3 nS and $\sigma = 0.001$ (**d**) different noise (σ) levels, as indicated, with $g_e = 0.3$ nS and $g_i = 0.75$ nS

periods. As gap-junctional strength g_e is increased (other colors, see legend), there is a sharp transition to a maximally synchronous state with no decay of the envelope of synchrony (green trace). With gap junctions only (Fig. 20.6c, blue), there is no progressive decay in the envelope of synchrony, but the amplitude is relatively low. As inhibition increases (other colors as indicated), the peak amplitude increases, again with a saturating threshold effect. Finally, in simulations with both inhibition and gap junctions (Fig. 20.6d), there is a sharp onset of asynchrony as noise level is increased. Again, as in the case of zero inhibition (Fig. 20.6b), the initial synchrony in the first period or two is fairly high, but the envelope decays rapidly in strong noise (Fig. 20.6d, green). Thus, synchronous oscillations of cells coupled with these SPRFs, can be organized very quickly even in the face of strong noise, but their persistence depends sensitively on g_e and their amplitude is strongly modulated by g_i. In view of the predominant role of the FS cell network in producing gamma oscillations, these findings may help to explain how gamma oscillations are transiently created and destroyed in the cortex.

In conclusion, the synaptic phase-resetting function of FS cells firing at gamma frequencies, as characterized here, is very well-suited to achieving rapid synchronization, and demonstrates complementary roles of the two types of synaptic connection in determining the frequency range of synchronization. It provides a simple yet surprisingly accurate model for predicting synchronization of these cells, and should be a useful component in network models aimed at understanding the complex spatiotemporal properties of locally synchronized gamma frequency firing in the cortex.

Acknowledgments Supported by grants from the BBSRC, EC, and the Daiwa Foundation (HPCR), and a Grant-in-Aid for Scientific Research on Priority Areas 17022012 from MEXT (KA). NWG was supported by a Churchill Foundation scholarship. We thank Boris Gutkin for useful discussions.

References

Amitai Y, Gibson JR, Beierlein M, Patrick SL, Ho AM, Connors BW, and Golomb D. The spatial dimensions of electrically coupled networks of interneurons in the neocortex. *J Neurosci* 22: 4142–4152, 2002.

Bacci A, and Huguenard JR. Enhancement of spike-timing precision by autaptic transmission in neocortical inhibitory interneurons. *Neuron* 49: 118–130, 2006.

Bartos M. Fast synaptic inhibition promotes synchronized gamma oscillations in hippocampal interneuron networks. *Proc Natl Acad Sci USA* 99: 13222–13227, 2002.

Bennett MV, and Zukin RS. Electrical coupling and neuronal synchronization in the Mammalian brain. *Neuron* 41: 495–511, 2004.

Buszaki G, and Draguhn A. Neuronal oscillations in cortical networks. *Science* 304: 1926–1929, 2004.

Cardin JA, Carlen M, Meletis K, Knoblich U, Zhang F, Deisseroth K, Tsai LH, and Moore CI. Driving fast-spiking cells induces gamma rhythm and controls sensory responses. *Nature* 459: 663–667, 2009.

Di Bernardo M, Budd CJ, and Champneys AR. Grazing and border-collision in piecewise-smooth systems: a unified analytical framework. *Phys Rev Lett* 86: 2553–2556, 2001.

Erisir A, Lau D, Rudy B, and Leonard CS. Function of specific K(+) channels in sustained high-frequency firing of fast-spiking neocortical interneurons. *J Neurophysiol* 82: 2476–2489, 1999.

Ermentrout B, and Wechselberger M. Canards, clusters and synchronization in a weakly coupled interneuron model. *SIAM J Appl Dynam Syst* 8: 253–278, 2009.

Ermentrout GB, and Kopell N. Multiple pulse interactions and averaging in systems of coupled neural oscillators. *J Math Biol* 29: 195–217., 1991.

Fisahn A, Pike FG, Buhl EH, and Paulsen O. Cholinergic induction of network oscillations at 40 Hz in the hippocampus in vitro. *Nature* 394: 186–189, 1998.

Galarreta M, and Hestrin S. Electrical and chemical synapses among parvalbumin fast-spiking GABAergic interneurons in adult mouse neocortex. *Proc Natl Acad Sci USA* 99: 12438–12443, 2002.

Galarreta M, and Hestrin S. Electrical synapses between GABA-releasing interneurons. *Nat Rev Neurosci* 2: 425–433, 2001.

Galarreta M, and Hestrin S. A network of fast-spiking cells in the neocortex connected by electrical synapses. *Nature* 402: 72–75, 1999.

Gibson JR, Beierlein M, and Connors BW. Two networks of electrically coupled inhibitory neurons in neocortex. *Nature* 402: 75–79, 1999.

Gouwens N, Zeberg H, Tsumoto K, Tateno T, Aihara K, and H.P.C. R. Synchronization of firing in cortical fast-spiking interneurons at gamma frequencies: a phase-resetting analysis. *PLoS Comput Biol* 6: e1000951, 2010.

Gutkin BS, Ermentrout GB, and Reyes AD. Phase-response curves give the responses of neurons to transient inputs. *J Neurophysiol* 94: 1623–1635, 2005.

Hasenstaub A, Shu Y, Haider B, Kraushaar U, Duque A, and McCormick DA. Inhibitory postsynaptic potentials carry synchronized frequency information in active cortical networks. *Neuron* 47: 423–435, 2005.

Holt GR, Softky WR, Koch C, and Douglas RJ. Comparison of discharge variability in vitro and in vivo in cat visual cortex neurons. *J Neurophysiol* 75: 1806–1814, 1996.

Hu H, Martina M, and Jonas P. Dendritic mechanisms underlying rapid synaptic activation of fast-spiking hippocampal interneurons. *Science* 327: 52–58, 2009.

Inoue J, Doi S, and Kumagai S. Numerical analysis of spectra of the Frobenius-Perron operator of a noisy one-dimensional mapping: Toward a theory of stochastic bifurcations. *Phys Rev* 64: 056219, 2001.

Kawaguchi Y, and Kubota Y. Correlation of physiological subgroupings of nonpyramidal cells with parvalbumin- and calbindinD28k-immunoreactive neurons in layer V of rat frontal cortex. *J Neurophysiol* 70: 387–396, 1993.

Kawaguchi Y, and Kubota Y. GABAergic cell subtypes and their synaptic connections in rat frontal cortex. *Cereb Cortex* 7: 476–486, 1997.

Kopell N, and Ermentrout GB. Chemical and electrical synapses perform complementary roles in the synchronization of interneuronal networks. *Proc Natl Acad Sci USA* 101: 15482–15487, 2004.

Lytton WW, and Sejnowski TJ. Simulations of cortical pyramidal neurons synchronized by inhibitory interneurons. *J Neurophysiol* 66: 1059–1079, 1991.

Mancilla JG, Lewis TJ, Pinto DJ, Rinzel J, and Connors BW. Synchronization of electrically coupled pairs of inhibitory interneurons in neocortex. *J Neurosci* 27: 2058–2073, 2007.

Martina M, Royer S, and Pare D. Cell-type-specific GABA responses and chloride homeostasis in the cortex and amygdala. *J Neurophysiol* 86: 2887–2895, 2001.

Martina M, Vida I, and Jonas P. Distal initiation and active propagation of action potentials in interneuron dendrites. *Science* 287: 295–300, 2000.

Morita K, Kalra R, Aihara K, and Robinson HPC. Recurrent Synaptic Input and the Timing of Gamma-Frequency-Modulated Firing of Pyramidal Cells during Neocortical "UP" States. *J Neurosci* 28: 1871–1881, 2008.

Netoff TI, Banks MI, Dorval AD, Acker CD, Haas JS, Kopell N, and White JA. Synchronization in hybrid neuronal networks of the hippocampal formation. *J Neurophysiol* 93: 1197–1208, 2005.

Pikovsky A, Rosenblum M, and Kurths J. *Synchronization: a universal concept in nonlinear sciences*. Cambridge: Cambridge University Press, 2001.

Reyes AD, and Fetz EE. Effects of transient depolarizing potentials on the firing rate of cat neocortical neurons. *J Neurophysiol* 69: 1673–1683, 1993.

Rinzel JM, and Ermentrout GB. Analysis of neuronal excitability. In: *Methods in Neuronal Modelling*, edited by C. K, and Segev I. Cambridge MA: MIT, 1998.

Robinson HPC. A scriptable DSP-based system for dynamic conductance injection. *J Neurosci Meth* 169: 271–281, 2008.

Robinson HPC. Synaptic conductances and spike generation in cortical cells. In: *Dynamic Clamp: From Principles to Applications*, edited by Destexhe A, and Bal T Springer, 2009, p. 49–72.

Robinson HPC, and Kawai N. Injection of digitally synthesized synaptic conductance transients to measure the integrative properties of neurons. *J Neurosci Meth* 49: 157–165, 1993.

Sharp AA, O'Neil MB, Abbott LF, and Marder E. Dynamic clamp: computer-generated conductances in real neurons. *J Neurophysiol* 69: 992–995, 1993.

Singer W. Neuronal synchrony: a versatile code for the definition of relations? *Neuron* 24: 49–65, 1999.

Stoop R, Schindler K, and Bunimovich LA. When pyramidal neurons lock, when they respond chaotically, and when they like to synchronize. *Neurosci Res* 36: 81–91, 2000.

Tamas G, Buhl EH, Lorinz A, and Somogyi P. Proximally targeted GABAergic synapses and gap junctions synchronize cortical interneurons. *Nat Neurosci* 3: 366–371, 2000.

Tateno T, Harsch A, and Robinson HP. Threshold firing frequency-current relationships of neurons in rat somatosensory cortex: type 1 and type 2 dynamics. *J Neurophysiol* 92: 2283–2294, 2004.

Tateno T, and Robinson HP. Phase resetting curves and oscillatory stability in interneurons of rat somatosensory cortex. *Biophys J* 92: 683–695, 2007.

Tateno T, and Robinson HPC. Integration of broadband conductance input in rat somatosensory cortical inhibitory interneurons: an inhibition-controlled switch between intrinsic and input-driven spiking in fast-spiking cells. *J Neurophysiol* 101: 1056–1072, 2008.

Traub RD. Gap junctions between interneuron dendrites can enhance synchrony of gamma oscillations in distributed networks. *J Neurosci* 21: 9478–9486, 2001.

Vida I, Bartos M, and Jonas P. Shunting inhibition improves robustness of gamma oscillations in hippocampal interneuron networks by homogenizing firing rates. *Neuron* 49: 107–117, 2006.

Whittington MA, Stanford IM, Colling SB, Jefferys JG, and Traub RD. Spatiotemporal patterns of gamma frequency oscillations tetanically induced in the rat hippocampal slice. *J Physiol* 502 (Pt 3): 591–607, 1997.

Whittington MA, Traub RD, and Jefferys JG. Synchronized oscillations in interneuron networks driven by metabotropic glutamate receptor activation. *Nature* 373: 612–615, 1995.

Yoshimura Y, and Callaway EM. Fine-scale specificity of cortical networks depends on inhibitory cell type and connectivity. *Nat Neurosci* 8: 1552–1559, 2005.

Index

N.W. Schultheiss et al. (eds.), *Phase Response Curves in Neuroscience: Theory, Experiment, and Analysis*, Springer Series in Computational Neuroscience 6,
DOI 10.1007/978-1-4614-0739-3, © Springer Science+Business Media, LLC 2012

GPSR Compliance
The European Union's (EU) General Product Safety Regulation (GPSR) is a set of rules that requires consumer products to be safe and our obligations to ensure this.

If you have any concerns about our products, you can contact us on

ProductSafety@springernature.com

In case Publisher is established outside the EU, the EU authorized representative is:

Springer Nature Customer Service Center GmbH
Europaplatz 3
69115 Heidelberg, Germany

www.ingramcontent.com/pod-product-compliance
Lightning Source LLC
Chambersburg PA
CBHW071256180926
51613CB00001B/42
* 9 7 8 1 4 6 1 4 0 7 3 8 6 *